ANALYTICAL MECHANICS

Analytical Mechanics provides a detailed introduction to the key analytical techniques of classical mechanics, one of the cornerstones of physics. It deals with all the important subjects encountered in an undergraduate course and prepares the reader thoroughly for further study at the graduate level.

The authors set out the fundamentals of Lagrangian and Hamiltonian mechanics early on in the book and go on to cover such topics as linear oscillators, planetary orbits, rigid-body motion, small vibrations, nonlinear dynamics, chaos, and special relativity. A special feature is the inclusion of many "e-mail questions," which are intended to facilitate dialogue between the student and instructor.

Many worked examples are given, and there are 250 homework exercises to help students gain confidence and proficiency in problem solving. It is an ideal textbook for undergraduate courses in classical mechanics and provides a sound foundation for graduate study.

Louis N. Hand was educated at Swarthmore College and Stanford University. After serving as an assistant professor at Harvard University during the 1964 academic year, he came to the Physics Department of Cornell University where he has remained ever since. He is presently researching in the field of accelerator physics.

Janet D. Finch, teaching associate in the Physics Department of Cornell University, earned her BS in engineering physics from the University of Illinois, and her MS in theoretical physics and her MA in teaching from Cornell. In 1994 she began working with Professor Hand on the Classical Mechanics course from which this book developed. She was the e-mail tutor for the course during the first-time implementation of this innovation.

ANALYTICAL MECHANICS

LOUIS N. HAND
and
JANET D. FINCH

CAMBRIDGE UNIVERSITY PRESS
Cambridge, New York, Melbourne, Madrid, Cape Town, Singapore, São Paulo

Cambridge University Press
40 West 20th Street, New York, NY 10011–4211, USA
www.cambridge.org
Information on this title:www.cambridge.org/9780521573276

© Cambridge University Press 1998

This publication is in copyright. Subject to statutory exception
and to the provisions of relevant collective licensing agreements,
no reproduction of any part may take place without
the written permission of Cambridge University Press.

First published 1998
Reprinted 2001, 2003, 2004, 2005

Printed in the United States of America

A catalogue record for this book is available from the British Library.

Library of Congress Cataloguing in Publication data
Hand, Louis, N., 1933—
Analytical mechanics / Louis N. Hand, Janet D. Finch.
p. cm.
Includes bibliographical references and index.
ISBN 0-521-57327-0 — ISBN 0-521-57572-9 (pbk.)
1. Mechanics, Analytic. I. Finch, Janet D., 1969—
II. Title.
QA805.H26   1998
531'.01'515352—dc21   97-43334
                                CIP

ISBN-13 978-0-521-57327-6 hardback
ISBN-10 0-521-57327-0 hardback

ISBN-13 978-0-521-57572-0 paperback
ISBN-10 0-521-57572-9 paperback

Cambridge University Press has no responsibility for
the persistence or accuracy of URLs for external or
third-party Internet Web sites referred to in this publication
and does not guarantee that any content on such
Web sites is, or will remain, accurate or appropriate.

CONTENTS

Preface		xi
1	**LAGRANGIAN MECHANICS**	1
	1.1 Example and Review of Newton's Mechanics: A Block Sliding on an Inclined Plane	1
	1.2 Using Virtual Work to Solve the Same Problem	3
	1.3 Solving for the Motion of a Heavy Bead Sliding on a Rotating Wire	7
	1.4 Toward a General Formula: Degrees of Freedom and Types of Constraints	10
	1.5 Generalized Velocities: How to "Cancel the Dots"	14
	1.6 Virtual Displacements and Virtual Work – Generalized Forces	14
	1.7 Kinetic Energy as a Function of the Generalized Coordinates and Velocities	16
	1.8 Conservative Forces: Definition of the Lagrangian L	18
	1.9 Reference Frames	20
	1.10 Definition of the Hamiltonian	21
	1.11 How to Get Rid of Ignorable Coordinates	22
	1.12 Discussion and Conclusions – What's Next after You Get the EOM?	23
	1.13 An Example of a Solved Problem	24
	Summary of Chapter 1	25
	Problems	26
	Appendix A. About Nonholonomic Constraints	36
	Appendix B. More about Conservative Forces	41
2	**VARIATIONAL CALCULUS AND ITS APPLICATION TO MECHANICS**	44
	2.1 History	44
	2.2 The Euler Equation	46
	2.3 Relevance to Mechanics	51
	2.4 Systems with Several Degrees of Freedom	53
	2.5 Why Use the Variational Approach in Mechanics?	54
	2.6 Lagrange Multipliers	56

2.7	Solving Problems with Explicit Holonomic Constraints	57
2.8	Nonintegrable Nonholonomic Constraints – A Method that Works	62
2.9	Postscript on the Euler Equation with More Than One Independent Variable	65
	Summary of Chapter 2	65
	Problems	66
	Appendix. About Maupertuis and What Came to Be Called "Maupertuis' Principle"	75

3 LINEAR OSCILLATORS 81

3.1	Stable or Unstable Equilibrium?	82
3.2	Simple Harmonic Oscillator	87
3.3	Damped Simple Harmonic Oscillator (DSHO)	90
3.4	An Oscillator Driven by an External Force	94
3.5	Driving Force Is a Step Function	96
3.6	Finding the Green's Function for the SHO	99
3.7	Adding up the Delta Functions – Solving the Arbitary Force	103
3.8	Driving an Oscillator in Resonance	105
3.9	Relative Phase of the DSHO Oscillator with Sinusoidal Drive	110
	Summary of Chapter 3	113
	Problems	114

4 ONE-DIMENSIONAL SYSTEMS: CENTRAL FORCES AND THE KEPLER PROBLEM 123

4.1	The Motion of a "Generic" One-Dimensional System	123
4.2	The Grandfather's Clock	125
4.3	The History of the Kepler Problem	130
4.4	Solving the Central Force Problem	133
4.5	The Special Case of Gravitational Attraction	141
4.6	Interpretation of Orbits	143
4.7	Repulsive $\frac{1}{r^2}$ Forces	151
	Summary of Chapter 4	156
	Problems	156
	Appendix. Tables of Astrophysical Data	167

5 NOETHER'S THEOREM AND HAMILTONIAN DYNAMICS 170

5.1	Discovering Angular Momentum Conservation from Rotational Invariance	170
5.2	Noether's Theorem	172
5.3	Hamiltonian Dynamics	175
5.4	The Legendre Transformation	175
5.5	Hamilton's Equations of Motion	180
5.6	Liouville's Theorem	184
5.7	Momentum Space	189

	5.8	Hamiltonian Dynamics in Accelerated Systems	190
		Summary of Chapter 5	195
		Problems	196
		Appendix A. A General Proof of Liouville's Theorem Using the Jacobian	202
		Appendix B. Poincaré Recurrence Theorem	204
6	**THEORETICAL MECHANICS: FROM CANONICAL TRANSFORMATIONS TO ACTION–ANGLE VARIABLES**		**207**
	6.1	Canonical Transformations	208
	6.2	Discovering Three New Forms of the Generating Function	213
	6.3	Poisson Brackets	217
	6.4	Hamilton–Jacobi Equation	218
	6.5	Action–Angle Variables for 1-D Systems	230
	6.6	Integrable Systems	235
	6.7	Invariant Tori and Winding Numbers	237
		Summary of Chapter 6	239
		Problems	240
		Appendix. What Does "Symplectic" Mean?	248
7	**ROTATING COORDINATE SYSTEMS**		**252**
	7.1	What Is a Vector?	253
	7.2	Review: Infinitesimal Rotations and Angular Velocity	254
	7.3	Finite Three-Dimensional Rotations	259
	7.4	Rotated Reference Frames	259
	7.5	Rotating Reference Frames	263
	7.6	The Instantaneous Angular Velocity $\vec{\omega}$	264
	7.7	Fictitious Forces	267
	7.8	The Tower of Pisa Problem	267
	7.9	Why Do Hurricane Winds Rotate?	271
	7.10	Foucault Pendulum	272
		Summary of Chapter 7	275
		Problems	276
8	**THE DYNAMICS OF RIGID BODIES**		**283**
	8.1	Kinetic Energy of a Rigid Body	284
	8.2	The Moment of Inertia Tensor	286
	8.3	Angular Momentum of a Rigid Body	291
	8.4	The Euler Equations for Force-Free Rigid Body Motion	292
	8.5	Motion of a Torque-Free Symmetric Top	293
	8.6	Force-Free Precession of the Earth: The "Chandler Wobble"	299
	8.7	Definition of Euler Angles	300
	8.8	Finding the Angular Velocity	304
	8.9	Motion of Torque-Free Asymmetric Tops: Poinsot Construction	305

8.10	The Heavy Symmetric Top	313
8.11	Precession of the Equinoxes	317
8.12	Mach's Principle	323
	Summary of Chapter 8	325
	Problems	326
	Appendix A. What Is a Tensor?	333
	Appendix B. Symmetric Matrices Can Always Be Diagonalized by "Rotating the Coordinates"	336
	Appendix C. Understanding the Earth's Equatorial Bulge	339

9 THE THEORY OF SMALL VIBRATIONS — 343

9.1	Two Coupled Pendulums	344
9.2	Exact Lagrangian for the Double Pendulum	348
9.3	Single Frequency Solutions to Equations of Motion	352
9.4	Superimposing Different Modes; Complex Mode Amplitudes	355
9.5	Linear Triatomic Molecule	360
9.6	Why the Method Always Works	363
9.7	N Point Masses Connected by a String	367
	Summary of Chapter 9	371
	Problems	373
	Appendix. What Is a Cofactor?	380

10 APPROXIMATE SOLUTIONS TO NONANALYTIC PROBLEMS — 383

10.1	Stability of Mechanical Systems	384
10.2	Parametric Resonance	388
10.3	Lindstedt–Poincaré Perturbation Theory	398
10.4	Driven Anharmonic Oscillator	401
	Summary of Chapter 10	411
	Problems	413

11 CHAOTIC DYNAMICS — 423

11.1	Conservative Chaos – The Double Pendulum: A Hamiltonian System with Two Degrees of Freedom	426
11.2	The Poincaré Section	428
11.3	KAM Tori: The Importance of Winding Number	433
11.4	Irrational Winding Numbers	436
11.5	Poincaré–Birkhoff Theorem	439
11.6	Linearizing Near a Fixed Point: The Tangent Map and the Stability Matrix	442
11.7	Following Unstable Manifolds: Homoclinic Tangles	446
11.8	Lyapunov Exponents	449
11.9	Global Chaos for the Double Pendulum	451
11.10	Effect of Dissipation	452
11.11	Damped Driven Pendulum	453

	11.12	Fractals	463
	11.13	Chaos in the Solar System	468
		Student Projects	474
		Appendix. The Logistic Map: Period-Doubling Route to Chaos; Renormalization	481
12	**SPECIAL RELATIVITY**	493
	12.1	Space–Time Diagrams	495
	12.2	The Lorentz Transformation	498
	12.3	Simultaneity Is Relative	501
	12.4	What Happens to y and z if We Move Parallel to the X Axis?	503
	12.5	Velocity Transformation Rules	504
	12.6	Observing Light Waves	505
	12.7	What Is Mass?	512
	12.8	Rest Mass Is a Form of Energy	513
	12.9	How Does Momentum Transform?	517
	12.10	More Theoretical "Evidence" for the Equivalence of Mass and Energy	519
	12.11	Mathematics of Relativity: Invariants and Four-Vectors	521
	12.12	A Second Look at the Energy–Momentum Four-Vector	526
	12.13	Why Are There Both Upper and Lower Greek Indices?	529
	12.14	Relativistic Lagrangian Mechanics	530
	12.15	What Is the Lagrangian in an Electromagnetic Field?	533
	12.16	Does a Constant Force Cause Constant Acceleration?	535
	12.17	Derivation of the Lorentz Force from the Lagrangian	537
	12.18	Relativistic Circular Motion	539
		Summary of Chapter 12	540
		Problems	541
		Appendix. The Twin Paradox	554
Bibliography		559
References		563
Index		565

PREFACE

PREREQUISITES

The Physics Department at Cornell offers two intermediate-level undergraduate mechanics courses. This book evolved from lecture notes used in the more advanced of the two courses. Most of the students who took this course were considering postgraduate study leading to future careers in physics or astronomy. With a few exceptions, they had previously taken an introductory honors course in mechanics at the level of Kleppner and Kolenkow.* Many students also had an Advanced Placement physics course in high school. Since we can assume that a solid background in introductory college-level physics already exists, we have not included a systematic review of elementary mechanics in the book, other than the brief example at the beginning of Chapter 1.

Familiarity with a certain few basic mathematical concepts is essential. The student should understand Taylor series in more than one variable, partial derivatives, the chain rule, and elementary manipulations with complex variables.† Some elementary knowledge of matrices and determinants is also needed.‡ Almost all of the students who took the honors analytic mechanics course at Cornell have either completed, or were concurrently registered in, a mathematical physics course involving vector analysis, complex variable theory, and techniques for solving ordinary and partial differential equations. However, a thorough grounding in these subjects is not essential – in fact some of this material can be learned by taking a course based on this book.

INTRODUCTION

Our intention in writing this book is to reduce the gap between undergraduate and graduate physics training. Graduate students often complain that their undergraduate training did not prepare them for the rigors of graduate school. For that reason we have

---

* *An Introduction to Mechanics*, D. Kleppner and R. J. Kolenkow, McGraw-Hill, 1973.
† At the level of *Advanced Calculus*, 2d ed., W. Kaplan, Addison-Wesley, 1984.
‡ For linear algebra, we recommend a book on the level of *Linear Algebra with Applications*, 2d edition, S. J. Leon, Macmillan, 1986, or one of the many other suitable texts at the intermediate level.

written a text that emphasizes those concepts that will be useful to know later. We feel that only tradition stands in the way of teaching Lagrangian and Hamiltonian mechanics at an earlier stage than has been the case in the education of physicists. In addition to advancing the stage when these basic concepts are encountered, we have a second purpose in mind. In many colleges and universities, quantum mechanics is now taught at the intermediate undergraduate level. As a result of this recent trend, there often is a mismatch between these courses and the preparatory courses, and student preparation is often inadequate in the upper-level undergraduate courses. This places a heavy burden on the instructor in the courses introducing quantum mechanics and modern physics to juniors and seniors. Many important topics can be more easily visualized if taught in a classical mechanics course instead. The use of eigenvectors and eigenvalues to solve physical problems is a prime example.

Classical mechanics is an excellent way to introduce the basic tools of theoretical physics. Lagrangian methods can be used to simplify problems that would be difficult to solve by other means. Mechanics problems can be written and solved in a few lines using these powerful techniques. It is usually easier to work with more "advanced" techniques than with the more complicated "elementary" methods. Deeper insight into the motion of a mechanical system is obtained with these more sophisticated methods. We cite the role of conserved quantities derived from symmetries via Noether's theorem as one example.

A course in classical mechanics need not be justified entirely on the grounds that it provides a path to something else more glamorous and fashionable. To quote Gutzwiller:

> Elementary mechanics, both classical and quantum, has become a growth industry in the last decade. A newcomer to this flourishing field must get acquainted with some unfamiliar concepts and get rid of some cherished assumptions. The change in orientation is necessary because physicists have finally realized that most dynamical systems do not follow simple, regular, and predictable patterns, but run along a seemingly random, yet well-defined, trajectory. The generally accepted name for this phenomenon is *chaos*, a term that accurately suggests that we have failed to come to grips with the problem.*

Far from being a dead subject, classical mechanics has reemerged in the forefront of modern physics research! Deterministic chaos is a special topic discussed in Chapter 11. Since the 1960s chaos has developed as a new branch of classical mechanics. It has wide applicability to other fields outside of physics as well, although we do not consider this. Here we confine ourselves to treating two simple dynamical systems, one a conservative system and the other a dissipative system. This text is intended to give a purely introductory, rather than comprehensive, treatment of chaos. It is not a substitute for textbooks devoted entirely to this subject but might stimulate the student to investigate the subject more in future courses.

We have included a final chapter on special relativity. There are many excellent introductions to this subject, some of which are listed in the bibliography. However we felt that there may be value in having an introduction that applies the lessons of the previous

---

* Martin C. Gutzwiller, in *Chaos in Classical and Quantum Mechanics*.

chapters. The case of special relativity illustrates the modern viewpoint and shows the power of Lagrangian mechanics to transcend its original role, which was a reformulation of Newton's Laws of Motion.

One word of warning: We believe there is more material in this book than can be reasonably presented in a one-semester undergraduate course. The teacher must select those parts which he or she feels comprise a unified course in mechanics. In particular, there is a natural choice between teaching the material in Chapter 6 (advanced theory) and Chapter 11 (chaos) on the one hand and teaching special relativity (Chapter 12) on the other.

This book will be much more effective if it serves as part of an experience that directly involves the student in an active learning process. In a separate publication,* we will discuss the way in which the course was taught at Cornell, in particular the innovations made possible by the Sloan Foundation: the seminars and the e-mail questions.† Let us only say here that a dialogue with the individual students, separately and in groups, is essential for success in teaching the material. In our previous experience with teaching mechanics, it was found that relying only on standard lectures, with spontaneous student questions and discussion in lecture, did not stimulate the kind of individual thought process needed. The need to cover the material in the lecture tends to reduce thoughtful discussion to a minimum. Reading the text alone did not fill this gap. Students tended to read the text rather superficially, accepting what was said, rather than questioning it and working examples. Problem sets were done "just in time." This book was designed to provoke a more thoughtful reading experience, and more continuity in the study process, when it was used in combination with the seminar and the e-mail questions.

Support by the Sloan Foundation, which has an interest in acceleration of the undergraduate phase of education, allowed us the freedom to depart from the standard lecture format to develop the alternative ways to teach this material. Judging by the student reaction, we were successful in most cases. The use of undergraduate teaching assistants (TAs), students who had taken the course in the previous year, was particularly successful and helped provide role models.

ADVICE TO STUDENTS WHO USE THIS BOOK

Learning physics is an active experience.

If you were learning to ride a bicycle or to play the violin, you would expect to practice the technique until it became second nature. Falling off the bicycle gives you immediate feedback – it tells you that you need more practice. Exactly the same thing is true of learning physics. It must become part of you, something intuitive. Almost everyone has to work very hard to achieve this. Until you can solve problems, your understanding is not sufficiently deep. It is one thing to watch a lecturer solve a problem, where every step

---

* J. D. Finch and L. N. Hand, "Using an Email Tutorial and Student Seminars to Improve an Intermediate-Level Undergraduate Physics Course," *American Journal of Physics*, to be published in 1998.
† The seminar problems are denoted by a "*" in the homework problem sections. The e-mail questions are distributed throughout the text.

seems to be a logical one, and quite another to tackle a real problem on your own. Do not think of learning physics as "art appreciation." It is a "do-it-yourself" activity.

The key to success is how one studies the subject outside of class. A last minute "all-nighter" to solve a problem set is an exercise in self-delusion. You are strongly advised against trying to learn physics in this way, because it inhibits the crucial transition from short-term to long-term memory. The new concepts have to soak into your consciousness. Remember that it took about 150 years to develop Hamiltonian dynamics. It can't be learned adequately in one night. You should put aside a regular time for studying this material and concentrate on it without distraction. Do the reading early in the week it is assigned. Think about the problems more than one day before they are due. Try to isolate the points you don't understand. Read the material again. Most important: *Discuss it with other students.* Don't hesitate to ask others for an explanation and don't be satisfied until you get one. Another tip: Make the effort to memorize what the notation means. By experience, we have often observed that lack of familiarity with the symbols and what they stand for is one difference between strong and weak students. Memorizing the meaning of symbols becomes automatic with trained physicists. Acquiring the skill of learning physics will serve you well in later years, when most learning must be self-taught.

If you don't fully understand what your are reading, try to construct a simple example for yourself. But don't let the lack of understanding remain. Pester someone – your teaching assistant, your colleague, or your professor – until your questions are answered. And don't assume that it is clear to everyone except yourself.

The questions scattered throughout the text are intended to test your comprehension as you read the material. Some of them were assigned to our students, and answers were given and graded by e-mail. Repeated improvements in the answer, following comments by the teaching assistant, led to repeated improvements in the grade for that particular question. Most people ended up with nearly perfect scores after a useful dialogue with the TA via e-mail.

ACKNOWLEDGMENTS

First and foremost, we would like to thank the students who took this course over a five-year period while the content and format were being developed. The course was taught by one of us (L. H.). The other (J. F.) joined for the last three years, first as an e-mail tutor and then, later, as an editor and coauthor on the several versions of the lecture notes that preceded this book.

Without the generous encouragement and support of Dr. Frank Mayadas of the Sloan Foundation, this book would not have been possible, nor would it have been possible to try the principal innovations in teaching classical mechanics at Cornell: a weekly two-hour problem-solving seminar and the e-mail tutorial which engages the students in a dialogue while they are doing the assigned reading. The importance of the Sloan support cannot be overstated. Nowadays, there are very few remaining sources of support for improving the quality of undergraduate physics education.

Thanks are also due to Prof. Kurt Gottfried of Cornell, who was Department Chairman at the time this course made the transition into a teaching experiment. His leadership and constant encouragement played a major role in bringing the project to fruition.

Special thanks should go to Dr. Steven Townsend, who not only did a superb job as an e-mail tutor, but later helped us revise the quiz questions and select those which had been helpful to the students. His insightful comments on the physics of these questions were of very great pedagogical value. We have tried to incorporate them here.

Paul Shocklee helped with proofreading the first edition of these notes. Alex Khein contributed many problems as well as a set of notes connecting Hamilton–Jacobi theory with quantum mechanics. Thomas Haeusser helped proofread, finding several errors which we corrected. The graduate student TAs: Alex Khein, Ard Louis, Thomas Haeusser, and Katrin Schenk provided solutions to assigned homework problems. The undergraduate TAs have been essential to the success of the weekly problem solving seminars we have conducted at Cornell. Thanks are due to Paul Shocklee, Robin Madryk, Abe Stroock, Smitha Vishveshwara and Jonathan Levine for their contributions as Teaching Assistants in these seminars.

We also would like to thank Kevin Hodgson of Triple C, who did a superb job of translating our rough sketches into figures which add immeasurably to the quality and professionalism of the book.

Prof. Saul Teukolsky provided some valuable information about the most recent numerical studies of the stability of the solar system. Dr. Mark Scheel contributed a critical reading of Chapter 12 on special relativity. We have adopted several of his excellent suggestions. Prof. Eberhardt Bodenshatz kindly allowed us to become part of his network of NEXT computers. Without his support, and the help of his graduate student Brendan Plapp, we would have found writing the book far more difficult.

The original suggestion that the lecture notes be turned into a book came from Dr. Robert Lieberman, who has been our agent. The book would not have been possible without him and his constant encouragement. We also wish to thank Dr. Philip Meyler of Cambridge University Press for his patience and constant support.

CHAPTER ONE

LAGRANGIAN MECHANICS

OVERVIEW OF CHAPTER 1

Joseph Louis Lagrange* reformulated Newton's Laws in a way that eliminates the need to calculate forces on isolated parts of a mechanical system. Any convenient variables obeying the constraints on a system can be used to describe the motion. If Lagrangian mechanics rather than Newtonian mechanics is used, it is only necessary to consider a single function of the dynamical variables that describe the motion of the entire system. The differential equations governing the motion are obtained directly from this function without any vector force diagrams. Lagrangian mechanics is extremely efficient: There are only as many equations to solve as there are physically significant variables.

Lagrange did not introduce new physical principles to mechanics. The physical concepts are due to Newton and Galileo. But he succeeded in giving a more powerful and sophisticated way to formulate the mathematical equations of classical mechanics, an approach that has spread its influence over physics far beyond the purely mechanical problems.

We will begin by solving some examples that lead us toward this new formulation of mechanics. We plan to use the concept of *virtual work* to derive this. We will consider extended rigid bodies to be made up of collections of massive point particles. Summing over the constituent particles will lead to an efficient and general method for obtaining the differential equations of motion for any frictionless mechanical system.

To use Lagrange's method we have to express the difference between kinetic and potential energies in terms of the variables we choose to describe the motion. This gives a single function, called the *Lagrangian*. After that, there is a straightforward procedure to follow to find the differential equations of the motion.

1.1 EXAMPLE AND REVIEW OF NEWTON'S MECHANICS: A BLOCK SLIDING ON AN INCLINED PLANE

A block of mass m slides on a frictionless inclined plane, which itself has a mass M and rests on a flat surface, without any friction between this surface and the inclined

---

* (1736–1813) His treatise on mechanics, *Mécanique Analytique*, was published in 1788.

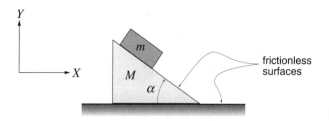

FIGURE 1.1

plane, as shown in Figure 1.1. There are then two types of motion, or *degrees of freedom*. The small block can slide down the plane, and the inclined plane can move horizontally. Whatever happens, we will assume the block always remains constrained to move along the upper surface of the inclined plane. A practical question is: If the small block starts at the top of the inclined plane, how long will it take to reach the bottom?

We assume the reader is well-versed in the standard methods of Newtonian mechanics. We are taught to isolate each body and then to consider all of the forces acting on that body. If we first consider the forces on the *SB* (small block), there is a constraint force $\vec{F}_1$ exerted by the *IP* (inclined plane) on the block. This force is considered a *constraint force* because it keeps the motion of the small block constrained to be along the upper surface of the *IP*. Therefore, the force $\vec{F}_1$ must be perpendicular to the tilted plane. (In the absence of friction, $\vec{F}_1$ is the only force exerted by the *IP* on the *SB*.) Let the horizontal direction be defined as the *X* axis, with a unit vector along the *X* axis being denoted by $\hat{\imath}$. The vertical direction will be the *Y* axis, with the positive direction upwards. A unit vector pointing up is denoted by $\hat{\jmath}$. With this notation, the vector $\vec{F}_1$ is $\vec{F}_1 = F_1(\sin\alpha\,\hat{\imath} + \cos\alpha\,\hat{\jmath})$. ($F_1$ without the arrow above it stands for the magnitude of $\vec{F}_1$.) Although the force $\vec{F}_1$ is actually distributed along the bottom of the small block, we will make the usual assumption that we can treat the block as a point mass so that this force acts through the center of the block, as does the force of gravity, $\vec{F}_{\text{gravity}}(SB) = -mg\hat{\jmath}$ (where $g \approx 9.8$ m/s² is the acceleration of gravity). Let $\vec{A}$ be the acceleration of the *IP* in an inertial reference system. This is any reference system moving at a constant velocity and therefore not accelerated itself. With the *X* axis oriented as we have done, $\vec{A}$ has only an *x* component, A_x. Let $\vec{a} + \vec{A}$ be the acceleration of the small block in the same reference system. Then $\vec{a}$ is the acceleration of the block relative to the inclined plane. Newton's Second Law for the small block is then, according to the law of addition of forces,

$$\vec{F}(SB) = \vec{F}_1 + \vec{F}_{\text{gravity}}(SB) = m(\vec{a} + \vec{A}). \qquad (1.1)$$

Equation (1.1) is really two equations. The components of $\vec{F}$ parallel to the top of the inclined plane ($F_\parallel$) and perpendicular to it ($F_\perp$) are

$$\begin{aligned} F_\parallel &= mg\sin\alpha = ma_\parallel + mA_x\cos\alpha, \\ F_\perp &= F_1 - mg\cos\alpha = ma_\perp + mA_x\sin\alpha. \end{aligned} \qquad (1.2)$$

Since the constraint that the small block move along the inclined plane's top surface means that $a_\perp = 0$, we can solve the second equation in (1.2) for F_1, the magnitude of the constraint force:

$$F_1 = mg\cos\alpha + mA_x\sin\alpha. \qquad (1.3)$$

1.2 USING VIRTUAL WORK TO SOLVE THE SAME PROBLEM

The forces on the inclined plane are a bit more complicated. According to Newton's Third Law of action and reaction, there is a reaction force $-\vec{F}_1$ exerted by the block on the inclined plane. A second constraint force is needed to prevent any downward motion of the *IP*. This is $\vec{F}_2 = F_2\,\hat{\jmath}$. Newton's Second Law becomes

$$\vec{F}(IP) = -\vec{F}_1 + \vec{F}_2 + \vec{F}_{\text{gravity}}(IP) = M\vec{A}. \tag{1.4}$$

$F_{\text{gravity}}(IP) = -Mg\hat{\jmath}$, and $F_2 - Mg$ just cancels the downward component of $-\vec{F}_1$,

$$F_2 - Mg = -F_1 \cos\alpha. \tag{1.5}$$

The horizontal component of $-\vec{F}_1$ produces a horizontal acceleration of the *IP*:

$$MA_x = -F_1 \sin\alpha. \tag{1.6}$$

Solve the linear equation obtained by combining (1.6) with (1.3):

$$A_x = -g\left(\frac{\sin\alpha \cos\alpha}{\sin^2\alpha + \frac{M}{m}}\right). \tag{1.7}$$

Equation (1.7) determines the motion of the inclined plane.

The acceleration of the small block along the plane, $a_\parallel$, can be obtained by solving (1.2):

$$a_\parallel = g\sin\alpha - A_x \cos\alpha. \tag{1.8}$$

The vertical component of the acceleration of the small block is

$$a_y = -a_\parallel \sin\alpha = -g\sin^2\alpha\left(\frac{M+m}{M+m\sin^2\alpha}\right). \tag{1.9}$$

If the inclined plane has a height h, a block starting at this height will reach the bottom in a time t given by the solution to $-h = a_y \frac{t^2}{2}$, that is, $t = \sqrt{-2h a_y}$.

This completes the solution of the problem using the standard methods of Newtonian mechanics. There were four equations, two for each object, and four unknowns: $F_1, F_2, a_\parallel,$ and A_x. Next we will see how to use the special properties of the constraint forces $\vec{F}_1, \vec{F}_2$ in order to eliminate them from the problem completely.

1.2 USING VIRTUAL WORK TO SOLVE THE SAME PROBLEM

What are the natural variables in the problem solved in the previous section? One useful choice is X, the horizontal position of the inclined plane, and d, the distance of the small block from the top of the inclined plane. Notice that d is a coordinate defined with respect to a noninertial reference frame, since the point $d = 0$ is accelerated. The first task is to find expressions in terms of these variables for the kinetic energies of the block and the plane. Throughout this book, we will denote kinetic energies by the letter T. As always in nonrelativistic mechanics, the kinetic energy of a point mass is $\frac{1}{2}mv^2$, where v is the magnitude of the velocity as measured in an inertial reference frame. We will treat both objects as point masses, since they do not rotate. (The proof that this is justified appears in Chapter 8.) For the inclined plane we have

$$T_{IP} = \frac{1}{2}M\dot{X}^2, \tag{1.10}$$

$$v_x(SB) = \dot{X} + \dot{d}\cos\alpha, \quad v_y(SB) = -\dot{d}\sin\alpha. \tag{1.11}$$

(The reader should check that he or she agrees with (1.11).) For the small block we have

$$T_{SB} = \frac{1}{2}m(v_x^2 + v_y^2) = \frac{1}{2}m(\dot{X}^2 + 2\dot{d}\dot{X}\cos\alpha + \dot{d}^2). \tag{1.12}$$

The total kinetic energy is

$$T = T_{SB} + T_{IP} = \frac{1}{2}(m+M)\dot{X}^2 + \frac{1}{2}m(\dot{d}^2 + 2\dot{d}\dot{X}\cos\alpha). \tag{1.13}$$

With the calculation of the total kinetic energy in terms of our chosen variables finished, we can now concentrate on the method of virtual work. Imagine that, at some fixed time, we make a small *virtual* displacement δd in d and another virtual displacement δX in X. The displacements are called "virtual" because they are purely imaginary. The requirements on acceptable virtual displacements are:

1. The time is held fixed.
2. The displacements are infinitesimal.
3. There is no change in time derivatives $\dot{d}$, $\dot{X}$.
4. There are as many possible virtual displacements as there are variables needed to describe the motion.
5. The displacements obey the constraints on the motion. This is true in our case because d and X are defined so that the block always slides on the upper surface of the plane as d and X change with time.

The concept of a virtual displacement was discovered in 1703 by Jacques Bernoulli as a tool for understanding static equilibrium but was developed by d'Alembert beginning in 1743 for use in problems involving motion.

It will be useful to express these virtual displacements as infinitesimal vectors in the same Cartesian coordinate system we used in the previous section:

$$\delta\vec{r}_{SB} = (\delta X + \delta d\cos\alpha)\hat{\imath} - \delta d\sin\alpha\,\hat{\jmath}, \tag{1.14}$$

$$\delta\vec{R}_{IP} = \delta X\,\hat{\imath}. \tag{1.15}$$

Equation (1.14) can be obtained by multiplying (1.11) by dt.

In mechanics, work is defined to be the line integral $W \equiv \int \vec{F} \cdot d\vec{r}$. In words, work is the integral of the component of force along the direction of displacement, integrated with

1.2 USING VIRTUAL WORK TO SOLVE THE SAME PROBLEM

respect to the displacement. Virtual work is defined in exactly the same way, except that the displacement is a virtual one:

$$\delta W \equiv \vec{F} \cdot \delta \vec{r}. \qquad (1.16)$$

No integration is necessary, because the displacement is an infinitesimal one.

In calculating the virtual work, we do not have to take into account the constraint forces $\vec{F}_1, \vec{F}_2$ we considered in the previous section, because constraint forces always act in a direction perpendicular to the possible displacements. The only nonconstraint forces are the gravitational forces on the two bodies ($-mg\hat{j}$ and $-Mg\hat{j}$). Only these two forces could influence the expression of the virtual work in terms of virtual displacements. However, only the gravitational force on the small block does any work when inserting (1.14) and (1.15) into (1.16):

$$\delta W(SB) = mg \sin\alpha \delta d, \quad \delta W(IP) = 0. \qquad (1.17)$$

The dynamics arises through a connection between the virtual work and partial derivatives of the kinetic energy with respect to $\dot{X}$ and $\dot{d}$. To find these partial derivatives, we treat $\dot{X}, \dot{d}$ as formal variables that are independent of X, d. It is perfectly straightforward to take partial derivatives of Equation (1.13) with respect to $\dot{d}, \dot{X}$.

As always in classical mechanics, the heart of the dynamics lies in the expression of Newton's Second Law: $\vec{F} = \dot{\vec{p}}$. ($\vec{p} \equiv m\vec{v}$. $m\vec{a} \neq \dot{\vec{p}}$ if the system's mass is changing, as in a rocket problem.) Rewriting (1.16) we obtain

$$\boxed{\delta W - \dot{\vec{p}} \cdot \delta \vec{r} = 0, \qquad (1.18)}$$
$$\text{d'Alembert's Principle}$$

$$\dot{\vec{p}} \cdot \delta \vec{r} = \frac{d(\vec{p} \cdot \delta \vec{r})}{dt} - \vec{p} \cdot \frac{d(\delta \vec{r})}{dt}. \qquad (1.19)$$

Equation (1.19) is a mathematical identity. It may seem like a contradiction that in making virtual displacements, $\delta \dot{d} = 0, \delta \dot{X} = 0$, while $\frac{d(\delta \vec{r})}{dt} \equiv \delta \dot{\vec{r}}$ could be nonzero. It would arise if the Cartesian vector virtual displacements $\delta \vec{r}$ in Equations (1.14) and (1.15) had depended on d or X as well as δd and δX. The time derivative of $\delta \vec{r}$ would not then vanish in general. In other words, the velocity of each body is a function of the coordinates as well as the time derivatives of the coordinates. One can see that the kinetic energy T will be a function of d and X as well as $\dot{d}, \dot{X}$ if this had actually happened. We *can* set $\delta \dot{\vec{r}} = 0$ for the particular case here, treating $\delta \vec{r}$ as a time-independent constant. Later we will extend our treatment to include the more general case.

To find the equations of motion, we need to obtain expressions for the total $\vec{p} \cdot \delta \vec{r}$, adding together the contributions of the small block and the inclined plane. The position of the small block, expressed as a vector $\vec{r}_{SB}$ is a function of both d and X. (We drop the subscript SB, since what we will say applies equally well to the inclined plane, although the position is only a function of X in that case.) Using the chain rule for differentiation of a function $\vec{r}(d, X)$ we get

$$\vec{p} \cdot \delta\vec{r} = \vec{p} \cdot \frac{\partial \vec{r}}{\partial d} \delta d + \vec{p} \cdot \frac{\partial \vec{r}}{\partial X} \delta X. \tag{1.20}$$

The velocity $\vec{v} = \dot{\vec{r}}$ is given by the general functional relation, again from calculus:

$$\dot{\vec{r}} = \frac{\partial \vec{r}}{\partial d} \dot{d} + \frac{\partial \vec{r}}{\partial X} \dot{X}. \tag{1.21}$$

Equation (1.11) is Equation (1.21) written out explicitly in component form for the small block.

Now replace $\frac{\partial \vec{r}}{\partial d}, \frac{\partial \vec{r}}{\partial X}$ in Equation (1.20) by $\frac{\partial \dot{\vec{r}}}{\partial \dot{d}}, \frac{\partial \dot{\vec{r}}}{\partial \dot{X}}$, respectively. This is a purely formal operation, which assumes that the variables and time derivatives of variables can be treated as formally independent of each other in Equation (1.21). For either body, Equation (1.20) becomes

$$\vec{p} \cdot \delta\vec{r} = \vec{p} \cdot \frac{\partial \dot{\vec{r}}}{\partial \dot{d}} \delta d + \vec{p} \cdot \frac{\partial \dot{\vec{r}}}{\partial \dot{X}} \delta X. \tag{1.22}$$

The vector $\vec{p} = m\dot{\vec{r}}$ can be expressed in Cartesian coordinates as

$$p_x = \frac{\partial T}{\partial \dot{x}}, \quad p_y = \frac{\partial T}{\partial \dot{y}}, \quad p_z = \frac{\partial T}{\partial \dot{z}}. \tag{1.23}$$

Inserting these three expressions into (1.22) and using the chain rule* together with an analogous relation for $\frac{\partial T}{\partial \dot{X}}$, we obtain an expression for $\vec{p} \cdot \delta\vec{r}$:

$$\boxed{\vec{p} \cdot \delta\vec{r} = \frac{\partial T}{\partial \dot{d}} \delta d + \frac{\partial T}{\partial \dot{X}} \delta X.} \tag{1.24}$$

The formula (1.24) is "boxed" because, when generalized to any number of variables, it plays a fundamental role in analytical mechanics. Equation (1.24) applies to any mechanical system described by two variables. Since the kinetic energy is an additive quantity, we can use (1.24) for either body or else for the total $(\vec{p} \cdot \delta\vec{r})_{SB} + (\vec{p} \cdot \delta\vec{r})_{IP}$. All that remains is to equate the time derivative of the total $\vec{p} \cdot \delta\vec{r}$ to the total virtual work, adding together both bodies. Inserting the specific form of the kinetic energy and the virtual work for this

---

* $\frac{\partial T}{\partial \dot{d}} = \frac{\partial T}{\partial \dot{x}} \frac{\partial \dot{x}}{\partial \dot{d}} + \frac{\partial T}{\partial \dot{y}} \frac{\partial \dot{y}}{\partial \dot{d}} + \frac{\partial T}{\partial \dot{z}} \frac{\partial \dot{z}}{\partial \dot{d}}.$

1.3 SOLVING FOR THE MOTION OF A HEAVY BEAD SLIDING ON A ROTATING WIRE

problem will lead us directly to the equations of motion. Returning to the specific problem of the small block and the inclined plane, use the expressions (1.24), (1.17), and (1.13) in (1.18) to obtain

$$mg \sin\alpha \delta d = \frac{d\left(\frac{\partial T}{\partial \dot{d}}\right)}{dt}\delta d + \frac{d\left(\frac{\partial T}{\partial \dot{X}}\right)}{dt}\delta X$$
$$= [(m+M)\ddot{X} + m\ddot{d}\cos\alpha]\delta X + m[\ddot{X}\cos\alpha + \ddot{d}]\delta d. \quad (1.25)$$

The virtual displacements are independent of each other, so it is possible to obtain two equations of motion from (1.25) by equating the coefficients of δd and δX separately to zero. When doing so, one obtains (1.7, 1.8).

QUESTION 1: *Virtual Work* The reader should test his or her understanding by using (1.25) to get the equations of the motion (1.7, 1.8) in order to prove that this solution agrees with that of the previous section.

At this point, this procedure may seem more indirect and elaborate than a straightforward application of Newtonian mechanics. But most of the effort was spent in justifying the steps involving partial derivatives of the kinetic energy. In future problems, all that will be required is to express the kinetic energy in terms of the chosen set of variables and their time derivatives and to find the virtual work, which is usually an easy task. The derivation presented here needs to be extended to cover the case where the velocities depend on the dynamical variables* as well as on their time derivatives. Otherwise we have derived an approach to mechanics that is very close to the spirit of Lagrangian mechanics.

1.3 SOLVING FOR THE MOTION OF A HEAVY BEAD SLIDING ON A ROTATING WIRE

This example will illustrate the method of the previous section, but it contains a new feature: The kinetic energy depends on both the dynamical variable and on its time derivative, instead of on the time derivative alone.

A bead slides without friction on a thin wire which is rotated about a vertical axis by a motor at a constant angular frequency ω as shown in Figure 1.2. The wire is tilted away from the Z axis by a fixed angle α. The bead is constrained to move on the wire. To describe the motion of the bead, it is only necessary to specify one dynamical variable. We choose this to be the distance q from the origin. (See Figure 1.2.)

The time dependence of $q(t)$ depends on the solution to a differential equation which we will derive. To specify this solution we must know the initial position $q(0)$ and velocity of the bead $\dot{q}(0)$ along the wire. There are two initial conditions. The position of the wire

---

* *Dynamical variables* are the time-dependent variables needed to describe the motion, such as $d(t)$ and $X(t)$ for the small block on the inclined plane.

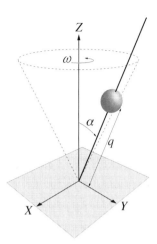

FIGURE 1.2
Bead on a wire.

is not a dynamical variable because it is a given function of time that does not depend on specific initial conditions.

Since only one dynamical variable is needed to describe the motion of the bead, this system is said to have only one *degree of freedom*. The system of the small block sliding on the movable inclined plane has two degrees of freedom. If the bead were unconstrained in free space, it would have three degrees of freedom. The wire provides two constraints, which reduce the number of degrees of freedom to $3 - 2 = 1$.

To describe the position of the bead in space, use a fixed set of Cartesian coordinates:

$$\vec{r} \equiv (x, y, z), \tag{1.26}$$

$$\begin{aligned} x(t) &= q(t) \sin \alpha \cos \omega t, \\ y(t) &= q(t) \sin \alpha \sin \omega t, \\ z(t) &= q(t) \cos \alpha. \end{aligned} \tag{1.27}$$

The number of degrees of freedom is an intrinsic property of the system, so there must be two relations between the three equations above (1.27) to reduce the number to one. These are $q^2 = x^2 + y^2 + z^2$ and $\tan \alpha = \sqrt{\frac{x^2+y^2}{z}}$.

Calculate the bead velocity in terms of q and $\dot{q}$ by differentiating the formulas (1.27) with respect to time:

$$\vec{v}(t) = \dot{\vec{r}} = (\dot{x}, \dot{y}, \dot{z}), \tag{1.28}$$

$$\begin{aligned} \dot{x}(t) &= \dot{q}(t) \sin \alpha \cos \omega t - \omega q(t) \sin \alpha \sin \omega t, \\ \dot{y}(t) &= \dot{q}(t) \sin \alpha \sin \omega t + \omega q(t) \sin \alpha \cos \omega t, \\ \dot{z}(t) &= \dot{q}(t) \cos \alpha. \end{aligned} \tag{1.29}$$

Assume that the XYZ reference system is an inertial frame. The kinetic energy of the bead

1.3 SOLVING FOR THE MOTION OF A HEAVY BEAD SLIDING ON A ROTATING WIRE

is $\frac{1}{2}mv^2$:

$$T \equiv \frac{1}{2}m\dot{\vec{r}} \cdot \dot{\vec{r}} = \frac{1}{2}m(\dot{x}^2 + \dot{y}^2 + \dot{z}^2) = \frac{1}{2}m(\omega^2 q^2 \sin^2\alpha + \dot{q}^2). \quad (1.30)$$

Add a virtual displacement δq to q at a fixed time, as before. Define $\delta\vec{r}$:

$$\delta\vec{r} \equiv \vec{r}(q + \delta q, t) - \vec{r}(q, t) = \frac{\partial \vec{r}}{\partial q}\delta q. \quad (1.31)$$

Likewise

$$\delta\dot{\vec{r}} \equiv \dot{\vec{r}}(q + \delta q, \dot{q}, t) - \dot{\vec{r}}(q, \dot{q}, t) = \frac{\partial \dot{\vec{r}}}{\partial q}\delta q. \quad (1.32)$$

Note that in this case, $\delta\dot{\vec{r}} \neq 0$. The reason is the appearance of q in Equations (1.29) above.

Again denote the virtual work by δW using (1.18). What are the forces acting on the bead? There are two distinct types of force. One is the downward force of gravity: $-mg\hat{z}$ ($\hat{z}$ is the unit vector in the z direction); the other is the time-dependent constraint force exerted by the wire on the bead. Only the force of gravity contributes to the virtual work:

$$\delta W = -mg\cos\alpha\,\delta q. \quad (1.33)$$

According to the identity, Equation (1.19), (1.18) can be written as a total time derivative of $\vec{p} \cdot \delta\vec{r}$ minus a second term:

$$\delta W = \frac{d}{dt}(\vec{p} \cdot \delta\vec{r}) - \vec{p} \cdot \delta\dot{\vec{r}}. \quad (1.34)$$

Using formula (1.24) for a system with only one degree of freedom we have

$$\vec{p} \cdot \delta\vec{r} = \frac{\partial T}{\partial \dot{q}}\delta q = m\dot{q}\,\delta q. \quad (1.35)$$

The chain rule implies that, if the kinetic energy ($T \equiv \frac{1}{2}m\dot{\vec{r}} \cdot \dot{\vec{r}}$) is a function of q, then

$$\frac{\partial T}{\partial q}\delta q = m\dot{\vec{r}} \cdot \frac{\partial \dot{\vec{r}}}{\partial q}\delta q = \vec{p} \cdot \delta\dot{\vec{r}}. \quad (1.36)$$

(We used (1.32) in the last step.) Using the explicit form of T from (1.30), we obtain

$$\vec{p} \cdot \delta\dot{\vec{r}} = m\,\omega^2 \sin^2\alpha\, q\,\delta q. \quad (1.37)$$

The differential equation of the motion (abbreviated "EOM") can be obtained by using (1.33), (1.35), and (1.37) in (1.34). (Recall that δq is a constant.) The EOM of the bead obtained in this way is

$$\ddot{q} - \omega^2 \sin^2\alpha\, q = -g\cos\alpha. \quad (1.38)$$

A homework problem asks you to investigate what this equation implies about the motion of the bead.

1.4 TOWARD A GENERAL FORMULA: DEGREES OF FREEDOM AND TYPES OF CONSTRAINTS

We want to derive general equations of motion from d'Alembert's Principle. There are some preliminary things to discuss before we do this.

Degrees of Freedom

The most general definition of *dynamical variables* is any set of variables that can describe completely the configuration of a mechanical system (i.e., the position of its parts) and that can change under the action of the forces. In what follows we will always make the assumption that any one of the q_k may be varied independently from the others.* The motion of the system is completely determined by specifying the dynamical variables as functions of time. Newton's Laws imply that these functions are found by solving a set of second-order differential equations. Initial conditions for all of the initial values $q_k(0)$ and $\dot{q}_k(0)$ must be known in order to determine a specific solution. (Often we will write q_k and $\dot{q}_k$ without explicitly indicating that they are functions of time.)

Usually the dynamical variables are chosen to be positions and/or angles. A mass point has three degrees of freedom. The motion of a point mass in space, $\vec{r}(t)$, is therefore described by three different, independent, dynamical variables. Often they are the Cartesian coordinates, $\vec{r}(t) = (x(t), y(t), z(t))$. Cylindrical polar coordinates could instead be used, with the variables (r, ϕ, z), or else spherical polar coordinates, with the variables (r, θ, ϕ), or perhaps some other convenient definition of coordinates.

Unspecified dynamical variables q_k are referred to as *generalized coordinates*. The number N of degrees of freedom is an intrinsic property of the system, but the actual specific choice of dynamical variables is up to us, as long as the number of independent coordinates is N.

Constraints reduce the number of degrees of freedom. A box confined to the top of a table has three degrees of freedom – two possible translations and one degree of freedom for arbitrary rotations about an axis perpendicular to the table top. A mass point confined to the same table top will have only two degrees of freedom, since the rotational motion is now not significant.

Consider a collection of M massive point particles. We might think of this collection as a single dynamical system with $3M$ degrees of freedom. If there are j *independent* constraints, the system has only $N = 3M - j$ degrees of freedom. Some examples of $M = 2, 3$ are shown in Figure 1.3. A rigid body with three or more mass points has only six degrees of freedom, like Figure 1.3C.

QUESTION 2: *Degrees of Freedom 1*   a) Prove that a rigid body, with three or more mass points, has six degrees of freedom. Remember to point out what is so special

---

* Situations in which independent variation of the dynamical variables is not possible would mean that the number of variables would be greater than the number of degrees of freedom. We will discuss this in Chapter 2.

1.4 TOWARD A GENERAL FORMULA: DEGREES OF FREEDOM AND TYPES OF CONSTRAINTS

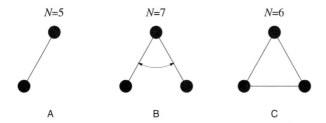

FIGURE 1.3
Systems with many degrees of freedom. All systems are point masses connected by rigid massless rods. Motion can be in three dimensions. System B has a flexible joint.

about a rigid body and why we need only six dynamical variables to describe the motion. b) What are these six degrees of freedom explicitly? c) Is the choice unique? With regard to uniqueness, you should consider the choice of coordinate system as well as the choice of origin.

QUESTION 3: *Degrees of Freedom 2* Identify how many degrees of freedom there are in the following mechanical systems: a) a compact disc in use (i.e., a rigid body rotating about a fixed axis – all the points in the body that lie along the axis are not moving), b) a spinning top (i.e., a rigid body rotating about a fixed point – all the points of the body can rotate except for the point on the body that is fixed), c) a bug crawling on the ground (i.e., a particle moving on a given surface), d) a dumbbell (i.e., two particles kept at a constant distance from each other), e) two components of a double star that revolve in the same plane (note that a double star is in three dimensions, but the force of gravity between them causes them to be coplanar; also note that a star should be considered to be a rigid body rather than a point mass), and *f*) the Earth–Moon–Sun system (note that a planet should also be considered to be a rigid body rather than a point mass).

Model of a Rigid Body as a Collection of Mass Points

It must be possible to express the spatial position $\vec{r}_i$ of each part of a system as a function of a finite number of dynamical variables. The subscript "i" is a label that will identify a specific part of the system, as shown in Figure 1.4. This requirement is indicated by writing

$$\vec{r}_i = \vec{r}_i(q_1, \ldots, q_k, \ldots, q_N, t). \tag{1.39}$$

N is an integer indicating the number of degrees of freedom: $1 \leq k \leq N$. Besides the time dependence that arises because the q_k depend on the time, there can sometimes be an explicit time dependence due to moving constraints, which we have indicated in Equation (1.39).

If there is no explicit time dependence, we may indicate this by writing, instead of (1.39),

$$\vec{r}_i = \vec{r}_i(q_1, \ldots, q_k, \ldots, q_N). \tag{1.40}$$

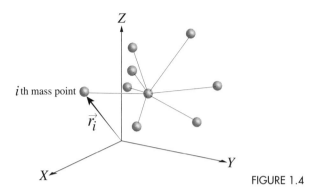

FIGURE 1.4

Equation (1.39) expresses the fact that at any instant we can specify the location of the system in space by specifying the value of all the q_k and the time. Strictly speaking, this is complete information only if we consider a system consisting of a finite number of point masses. For truly rigid bodies, we can, in our minds, subdivide the body down to the atomic level if we wish, still treating the system as a finite number of point masses. As noted above, there will be only six degrees of freedom. If the body is not a rigid one, the location of its parts will be considerably more complicated, depending on an infinite number of dynamical variables. We exclude this possibility from consideration; it lies in the realm of continuum mechanics.

For example, a chair may have 10^{24} molecules, hence 3×10^{24} coordinates. In practical situations we need only a maximum of six coordinates to describe the motion of the chair. We imagine that we could know the position of each molecule as a function of these six variables:

$$\vec{r}_i = \vec{r}_i(q_1, \ldots, q_6), \quad i = 1, \ldots, M = 10^{24}. \tag{1.41}$$

We don't actually have to know the 10^{24} individual molecular positions, just that it is (*in principal*) possible to write a function for each molecule that only depends on the degrees of freedom for the entire chair.

The Two Types of Constraints: Time-Independent or Time-Dependent

A spherical pendulum can be constructed by suspending a mass from a pivot anchored in the ceiling. The pivot allows the mass to swing in any direction. The length of the pendulum remains constant. The mass is free to move on the inside surface of an imaginary sphere. How many dynamical variables does it take to describe the motion of the pendulum? The answer to this question is two, which could be θ and ϕ, the polar and azimuthal angles. There are two *degrees of freedom*. It is similar to describing your location on the face of the Earth. Again two numbers are required to do this: latitude and longitude.

For the spherical pendulum, the constraint that the distance from the mass to the pivot remain constant is a *time-independent* constraint. Not all constraints are independent of time. The bead on a rotating wire is an example of an *explicit time dependence* in the constraints.

1.4 TOWARD A GENERAL FORMULA: DEGREES OF FREEDOM AND TYPES OF CONSTRAINTS

The difference between the two types of constraints is very important for the physics. We have already seen that it is necessary to find not only $\frac{\partial T}{\partial \dot{q}}$ but also $\frac{\partial T}{\partial q}$ if the constraints introduce a q dependence into the kinetic energy. For example, one could have a particle constrained to slide on the surface of a sphere. This has two degrees of freedom, since the distance r of the particle from the center of the sphere is constrained to be the radius of the sphere R. But suppose the radius of this sphere is changing in a predetermined way with time: $R = R(t)$. You would expect to get a completely different answer for the motion of the particle in this situation.

When *either* Equations (1.40) or (1.39) hold for all times, we say the constraints are *holonomic constraints*. If (1.40) is the relevant one, the constraints are time-independent. They are then called *scleronomic constraints*. For this case, the parts of the system move only because the dynamical variables change with time.

It could happen that the functional equation between the $\vec{r}_i$ and the set of generalized coordinates q_k involves the time explicitly, as indicated symbolically by Equation (1.39). Time-dependent constraints are called *rheonomic constraints*. Note that an explicit time dependence means that you can *see* the time in the constraint equations. For the example of the bead sliding on a rotating wire, the position in space of the bead was expressed as a function of the generalized coordinate q as well as t in Equations (1.27). Unlike in the case of the small block sliding on the inclined plane or the spherical pendulum, the constraint equations for the bead involve the time *explicitly* as well as implicitly.

Both rheonomic and scleronomic constraints share a common property: All positions of all parts of the system are described by functions of the dynamical variables and, in the rheonomic case, the time. Both types of constraints are holonomic. There are constraints that are not holonomic. Sometimes we have constraints on the velocities, such as for a ball rolling without slipping on a table top. We will deal with this situation later. For now we will only focus on holonomic constraints. The Greek meanings of the two names for the types of holonomic constraints are

> scleronomic: "rigid law,"
> rheonomic: "running law."

QUESTION 4: *Rheonomic vs. Scleronomic* a) Identify whether or not the following examples need to be described by generalized coordinates with rheonomic constraints or scleronomic constraints: i) a point mass sliding on the surface of a bowl, ii) a pendulum whose support point is driven vertically up and down, iii) a top spinning on a table, and iv) a spinning top in free fall. b) Will the kinetic and/or potential energy contain an explicit time dependence for scleronomic constraints? How about for rheonomic constraints? Explain.

Some examples of nonholonomic constraints are given in Appendix A of this chapter.

1.5 GENERALIZED VELOCITIES: HOW TO "CANCEL THE DOTS"

By *generalized velocities*, we mean the set of all first time derivatives of the q_k. We can regard $q_k(t)$ and $\dot{q}_k(t)$ as independent variables. They are physically independent since we have independent control over the starting values of $q_k(0)$ and $\dot{q}_k(0)$.

Changing the dynamical variables can mean moving all parts of the system. The positions $\vec{r}_i$ will change as q_k varies. If one particular q_k is varied and all the other $N-1$ q_ks are held constant as well as time, the partial derivative of $\vec{r}_i$ with respect to q_k is

$$\frac{\partial \vec{r}_i}{\partial q_k}. \tag{1.42}$$

We relate the "real" velocity of the ith particle, $\dot{\vec{r}}_i$, to the generalized velocities, $\dot{q}_k$. By the chain rule applied to differentiation of Equation (1.39) or (1.40), the velocity of the ith mass point is

$$\vec{v}_i = \dot{\vec{r}}_i = \sum_{k=1}^{N} \frac{\partial \vec{r}_i}{\partial q_k} \dot{q}_k + \frac{\partial \vec{r}_i}{\partial t}. \tag{1.43}$$

The last term is absent if we have scleronomic constraints (i.e., constraint equations of the form of Equation (1.40)). Take the partial derivative with respect to $\dot{q}_k$ of the expression above to obtain the relation

$$\frac{\partial \dot{\vec{r}}_i}{\partial \dot{q}_k} = \frac{\partial \vec{r}_i}{\partial q_k}. \tag{1.44}$$

It is as if the "dots" can be cancelled out. This useful identity holds for both types of holonomic constraints.

QUESTION 5: *Dot Cancellation* a) Why isn't Equation (1.44) true if the holonomic constraint equations represented by Equations (1.39, 1.40) also would have contained the generalized velocities? What would an application of the chain rule look like in this case? b) See if "dot cancellation" works for two simple examples: i) a mass sliding without friction down a stationary inclined plane and ii) the rotating bead on a wire. These are prototypes of scleronomic and rheonomic constraints respectively. These are really simple examples, because the coordinate representing the degree of freedom appears linearly in the equations for the position. But the "dots cancel" even in more complicated functional relations, as proved above.

1.6 VIRTUAL DISPLACEMENTS AND VIRTUAL WORK – GENERALIZED FORCES

In this section we will discuss virtual displacements and virtual work from a completely general point of view.

Observe the motion of a mechanical system. Now "freeze" the motion at some particular moment in time. Displace only one of the degrees of freedom (q_ks) an infinitesimal

1.6 VIRTUAL DISPLACEMENTS AND VIRTUAL WORK – GENERALIZED FORCES

amount δq_k. (Each coordinate can be varied independently of the other generalized coordinates.) Look at a particular part of the system, called the ith part. The position of this part, $\vec{r}_i$, will change in space by an amount

$$\delta \vec{r}_i = \frac{\partial \vec{r}_i}{\partial q_k} \delta q_k. \qquad (1.45)$$

Due to the nonconstraint forces being exerted on the different parts of the system, virtual work will be done by this virtual displacement of one of the degrees of freedom. Now consider the various possible virtual displacements as acting simultaneously and independently. If more than one of the q_ks is changed, then the virtual work is the sum of the virtual work done by each generalized coordinate variation. Call this total virtual work δW. By definition:

$$\delta W \equiv \sum_i \vec{F}_i \cdot \delta \vec{r}_i \quad \left(\sum_i \equiv \sum_{i=1}^{i=M} \right). \qquad (1.46)$$

Using the formula (1.45), we can obtain partial derivatives connecting the virtual displacements of the different parts with the virtual displacement of q_k:

$$\delta W_k \equiv \left(\sum_i \vec{F}_i \cdot \frac{\partial \vec{r}_i}{\partial q_k} \right) \delta q_k,$$
$$\delta W = \sum_k \delta W_k \quad \left(\sum_k \equiv \sum_{k=1}^{k=N} \right). \qquad (1.47)$$

Recall that since constraint forces always act to maintain the constraint, they point in a direction perpendicular to the movement of the parts of the system. This means that the constraint forces do not contribute anything to the virtual work. Equation (1.47) can be rewritten in terms of the nonconstraint forces as a sum over all parts of the system,* followed by a sum over the N generalized coordinates:

$$\delta W = \sum_k \left(\sum_i \vec{F}_i^{\,nc} \cdot \frac{\partial \vec{r}_i}{\partial q_k} \right) \delta q_k. \qquad (1.48)$$

The *generalized force* $\mathcal{F}_k$ associated with the kth degree of freedom is defined:

$$\boxed{\mathcal{F}_k \equiv \sum_i \vec{F}_i \cdot \frac{\partial \vec{r}_i}{\partial q_k} = \frac{\delta W}{\delta q_k}.} \qquad (1.49)$$

We have suppressed the superscript "nc" on $\vec{F}_i$. From now on we will refer only to the forces that do virtual work. The dimensions of $\mathcal{F}_k$ depend on the dimensions of q_k, as can be seen from the defining equation (1.49). They are not the same as ordinary force unless q_k happens to have the dimensions of length.

---

* The notation $\vec{F}_i^{\,nc}$ stands for the nonconstraint force on part i.

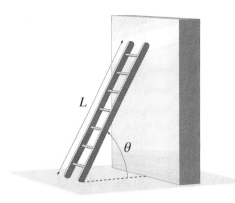

FIGURE 1.5

1.7 KINETIC ENERGY AS A FUNCTION OF THE GENERALIZED COORDINATES AND VELOCITIES

Kinetic energy is additive. The most general form for the kinetic energy for any holonomic system is

$$T = \frac{1}{2}\sum_{i=1}^{M} m_i \dot{\vec{r}}_i \cdot \dot{\vec{r}}_i = T(q_1, \ldots, q_N, \dot{q}_1, \ldots, \dot{q}_N, t). \qquad (1.50)$$

The sum over the index i goes from 1 to M (i.e., all parts of the system), but the kinetic energy is only a function of the $2N$ variables representing the generalized coordinates and their first time derivatives. This function can be found for a particular case by summing over the parts of the system. Even for rheonomic constraints, it is often true that the time does not appear explicitly in the equation for T. This was true for the bead on a rotating wire, for example.

> **QUESTION 6: Kinetic Energy** A ladder of length L and mass M is leaning against a wall as shown in Figure 1.5. Assuming the wall and the floor are frictionless, the ladder will slide down the wall and along the floor until the left end loses contact with the wall. Before the ladder loses contact with the wall there is one degree of freedom (θ, see Figure 1.5). Express the kinetic energy of the ladder in terms of θ and $\dot{\theta}$. Hint: Divide the ladder into infinitesimal slices of thickness ds, a distance s away from the origin. Then note that the mass dm of each slice is $dm = \frac{M}{L}ds$. The total kinetic energy is the sum (i.e., integral) of the kinetic energy of each slice.

> **QUESTION 7: Virtual Work, Generalized Forces** Find the virtual work and generalized force for the example of the ladder sliding down the frictionless wall. Compare to what you know about the ladder from Newtonian mechanics.

We need to know partial derivatives of the kinetic energy. Using the definition of T from Equation (1.50) and the chain rule we have

$$\frac{\partial T}{\partial q_k} = \sum_i m_i \dot{\vec{r}}_i \cdot \frac{\partial \dot{\vec{r}}_i}{\partial q_k} = \sum_i \vec{p}_i \cdot \frac{\partial \dot{\vec{r}}_i}{\partial q_k}, \qquad (1.51)$$

1.7 KINETIC ENERGY AS A FUNCTION OF THE GENERALIZED COORDINATES AND VELOCITIES

$$\frac{\partial T}{\partial \dot{q}_k} = \sum_i m_i \dot{\vec{r}}_i \cdot \frac{\partial \dot{\vec{r}}_i}{\partial \dot{q}_k} = \sum_i \vec{p}_i \cdot \frac{\partial \vec{r}_i}{\partial q_k}. \tag{1.52}$$

The second part of (1.52) is only true for holonomic constraints that allow "dot cancellation" (1.44).

The virtual work and thus $\mathcal{F}_k$ can be related by d'Alembert's Principle to the partial derivatives of the kinetic energy on the left sides of Equations (1.51, 1.52). This is the crucial step in the development of analytical mechanics. Take the total time derivative of $\frac{\partial T}{\partial \dot{q}_k}$ as expressed in Equation (1.52). Work it out using the rule for differentiating a product:

$$\frac{d}{dt}\left(\frac{\partial T}{\partial \dot{q}_k}\right) = \sum_i \dot{\vec{p}}_i \cdot \frac{\partial \vec{r}_i}{\partial q_k} + \sum_i \vec{p}_i \cdot \frac{\partial \dot{\vec{r}}_i}{\partial q_k}. \tag{1.53}$$

In the second term on the right-hand side of Equation (1.53), the order of the total time derivatives and the partial derivatives have been exchanged.

The last step is where the physics enters – use Newton's Second Law to replace $\dot{\vec{p}}_i$ by the force on the part i: Only if the $\vec{r}_i$ are referred to an *inertial* frame, can Newtonian dynamics be used. The first term on the right-hand side of Equation (1.53) is the generalized force $\mathcal{F}_k$, as can be seen by comparison with Equation (1.49) above. From Equation (1.51), the last term on the right of Equation (1.53) was already shown to equal $\frac{\partial T}{\partial q_k}$. We have discovered the basic formula of analytical mechanics, in its most general form:

> **GOLDEN RULE #1**
>
> $$\mathcal{F}_k = \frac{d}{dt}\left(\frac{\partial T}{\partial \dot{q}_k}\right) - \frac{\partial T}{\partial q_k}, \tag{1.54}$$
>
> generalized equations of motion
>
> $k = 1, \ldots, N.$

Only a single scalar function of the generalized coordinates and velocities, the kinetic energy, and the functions $\mathcal{F}_k$ of the same coordinates are needed to use the N Equations (1.54). It is unnecessary to find the constraint forces. Newton's Laws are not used explicitly, but they are contained in (1.54).

We have obtained a set of general equations that will give the differential equations of motion for any holonomic system. A major advantage of this method is that the number of unknowns equals the number of degrees of freedom, unlike for Newton's vector mechanics, in which the constraint forces are additional unknowns and more equations must be solved.

If the forces are conservative forces, it is possible to simplify things even more and eliminate the need to find $\mathcal{F}_k$ explicitly. We will do this in the next section.

QUESTION 8: *Generalized EOM 1* Using the formula for the generalized EOM (1.54), find the EOM for a mass sliding without friction down an inclined plane. Assume that gravity is present, and explicitly find the generalized forces. Compare your EOMs to what you would obtain using Newtonian mechanics.

QUESTION 9: *Generalized EOM 2* Using the formula for the generalized EOM (1.54), find the EOM for the ladder example (see Figure 1.5). Assume that gravity is present, and explicitly find the generalized forces. Compare your EOMs to what you would obtain using Newtonian mechanics.

1.8 CONSERVATIVE FORCES: DEFINITION OF THE LAGRANGIAN *L*

If the nonconstraint forces are conservative, then there exists, by definition, a potential energy function $V(q_1, \ldots, q_N)$. The nonconstraint force on the ith point mass part of the system is defined as the gradient of the potential energy with respect to the coordinates of this part of the system:

$$\vec{F}_i = -\vec{\nabla}_i V(\vec{r}_1, \ldots, \vec{r}_i, \ldots, \vec{r}_M). \tag{1.55}$$

For example, the x component of this force is

$$F_{i,x} = -\frac{\partial V}{\partial x_i}. \tag{1.56}$$

Equation (1.55) can be integrated to calculate the work done by a change in the q_k:

$$W = \int \sum_i \vec{F}_i \cdot d\vec{r}_i = -\sum_i \int \vec{\nabla}_i V(\vec{r}_1, \ldots, \vec{r}_i, \ldots, \vec{r}_M) \cdot d\vec{r}_i$$

$$= -\sum_k \int_{q_k(0)}^{q_k(1)} \left(\sum_i \vec{\nabla}_i V \cdot \frac{\partial \vec{r}_i}{\partial q_k} \right) dq_k = -\sum_k \int_{q_k(0)}^{q_k(1)} \frac{\partial V}{\partial q_k} dq_k. \tag{1.56}$$

In the last line, we have changed from viewing V as a function of the individual parts to viewing the same V as a function of the generalized coordinates. This is possible only for holonomic constraints. The work done is a function of the q_k at the end points of the motion, which are denoted $q_k(0), q_k(1)$ in the formula above. This work will not depend on the specific initial conditions and the actual path $q(t), \dot{q}(t)$.* If V did depend on the generalized velocities, there would be a path dependence. We must exclude this possibility.

We can also integrate (1.48), and use (1.49) to obtain

$$W = \int \sum_k \mathcal{F}_k \, \delta q_k. \tag{1.57}$$

Comparing this to (1.56) and noting that we can exchange the sum and integral sign, we conclude that the generalized force is given by

$$\mathcal{F}_k = -\frac{\partial V}{\partial q_k}. \tag{1.58}$$

---

* See Appendix B for a more detailed discussion of this important point.

1.8 CONSERVATIVE FORCES: DEFINITION OF THE LAGRANGIAN L

This result tells us that, if there are only conservative forces and constraint forces, we can bypass the virtual work method and calculate $\mathcal{F}_k$ directly from the partial derivative of the potential energy expressed in terms of the generalized coordinates: $V(q_1, \ldots, q_N)$.

QUESTION 10: *Component of Force from V* Suppose V is a known function of cylindrical polar coordinates r, ϕ, z. Find all three x, y, z components of the force in terms of the partial derivatives of V: $\frac{\partial V}{\partial r}, \frac{\partial V}{\partial \phi}, \frac{\partial V}{\partial z}$.

QUESTION 11: *Gravitational Potential Energy* Prove that the gravitational potential energy for a body of arbitrary shape and mass depends only on the height of the center of mass.

Since we have assumed the potential V is velocity-independent, $\frac{\partial V}{\partial \dot{q}_k} = 0$. Define the *Lagrangian L*:

$$L \equiv T - V. \tag{1.59}$$

Substituting this into the general form of the EOM, Equation (1.54), and using (1.58), we obtain the *Euler–Lagrange equations* in their final form:

> **GOLDEN RULE #2**
>
> $$\frac{d}{dt}\left(\frac{\partial L}{\partial \dot{q}_k}\right) - \frac{\partial L}{\partial q_k} = 0, \tag{1.60}$$
>
> for $k = 1, \ldots, N$.
>
> Euler–Lagrange equations

This is the second and most useful form of the equations of motion. These are the fundamental equations of Lagrangian mechanics. The number of these equations equals the number of degrees of freedom, N. Thus we always have exactly the minimum number of equations needed to solve the problem. What is more, we have reduced the entire problem of solving the motion to finding a single scalar function. Of course, the EOM still need to be solved, but this is a mathematics problem. The physics of the problem is entirely contained in the Lagrangian function.

How often do we encounter conservative forces? Almost 100% of the cases in mechanics texts involve conservative forces. The exception in this book is when we consider damped simple harmonic oscillators in Chapter 3. In real life, it is very often a useful approximation to consider that the forces are conservative as a first approximation. Often that is all you will need, as long as friction plays a minor role.

QUESTION 12: *Euler–Lagrange EOM 1* Using the formula for the Euler–Lagrange EOM (1.60), find the Lagrangian and then the EOM for a mass sliding without friction down an inclined plane. Assume that gravity is present. What happens if there is friction present between the mass and the plane? Why can't you use the Lagrange method?

QUESTION 13: *Euler–Lagrange EOM 2* Using the formula for the Euler–Lagrange EOM (1.60), find the Lagrangian and then the EOM for the ladder example (see Figure 1.5). Assume that gravity is present.

1.9 REFERENCE FRAMES

The kinetic energy T and potential energy V must be evaluated in an inertial reference frame, where Newton's Laws hold. However, the variables by which we choose to express T and V can be variables with respect to any frame, even a noninertial accelerated reference frame.

As a simple example of this, consider observing the motion of a ball thrown upwards from a rapidly ascending or descending elevator, one accelerated with $a = a_{\text{elevator}}$. A person in an *inertial* reference frame will see the height of the ball, $y(t)$, obey the Lagrangian

$$L = \frac{1}{2}m\dot{y}^2 - mgy. \tag{1.61}$$

The solution is

$$y(t) = v_0 t - \frac{1}{2}gt^2, \tag{1.62}$$

where v_0 is the initial velocity of the ball.

We can also chose to express the motion as seen from the elevator, a noninertial reference frame. If $\tilde{y}$ is the height of the ball as seen by the person in the elevator, the differential equation for the motion is

$$\frac{d^2\tilde{y}}{dt^2} = -(g+a). \tag{1.63}$$

The physical trajectory $\tilde{y}(t)$, as measured with respect to the accelerated floor of the elevator, is

$$\tilde{y}(t) = v_0 t - \frac{1}{2}(g+a)t^2. \tag{1.64}$$

If you simply calculate T and V using the noninertial frame variables, as in the wrong column of Table 1.1, the answer will be incorrect, since we derived the Lagrangian formalism from Newton's laws, which are only valid in inertial frames. Therefore, we must continue to use the Lagrangian (1.61), and substitute into this expression the noninertial variables as shown in the correct column of Table 1.1. This will give the correct EOM for $y(t)$ or $\tilde{y}(t)$ (1.63).

TABLE 1.1 COMPARISON OF RIGHT AND WRONG APPROACHES FOR A UNIFORMLY ACCELERATED REFERENCE FRAME

Correct	Wrong
$T = \frac{1}{2}m(\dot{\tilde{y}} + at)^2 = \frac{1}{2}m\dot{y}^2$	$T = \frac{1}{2}m\dot{\tilde{y}}^2$
$V = mg(\tilde{y} + \frac{1}{2}at^2) = mgy$	$V = mg\tilde{y}$

1.10 DEFINITION OF THE HAMILTONIAN

A function of the dynamical variables and their time derivatives is said to be "conserved" if it is a constant of the motion, that is, constant in time as the motion of the system proceeds. We say that the energy of a system is conserved if the total energy $E \equiv T + V$ is constant.

Without even solving the EOM, we can know that a certain quantity, called the *Hamiltonian* and denoted by H, is conserved under very general conditions. The definition of H is

$$H \equiv \sum_k \dot{q}_k \frac{\partial L}{\partial \dot{q}_k} - L. \tag{1.65}$$

To prove H is constant, take $\frac{dH}{dt}$:

$$\frac{dH}{dt} = \sum_k \ddot{q}_k \frac{\partial L}{\partial \dot{q}_k} + \dot{q}_k \frac{d}{dt}\left(\frac{\partial L}{\partial \dot{q}_k}\right) - \frac{dL}{dt}. \tag{1.66}$$

According to the chain rule for differentiating an implicit function of time:

$$\frac{dL}{dt} = \sum_k \frac{\partial L}{\partial q_k}\dot{q}_k + \sum_k \frac{\partial L}{\partial \dot{q}_k}\ddot{q}_k + \frac{\partial L}{\partial t}. \tag{1.67}$$

Putting these equations together, we see that there are cancellations after we use the Euler–Lagrange equations (1.60) to eliminate $\frac{d}{dt}(\frac{\partial L}{\partial \dot{q}_k})$. You should prove for yourself that we get the simple and completely general result:

$$\frac{dH}{dt} = -\frac{\partial L}{\partial t}. \tag{1.68}$$

If the time doesn't appear *explicitly* in the Lagrangian, $\frac{\partial L}{\partial t} = 0$. Then and only then is H conserved. This important quantity H may be calculated once the Lagrangian is known. There will be much more about the Hamiltonian in Chapter 5.

For scleronomic constraints, the Lagrangian cannot have any explicit time dependence. Furthermore, it can be shown (as a homework problem) that in this case T is a quadratic function of the generalized velocities (*quadratic form*) and thus $H = E = T + V$. *Total energy is always conserved for scleronomic constraints.* For rheonomic constraints, the Lagrangian may or may not have an explicit time dependence. If it does not, as in the rotating bead example, then H must be a constant of the motion, but it need not equal the total energy E. If the energy is not conserved it is because there is an external source of energy in the problem.

When H is a constant and there is only one degree of freedom, the equations of motion can be integrated once for "free." Then only a first-order differential equation needs to be solved. This can make a big difference, since a first-order equation can always be solved in terms of quadratures (integrals), whereas that is not true of the original second-order EOM.

QUESTION 14: Hamiltonian Consider the Lagrangian for a bead on a rotating horizontal wire: $L = \frac{m}{2}(\dot{q}^2 + \omega^2 q^2)$. a) What is H? Is it constant? b) What if angular speed of rotation ω were not a constant? If $\omega = \omega(t)$ what is H? Would it be constant? c) In either case does $H = E$, the total energy?

1.11 HOW TO GET RID OF IGNORABLE COORDINATES

If only the time derivative of a particular coordinate appears in the Lagrangian, but not the coordinate itself, we say that this coordinate is *ignorable*.* For example, for a free particle, the position of the particle is ignorable because it does not appear in the Lagrangian, which is $\frac{1}{2}mv^2$. Assuming only the time derivative of a coordinate q appears in the Lagrangian, then from the Euler–Lagrange equations for this coordinate we have

$$\frac{d}{dt}\left(\frac{\partial L}{\partial \dot{q}}\right) = 0. \qquad (1.69)$$

This means that

$$p \equiv \frac{\partial L}{\partial \dot{q}} \qquad (1.70)$$

is a constant of the motion. We can solve Equation (1.70) as an implicit equation for $\dot{q}$ in terms of p and q, plus all the other degrees of freedom and their time derivatives: $\dot{q} = \dot{q}(p, \ldots)$.

The term p is called the *canonically conjugate momentum* to the coordinate q, whether or not q is ignorable. We will see later that this generalization of the idea of momentum plays an important role in mechanics. For a free particle, the momentum conjugate to x is $m\dot{x}$, and thus canonically conjugate momentum reduces to the usual definition of momentum. This is not always true, however. Canonically conjugate momentum does not equal ordinary momentum for a charged particle in a magnetic field, for example.

---

* In some texts, the term "cyclic" is used instead of "ignorable."

It is possible to eliminate ignorable coordinates from the Lagrangian, but the Lagrangian must be modified to do this. One forms a new Lagrangian, called the *Routhian*, which is

$$R \equiv L - p\dot{q}. \tag{1.71}$$

It isn't obvious from what has been said, but R can be treated just like a Lagrangian, with $N - 1$ degrees of freedom.

1.12 DISCUSSION AND CONCLUSIONS – WHAT'S NEXT AFTER YOU GET THE EOM?

The basic concepts of virtual displacement and virtual work were used to derive a general equation of motion for holonomic systems of either type: scleronomic or rheonomic (Equation (1.54)). The Euler–Lagrange equations (1.60) follow after the additional assumption of conservative forces. A single scalar function, the Lagrangian, contains all of the information about the possible motions of any system we want to study. After we determine the kinetic and potential energy functions, we treat the system as a whole. The coordinate system can be chosen so that it best fits the particular problem being solved.

Euler found the equations we call his in 1760, but they were discovered as pure mathematics, as a tool for solving the calculus of variations. It was Lagrange who first applied them to physics. He published a general theory of mechanics in the treatise *Mécanique Analytique* which he wrote in 1788, about 100 years after Newton's work. Lagrange was very proud of the fact that not a single diagram appeared in his book: Mechanics was "reduced" to pure mathematics. What Lagrange did *is* purely mathematical. It is not only beautiful in its mathematical austerity, but quite useful in solving problems, too. Perhaps the most significant conceptual advance is that we treat the mechanical system as a whole. Lagrange's formulation of mechanics has become the starting point in practically all branches of theoretical physics.

The physical content of Newton and Galileo's mechanics remains intact after Lagrange. His physics is completely equivalent to Newton's but mathematically much more powerful. The physics of mechanics did not change until Einstein's theory of special relativity was formulated in 1905. The change involves our most basic ideas about time and space.

However, Lagrangian mechanics reveals many things that weren't obvious in Newtonian mechanics. For example, we see that conservation of energy is a direct consequence of having a time-independent Lagrangian function: Physical laws are invariant to time translations. In general you'll see that conservation laws, such as momentum or angular momentum conservation, follow from symmetries of the Lagrangian function.

Newton's original version of mechanics can be thought of as "vectorial mechanics." Lagrange's mechanics is called "analytical mechanics." Probably the most important reason for studying analytical mechanics is that it plays an essential role in quantum mechanics. In particular, the whole theory of quantum mechanics is founded on the notion of canonically conjugate momentum. Quantum dynamics is based on the Hamiltonian, rather than the Lagrangian.

In this chapter it has been proved that $L \equiv T - V$ plays a key role in mechanics. In the next chapter we will show that the Euler–Lagrange equations imply that the action $S = \int L\, dt$ is an extremum (minimum or maximum) for the physical path. Indeed, one could completely avoid using Newton's Laws and make a fresh start in mechanics by postulating the existence of such a minimum for the action along the physical path. Instead, in Chapter 2 we follow a more traditional route of showing how this minimum principle is equivalent to the use of Newton's Laws of motion.

The Euler–Lagrange equations are differential equations that allow us to predict the motion of a system, that is, know all of the $q_k(t)$ for all times. Solving a differential equation always involves specifying the initial conditions. For a second-order differential equation, not only the starting point $q_k(0)$ must be specified, but also the starting velocities $\dot{q}_k(0)$. To see this, think of a Taylor series expansion* in the time dependence of $q_k(t)$ near to $t = 0$:

$$q_k(t) = q_k(0) + t\dot{q}_k(0) + \frac{t^2}{2}\ddot{q}_k(0) + \cdots . \tag{1.72}$$

Since the second-order differential equation gives us $\ddot{q}_k$ in terms of q_k, $\dot{q}_k$, we know $\ddot{q}_k(0)$ once these initial conditions are specified at $t = 0$. By differentiating the equation once we can get the third derivative at $t = 0$ and so on for the higher derivatives. So we know all the terms in the Taylor series from the equation of motion once the $q_k(0)$ and $\dot{q}_k(0)$ are given. In principle this gives us the function $q_k(t)$ for all times.

In practice nobody does this. Either we recognize a differential equation for which we know the solution, or else we solve it on the computer. Analytic solutions are much more useful than numerical solutions, because it is easier to visualize what is happening in general. Insight comes more quickly, as a rule, from analytic solutions, which explains why we use numerical techniques only as a last resort. But very few differential equations have analytic solutions. The most important case is the simple harmonic oscillator. Here we are very fortunate that the most common situation encountered in practice can be approximated to this: We can write an analytic solution in terms of sines and cosines. In Chapter 3 we will study this system in detail.

1.13 AN EXAMPLE OF A SOLVED PROBLEM

> Example: Two-dimensional Central Force
>
> A particle of mass m moves in a plane. Suppose that the potential energy depends only on the distance from the origin: $V = V(r)$. By taking the gradient it can be shown that this leads to a force directed along the radius vector from the origin to the particle. In fact this is the definition of a *central force*. The kinetic energy is
>
> $$T = \frac{1}{2}m\left(\frac{ds}{dt}\right)^2, \tag{1.73}$$

---

* $f(x) = f(x_0) + (x - x_0)f'(x_0) + \frac{(x-x_0)^2}{2}f''(x_0) + \cdots .$

where $ds^2 = dr^2 + r^2 d\phi^2$ is the differential of arc length in plane polar coordinates (r, ϕ). Putting the expression for T, Equation (1.73), together with the potential energy, we get the Lagrangian

$$L = T - V = \frac{m}{2}[\dot{r}^2 + r^2\dot{\phi}^2] - V(r). \tag{1.74}$$

Since there are two degrees of freedom, there will be two Euler–Lagrange equations:

$$\frac{d}{dt}(m\dot{r}) - mr(\dot{\phi})^2 + \frac{dV}{dr} = 0, \tag{1.75}$$

and

$$\frac{d}{dt}(mr^2\dot{\phi}) = 0. \tag{1.76}$$

The second Euler–Lagrange equation, (1.76), shows that ϕ is an ignorable coordinate and has the solution $mr^2\dot{\phi} =$ a constant $= l_z$. To find the motion we have to integrate the one-dimensional equation

$$m\ddot{r} = \frac{l_z^2}{mr^3} - \frac{dV}{dr}. \tag{1.77}$$

The first term on the right side of Equation (1.77) is really a kinematic term – it is part of the kinetic energy. It acts like a "fictitious" repulsive force that depends on the constant l_z. We will discuss this example in more detail in Chapter 4.

The generalized forces are computed from the potential energy to be

$$\mathcal{F}_\phi = -\frac{\partial V}{\partial \phi} = 0 \tag{1.78}$$

and

$$\mathcal{F}_r = -\frac{dV}{dr}. \tag{1.79}$$

The dimensions of $\mathcal{F}_\phi$ and $\mathcal{F}_r$ are different, the former having the dimensions of torque, and the latter of force. In both cases $\mathcal{F}_k dq_k$ has the dimensions of work, (i.e., energy). With this method, there is no need to work out acceleration in polar coordinates as we would need to do in the case of vectorial mechanics.

SUMMARY OF CHAPTER 1

- Types of holonomic constraints (N degrees of freedom, $\vec{r}_i$ is location of ith part):

$$\vec{r}_i = \vec{r}_i(q_1, \ldots, q_N) \quad \text{scleronomic,} \tag{1.80}$$

$$\vec{r}_i = \vec{r}_i(q_1, \ldots, q_N, t) \quad \text{rheonomic.} \tag{1.81}$$

- Dot cancellation works for holonomic constraints of both types:

$$\frac{\partial \dot{\vec{r}}_i}{\partial \dot{q}_k} = \frac{\partial \vec{r}_i}{\partial q_k}. \quad (1.82)$$

- Generalized force $\mathcal{F}_k$:

$$\mathcal{F}_k \equiv \frac{\delta W}{\delta q_k} \quad \text{at fixed time, } \delta W \text{ is virtual work,} \quad (1.83)$$

$$\mathcal{F}_k = \sum_{i=1}^{N} \vec{F}_i \cdot \frac{\partial \vec{r}_i}{\partial q_k}. \quad (1.84)$$

- Generalized equations of motion:

$$\mathcal{F}_k = \frac{d}{dt}\left(\frac{\partial T}{\partial \dot{q}_k}\right) - \frac{\partial T}{\partial q_k}. \quad (1.85)$$

- Generalized force for a system with a potential energy function $V(q_1, \ldots, q_N)$:

$$\mathcal{F}_k = -\frac{\partial V}{\partial q_k}. \quad (1.86)$$

- Definition of the Lagrangian for a holonomic conservative system:

$$L(q_1, \ldots, q_N, \dot{q}_1, \ldots, \dot{q}_N, t) \equiv T - V. \quad (1.87)$$

- Euler–Lagrange equations for a conservative system with $\frac{\partial V}{\partial \dot{q}_k} = 0$:

$$\frac{d}{dt}\left(\frac{\partial L}{\partial \dot{q}_k}\right) - \frac{\partial L}{\partial q_k} = 0, \quad k = 1, \ldots, N. \quad (1.88)$$

- Definition of the Hamiltonian:

$$H \equiv \sum_k \dot{q}_k \frac{\partial L}{\partial \dot{q}_k} - L. \quad (1.89)$$

- H is conserved if the Lagrangian does not explicitly contain time.
- H is the total energy $E = T + V$ for scleronomic constraints

PROBLEMS

Try to write your problem solutions as an explanation to a colleague, rather than to a grader who already understands the solution. Imagine that you will want to read through your solutions before the course exam, or perhaps even several years later. Indicate your

PROBLEMS

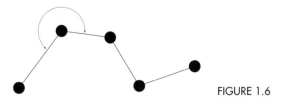

FIGURE 1.6

reasoning at every step. It is a good habit to get into. The object is not only to solve the problem, but also to be able to explain it. The extra time invested is worth it!

Degrees of Freedom

Problem 1: *(Bicycle)* Make a simplified model of a bicycle. How many degrees of freedom are there? Restrict your model to the most important degrees of freedom.

Problem 2: *(Flexible chain)* A flexible chain of M massive point particles has rigid weightless rods as $M - 1$ links as shown in Figure 1.6. Each joint is free to move in any direction. How many degrees of freedom does the chain have? If you place the chain on a flat table, how many degrees of freedom does it then have? Finally, suppose the chain is lifted off the table and is closed by one more link. How many degrees of freedom are there then?

Dot Cancellation

Problem 3: *(Spherical polar coordinates)* Prove that the relation $\frac{\partial \vec{r}_i}{\partial q_k} = \frac{\partial \dot{\vec{r}}_i}{\partial \dot{q}_k}$ (Equation (1.44)) holds if you have a one particle system described by spherical polar coordinates: Choose for q_1, q_2, q_3 the parameters r, θ, ϕ.

Kinetic Energy

Problem 4: *(Spherical pendulum)* Consider the spherical pendulum, which consists of a mass m suspended by a string from the ceiling. The mass is free to swing in both directions but maintains a constant distance from the point of suspension. Choose spherical polar coordinates θ, ϕ as generalized coordinates for this problem. What is $T(\theta, \dot{\theta}, \phi, \dot{\phi})$?

Virtual Work

Problem 5: *(Spring pendulum)* Imagine that you have a pendulum made of a mass hanging from a spring. Unlike the previous problem, restrict all motion to take place in a vertical plane here. At rest the pendulum has a length l_0. The spring constant is k. There are two degrees of freedom, which you can take as θ, the angle from the vertical of the pendulum, and x, the extension of the spring. (When the spring is extended, the pendulum length is $l_0 + x$.) Find the generalized forces $\mathcal{F}_\theta$, and $\mathcal{F}_x$ using the principle of virtual work.

Invariance

Problem 6*:† (*Physically equivalent Lagrangians*)

a) Prove that adding a constant to the Lagrangian L or else multiplying the Lagrangian by a constant produces a new Lagrangian L' that is physically equivalent to L. What we mean by physically equivalent is that the Euler–Lagrange equations for the $q(t)$ remain the same (i.e., are invariant) under this change of Lagrangian.

b) There is even more freedom to change the Lagrangian *without changing the physics it describes*. A total time derivative of an arbitrary function of the dynamical variables can be added to the Lagrangian to produce a completely equivalent Lagrangian. Consider a new Lagrangian L' which is produced as follows:

$$L \to L' = L + \frac{dF}{dt}. \tag{1.90}$$

We assume that F is an arbitrary function of the qs and t but is not a function of the $\dot{q}$s. Prove that the Euler–Lagrange equations for $q(t)$ are invariant under this change of Lagrangian. Since one can always make transformations of this sort, the Lagrangian for a given physical system is not unique.

Problem 7*: (*Guessing the Lagrangian for a free particle*) Assume that you do not know about kinetic energy or Newton's Laws of motion. Suppose instead of deriving the Euler–Lagrange equations, we postulated them. We define the basic law of mechanics to be these equations and ask ourselves the question: What is the Lagrangian for a free particle? (This is a particle in empty space with no forces acting on it. Be sure to set up an inertial reference system.)

a) Explain why, on very general grounds, L cannot be a function of x, y, or z. It also cannot depend on the individual coordinates of velocity in any way except as a function of the magnitude of the velocity: $v^2 = v_x^2 + v_y^2 + v_z^2$. On what assumption about the properties of space does this depend?

b) The simplest choice might be to guess it must be proportional to v^2, where $\vec{v}$ is the particle velocity in an inertial frame K. Take $L = v^2$. A second inertial frame K' moves at the constant velocity $-\vec{V}_0$ with respect to K, so that the transformation law of velocities is

$$\vec{v}' = \vec{v} + \vec{V}_0. \tag{1.91}$$

Prove that $L' = v'^2$ is a possible choice for the Lagrangian in the frame K'. Explain how this proves that all inertial frames are equivalent. You will have to

---

† The symbol "*" will be used to denote problems used in a weekly student seminar that was part of the course taught at Cornell in 1994–1996. In the seminar, student groups had an hour to solve an assigned problem, after which they presented the solution to the class.

make use of the result of the previous problem to show this. With this approach we prove the equivalence of inertial frames from the form of the Lagrangian, instead of postulating this equivalence at the start, which is the usual way of doing things.

c) Instead of proving it, adopt the equivalence of inertial frames as a postulate, in addition to the Euler–Lagrange equations. Explain why this means that

$$L'(v + V_0) = L(v) + \frac{dF(x,t)}{dt}. \tag{1.92}$$

$L(v)$ is an unknown function for the free particle that we are trying to determine from these principles. (Work in one dimension to make things easier.) Let V_0 be an infinitesimal quantity. Expand the left side of Equation (1.92) in a Taylor series and keep only the first two terms. From this prove $L(v) \sim v^2$.

Problem 8: *(Potentials with scaling properties)* Let $V(\vec{r}_1, \ldots, \vec{r}_M)$ be the potential energy of a system of M massive particles which has the scaling property

$$V(\alpha \vec{r}_1, \ldots, \alpha \vec{r}_M) = \alpha^k V(\vec{r}_1, \ldots, \vec{r}_M) \tag{1.93}$$

(k is usually an integer, α an arbitrary constant.) Prove that, if the Lagrangian is to remain invariant (except for multiplication by a constant), and all distances are scaled by a factor α, then the time must be scaled by a factor $\beta = \alpha^{1-\frac{k}{2}}$. Applications of this include:

a) If $k = 1$, the force is constant, like gravity. Prove that distances scale like t^2.
b) If $k = 2$, the force is like that of a harmonic oscillator or a system of harmonic oscillators coupled to each other. Prove that the frequency or frequencies of such a system are independent of the amplitude of oscillation.
c) If $k = -1$, we have the Kepler problem (inverse square force law). Prove Kepler's third law from this scaling law above. (That is, prove $d^3 = t^2$, where d could be any distance in the problem. Normally it is the mean distance of a planet from the sun.)

Hamiltonian Concept/Energy

Problem 9: *(Quadratic forms)* Prove that, if the constraints are scleronomic (i.e., time-independent), T is a *quadratic* function (*quadratic form*) of the generalized velocities. Then prove this implies

$$\sum_k \dot{q}_k \frac{\partial T}{\partial \dot{q}_k} = 2T. \tag{1.94}$$

Assuming that the kinetic energy *is* a quadratic form in the generalized velocities so that the formula above is correct, prove that the Hamiltonian H (Equation (1.65)) is the total energy ($H = T + V = E$).

Problem 10: *(Bead on a wire of arbitrary shape)* A bead slides without friction down a wire that has the shape $y = f(x)$ (Y is vertical, X is horizontal).

a) Prove that the EOM is

$$(1 + f'^2)\ddot{x} + f'f''\dot{x}^2 + gf' = 0 \tag{1.95}$$

(where $f' \equiv \frac{df}{dx}$, $f'' \equiv \frac{d^2f}{dx^2}$).

b) Since the Hamiltonian is a constant in this problem, it always equals its value at $t = 0$. Use this fact to solve for $\dot{x}(t)$.

c) Let τ be the time to slide down the wire between two heights $y_1 = f(1)$ and $y_0 = f(0)$. Show that this leads to a solution of the form $\sqrt{g}\,\tau = \int_0^1 h(x)\,dx$, where $h(x)$ is the function you should find in terms of $f(x)$ and its derivatives.

Problem 11: *(Comparing H and E)* Invent a concrete example of each type of the situation described below:

a) H is conserved, but $H \neq E$.
b) $H = E$, but $\frac{dH}{dt} \neq 0$, so H is not conserved.

Lagrangian/EOM

Problem 12: *(L for free particle in plane polar coordinates)* Express the Lagrangian for a free particle moving in a plane in plane polar coordinates. From this prove that, in terms of radial and tangential components, the acceleration in polar coordinates is

$$\vec{a} = (\ddot{r} - r\dot{\theta}^2)\,\hat{\mathbf{e}}_r + (r\ddot{\theta} + 2\dot{r}\dot{\theta})\,\hat{\mathbf{e}}_\theta \tag{1.96}$$

(where $\hat{\mathbf{e}}_r$ and $\hat{\mathbf{e}}_\theta$ are unit vectors in the positive radial and tangential directions).

Problem 13: *(Bead on a wire)* Discuss the motion of the bead according to the equation of motion (1.38) as completely as you can. Find explicit solutions. What will happen if the bead is slightly displaced from the point where it has no acceleration?

Problem 14*: *(L for charged particle in a magnetic field)* The Lagrange method *does* work for some velocity-dependent potentials. A very important case is a charged particle moving in a magnetic field. The magnetic field $\vec{B}$ can be represented as the curl of a "vector potential" $\vec{A}$: $\vec{B} = \vec{\nabla} \times \vec{A}$. A uniform magnetic field $\vec{B}_0$ corresponds to a vector potential $\vec{A} = \frac{1}{2}\vec{B}_0 \times \vec{r}$.

a) Check that $\vec{B}_0 = \vec{\nabla} \times \vec{A}$.
b) From the Lagrangian

$$L = \frac{1}{2}mv^2 + e\vec{v} \cdot \vec{A} \quad \text{(MKSI units)} \tag{1.97}$$

(where e is the charge and m is the mass) show that the EOM derived from the Euler–Lagrange equations is identical with the result from Newtonian mechanics plus the Lorentz force on a moving charged particle in a magnetic field:

$$\vec{F} = e\vec{v} \times \vec{B}. \tag{1.98}$$

Hint: Use the fact that you can exchange the dot and cross in $\vec{v} \cdot \vec{B}_0 \times \vec{r}$ to find the gradient of this expression. Also use the fact that $\frac{d\vec{A}}{dt} = \frac{1}{2}\vec{B}_0 \times \vec{v}$. This result is a very general one – the Lagrangian takes the simple form of Equation (1.97) even with an arbitrary magnetic and electric field (in the absence of an electrostatic potential).

Problem 15*: *(Double Atwood machine)* Figure 1.7 shows a "double Atwood machine." The center pulley is free to move vertically and it has a mass M. The string connecting the three masses shown is weightless. Masses m_1 and m_2 hang on the left and right respectively from the fixed pulleys. The acceleration of gravity is g. All three pulleys are frictionless, so that the string slides freely over them.

Prove that, if the virtual work vanishes for the independent variation of the two degrees of freedom, the conditions for static equilibrium are obtained:

$$\frac{M}{2} = m_1, \quad \frac{M}{2} = m_2. \tag{1.99}$$

This means that $m_1 = m_2$ and $M = m_1 + m_2$, a simple result.

Solve for the dynamics in the event the static equilibrium condition is not satisfied. Find the Lagrangian, and prove that the equations of motion are

$$\left(m_1 + \frac{M}{4}\right)\ddot{x}_1 + \frac{M}{4}\ddot{x}_2 = g\left(\frac{M}{2} - m_1\right),$$

$$\frac{M}{4}\ddot{x}_1 + \left(m_2 + \frac{M}{4}\right)\ddot{x}_2 = g\left(\frac{M}{2} - m_2\right). \tag{1.100}$$

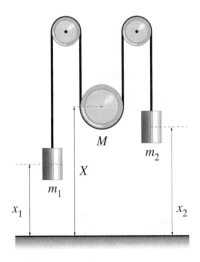

FIGURE 1.7
Double Atwood machine.

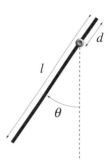

FIGURE 1.8
Physical pendulum.

Equations (1.100) are two simultaneous equations in the two unknown constants $\ddot{x}_1, \ddot{x}_2$. It can be seen that the static equilibrium condition is satisfied if the condition on the masses (1.99) is obeyed.

Problem 16: *(A simple oscillator: the physical pendulum)* A meter stick of total length l is pivoted a distance d from one end on a frictionless bearing. The stick is suspended so that it becomes a pendulum as shown in Figure 1.8. This is called a "physical" pendulum because the mass is distributed over the body of the stick. Assume the total mass is m and the mass density of the stick is uniform. The acceleration of gravity is g. Find T and V as functions of the generalized coordinate θ and velocity $\dot{\theta}$. Do this by considering the stick to be divided into infinitesimal parts of length dl and integrating to find the total kinetic and potential energy. Set up the Lagrangian and find the equation of motion.

Problem 17: *(Stick on a table)* A stick of length l and mass m lies on a frictionless table. A force parallel to the table top, $\vec{F}(t)$, is applied to one end of the stick for a very short time: $\int \vec{F} dt = \vec{I}$, the "impulse." Choose a convenient set of two generalized coordinates. In terms of your choice of generalized coordinates, what are the generalized forces when $F \neq 0$? What is the subsequent motion of the stick?

Problem 18: *(Governor)* A common type of *governor* for regulating motor speed is sketched in part A of Figure 1.9. Two masses, each equal to m, are located on arms

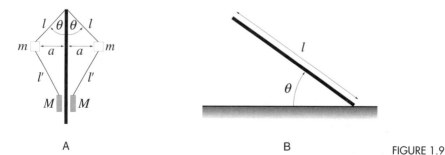

A B FIGURE 1.9

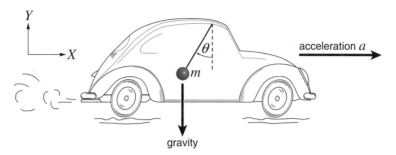

FIGURE 1.10

that pivot at the top and bottom so that the angle θ is a function of the angular speed ω of the shaft. Find the function $\theta(\omega)$ using Lagrangian methods. The ring mass M at the bottom can slide up and down on the shaft.

Problem 19: *(Inclined stick on a table)* A stick is initially held at a vertical angle θ as shown in part B of Figure 1.9. First consider a table where the bottom of the stick is fixed (with a frictionless bearing) to the table top. Is this problem holonomic? Solve for the motion of the stick after it is released. Next assume that the bottom end is on a frictionless table instead. How many degrees of freedom are there? Is the problem holonomic? Again solve for the motion of the stick after it is released.

Problem 20: *(Pendulum in an accelerated reference frame)* A pendulum with a weightless string of length D and mass m is attached to a moving car, as shown in Figure 1.10. The car is continuously accelerated along a horizontal track with constant acceleration a, starting from an initial horizontal velocity v_0. Gravity acts in the vertical direction with acceleration g.

Assume that the (x, y) coordinate system shown is at rest with respect to the ground (not located in the car). *The car is not an inertial frame of reference.* There is one degree of freedom. Use θ (see Figure 1.10, which shows $\theta < 0$) as the generalized coordinate/dynamical variable. The goal is to find out how the acceleration of the pendulum support affects the motion of the pendulum as seen by a person in the car.

a) Find the components of the velocity of the pendulum bob in the *laboratory frame* (i.e., the frame at rest with respect to the moving car). Find the kinetic energy as a function of θ, $\dot{\theta}$, and the other variables, all of which are known functions.
b) Find the Lagrangian $L(\theta, \dot{\theta}, t)$. Does L depend explicitly on the time?
c) Prove that the equation of motion for the pendulum is

$$\ddot{\theta} + \frac{g}{D} \sin\theta + \frac{a}{D} \cos\theta = 0. \qquad (1.101)$$

Notice that the velocity of the car is not detectable by observing the pendulum from inside the car, but the car acceleration *is* detectable by the "tilt" of the pendulum when it is at rest with respect to its support point. This is in accordance with the Galilean principle of relativity. Explain why this is true.

d) What is the angle of the pendulum when it remains at rest in stable equilibrium? Give an expression for the tangent of this angle (call it θ_{eq}).

e) Set $\theta(t) = \theta_{eq} + \eta(t)$, that is, measure the motion with respect to the equilibrium point. Use a Taylor series to find the equation of motion for η for small oscillations around θ_{eq}. If η is sufficiently small, show that the equation for η is

$$\ddot{\eta} + \omega^2 \eta = 0. \tag{1.102}$$

The solutions to this equation are $\sin \omega t$ and $\cos \omega t$, which means that the pendulum makes simple harmonic oscillations about the equilibrium angle with frequency ω. Prove that the angular frequency of these small oscillations is

$$\omega = \sqrt{\frac{\sqrt{a^2 + g^2}}{D}}.$$

Problem 21: *(Simple pendulum with driven support)* A simple pendulum with a point mass m suspended from a weightless rod of length l has its support point driven rapidly up and down with an amplitude of vertical motion

$$A \cos \omega t, \tag{1.103}$$

where A and ω are independently adjustable constants. Find the Lagrangian for this system using θ, the angle the pendulum makes with the vertical, as the generalized coordinate. Is H constant? Is H the total energy?

Problem 22*: *(Box sliding horizontally)* A box of mass M slides horizontally on a frictionless surface. The distance of the box's center of mass from the origin is denoted by X. Suspended from inside the center of the box is a pendulum of length l at the bottom of which is a mass m. All the motion takes place in the XY plane. What is the Lagrangian for this system? What are the EOMs?

Problem 23: *(Bead on a rotating circular hoop)* Imagine a vertical circular hoop of radius R rotating about a vertical axis with constant angular velocity Ω as shown in Figure 1.11. A bead of mass m is threaded on the hoop, so that it can move without friction, but is confined to move on the hoop. (Define the angle θ to be the angle from the vertical line through the center of the hoop to the bead.) Find the Lagrangian and the equations of motion. Find the Hamiltonian H explicitly. Is H a constant of the motion? Is energy $(T + V)$ constant?

Conjugate Momentum/Routhian

Problem 24: *(Physical pendulum)* For the physical pendulum (see Figure 1.8, using θ as the generalized coordinate, what is the canonically conjugate momentum to θ, i.e., p_θ? What is another name for p_θ in this case?

PROBLEMS

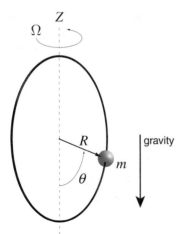

FIGURE 1.11

Problem 25: *(2-D central force motion)* Find the canonically conjugate momentum p_ϕ from the Lagrangian for the example of the two-dimensional central force motion worked out in the text (1.74). Try to eliminate $\dot\phi$ from the problem, using the fact that the momentum conjugate to ϕ is a constant of the motion. Does it work to just substitute it for $\dot\phi$ in terms of p_ϕ and use the result as a new Lagrangian? Find the Routhian and use it to get equations of motion.

Nonholonomic

Problem 26: *(Particle on a bowling ball)* A pointlike particle of mass m slides without friction down the surface of a bowling ball of radius R. The particle starts at rest, arbitrarily close to, but not exactly at the top, of the bowling ball.

a) Find out where the particle leaves the surface of the ball and its velocity at that point. Hint: The motion takes place in a plane. Why can you assume this?

b) Discuss how the problem would be changed if the bowling ball had a mass M and were free to move by rolling without slipping along the horizontal table. Are the constraints holonomic?

Problem 27: *(Ball on a table)* As a rigid body, a spherical ball must have six degrees of freedom. Placed on a flat table top, the number of degrees of freedom of the ball is reduced to five. These could be taken to be the two coordinates of the center of the ball and the three angles expressing the ball's orientation. These angles could be the polar and azimuthal angles of an arbitrary point on the ball, plus an angle of rotation about this point.

a) Suppose the table provides enough friction so that the ball rolls without slipping. Prove that this must reduce the number of degrees of freedom to three. (Hint: What is the velocity of the point where the ball contacts the table?)

b) These three remaining degrees of freedom can be chosen to be x_c, y_c (location of the ball center), and a rotation ϕ about the vertical axis, perpendicular to the table surface. Experiment with an actual "ball" on a table to show that the orientation of the ball at a particular x_c, y_c depends on the previous history of rolling the ball, say around different closed loops, and not just on x_c, y_c, ϕ. This demonstration proves that it is impossible to use these three quantities as dynamical variables. In actual practice since the motion is nonholonomic, we have to embed the three degrees of freedom in a five-dimensional space. We can never eliminate the other two parameters.

c) Finally, prove that if the ball rolls without slipping, it must be true that

$$\vec{v}_c + R\vec{\omega} \times \hat{n} = 0, \qquad (1.104)$$

where $\vec{\omega}$ is parallel to the instantaneous axis of rotation, $|\vec{\omega}|$ is the rate of rotation around the axis, and $\hat{n}$ is a unit vector along the z direction, perpendicular to the table. Since this problem involves a constraint on the velocities, the constraints are not holonomic.

APPENDIX A

ABOUT NONHOLONOMIC CONSTRAINTS

We have emphasized mechanics problems involving holonomic constraints, either scleronomic (time-independent) or rheonomic (time-dependent). Nonholonomic systems require special treatment.

How Can Constraints be Nonholonomic?

There are at least three types of nonholonomic constraints. They are all characterized by the fact that we cannot write simple equations connecting the number of degrees of freedom with the location of all parts of the system. In some fundamental way, we need to know the previous history of the system in addition to the possible ways it can move at any given moment.

Conditions on velocities often lead to equations that cannot be integrated until the solution is known. These are called *nonintegrable* constraints. This type of nonholonomic constraint involves M nonintegrable conditions of the form:

$$0 = \sum_{j=1}^{N_C} \omega_{ij}(q_1, \ldots, q_{N_C}) \dot{q}_j, \quad i = 1, \ldots, M. \qquad (1.105)$$

The coefficients ω_{ij} can be functions of the generalized coordinates. $N_C = N + M$. The number of generalized coordinates is now greater than N, the number of degrees of freedom. The problem cannot be described by N-dimensional space, but must be embedded in an N_C-dimensional configuration space.*

---

* This is the term for the abstract N-dimensional space spanned by the $\{q_k\}$.

APPENDIX A ABOUT NONHOLONOMIC CONSTRAINTS

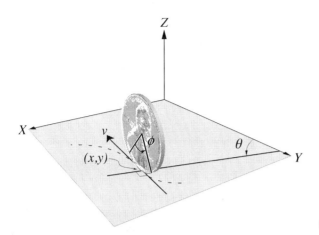

FIGURE 1.12

One example of this type of problem is a vertical penny rolling on its edge on a flat plane, as shown in Figure 1.12. The orientation of the penny, which has a radius r, is specified by two coordinates θ, ϕ, so there are two degrees of freedom. The position of the center of the coin, projected onto the XY plane is specified by (x, y). This is also the point of contact with the XY plane, since the coin is assumed to be vertical. We have the equations:

$$\begin{aligned} v &= r\dot{\phi}, \\ \dot{x} &= -v\cos\theta, \\ \dot{y} &= -v\sin\theta. \end{aligned} \qquad (1.106)$$

If multiplied by dt, Equations (1.106) give the two differential relations with the form of Equation (1.105):

$$dx + r\cos\theta \, d\phi = 0 \qquad (1.107)$$

and

$$dy + r\sin\theta \, d\phi = 0. \qquad (1.108)$$

If these could be integrated, they would give two equations $f_1(x, \theta, \phi) = 0$, $f_2(y, \theta, \phi) = 0$. You might guess that these equations could then be solved to obtain $x(\theta, \phi)$, $y(\theta, \phi)$. Specifying θ, ϕ would completely determine x, y. But this can't be done!

To see this, imagine a motion for the penny in which ϕ and θ are varied but eventually return to the same ϕ, θ point. In general, x and y will *not* return to the starting point after a closed path in ϕ, θ space, as shown in Figure 1.13.

This proves that x and y cannot be eliminated from the problem, because the values of x and y depend not only on θ, ϕ, but also on the history of the system. The "configuration space" has two degrees of freedom but it must be embedded in a four-dimensional space and cannot be projected onto either the (x, y) or the (ϕ, θ) planes. The differential Equations (1.107, 1.108) cannot be integrated unless the solution is known. This is typical of rolling motion.

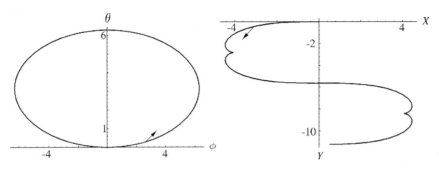

FIGURE 1.13

As a mathematical proof, if (1.107) were the differential of a function $f_1(x, \theta, \phi)$, it would be of the form

$$\frac{\partial f_1}{\partial x} dx + \frac{\partial f_1}{\partial \theta} d\theta + \frac{\partial f_1}{\partial \phi} d\phi = 0. \quad (1.109)$$

From the original equation (1.107), we conclude that $\frac{\partial f_1}{\partial \theta} = 0$ and $\frac{\partial f_1}{\partial \phi} = r \cos \theta$. For any well-behaved function f_1, the order of taking partial derivatives is not significant: $\frac{\partial^2 f_1}{\partial \theta \partial \phi} = \frac{\partial^2 f_1}{\partial \phi \partial \theta}$. For the expressions above, $\frac{\partial^2 f_1}{\partial \theta \partial \phi} \neq \frac{\partial^2 f_1}{\partial \phi \partial \theta}$, which proves that f_1 does not exist.

Besides rolling motion, there are two other types of constraints that are not holonomic. One of them involves inequalities. The classic example is a point mass sliding without friction on a bowling ball, starting from near the top. This example will be worked out in detail next.

Problems with friction constitute the last type of nonholonomic problem. These are most often solved by reverting to Newtonian dynamics. Some examples are the damped harmonic oscillator and the baseball home run, which is strongly affected by air resistance.

Point Mass Sliding on a Spherical Bowling Ball

This problem is nonholonomic because it involves the constraint $r \geq R$, where r is the position of the point of contact of the bead with the sphere of radius R, as shown in Figure 1.14. Let's first assume that the bead starts out in the region $r = R$ and find the

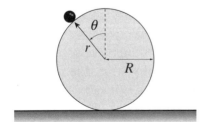

FIGURE 1.14

APPENDIX A ABOUT NONHOLONOMIC CONSTRAINTS

equation of motion. We have

$$T = \frac{1}{2}mR^2\dot\theta^2,$$
$$V = -mgR(1 - \cos\theta), \qquad (1.110)$$
$$L = T - V.$$

From the Euler–Lagrange equations, the EOM is

$$\ddot\theta - \frac{g}{R}\sin\theta = 0. \qquad (1.111)$$

Use energy conservation to solve this equation. The kinetic energy T is a quadratic form in the generalized coordinate θ, so the total energy E is

$$E = T + V = \frac{1}{2}mR^2\dot\theta^2 - mgR(1 - \cos\theta). \qquad (1.112)$$

The total energy must be a constant because the Lagrangian does not contain the time explicitly, and the constraint is time-independent. The equations are true as long as the bead remains on the sphere. Assume that the starting point at $t = 0$ has $\theta = \theta_0$ and $\dot\theta_0 = 0$. Since the energy is constant, we can evaluate it at $t = 0$, where $E = -mgR(1 - \cos\theta_0)$.

We want to know how long it takes the bead to leave the surface of the sphere. Solving Equation (1.112) for $\dot\theta$,

$$\dot\theta = \sqrt{\frac{2g}{R}}\sqrt{[\cos\theta_0 - \cos\theta]}, \qquad (1.113)$$

and integrating with respect to the time, we get

$$\tau = \int dt = \sqrt{\frac{R}{2g}}\int_{\theta_0}^{\theta_{\max}} \frac{d\theta}{\sqrt{\cos\theta_0 - \cos\theta}} \qquad (1.114)$$

(where τ is the time interval between releasing the bead and when it leaves the sphere.) To evaluate this, we need to know $\theta_{\max}$, the point at which the bead leaves the surface of the sphere. This will happen when the radial constraint force vanishes.

But here we hit a snag. The elimination of the constraint forces is the source of the power of the methods we have developed. This is a blessing if we have holonomic constraints, but in this problem it prevents further progress. Blindly solving the equation of motion (1.111) would lead to the ridiculous conclusion that the bead continues to move on the bottom part of the sphere instead of falling off, as we know to be the actual case. This problem remains holonomic only as long as $r = R$. We could solve this problem as two different holonomic problems that must be "joined" by imposing continuity on the dynamical variables and their time derivatives. The transition occurs at the moment when

TABLE 1.2 SUMMARY OF ALL TYPES OF CONSTRAINTS DISCUSSED IN THIS CHAPTER

Constraint Type	Example	# of DOF
Holonomic		
Scleronomic	Spherical pendulum	2
Rheonomic	Bead on a rotating wire	1
Nonholonomic		
Inequality	Particle sliding down a bowling ball ($r \geq R$)	2
Rolling motion	Coin spinning on an inclined plane	2
Most velocity-dependent forces	Problems involving friction	?

the point mass leaves the surface of the sphere. This occurs when the radial force on the bead $\mathcal{F}_r = 0$. Lagrangian mechanics without Lagrange multipliers (see Chapter 2) cannot reveal this information.

Newtonian mechanics can be used to find $\theta_{\max}$. When the constraint force exerted by the bowling ball vanishes, the centripetal acceleration must be equal to the component of the acceleration of gravity along the radial direction

$$R\dot{\theta}^2\Big|_{t=\tau} = g\cos\theta_{\max}. \tag{1.115}$$

Inserting (1.113) into (1.115) to eliminate $\dot{\theta}$ in the expression (1.112) at $t = \tau$ we obtain

$$\cos\theta_{\max} = \frac{2}{3}\cos\theta_0. \tag{1.116}$$

The bead leaves the surface of the sphere when $\cos\theta_{\max} = \frac{2}{3}\cos\theta_0$. The time τ can then be calculated using (1.114) and the bead velocity for $t = \tau$ using (1.113). The second half of the motion is free fall with these initial conditions. It can also be proved that $\tau(\theta_0)$ increases logarithmically as $\theta_0 \to 0$, so the bead takes an infinite time to fall off the sphere if started exactly at the top ($\theta_0 = 0$). This problem remains holonomic only while the bead is in contact with the surface of the sphere. Afterwards, a different EOM is obeyed, with a different Lagrangian.

QUESTION 15: *Holonomic vs. Nonholonomic* Consider the ladder sliding down a frictionless vertical wall (see Figure 1.5). The base of the ladder also slides on a frictionless floor. What happens after the ladder slides far enough to lose contact with the wall? Find the generalized coordinates before and after contact. In what sense is this situation nonholonomic? In what sense is it holonomic?

APPENDIX B

MORE ABOUT CONSERVATIVE FORCES

By definition, if the forces are conservative forces, the net work done by taking the system from one configuration to another and back again will always be zero. Using the definition of work, this means that

$$\Delta W \equiv \sum_{i=1}^{M} \left(\oint \vec{F}_i \cdot d\vec{r}_i \right) = 0. \tag{1.117}$$

definition of a conservative force

(The symbol "$\oint$" stands for the line integral around a closed path. This is taken in real space for each part of the system and then summed over all of the parts.)

Imagine that you make a series of arbitrary virtual displacements of the generalized coordinates which change the configuration of the system but eventually restore it to the starting point. Using the chain rule for differentiating an implicit function, since the $\vec{r}_i$ are functions of the set q_k* we get

$$\vec{F}_i \cdot d\vec{r}_i = \vec{F}_i \cdot \sum_{k=1}^{N} \frac{\partial \vec{r}_i}{\partial q_k} dq_k. \tag{1.118}$$

Interchanging the order of summing over $i = 1, \ldots, M$ parts and over $k = 1, \ldots, N$ degrees of freedom and using (1.49) yields

$$\sum_{i=1}^{M} \vec{F}_i \cdot d\vec{r}_i = \sum_{k=1}^{N} \mathcal{F}_k dq_k. \tag{1.119}$$

Since we can vary the q_k independently one at a time, it must be possible to obtain special cases of the general formula from Equations (1.117, 1.119) above that involve changing only a single degree of freedom:

$$\oint \mathcal{F}_k \, dq_k = 0, \quad k = 1, \ldots, N. \tag{1.120}$$

conservative force

---

* Time is held constant, and we assume that the constraints are holonomic.

The function defined by

$$V(q_1(1), \ldots, q_N(1)) \equiv -\sum_k \int_{q_k(0)}^{q_k(1)} \mathcal{F}_k \, dq_k \quad (1.121)$$

is uniquely defined as a function only of the end points $q_k(0)$ and $q_k(1)$, even for arbitrary variations of the integration path. We usually consider the upper limits to be variables, while the lower limits are fixed constants. The uniqueness of V for arbitrary paths follows because the difference between any two different paths from the same starting point to the same upper limit is just an integral around a closed path, and hence equals zero. For that reason, V is a function only of the upper limit of the integral and not also of the path followed to get there:

$$\mathcal{F}_k = -\frac{\partial V}{\partial q_k}. \quad (1.122)$$

Conservative force:
generalized force is negative gradient of V.

⊃ Example: Rubber Band

Consider the force required to stretch a rubber band. A graph of the applied force as a function of the stretch distance x will look something like Figure 1.15. (You can try this experiment yourself.)

The work done in a complete cycle of stretching and releasing is $\int_{x_0}^{x_0+\Delta x} F \, dx + \int_{x_0+\Delta x}^{x_0} F \, dx$. This is the shaded area on the graph. In this case, the integral around a closed cycle of the coordinate, $\oint F \, dx$, is greater than zero. The work you do to stretch the rubber band is more than the work the rubber band does on you when it is released. Work is a form of energy, so it takes energy to cycle the rubber band and this energy appears as heat. The rubber band will get hot if you repeat the experiment rapidly. The heat is generated by the internal friction in the rubber.

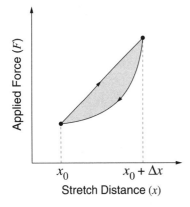

FIGURE 1.15

APPENDIX B MORE ABOUT CONSERVATIVE FORCES

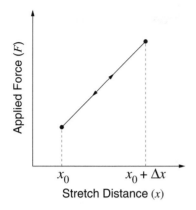

FIGURE 1.16

The rubber band is an example of a nonconservative force, since the motion is not reversible. It would make no sense to try to define a function

$$V(\Delta x) \equiv -\int_{x_0}^{x_0+\Delta x} F\,dx \qquad (1.123)$$

since the value of the integral would depend on the history of how the rubber is stretched and not just on the end points x_0, $x_0 + \Delta x$.

In contrast, a conservative force versus displacement has a graph that looks like Figure 1.16. For a conservative force, (1.123) makes perfect sense, since the function defined is unique. It does not depend on the rate at which we change x. The minus sign is due to using the force *on* the object (not the reaction force exerted *by* the object).

Think of lifting a brick. The force on the brick is the downward force of gravity: $-mg\,\hat{z}$ (where $\hat{z}$ is a unit vector pointing straight up). So when you lift a brick, work is done on the brick, and its potential energy increases. Notice that potential energy is a scalar. For the gravitational force it is just mg times the distance the brick is raised and does not depend on the path we take to raise it. Only *changes* in V are physically significant. $V = -W$, where W is the work done by the object to change V. To raise the brick we do positive work, that is, the brick does negative work on us.

CHAPTER TWO

VARIATIONAL CALCULUS AND ITS APPLICATION TO MECHANICS

OVERVIEW OF CHAPTER 2

Nature is found to conspire in just such a way that the time integral of the Lagrangian is smallest if the motion obeys Newton's Laws. Mechanics can be based on the single principle: Minimize the time integral of the Lagrangian. Three laws of motion can be condensed into one universal principle!

The mathematical language needed to provide the framework for this is called variational calculus. The variational calculus can be used as a powerful tool in solving mechanics problems with explicit constraints. It is also the most general means of solving nonholonomic problems with constraints on the velocities such as for rolling motion. This type of problem cannot be solved by choosing coordinates equal to the number of degrees of freedom but must be embedded in a higher-dimensional space.

The well-known theoretical physicist E. P. Wigner refers to the "unreasonable effectiveness of mathematics in theoretical physics." Mathematical beauty is and should be the chief guiding principle of theorists, according to P. A. M. Dirac, one of the inventors of quantum mechanics. Although it is hard to define exactly what mathematical beauty is, the search for beauty was the guiding principle in the invention of two major advances in physics in the twentieth century: relativistic quantum mechanics and general relativity. In this chapter, we will discover an elegant formulation of classical mechanics. The mathematical techniques uncovered here are not only beautiful, but they have become the language of modern theoretical physics.

2.1 HISTORY

In variational calculus (sometimes called "functional calculus"), we study "functions of functions," which are called *functionals*. The history of how the variational calculus was discovered is fascinating. In 1696, Jean Bernoulli managed to find a solution to what is now called the "brachistochrone problem."* Bernoulli then issued a challenge to

---

* From two Greek words "brachistos," meaning "shortest" and "chronos," meaning "time."

2.1 HISTORY

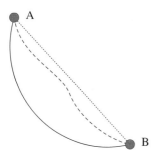

FIGURE 2.1
Brachistochrone problem.

the mathematicians of Europe to solve this problem. We give an excerpt from a translation from the Latin:*

NEW PROBLEM
WHICH MATHEMATICIANS ARE INVITED TO SOLVE

If two points A and B are given in a vertical plane, to assign to a mobile particle M the path AMB along which, descending under its own weight, it passes from the point A to the point B in the briefest possible time.

(See Figure 2.1.)

After six months, in which no one except Leibnitz managed to solve this problem, Bernoulli published his solution:

PROCLAIMATION
MADE PUBLIC AT GRONINGEN JANUARY 1697

(referring to his earlier publication) . . . six months from the day of publication . . . , at the end of which, if no one had brought a solution to light, I promised to exhibit my own . . .

He adds that Leibnitz had solved the problem in the interim. Daniel Bernoulli also solved it, but Jean didn't mention this at the time. It is a curious fact that Isaac Newton only heard about this problem from a friend in France. Receiving the letter on January 29, 1697, Newton had the answer by the next day. Newton had, in fact, come very close to inventing variational calculus earlier, in 1686, but he had not published. His 1697 solution was published anonymously. Upon reading it, Jean Bernoulli recognized the work as Newton's. He made the famous remark that one "recognizes a lion from his claw marks." At the time, there was an intense rivalry between Newton and Leibnitz. This may have been the reason Newton chose to publish anonymously. In all, five people eventually solved the problem: Jean and Daniel Bernoulli, l'Hospital, Leibnitz, and Newton.

---

* From *Acta Eruditorium*, Leipzig, June 1696, as discussed in *A Source Book in Mathematics* by D. E. Smith, Dover Press.

The modern form of the variational calculus, derived by Euler, didn't appear until 1744. Lagrange was the first to apply it to mechanics, and Hamilton's Principle was formulated by Hamilton 137 years after Bernoulli's challenge. What we refer to as *Hamilton's Principle* is also called the *Principle of Least Action*. It is impossible to overestimate the importance of the variational calculus as a mathematical framework for much of modern theoretical physics. We begin with pure mathematics and later apply this mathematics to mechanics.

2.2   THE EULER EQUATION

The problem Euler considered was a generalization of the brachistochrone problem. He considered it as a problem in pure mathematics: Given a known function $F(y, \frac{dy}{dx}, x)$ of an unknown curve $y(x)$ and its first derivative $\frac{dy}{dx}$, find the curve which makes the integral I over the independent variable x an extremum. Since the value of $I[y]$ depends on the choice of the curve $y(x)$, $I[y]$ is a functional of y:

$$I[y] \equiv \int_{x_0}^{x_1} F\left(y, \frac{dy}{dx}, x\right) dx. \tag{2.1}$$

To say that the integral is an extremum means that it is either a maximum or a minimum for a unique curve $y(x)$. This means that small variations in this curve, once it is found, can produce only second-order variations in the integral. It is analogous to the statement in calculus that if the tangent to a curve is zero, there is a maximum or a minimum. These two possibilities cannot be distinguished without examining the second or sometimes higher derivative. The solution to this problem does not exist unless $y(x)$ is required to take on fixed values at the end points x_0 and x_1.

For any reasonable* arbitrary curve $y(x)$, we can insert it and its first derivative into the known function F and then calculate the integral $I[y]$. To find the extremum, a unique curve $y^*(x)$ must be selected from an infinity of possible curves that could connect the fixed end points.

Euler found the general solution in the form of a differential equation for $y^*(x)$. It is

$$\frac{\partial F}{\partial y} = \frac{d}{dx}\left(\frac{\partial F}{\partial \frac{dy}{dx}}\right). \tag{2.2}$$

Note carefully the structure of this equation. It involves partial derivatives of F with respect to both y and $\frac{dy}{dx}$. One treats $\frac{dy}{dx}$ as an independent variable, on the same footing as y, when taking partial derivatives of F. For the partial derivative $\frac{\partial F}{\partial \frac{dy}{dx}}$, a *total* derivative with respect to the independent variable x must then be taken to obtain the right side of (2.2). Finally, we solve the equation thus obtained for $y^*(x)$.

---

* "Reasonable" in this context means both continuous and differentiable at almost every x value.

2.2 THE EULER EQUATION

It is very surprising that a local differential equation gives the global result that its solution $y^*(x)$ has $I[y^*] \leq I[y]$ for all curves $y(x)$ (in the case that the extremum is a minimum). We will prove below that Equation (2.2) indeed does give the solution curve leading to the extremum of Equation (2.1). Equation (2.2) can be used to solve a wide variety of problems.

Derivation of the Euler Equation

The integrand in Equation (2.1) is considered to be a *known, but unspecified* function $F(y, \frac{dy}{dx}, x)$. The problem is to find the unknown function $y^*(x)$ that minimizes or maximizes this integral. Here x is the *independent variable* over which we integrate. With respect to x, y is a *dependent variable*. Insofar as the value of F depends on y and the derivative of y in a way we assume is known, $(y, \frac{dy}{dx})$ will be treated as independent variables in the function F.

Consider the effect on F of making small variations in y and $\frac{dy}{dx}$ at a *fixed value of* x. These small variations in the functions $y(x)$, $\frac{dy}{dx}(x)$ can depend on x. Call them δy and $\delta \frac{dy}{dx} \equiv \frac{d}{dx}\delta y$. If these quantities are varied, the value of the function F will change, even if x is held constant. By assumption, the values of $y(x)$ must be held constant at the end points; thus we require that $\delta y(x_0) = \delta y(x_1) = 0$. Since the variation $\delta y(x)$ can be chosen to be (almost) arbitrarily different at different x values between the end points, it can be considered to be an arbitrary function of x, provided that it has a derivative with respect to x almost everywhere.

It may help to think of the curve $y(x)$ as a rigid wire, whereas the varied curve $y(x) + \delta y(x)$ is a flexible string attached to the end points of the wire. The string can be deformed in an arbitrary, but small way from the shape of the wire. For each particular deformation we can use the fact that the difference between the string and the wire is small to calculate the change in F and hence I, the integral of F over x.

Consider the change in Equation (2.1) for the "varied path" $y + \delta y$:

$$\delta I \equiv I[y + \delta y] - I[y] = \int F\left(y + \delta y, \frac{dy}{dx} + \delta \frac{dy}{dx}, x\right) dx$$
$$- \int F\left(y, \frac{dy}{dx}, x\right) dx. \quad (2.3)$$

A Taylor's series in the two variables y, $\frac{dy}{dx}$ can be used to calculate how much the function F changes when the curve becomes $y + \delta y$ instead of y. The Taylor series is a power series in both δy and $\delta \frac{dy}{dx}$. Because we expect to go to the limiting case where δy, $\delta \frac{dy}{dx}$ are infinitesimally small, we keep only the terms of first order in δy and $\delta \frac{dy}{dx}$.

> ➤ **Mathematical reminder:** *We know from calculus* that if we have an arbitrary well-behaved function of two variables, $F(x, y)$, and we expand in a Taylor series around*

---

* Kaplan, *Advanced Calculus*, third edition, p. 445 ff.

fixed points x_0, y_0, we get

$$F(x, y) = F(x_0, y_0) + (x - x_0)\frac{\partial F}{\partial x}\bigg|_{x=x_0, y=y_0}$$
$$+ (y - y_0)\frac{\partial F}{\partial y}\bigg|_{x=x_0, y=y_0} + O[\Delta^2]. \quad (2.4)$$

The symbol $O[\Delta^2]$ stands for other terms proportional to the squares $(x - x_0)^2$, $(x - x_0)(y - y_0), (y - y_0)^2$. If the point x, y is sufficiently close to (x_0, y_0), we can truncate this series after the first order, as shown in Equation (2.4).

Use the Taylor series expansion above at a fixed value of x. For small variations close to the curve $y(x)$:

$$F\left(y + \delta y, \frac{dy}{dx} + \delta\frac{dy}{dx}, x\right) = F\left(y, \frac{dy}{dx}, x\right) + \frac{\partial F}{\partial y}\delta y + \frac{\partial F}{\partial \frac{dy}{dx}}\delta\frac{dy}{dx}$$
$$+ \text{ higher order terms in } \delta y, \delta\frac{dy}{dx}. \quad (2.5)$$

Integrate F along the varied curve over x and take the difference between this and the result for the unvaried curve. Obtain the variation in $I \equiv \delta I \equiv I[y + \delta y] - I[y]$ by substituting (2.5) into (2.3) and taking the limit where $\delta y, \delta\frac{dy}{dx} \to 0$ but do not actually vanish. Since the variations $\delta y, \delta\frac{dy}{dx}$ become arbitrarily small, higher-order terms in the Taylor series can be neglected. We obtain

$$\delta I = \int \left(\frac{\partial F}{\partial y}\delta y + \frac{\partial F}{\partial \frac{dy}{dx}}\delta\frac{dy}{dx}\right) dx. \quad (2.6)$$

The term proportional to $\delta\frac{dy}{dx}$ is now integrated by parts (remember that $\delta\frac{dy}{dx} = \frac{d}{dx}\delta y$) to yield

$$\int \frac{\partial F}{\partial \frac{dy}{dx}}\frac{d}{dx}\delta y \, dx = \frac{\partial F}{\partial \frac{dy}{dx}}\delta y(x)\bigg|_{x_0}^{x_1} - \int \frac{d}{dx}\frac{\partial F}{\partial \frac{dy}{dx}}\delta y \, dx. \quad (2.7)$$

Contributions at the end points vanish because we have required δy to vanish there. Inserting the expression (2.7) into the integral over the varied curve (2.6), the variation in the integral I is

$$\boxed{\delta I = \int \left[\frac{\partial F}{\partial y} - \frac{d}{dx}\left(\frac{\partial F}{\partial \frac{dy}{dx}}\right)\right]\delta y(x)\, dx.} \quad (2.8)$$

If the integral I for the path $y(x)$ is an extremum, it means that I does not change for infinitesimal variations in the curve: $y(x) + \delta y(x)$. This implies that $\delta I = 0$ for any

2.2 THE EULER EQUATION

choice of $\delta y(x)$ for which we can neglect terms of order $O[\delta y^2]$ and higher:

$$\delta I = 0 + O[\delta y^2]. \qquad (2.9)$$

condition for an extremum

To see why this is true, assume that, for a particular choice of variation $\delta y(x)$, we find $\delta I < 0$. This would contradict our assumption that y gives the minimum possible action, since it would mean $I[y + \delta y] < I[y]$. Now suppose we find $\delta I > 0$ instead. Recalculate the integral with a new variation that has the opposite sign: $\delta y \to -\delta y$. Then $\delta I(\text{new}) = -\delta I(\text{old})$, so there is still a contradiction with the assumption that $I[y]$ is the minimum value of the integral. The same argument can be used if I is a maximum instead of a minimum. It is the same argument used in elementary calculus to prove that the tangent of a curve has zero slope at a maximum or minimum of that curve.

The variation δI in (2.8) is expressed as an integral over x of the product of two functions, one of which is arbitrary (except for being small). Since (within the limits of integration) δy is an arbitrary function, for $\delta I = 0$ it must be true that

$$\delta I = 0 \Leftrightarrow \frac{\partial F}{\partial y} - \frac{d}{dx}\left(\frac{\partial F}{\partial \frac{dy}{dx}}\right) = 0. \qquad (2.10)$$

This is the proof of Euler's equation (2.2). The curve $y^*(x)$ is the solution to this equation.

⟩ Example

As a very simple example of the calculus of variations, suppose that you want to know the curve in the XY plane that minimizes the length between the points $(0, 0)$ and $(2, 1)$.

By the Pythagorean theorem for infinitesimal right triangles, if ds lies along the tangent to the curve in the XY plane then

$$ds^2 = dx^2 + dy^2. \qquad (2.11)$$

The element of arc length in Cartesian coordinates is $ds = \sqrt{dx^2 + dy^2}$, as shown in Figure 2.2. Assuming that x is chosen to be the independent variable, the length of any curve $y(x)$ with $y(0) = 0$ and $y(2) = 1$ is given by the integral

$$s[y] = \int_0^1 \sqrt{1 + \left(\frac{dy}{dx}\right)^2}\, dx. \qquad (2.12)$$

Use $F = \sqrt{1 + \frac{dy}{dx}^2}$ to get

$$\frac{\partial F}{\partial y} = 0, \qquad \frac{\partial F}{\partial \frac{dy}{dx}} = \frac{\frac{dy}{dx}}{\sqrt{1 + \frac{dy}{dx}^2}}. \qquad (2.13)$$

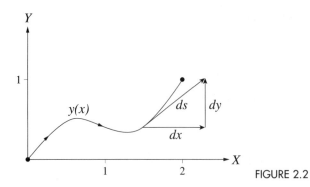

FIGURE 2.2

In this case, the Euler equation says that the total derivative with respect to x of a function of $\frac{dy}{dx}$ vanishes. Therefore that function must be a constant, which implies that $\frac{dy}{dx}$ is also a constant. The desired curve must be of the form $y(x) = ax + b$. The constants are chosen to satisfy the end point conditions: $a = .5, b = 0$. Since there is nothing special about the choice of end points, it has been proved that the extremum distance between any two points is a straight line. In this case, the extremum must be a minimum.

In the example it has been proved that, if $y(x)$ is a straight line between any two fixed points, then any other curve between these points will be longer:

$$s[y + \delta y] > s[y], \tag{2.14}$$

where $y + \delta y$ is defined in Figure 2.3.

QUESTION 1: Length Minimization 1 Keeping in mind that the length of the curve $s[y]$ involves an integral over x, what kinds of variations δy could we allow? Need they be continuous to minimize the length?

QUESTION 2: Length Minimization 2 For a very small deviation from the straight line $y(x)$, calculate the difference $s[y + \delta y] - s[y]$. How can you be certain it is always positive for an arbitrary, but very small deviation $\delta y(x)$?

In a curved space, such as the surface of the Earth for example, the infinitesimal form of the Pythagorean law (Equation 2.11)) is altered:

$$ds^2 = g_{11}\, du^2 + 2g_{12}\, du\, dv + g_{22}\, dv^2, \tag{2.15}$$

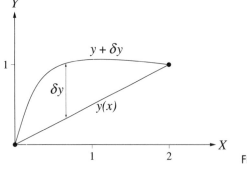

FIGURE 2.3

where du, dv are infinitesimal displacements of two arbitrary coordinates u, v which describe a position on the surface of the sphere (such as latitude θ and longitude ϕ). The g_{ij} are called the components of the "metric tensor." In general they are functions of u, v. For spherical polar coordinates on a unit sphere, $u = \theta$, $v = \phi$, and $g_{11} = 1$, $g_{12} = 0$, $g_{22} = \sin^2 \theta$.

A curve that is the shortest distance between two points in an arbitrarily curved space is called a "geodesic" curve. The geodesic curve on a sphere is a great circle. Flying along the great circle in one direction minimizes the distance; in the other direction, the distance is maximized, but in either case it is an extremum.

The curve that is the shortest distance between two points does not depend on the choice of coordinate system. It is an intrinsic property of the space in which the curve lies. In the example of the minimum distance on a plane surface, you could have chosen to use plane polar coordinates. The equation describing the straight line looks different in this coordinate system, but it is still a straight line.

2.3 RELEVANCE TO MECHANICS

What relevance does all this have to mechanics? It leads to a new and elegant way to formulate mechanics. A certain functional called the *action S* can be defined for any path $q(t), \dot{q}(t)$ taken by a physical system as the time integral of the Lagrangian along this path:

$$S[q] \equiv \int L(q(t), \dot{q}(t), t)\, dt. \qquad (2.16)$$

Here $S[q]$ is a functional of $q(t)$. We assume that $L(q, \dot{q}, t)$ is a known function, while $q(t)$ is a trial function that we may substitute into L and compute the action integral. Every trial function $q(t)$ gives a value for the action S. What distinguishes the physical path from all others is that this particular path minimizes the action. (Technically, it is the path that makes the action an extremum or, equivalently, "stationary.") This "least action" property of the physical path is known as *Hamilton's Principle*.

Hamilton's Principle can be adopted as the basic principle of classical mechanics. To prove this, we must show that Hamilton's Principle is equivalent to the Euler–Lagrange Equations we derived in Chapter 1. Since the Euler–Lagrange equations have already been proved to be the general form of Newton's Laws of Motion, we are free to adopt Hamilton's Principle (instead of Newton's three laws) as the basic principle, the starting point, of classical mechanics. All we need to solve any specific problem is the explicit form of the Lagrangian. Please keep in mind that everything is mathematics and not physics until we identify F with the Lagrangian L, the independent variable with time, and the integral I with the action S. Making these substitutions into (2.10) the physical path satisfies

$$\delta S = 0 \Leftrightarrow \frac{\partial L}{\partial q} - \frac{d}{dt}\left(\frac{\partial L}{\partial \dot{q}}\right) = 0. \qquad (2.17)$$

Now we have reproduced the Euler–Lagrange Equation (1.60) derived in Chapter 1, so the new way to formulate mechanics is completed by seeing that Hamilton's Principle is an alternative to using $\vec{F} = m\vec{a}$. The action for the physical curve is an extremum. Notice that the time derivative in (2.17) is a total time derivative, not a partial time derivative.

Suppose there are two runners who follow different paths to the same end point, arriving at the same time. If one of the runners follows the physical path, the one determined by Newton's Laws of Motion, and the other runner follows an arbitrary but nearby path, the first runner will have slightly less action than the second: $\delta S = S_1 - S_2 = 0 - \epsilon$, where ϵ is a very small positive quantity. This statement applies only in the limit where the path variation becomes arbitrarily small.

Example

As an example, imagine a cannon being fired at $t = t_0$. At $t = t_1$, the cannonball strikes the ground. The true path for the height $y(t)$ is a parabola as a function of the time. We will show that any small deviation from this parabolic path, $y + \delta y$, will increase the value of the action integral (2.16) owing to quadratic terms in the variation. The Lagrangian (ignoring horizontal motion, since it is independent of vertical motion here) is

$$L = \frac{1}{2}m\dot{y}^2 - mgy, \qquad (2.18)$$

where m is the mass, and g the acceleration of gravity. Then

$$\delta L \equiv L(y + \delta y) - L(y) = \frac{1}{2}m\left[2\dot{y}\left(\frac{d}{dt}\delta y\right) + \left(\frac{d}{dt}\delta y\right)^2\right] - mg\delta y. \qquad (2.19)$$

This last expression, (2.19), is exact because it includes a second-order term in δy, a term we have previously neglected. Next rewrite the first term in Equation (2.19) in terms of a total time derivative:

$$\delta L = \frac{1}{2}m\left[2\frac{d}{dt}(\dot{y}\delta y) - 2\ddot{y}\delta y + \left(\frac{d}{dt}\delta y\right)^2\right] - mg\delta y, \qquad (2.20)$$

$$\delta L = -m(\ddot{y} + g)\delta y + m\frac{d}{dt}(\dot{y}\delta y) + (\delta\dot{y})^2. \qquad (2.21)$$

The second term in Equation (2.21) is a total time derivative, which will not contribute to the action integral variation, because this will give a part that depends only on the end points. The other first-order terms proportional to δy in Equation (2.21) vanish if $y(t)$ obeys the equation of motion derived from the Euler–Lagrange equation:

$$\ddot{y} = -g. \qquad (2.22)$$

QUESTION 3: *Cannon Example* Using the expression for δL (2.19), show that, after integrating by parts, δS has one part that vanishes if $\ddot{y} = -g$ regardless of δy and one part that depends on $(\delta y)^2$ and thus is always positive or zero. How does this prove that requiring S to be a minimum implies that $\ddot{y} = -g$?

Definition of the Variational Derivative

The combination of total and partial derivatives that enters into the Euler equation is called the *variational derivative* of L with respect to q:

$$\delta S = \int \delta L \, dt \equiv \int \frac{\delta L}{\delta q} \delta q(t) \, dt. \tag{2.23}$$

Comparing with (2.8), we have

$$\boxed{\frac{\delta L}{\delta q} \equiv \frac{\partial L}{\partial q} - \frac{d}{dt}\left(\frac{\partial L}{\partial \dot q}\right).} \tag{2.24}$$

definition of a notation for variational derivative

Note that δ is quite different from ∂! $\frac{\delta L}{\delta q}$ is just a notation for the right side of Equation (2.24) – nothing else. It is *not* a derivative in the usual sense of partial or total derivative. The use of a symbol for the right side of Equation (2.24) is a great convenience, but the reader will have to be aware what is meant by $\frac{\delta L}{\delta q}$.

It has been proved that the vanishing of the variational derivative is a *sufficient* condition for an extremum in the action integral. This means that if the variational derivative vanishes, then it is implied that $\delta S = 0$. Is this condition also a necessary one? In other words, if $\delta S = 0$, does that imply that the variational derivative must vanish? The necessity of the variational dervative to vanish is guaranteed for the one dependent variable case we have considered here. With more than one dependent variable, necessity of this condition for an extremum means that if $\delta S = 0$, *all* of the variational derivatives must vanish. This can be true only if the variations $\delta q_i(t), i = 1, 2, \ldots$ are all independent. With more than one dependent variable, we are not always either at the bottom of a valley or at the top of a mountain peak. Theoretically there is the additional possibility of a "saddle point," which, like a mountain pass, is neither a maximum nor a minimum. The proper terminology is that the action integral is then said to be "stationary." A functional that depends on several functions $S[y, z, \ldots]$ is stationary if there is no variation in its value for small independent and arbitrary variations in the functions $y, z, \ldots$. We wouldn't know what kind of extremum it is unless we examine the second derivatives. Strangely, in physics this is commonly never done. Physicists really only care that the action is stationary, not whether it represents an actual minimum, maximum, or saddle point.

2.4 SYSTEMS WITH SEVERAL DEGREES OF FREEDOM

Suppose there are N degrees of freedom. Call them $q_k(t)$ ($k = 1, \ldots, N$). For the action integral variation, by a straightforward application of Taylor's series in N dependent variables, we have

$$\delta S = \int \left(\sum_{k=1}^{N} \frac{\delta L}{\delta q_k} \delta q_k \right) dt. \qquad (2.25)$$

(To save writing it out, we've used the notation for the variational derivative (2.24).) *If* all the δq_ks can be varied independently, $\delta S = 0$ implies that all of the variational derivatives must vanish. N Euler–Lagrange equations are then obtained from a single Lagrangian. This is only true for the case of holonomic constraints that have the same number of generalized coordinates as degrees of freedom. No matter how many degrees of freedom there are, the Euler–Lagrange equations are a consequence of requiring that the action integral be stationary with respect to *independent* arbitrary variations of the time-dependent generalized coordinates. These variations are arbitrary except that they are required to vanish at the (fixed) end points in time. One equation per degree of freedom is obtained in this way.

But what do we conclude if the q_ks *cannot* all be varied independently owing to constraints? This is the kind of situation where we resort to the use of Lagrange multipliers to restore the apparent independence of the δq_ks. Usually one tries to choose coordinates so that they are all independent, and each one expresses a different degree of freedom. However, there is one situation where Lagrange multipliers cannot be avoided: certian types of nonholonomic problems. We will discuss this technique later in this chapter.

Still another interesting situation is the case of more than one independent variable. A stretched string obeys a variational principle with two independent variables, the horizontal distance x and the time t. One then minimizes a double integral over x and t. This case is also briefly discussed later in the chapter.

2.5 WHY USE THE VARIATIONAL APPROACH IN MECHANICS?

If you had never heard of Newton's Laws, we could have introduced classical mechanics by postulating that the action be minimized by the physical path. To find the form of the Lagrangian for a free particle, we would then have only very general principles to guide us. Yet by imposing the Principle of Galilean Invariance and assuming that space is the same in all directions, we could show that $L \sim v^2$. For other problems we would have to guess the correct form of the potential energy. This seems a peculiar way to introduce classical mechanics, but it is perfectly rigorous. The notion of Galilean invariance (equivalence of inertial frames) is the essence of Newtonian mechanics. Mass could be defined as the proportionality constant (or rather twice this constant) of v^2 in the Lagrangian. We would probably have to include the additivity of mass in our postulates, but little else. Imposing very general invariance requirements on the Lagrangian is in fact what theoretical physicists do in elementary particle physics. All the symmetry principles we believe in restrict the possible Lagrangians so much that we end up with just a handful of possible Lagrangians, each of which can be tried and compared with experiment, until we find the best one that agrees with experimental data.

2.5 WHY USE THE VARIATIONAL APPROACH IN MECHANICS?

If we postulate the action principle instead of deriving it from Newton's Laws, is Hamilton's Principle a law of physics or is it "just" mathematics? The answer is mathematics. The physics is expressed in the form of the Lagrangian we choose. This Lagrangian must obey the principles we believe are physically correct. The principle of least action serves as an ideal mathematical platform for expressing these principles.

Historically, the first variational principle to be used in physics came from Fermat, who derived the laws of refraction by requiring that the "optical path" for light be minimized. Maupertuis tried to generalize these ideas to mechanics, arguing that the existence of a "minimal action" principle was evidence for the existence of God! His physics was not correct, and he was ridiculed for his religious claims, but it did generate interest in pursuing the idea of variational calculus as a tool for theoretical physics. Hamilton published the correct formulation in 1834, many years later. (See the appendix for a discussion of Maupertuis' Principle.)

Review of Results So Far

Hamilton's Principle can be taken as the fundamental principle of classical mechanics, provided that we specify the form of the Lagrangian. By requiring that Hamilton's Principle hold, we have given an alternative derivation of the basic equations of the mechanical motion: the Euler–Lagrange equations. The variational approach makes it clear that the Euler–Lagrange equations will have exactly the same form no matter what coordinate system we happen to choose to do the problem.

QUESTION 4: *Variational Approach* How exactly does the variational approach make it clear that the Euler–Lagrange equations will have exactly the same form no matter what coordinate system we happen to choose to do the problem?

In more precise mathematical language, we say that any invertible mapping between the coordinate systems $\{q_k\}$ and another set $\{Q_k\}$ is called a *diffeomorphism*. "Invertible" implies that one can always express one set of coordinates in terms of the other. The coordinate transformation can also depend on the time (but not on the velocities). Even though we did not explicitly state it, we have already shown that

$$\frac{\delta L}{\delta q_k} = 0 \Leftrightarrow \frac{\delta \bar{L}}{\delta Q_k} = 0. \qquad (2.26)$$

Here $\bar{L}$ is the new Lagrangian obtained by substituting the transformation equations into L.

There is some freedom as to what we choose for the Lagrangian in a given problem: We can add a constant, multiply by a constant, change the time scale by a multiplicative constant, or add the total time derivative of an arbitrary function of q and t: $F(q, t)$. Any of these transformations will lead to a Lagrangian that is perfectly satisfactory for describing the motion.

QUESTION 5: *Invariant Transformations* Prove the statements in the paragraph above. Why is it possible to change the time scale by a constant factor? Why can't the arbitrary function F also be a function of $\dot{q}$ as well as q and the time?

QUESTION 6: *Variational Calculus* Explain the meaning of the term "variational calculus."

2.6 LAGRANGE MULTIPLIERS

The method of Lagrange multipliers is sometimes useful or even essential to solve problems in mechanics using Hamilton's Principle. Lest you think that the method of Lagrange multipliers is necessarily only used in the calculus of variations, we begin here with an application in a calculus minimization problem.

⊃ Example: Lagrange Multipliers

The problem is to minimize $F(x, y) = x^2 + y^2$, subject to the constraint $y = 2x+1$ as shown in Figure 2.4. The straightforward way to do this problem is to eliminate y using the constraint and treat it as a problem in a single variable. In order to demonstrate the method of Lagrange multipliers we won't do this. Instead, consider modifying F to $F' = x^2 + y^2 + \lambda(y - 2x - 1)$, where λ is an initially unknown constant. Since λ multiplies something that vanishes according to the constraint, as long as the constraint is satisfied, F is minimized when F' is minimized. If x and y were independent variables, the minimum value of F' would necessarily imply $\frac{\partial F'}{\partial x} = \frac{\partial F'}{\partial y} = 0$ at this value of (x, y). We will choose λ so that both partial derivatives do vanish (even though x and y are not independent on the line $y = 2x + 1$). The three equations for the three unknowns x, y, and λ are:

$$\frac{\partial F'}{\partial x} = 2x - 2\lambda = 0,$$

$$\frac{\partial F'}{\partial y} = 2y + \lambda = 0, \quad (2.27)$$

$$y = 2x + 1.$$

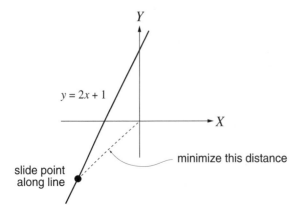

FIGURE 2.4

The unique solution is $x = -\frac{2}{5}$, $y = \frac{1}{5}$, $\lambda = -\frac{2}{5}$, $F_{\min} = \frac{1}{5}$. Check for yourself to see if this is the same answer you get if, instead, y is expressed in terms of x using the constraint equation after which the function $F(y(x), x)$ is minimized with respect to x. For a more formal discussion of Lagrange multipliers, see any text on advanced calculus.*

QUESTION 7: *Lagrange Multipliers* Find the equation for the minimum of $F = 3x^2 + 2y^2$ subject to the constraint that $x = \cos 2y$ using the method of the Lagrange multipliers. It will be a transcendental equation. Solve it numerically.

2.7 SOLVING PROBLEMS WITH EXPLICIT HOLONOMIC CONSTRAINTS

It may happen that the number of dynamical variables is greater than the number of degrees of freedom. One should then go back and try to come up with a better set of coordinates. For example, consider the pendulum of length l, which has only one degree of freedom. Instead of choosing θ as the generalized coordinate, suppose you, stubbornly and foolishly, insist on choosing x and y instead as your generalized coordinates. You also have the constraint equation $x^2 + y^2 = l^2 = \text{constant}$. If one looks at the variation of the action, one gets

$$\delta S = \int \left\{ \frac{\delta L}{\delta x} \delta x + \frac{\delta L}{\delta y} \delta y \right\} dt. \tag{2.28}$$

Nothing in the derivation of this expression depended on independent variations of $x(t)$ and $y(t)$. By Hamilton's Principle, it is still true that $\delta S = 0$. Unfortunately, since δx and δy are no longer independent, you cannot then conclude that each coefficient of the variations δx and δy (i.e., each variational derivative) vanishes by itself. If they did, you would obtain two Euler–Lagrange equations, one for x and the other for y. But they aren't independent. What should you do in this case?

Lagrange's method for dealing with constrained variables works here too. The method allows us to treat x and y as if they were independent, provided that we introduce an unknown constant, the Lagrange multiplier. It is well hidden in the mathematics, but secretly he is reintroducing the constraint forces we thought we had escaped. (In fact we *have* escaped them in most cases, but not in problems of this type with explicit constraint equations.) Assume there is a constraint equation of the type

$$G(x, y) = \text{constant} \equiv C. \tag{2.29}$$

Here G is assumed to be a known function of the coordinates but does *not* involve velocities – only coordinates and, possibly, also the time. For small variations in x and y, the relation from differentiating the equation for G is

$$\delta G = \frac{\partial G}{\partial x} \delta x + \frac{\partial G}{\partial y} \delta y = 0. \tag{2.30}$$

This is the explicit equation connecting variations in x with variations in y.

---

* e.g., Kaplan, *Advanced Calculus*, third edition, pp. 159–160.

Instead of substituting for $y(x)$ in the action principle, try this equivalent procedure: Multiply Equation (2.29) by an unknown "constant" λ. (Note that λ can be a function of t, the independent variable, but not x or y.) Since δG equals zero, we can add it inside the action integral (2.16) without changing anything. Then calculate the variation of S with the time integral of $\delta(L + \lambda(G - C))$ with respect to x and y:

$$\delta S = \int \left\{ \left(\frac{\delta L}{\delta x} + \lambda \frac{\partial G}{\partial x} \right) \delta x + \left(\frac{\delta L}{\delta y} + \lambda \frac{\partial G}{\partial y} \right) \delta y \right\} dt. \qquad (2.31)$$

Now since λ can be an arbitrary function of t, we can choose it to make the first term (coefficient of δx) vanish. Then the least action principle $\delta S = 0$ means that the coefficient of δy must also vanish, since δy is arbitrary. In this way, even though δx and δy are not independent, we can conclude that *both* coefficients vanish in Equation (2.31), just as in the earlier case (Equation (2.25)), where we drew the same conclusion for a different reason.

This implies that the Euler–Lagrange equations now become the two equations*

$$\begin{aligned} \frac{\delta L}{\delta x} + \lambda \frac{\partial G}{\partial x} &= 0, \\ \frac{\delta L}{\delta y} + \lambda \frac{\partial G}{\partial y} &= 0, \end{aligned} \qquad (2.32)$$

with the additional constraint equation

$$G(x, y) = \text{constant}. \qquad (2.33)$$

There are three equations in three unknowns: x, y, and $\lambda(t)$. In principle these can be solved. You can easily check that this is equivalent to solving for δx in terms of δy. The most general form of the method would involve more than one Lagrange multiplier. The number of λs you must introduce equals the number of constraint equations. You can often avoid complicated substitutions by using the Lagrange multiplier method. This becomes particularly useful when there are three or more variables. A kind of symmetry is maintained in this way between the dynamical variables. However, it can be shown that the number of equations that must be solved is equal to the number of coordinates plus the number of constraints, even though the number of degrees of freedom in the problem is the number of coordinates minus the number of constraints. Thus additional equations and variables are introduced by using the Lagrange multiplier method.

⊃ Example: Solving the Linear Pendulum Using Lagrange Multipliers

To illustrate the Lagrangian multiplier method, we will solve the pendulum shown in Figure 2.5 this way. This will also demonstrate that physically the Lagrange multipliers represent constraint forces. We choose the x and y coordinates of the mass as the dynamical variables instead of θ. The Lagrangian is

$$L = \underbrace{\frac{m}{2}(\dot{x}^2 + \dot{y}^2)}_{T} - \underbrace{mgy}_{V}. \qquad (2.34)$$

---

* Recall the definition of the variational derivative, Equation (2.24) above.

2.7 SOLVING PROBLEMS WITH EXPLICIT HOLONOMIC CONSTRAINTS

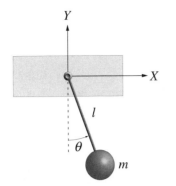

FIGURE 2.5

The constraint is that the length of the pendulum remain constant:

$$G(x, y) = \sqrt{x^2 + y^2} = \text{constant} = l. \qquad (2.35)$$

Notice that we use the radical sign in the constraint equation. This is purely a matter of choice, but it will affect the interpretation of the Lagrange multiplier.

QUESTION 8: *Pendulum Example* Using Lagrange multipliers, prove that the following equations are the EOM for the pendulum with the Lagrangian defined in (2.34) and the constraint defined in (2.35):

$$m\ddot{x} = \lambda \frac{x}{l},$$
$$m\ddot{y} + mg = \lambda \frac{y}{l}. \qquad (2.36)$$

There is a force $\equiv \mathcal{T}$ directed up the length of the pendulum. This is the constraint force that causes the pendulum mass to swing in the arc of a circle. There is a direct connection between the Lagrange multiplier, which we introduced for purely mathematical purposes, and the physical constraint force. The potential energy is $V = mgy$. The negative gradient of V is $(0, -mg)$. Using Newtonian mechanics, the components of the force are ($y < 0$):

$$\mathcal{F}_x = -\mathcal{T} \sin\theta = -\mathcal{T}\left(\frac{x}{l}\right), \qquad (2.37)$$

$$\mathcal{F}_y = -mg - \mathcal{T} \cos\theta = -mg - \mathcal{T}\left(\frac{y}{l}\right). \qquad (2.38)$$

Comparing Equations (2.37, 2.38) with the EOM derived using the Lagrange multipliers (2.36), we conclude that $\lambda = -\mathcal{T}$. The components of the constraint force are $-\lambda \frac{\partial G}{\partial x}$ and $-\lambda \frac{\partial G}{\partial y}$. Note that if we had chosen another form for the constraint equation other than (2.35), we would not have obtained the relation $\lambda = -\mathcal{T}$. But the Lagrange multiplier would still be proportional to this force.

It is not an accident that this connection between constraint forces and Lagrange multipliers exists. To prove it in the general case, we must generalize the example above to any number of nonindependent coordinates connected by one or more constraint equations.

Define the generalized constraint force as

$$\mathcal{N}_k \equiv \sum_{i=1}^{M} \vec{N}_i \cdot \frac{\partial \vec{r}_i}{\partial q_k}. \quad (2.39)$$

definition of generalized constraint force
$\vec{N}_i$ is constraint force on ith part of system

Even though the constraint forces do no work, we can't in this case conclude that $\mathcal{N}_k = 0$. Since, for example, δq_1 and δq_2 are not independent, we can only conclude that $\mathcal{N}_1 \delta q_1 + \mathcal{N}_2 \delta q_2 = 0$, as in the pendulum example above. Since the coordinates are not independently variable, we cannot eliminate the constraint forces associated with them from the equations of motion. Assume that you have N coordinates and N_C constraints. It is true from (1.53) that

$$\frac{\delta T}{\delta q_k} \equiv \frac{\partial T}{\partial q_k} - \frac{d}{dt}\left(\frac{\partial T}{\partial \dot{q}_k}\right) = -\sum_i \dot{\vec{p}} \cdot \frac{\partial \vec{r}_i}{\partial q_k}. \quad (2.40)$$

For conservative systems, from Newton's Law of Motion applied to the ith part, $\dot{\vec{p}}_i = -\vec{\nabla}_i V + \vec{N}_i$, where V is the potential energy function. Note the important mathematical identity $\frac{\partial V}{\partial q_k} = \sum_i \vec{\nabla}_i V \cdot \frac{\partial \vec{r}_i}{\partial q_k}$, which follows from the chain rule for differentiating implicit functions, since V is a function of the $\vec{r}_i$, which are in turn functions of the q_k. Replace the part of (2.40) involving the applied force by the negative gradient of the potential energy with respect to q_k, making use of this identity. This term is then moved to the left side of Equation (2.40). Keep the constraint force part on the right-hand side of the equation. Writing the left side in terms of the variational derivative of the Lagrangian (because $\frac{\partial V}{\partial \dot{q}_k} = 0$), we obtain

$$\frac{\delta L}{\delta q_k} = -\mathcal{N}_k, \quad k = 1, \ldots, N. \quad (2.41)$$

The effect of forces not derived from a potential energy function is to give a nonvanishing right-hand side to the usual Euler–Lagrange equations of motion. This new term is just the generalized constraint force as defined by Equation (2.39).

Equation (2.41) is also what we obtain by using the Euler–Lagrange equations, incorporating the Lagrange multipliers and the constraint equations $G_j = \text{constant}$ ($j = 1, \ldots, N_C$). Since the "augmented" Lagrangian is $L' = L + \sum_{j=1}^{N_C} \lambda_j G_j$, the action principle augmented by one or more Lagrange multipliers (multidimensional version of (2.31)) allows us to assume that all of the coefficients of the separate δq_k vanish in the variation of the action. The modified Euler–Lagrange equations (multidimensional verison of (2.32)) obtained after adding $\sum_{j=1}^{N_C} \lambda_j G_j$ can be written as

$$\frac{\delta L}{\delta q_k} = -\sum_{j=1}^{N_C} \lambda_j \frac{\partial G_j}{\partial q_k}. \quad (2.42)$$

2.7 SOLVING PROBLEMS WITH EXPLICIT HOLONOMIC CONSTRAINTS

Comparing Equation (2.42) with Equation (2.41) gives us

$$\mathcal{N}_k = \sum_{j=1}^{N_C} \lambda_j \frac{\partial G_j}{\partial q_k}. \tag{2.43}$$

This is the explicit connection between the constraint forces on the one hand and the Lagrange multipliers on the other.

The number of degrees of freedom is thus $N_D = N - N_C$. The number of equations that must be solved, including the constraints, is $N + N_C = N_D + 2N_C$. Since the constraints expressed by a suitable choice of generalized coordinates causes the number of equations to equal only the number of degrees of freedom, Lagrange multipliers should be introduced into this type of problem only as a last resort.

Frictional forces or other types of nonconservative forces excluded from the Lagrangian can also be incorporated into the equations of motion for generalized coordinates in a similar way. A generalized force can be defined in exactly the same way as was done in Equation (2.39) above.

> **Example**
>
> Lagrange multipliers can also be used in problems involving static equilibrium. Imagine a chain stretched across a river. The length of the chain is fixed as D, while the river has a width of $2a < D$. Taking the y coordinate to be vertical and the x coordinate to be horizontal, what curve does the chain follow in order to minimize the potential energy? (If the chain is at rest, there is no kinetic energy, so maximizing the Lagrangian is the same as minimizing the potential energy.) If the mass density of the chain is ρ, the potential energy is
>
> $$V = \rho g \int_{-a}^{a} y \, ds, \tag{2.44}$$
>
> where $ds = dx\sqrt{1 + (\frac{dy}{dx})^2}$ is the element of arc length. Using a Lagrange multiplier to express the fact that the length of the chain is constant: $D = \int_{-a}^{a} \sqrt{1 + (\frac{dy}{dx})^2} \, dx$, we make the expression below an extremum:
>
> $$\int_{-a}^{a} \left[y\sqrt{1 + \left(\frac{dy}{dx}\right)^2} + \lambda\sqrt{1 + \left(\frac{dy}{dx}\right)^2} \right] dx. \tag{2.45}$$
>
> Using the Euler equation, after some algebraic simplification, we arrive at the differential equation obeyed by the chain:
>
> $$1 + y'^2 = (y + \lambda)y'' \tag{2.46}$$
>
> ($y' \equiv \frac{dy}{dx}$). Rewrite this equation as
>
> $$\frac{2y'y''}{1 + y'^2} = 2\frac{y'}{y + \lambda}. \tag{2.47}$$

Integrating both sides and exponentiating, we obtain

$$1 + y'^2 = C(y + \lambda)^2, \tag{2.48}$$

where C is an integration constant. By choosing $C = \frac{1}{\lambda^2}$ after solving for y' in (2.48), the equation for x in terms of y is ($u = \frac{y}{\lambda}$)

$$x = \lambda \int_0^{\frac{y}{\lambda}} \frac{du}{\sqrt{(u+1)^2 - 1}}. \tag{2.49}$$

In writing Equation (2.49), we have assumed that $y = x = 0$ at the center of the river. The integral gives

$$\frac{x}{\lambda} = \log\left\{1 + \frac{y}{\lambda} + \sqrt{\frac{y}{\lambda}\left(2 + \frac{y}{\lambda}\right)}\right\}. \tag{2.50}$$

This equation can be inverted to find $y(x)$:

$$\frac{y}{\lambda} = \cosh\left(\frac{x}{\lambda}\right) - 1. \tag{2.51}$$

The shape of the chain is a *catenary* curve. λ is clearly a distance scale. $\frac{dy}{dx} = \sinh(\frac{x}{\lambda})$. From this, we obtain λ by calculating the length of the chain:

$$D = \lambda \int_{-\frac{a}{\lambda}}^{\frac{a}{\lambda}} \cosh u \, du = 2\lambda \sinh\left(\frac{a}{\lambda}\right). \tag{2.52}$$

The last equation is a transcendental equation that must be solved for λ in terms of the length of the chain and the width of the river, after which the description in Equation (2.51) is complete.

2.8 NONINTEGRABLE NONHOLONOMIC CONSTRAINTS – A METHOD THAT WORKS

The method we are about to describe for using Lagrange multipliers in nonholonomic problems was only discovered[*] in 1871. The crucial point is to express the constraint equations in terms of small variations in the coordinates at constant time, then use the variational principle. Just sending $L' \to L + \lambda G$ will *not* work in this case.

Stand a penny up on edge on a sloping table as shown in Figure 2.6. The penny is free to roll down the table, which it has a tendency to do. It is also free to spin on an axis passing through the point of contact with the table. But the table is rough, so the penny cannot skid.

---

[*] See the references in E. T. Whittaker's *Analytical Dynamics of Particles and Rigid Bodies*, p. 215.

2.8 NONINTEGRABLE NONHOLONOMIC CONSTRAINTS – A METHOD THAT WORKS

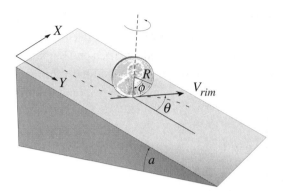

FIGURE 2.6

Here θ is the angle that measures the spinning of the coin around a vertical axis. $\theta = 0$ corresponds to having the coin lined up with the y (sloping) direction. Since the velocity of the rim $v_{\text{rim}} = R\dot\phi$, the velocity-dependent constraint equations are

$$\begin{aligned}\dot x &= R\dot\phi \sin\theta, \\ \dot y &= R\dot\phi \cos\theta.\end{aligned} \tag{2.53}$$

These constraints are relations between velocities, not coordinates. They do not lead to equations linking y with the other two coordinates. Just like Equations (1.107, 1.108), these are *nonintegrable* equations. Thus it is impossible to describe the penny with only two coordinates, despite the fact that the penny has only two degrees of freedom. It is a case of nonholonomic constraints.

If we freeze the motion, and make small virtual displacements in ϕ, $\delta\phi$, then since the rolling constraint (2.53) must still hold for every virtual displacement $\delta\phi$:

$$\delta x = R\sin\theta\, \delta\phi, \quad \delta y = R\cos\theta\, \delta\phi. \tag{2.54}$$

As always the Lagrangian is the difference of the kinetic and potential energies. It can be shown that the kinetic energy is the sum of two terms, one corresponding to rotation about a vertical axis through the center of the penny and a second term corresponding to rotations about an axis perpendicular to the plane of the penny through the point of contact between the penny and the surface of the plane. Without giving the derivation in detail (it is discussed in a problem at the end of the chapter), the result for the kinetic energy is

$$T = \frac{3}{4}mR^2\dot\phi^2 + \frac{1}{8}mR^2\dot\theta^2. \tag{2.55}$$

If the wedge angle of the plane is a, the potential energy due to gravity is

$$V = mgy\sin a. \tag{2.56}$$

The correct Lagrangian is

$$L = \frac{mR^2}{2}\left\{\frac{3}{2}\dot\phi^2 + \frac{1}{4}\dot\theta^2\right\} - mgy\sin a. \tag{2.57}$$

Notice that x doesn't appear in the Lagrangian at all with this particular choice of coordinates. There are two degrees of freedom, but three coordinates are needed in the Lagrangian: θ, ϕ, y. Since there is one relevant constraint equation, we need one Lagrange multiplier. Under the integral giving the variation of the action (2.25), add the vanishing quantity $\lambda(\delta y - R\cos\theta\,\delta\phi)$. Variation of the coordinates gives us the corresponding variation of the action:

$$\delta S = 0 = \int \left\{ \left(\frac{\delta L}{\delta y} + \lambda\right)\delta y + \left(\frac{\delta L}{\delta \phi} - \lambda R\cos\theta\right) \times \delta\phi + \left(\frac{\delta L}{\delta \theta}\right)\delta\theta \right\} dt. \tag{2.58}$$

Any variation of θ in the constraint equations would lead to quantities of the second order in the variations, so we can ignore this. Using the Lagrange multiplier to make all three coefficients of the virtual displacements vanish as before, we have three equations of motion from the action principle:

$$-mg\sin a + \lambda = 0,$$

$$\frac{3mR^2}{2}\ddot{\phi} - \lambda R\cos\theta = 0, \tag{2.59}$$

$$\frac{mR^2}{4}\ddot{\theta} = 0.$$

Notice that y has now been eliminated. These equations can be integrated to find the solution for the rolling and spinning penny. After the solutions are known, $x(t)$ and $y(t)$ are found by integrating the constraint Equations (2.54). The integration can be done both analytically and numerically. This problem is a good tutorial for solving systems with nonintegrable, nonholonomic constraints. You can also graph the rather amazingly complicated motion that can occur as the penny rolls down the inclined plane. This will be done in a problem at the end of the chapter.

Instead of using an "effective Lagrangian," $L + \lambda G$, which is only useful with velocity-independent (holonomic) constraints, we have multiplied by dt to convert linear relations constraining the velocities (2.53) into linear relations among the variations δq_k. Each of these constraint relations is then multiplied by a Lagrange multiplier* and added to δS directly *under the time integral*. We can then choose the Lagrange multipliers to allow us to act as if the different coordinates can be varied independently, thus obtaining a set of Euler–Lagrange equations by imposing Hamilton's Principle. We use these equations, plus the constraints on the velocities, to solve the problem. Although the details differ between handling holonomic and nonholonomic constraints, the net result is the same: We can treat all the q_ks as if they are independent after we insert the Lagrange multiplier terms into the integral for δS.

---

* Recall that the Lagrange multiplier can depend on the time.

2.9 POSTSCRIPT ON THE EULER EQUATION WITH MORE THAN ONE INDEPENDENT VARIABLE

In the mechanics we have discussed so far, time is the only independent variable. Consider, however, the case of a string stretched along the X axis. The string is clamped at both ends, but may be displaced vertically, so that $y \neq 0$. In fact, we may choose any point x and any time t at which to observe the deflection, so $y = y(x, t)$. Now both x and t are independent variables and the action for this system is a *double integral* over x, t. The Lagrangian is replaced in these integrals by a function $\mathcal{L} = \mathcal{L}(y, \frac{\partial y}{\partial x}, \frac{\partial y}{\partial t}, x, t)$. Exactly as before, one first does a Taylor series expansion in the small function δy and its derivatives. Since there are two (partial) derivatives of y, when y is varied, partial integrations are performed with respect to x for terms with $\frac{\partial \delta y}{\partial x}$ and with respect to t for terms proportional to $\frac{\partial \delta y}{\partial t}$. These partial integrations convert terms proportional to the derivatives of the variation δy into terms proportional to δy. The boundary conditions ensure the vanishing of the additional terms at the end points. A generalized Euler equation is the result:

$$\frac{\partial \mathcal{L}}{\partial y} - \frac{\partial}{\partial x}\left(\frac{\partial \mathcal{L}}{\partial \left(\frac{\partial y}{\partial x}\right)}\right) - \frac{\partial}{\partial t}\left(\frac{\partial \mathcal{L}}{\partial \left(\frac{\partial y}{\partial t}\right)}\right) = 0. \tag{2.60}$$

For more than two independent variables, one just subtracts more terms of the form above. There is no contradiction between the appearance of $\frac{d}{dt}$, the total time derivative in the original form of the Euler–Lagrange equations, and the partial derivatives $\frac{\partial}{\partial x}$ and $\frac{\partial}{\partial t}$ in (2.60). With one independent variable, we used the total time derivative to mean that we include the explicit time variation of the dependent variable when taking the derivative. We do the same thing here, but the notation means that t is constant when differentiating with respect to x and vice versa. The ordinary differential equation of motion obtained previously for a single independent variable becomes a partial differential equation if there is more than one independent variable. There remains only a single equation if there is a single *dependent* variable y. If there is more than one dependent variable y, instead say y and z, then we would have also a second equation like (2.60) with y replaced by z.

The function $\mathcal{L}$ is derived explicitly in the homework problem on the string. From this one can deduce the velocity of waves on a stretched string.

SUMMARY OF CHAPTER 2

- All of classical mechanics follows from

$$\boxed{\text{Hamilton's Principle: } \delta \int L \, dt = 0}$$

(with fixed start and finish times and no variation of path at end points).

- For each degree of freedom:

$$\delta \int L\, dt = 0 \Leftrightarrow \frac{\delta L}{\delta q} \equiv \frac{\partial L}{\partial q} - \frac{d}{dt}\left(\frac{\partial L}{\partial \dot{q}}\right) = 0 \qquad (2.61)$$

(Euler–Lagrange equations).
- If the number of coordinates exceeds the number of degrees of freedom, use the method of Lagrange multipliers. For N_C holonomic constraints of the form $G_j(q)$ = constant, use

$$L' \equiv L + \sum_{j=1}^{N_C} \lambda_j G_j \qquad (2.62)$$

as the "effective Lagrangian." If there are N_D degrees of freedom and N_C constraints, you can treat all $N = N_D + N_C$ coordinates as independent using this method. You solve for $N_D + 2N_C$ unknowns, using the N Euler–Lagrange equations and the N_C constraint equations. The unknowns are the N coordinates and the N_C Lagrange multipliers.
- The Lagrange multipliers λ_j for $j = 1, \ldots N_C$ are proportional to the constraint forces. They can be functions of time but not of the q_k or $\dot{q}_k$.
- Lagrange multipliers can also be used to solve nonholonomic problems with velocity-dependent constraints, such as a penny rolling on an inclined plane. This is done differently than in the holonomic case; the constraints on virtual displacements are multiplied by Lagrange multipliers and then added directly to the variation of the action under the time integral.
- If the orbit is needed without the time, use Maupertuis' Principle (see the appendix).

PROBLEMS

Variational Calculus

Problem 1: *(Straight line in polar coordinates)* Using plane polar coordinates and the variational calculus, find the minimum distance (i.e., the equation of the straight line) from the origin to the point (1, 1).

Problem 2: *(Geodesic on a sphere)* Use spherical coordinates and the variational calculus to find the *geodesic* (curve that is the shortest distance between two points) on a sphere. This is called by airplane pilots "the great circle route."

Problem 3: *(Geodesic on a cylinder)* Prove that the geodesic on a circular cylinder is a helix. Roll a piece of lined paper into a cylinder to see the helical curves. Notice

PROBLEMS

that rolling up the paper does not change the geodesic property of the curves on the surface. Why not? Hint: Set up a suitable coordinate system, and find an integral expression for the length of an arbitrary curve on the surface of this cylinder.

Problem 4: *(Geodesic on a cone)* Assume you are on the surface of a cone with a half angle α which is a surface of revolution about the Z axis. Find an equation in plane polar coordinates for the geodesic curves on this surface. Notice, as in the previous problem, that you can roll up a piece of paper into a cone and visualize these curves geometrically. Why can't you use the "paper roll" to also answer the question about geodesics on a sphere? What is the essential difference between a cone and cylinder on the one hand and a spherical surface on the other?

Problem 5: *(Variational Principle for quantum mechanics)* The quantum mechanics of a one-dimensional system is described by the Schrödinger equation for the complex wave function $\psi(x, t)$:

$$-\frac{\hbar^2}{2m}\frac{\partial^2 \psi}{\partial x^2} + V(x)\psi = i\hbar \frac{\partial \psi}{\partial t}, \qquad (2.63)$$

where $\hbar$ is Planck's constant $\frac{h}{2\pi}$, m the mass, and $V(x)$ the potential energy. Find a variational principle for quantum mechanics using the two dependent variables ψ, ψ^* (complex conjugate of ψ) and the two independent variables x, t. You can treat ψ, ψ^* as two independent generalized coordinates, since the real and imaginary parts are independent variables. Hint: You will try to make the variation of a double integral of the form below vanish:

$$0 = \delta \iint \mathcal{L}\left(\psi, \psi^*, \frac{\partial \psi}{\partial x}, \frac{\partial \psi}{\partial t}, \frac{\partial \psi^*}{\partial x}, \frac{\partial \psi^*}{\partial t}, x\right) dx\, dt = 0. \qquad (2.64)$$

Furthermore, you can assume that $\mathcal{L}$ is real. It might have pieces of the form $V(x)\psi^*\psi$ or $\frac{\partial \psi^*}{\partial x}\frac{\partial \psi}{\partial x}$, for example. See if you can guess the correct form for $\mathcal{L}$ such that the Euler–Lagrange equations lead to the Schrödinger equation and its complex conjugate. The potential energy $V(x)$ is a real function.

Problem 6*: *(One dependent and three independent variables: an electrostatics problem)*

a) Derive the form of the Euler–Lagrange equation for one dependent variable and three independent variables. You want to use x, y, z as independent variables and a function $\Phi(x, y, z)$ as the single dependent variable. Suppose there is a "known Lagrangian"

$$L(\Phi, \vec{\nabla}\Phi, x, y, z). \qquad (2.65)$$

You want to minimize the triple integral

$$I \equiv \iiint_V L\, dx\, dy\, dz, \qquad (2.66)$$

where V is the volume of integration. What equation should Φ satisfy to accomplish this? The main difficulty is to know how to do the partial integration we did so easily with one independent variable. There is a useful vector calculus identity you can use to do it:

$$\vec{\nabla} \cdot (\vec{F}G) = \vec{F} \cdot \vec{\nabla}G + G\vec{\nabla} \cdot \vec{F}, \qquad (2.67)$$

where $\vec{F}$ and G are arbitrary vector and scalar functions of x, y, z. We will assume the variation of Φ on the boundary of the volume V vanishes, in analogy with the case of one independent variable in the principle of least action. Also, the divergence theorem is useful here:

$$\iiint_V \vec{\nabla} \cdot \vec{F} \, dx \, dy \, dz = \iint_S \vec{F} \cdot d\vec{S}, \qquad (2.68)$$

where S is the surface of the volume V. This mathematical result holds for any vector function of x, y, z. Use the divergence theorem to derive the final form of the Euler–Lagrange equations for three independent variables in vector calculus notation.

b) In electrostatics, the energy stored in the electric field is proportional to $\iiint E^2 \, dV$, where $\vec{E}$ is the electric field and $dV = dx \, dy \, dz$ is the volume element. (We assume there is no free charge in the volume here.) Show that, if $\vec{E} = -\vec{\nabla}\Phi$, and the stored energy is minimized, while Φ (the electrostatic potential) is held constant on the boundaries, Φ must obey Laplace's equation ($\nabla^2 \Phi = 0$). Do this in Cartesian coordinates (x, y, z).

Fermat's Principle

Problem 7: *(Fermat's Principle and the bending of light)* A sugar solution with a nonuniform index of refraction $n[y]$ bends a ray of light passing through the solution, as shown in Figure 2.7. The index $n[y]$ is a decreasing function of y, $y(x)$ is the height of the light ray in the tank, and $0 \leq x \leq D$, the distance along the horizontal. The physical reason why a light wave is bent downwards can be seen by considering wavefronts of the light passing through the medium. Because the light velocity is higher at the top of the wave front than at the bottom, the upper portion travels faster and gets ahead of the lower part. Since the light ray is defined by the normal to the wave front, the light bends downward.

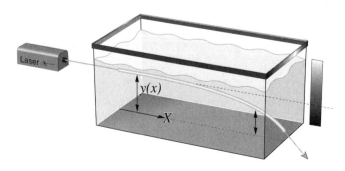

FIGURE 2.7

PROBLEMS

The motion of a light ray through an inhomogeneous medium can be mathematically described by Fermat's principle of least time, which says

$$\int dt = \text{a minimum}. \tag{2.69}$$

In other words, the light follows the path that minimizes the transit time between two fixed points ($x = 0$ and $x = D$). If we use this principle, we can calculate the light path taken as a function of the variable index of refraction $n[y]$ without using the wave theory of light at all. The paths of light inside fiber optics can be calculated in this way, for example. The light starts out at $x = y = 0$ and is initially horizontal: $\frac{dy}{dx}|_{x=0} = 0$. We know that

$$\text{velocity of light} = \frac{ds}{dt} = \frac{c}{n[y]}, \tag{2.70}$$

where ds is the arc length and $c = 3 \times 10^8$ m/s.

a) Use x as the independent variable, and show that Fermat's Principle is equivalent to finding the path $y(x)$ that minimizes

$$\int_0^D L \, dx, \tag{2.71}$$

where $L = L[y, \frac{dy}{dx}]$ is a function you must derive.

b) Assume that $n[y(0)] \equiv n_0$, and use the Euler–Lagrange equation to prove $y(x)$ is the solution to the equation ($n[y]$ is a known function)

$$\frac{d^2y}{dx^2} = \frac{d \ln n[y]}{dy}(1 + y'^2). \tag{2.72}$$

c) If $y'(0) = 0$, prove the solution to Equation (2.72) is given by

$$\left(\frac{n[y]}{n_0}\right)^2 = 1 + \left(\frac{dy}{dx}\right)^2 \tag{2.73}$$

or, equivalently, we get $x(y)$ by solving

$$x = \int_0^y \frac{dy}{\sqrt{\left(\frac{n[y]}{n_0}\right)^2 - 1}}. \tag{2.74}$$

We can then invert this function for $y(x)$ if we wish to.

Problem 8: *(More about bending light with a variable refractive index)* A tank of sugar dissolved as a highly concentrated solution in water has an index $n_0 = 1.5$. Suppose the tank is 30 cm long. Assume that $n[y] = n_0 e^{-\alpha y}$. A deflection of the light beam downwards is observed, and it emerges with $y(30) = -1$ cm. Find the numerical value of α and the approximate shape of the trajectory of the ray in the tank.

Problem 9: *(Brachistochrone 1)* Solve the Brachistochrone problem: Find the function $y(x)$ that connects two fixed points in the XY plane (as shown in Figure 2.1), such that a frictionless mass sliding down the curve arrives at the destination in the least possible time $T[y(x)]$. (Hint: Make the $+x$ direction downwards and the $+y$ direction to the right. This choice is to avoid having to solve for $x(y)$, which you could do instead of turning the axes.) An expression for T, given some trial function $y(x)$, is

$$T = \int dt = \int \frac{ds}{v} = \int \frac{dx}{v} \frac{ds}{dx}, \qquad (2.75)$$

where the velocity $v = \frac{ds}{dt}$ and $ds = $ the arc length.

a) Using energy conservation in a constant gravitational field, prove that the form of the functional integral we want is

$$\sqrt{2g}\, T[y] = \int_{x_0}^{x_1} \frac{dx}{\sqrt{x}} \sqrt{1 + \left(\frac{dy}{dx}\right)^2}. \qquad (2.76)$$

b) Prove that the curve $y(x)$ that minimizes $T[y]$ is

$$y(x) = -\sqrt{x(2r-x)} + 2r \arcsin\left(\frac{x}{2r}\right), \qquad (2.77)$$

where "r" is a constant of integration chosen so that the curve passes through the end point. It is really just a scale factor.

c) Plot the curve (using a computer would be helpful here). For $r = 1$, the curve runs between $x = 0$ and $x = 2$. Turn your head 90 degrees to see the Brachistochrone curve!

Problem 10: *(Brachistochrone 2)*

a) Derive the differential equation for $y(x)$ by minimizing the expression for the time using Equation (2.75).

b) Now assume that there is a parameter θ and that, in terms of θ, $x(\theta) = a(1 - \cos\theta)$. (We are still using rotated coordinates, with $+x$ vertically downward, so a is a negative constant.) Using the equation you have obtained for minimizing the time, prove that

$$y(\theta) = |a|(\theta - \sin\theta) \qquad (2.78)$$

is a solution, assuming that the particle starts from the origin $\theta = 0$.

c) Graph this curve, which is a cycloid curve. Explain why it is the curve traced out by a point on a wheel of radius $|a|$ rolling down the $+Y$ axis.
d) Calculate the time taken to slide down this curve, assuming that θ varies from 0 to $\frac{\pi}{2}$. Compare it to the time taken to slide down a straight line from the origin to this end point.

Problem 11: *(Ski race)* Imagine you are standing on top of a mountain. The altitude is given by the z coordinate. The shape of the surrounding hills is given by $z = f(x, y)$, where f is a known function. You are an Olympic skier in a race to get to the finish line located down in the valley at a point x_1, y_1.

a) What route should you choose to win the race? First find a set of differential equations for $x(t)$, $y(t)$, then explain how you would find the solution you want.
b) Solve the equations if $f(x, y)$ is the function

$$z = f(x, y) = (\sin^2 2\pi x)(\sin^2 2\pi y). \tag{2.79}$$

Start at $x = y = 0.25$ and ski down to $x = y = 0$. First make a 3-D plot of the surrounding hills with a computer and guess which route you should take. (Hint: Notice that this problem is symmetric under the exchange of x and y.)

Problem 12*: *(Snell's Law)* An open question for physics up to the start of the nineteenth century was about the nature of light: Does light consist of particles or waves? By observing the refraction of light at the interface between two media (say vacuum and glass), and measuring the speed of light in both media, it would have been possible to decide this question.

a) First assume that light consists of a stream of classical particles and that the light is bent towards the perpendicular to the interface as it passes from the vacuum into the glass. $v_{\text{vacuum}} = v_1$, $v_{\text{glass}} = v_2$ as in Figure 2.8. Could there be a transverse force (along the interface plane) exerted on the particles? If not, how is the change in direction related to the relative speed in the two different media? Derive an equation of the form

$$\frac{\sin \theta'}{\sin \theta} = f\left(\frac{v_2}{v_1}\right). \tag{2.80}$$

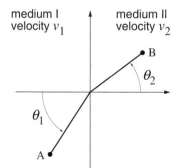

FIGURE 2.8

Find the form of the function f in Equation (2.80). It depends only on the velocity ratio. Do the particles speed up or slow down upon entering the glass if the ray is bent as shown in the figure?

b) Now take the point of view that light is a wave. Fermat's Principle states that the light ray will minimize the time it takes to go from point A to point B. Since you know that the light must travel in a straight line in a medium where the wave velocity is constant (why?), the only thing you can vary is the point where the ray intersects the interface. Find this point, assuming that the light velocity is given in terms of the index of refraction n, $v = \frac{c}{n}$. Prove Snell's Law and answer the question: If light is a wave and the ray is bent toward the normal direction as shown, is the velocity of light in glass, v_2, greater or less than the velocity in a vacuum, v_1? Compare with part a.

Hamilton's Principle

Problem 13: *(Stretched string)* A string is stretched under tension between two fixed end points at $x = 0$ and $x = L$. The string displacement is described by a single function of two independent variables: $y(x, t)$. ($y(0, t) = y(L, t) = 0$.) If the density of the string is ρ kg/m, the kinetic energy is

$$T = \frac{1}{2}\rho \int_0^L dx \left(\frac{\partial y}{\partial t}\right)^2. \tag{2.81}$$

To get the potential energy V, consider a small piece of the string of length dx as shown in Figure 2.9. The tension τ acts parallel to the string, so for small deflections

$$F_{y1} \approx \tau \tan\theta = \tau \frac{dy}{dx}\bigg|_x, \qquad F_{y2} \approx \tau \tan\theta' = \tau \frac{dy}{dx}\bigg|_{x+dx}, \tag{2.82}$$

$$dF_y = F_{y2} - F_{y1} = \tau \left(\frac{dy}{dx}\bigg|_{x+dx} - \frac{dy}{dx}\bigg|_x\right) \approx \tau dx \frac{d}{dx}\left(\frac{dy}{dx}\right) \tag{2.83}$$

Integrate the force from 0 to dy, holding x constant. The result for the total potential

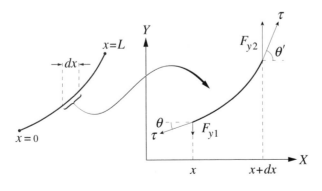

FIGURE 2.9

energy of the string is

$$V = \frac{1}{2}\tau \int_0^L dx \left(\frac{\partial y}{\partial x}\right)^2. \tag{2.84}$$

Find the equation of motion for the string. Notice that it has traveling wave solutions of the form $y(x, t) = f(x \pm ct)$, with f an arbitrary function. Find the wave velocity c.

Lagrange Multipliers

Problem 14*: *(Rolling hoop)* A hoop of mass M and radius R rolls without slipping down an inclined plane which makes an angle α with the horizontal. Gravity acts on the hoop in the vertical direction. You can assume that the potential energy of the hoop is the same as if all of its mass were concentrated at the center of the hoop. Using Lagrangian mechanics, find the equation of motion of the hoop.

This problem can be done in at least two different ways. Since there is only one degree of freedom, you can choose the angle ϕ through which the hoop has rolled and write the Lagrangian only in these terms. Or else you can use the distance along the hypotenuse of the plane d as well as ϕ plus a Lagrange multiplier that expresses the rolling constraint $d = R\phi$. In this example, the rolling constraint is holonomic because only one-dimensional motion is involved.

Problem 15: *(Rolling penny on an inclined plane)* Set up the Lagrangian for the problem of the penny on the inclined table (2.57). First calculate the kinetic energy (2.55) for rolling ($\dot{\phi} \neq 0$) and spinning ($\dot{\theta} \neq 0$). (Hint: Work out the kinetic energy as a function of $\dot{\phi}, \dot{\theta}$ for a ring of radius r; then integrate to get the kinetic energy for a uniform disk.) Then set up the Lagrangian and the constraint equations.

Problem 16*: *(Maximizing the area under a string of fixed length)* This problem involves an elementary application of the method of Lagrange multipliers. A string of fixed length l is placed with its ends on the X axis at $x = \pm a$ as shown in Figure 2.10. The problem is to find the curve $y(x)$ that maximizes the area between the curve and

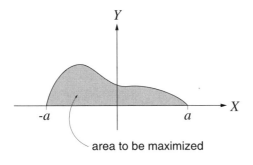

FIGURE 2.10

the X axis:

$$A = \int_{-a}^{a} y\, dx. \qquad (2.85)$$

The intuitive answer is fairly obvious. What is it? The length of the string is given by

$$l = \int_{-a}^{a} \sqrt{1 + y'^2}\, dx, \qquad (2.86)$$

where l is fixed (i.e., constrained) and $y' \equiv \frac{dy}{dx}$.

a) Since arbitrary variations $\delta y(x)$ are not possible (why not?), you can't use the calculus of variations directly. But there is a way to do the problem using Lagrange multipliers. Consider introducing an arbitrary constant λ and then maximizing the functional

$$K[y] \equiv A + \lambda l. \qquad (2.87)$$

If, for arbitrary variations $\delta y(x)$, you have $\delta K = 0$, then for the special variations δy that leave the string's length unchanged ($\delta l = 0$), it will be true that $\delta A = 0$. (Make sure you understand the logic of this last statement.) Find the differential equation from the variational derivative:

$$\frac{\delta}{\delta y}\left[y + \lambda\sqrt{1 + y'^2}\right] = 0. \qquad (2.88)$$

b) Integrate this equation once to find $y'(x)$ explicitly. Choose the integration constant so that $y'(0) = 0$. (Symmetry implies $y(x)$ is an even function of x.)
c) Integrate a second time to find the most general form of $y(x)$.
d) Evaluate the up-to-now unknown constant λ as a function of a and l. You may want to use the mathematical integral

$$\int_0^\alpha \frac{du}{\sqrt{1 - u^2}} = \arcsin \alpha. \qquad (2.89)$$

Did this solution agree with your intuition?

Problem 17: *(Particle in a constant magnetic field)* Inside a solenoid it is a good approximation to regard the magnetic field as constant and directed along the Z axis. Particle motion in such a field is a helical orbit, with particles that start from the axis eventually returning to the axis. Since in a magnetic field the kinetic energy must be constant, this would lead erroneously to the conclusion that Maupertuis' Principle would mean that $\delta \int ds = 0$. Explain why this is not true. Minimizing the arc length gives orbits that are straight lines instead of helices. What is wrong with this argument?

APPENDIX

ABOUT MAUPERTUIS AND WHAT CAME TO BE CALLED "MAUPERTUIS' PRINCIPLE"

Starting in 1747, there was a great controversy caused by Maupertuis, who stated the Principle of Least Action in a way we now know to be incomplete. He claimed that nature acted in such a way as to minimize the product of mass times velocity times distance. Furthermore, he said this principle had a divine origin, thus infuriating most scientists of his time who believed in science as a purely *logical* activity. As we said earlier, he was motivated by Fermat's Principle in optics, which asserted that light rays always follow the minimum optical path (minimum transit time for the light). Maupertuis tried to find an analogous principle that would apply to mechanics. Maupertuis was accused of a) being wrong and b) stealing the idea from Leibnitz. We quote Voltaire on Maupertuis:

> The assertion that the product of the distance and the velocity is always a minimum seems to us to be false, for this product is sometimes a maximum, as Leibnitz believed and as he has shown. It seems that the young author has only taken half of Leibnitz's idea; and, in this, we vindicate him of ever having had an idea of Leibnitz in its entirety.

Of course, the later work by Lagrange and Hamilton developed the correct approach without any mystical assumptions. Notice that the correct application of variational calculus to mechanics took place 87 years later, fully 138 years after Newton and others solved Bernoulli's challenge of the brachistochrone problem.

Nevertheless, Maupertuis has become immortalized by a principle that bears his name. Suppose we want to know the equation for the path of a particle without specifying the time. For example, in two-dimensional motion, instead of finding the parametric equations $y(t)$, $x(t)$, suppose that we would like a differential equation for $y(x)$. We can formulate a second kind of variational principle, making use of a new kind of variation Δ.

Note that the action S can be rewritten as (use the definition of canonical momentum p_k (1.70) and H (1.65))

$$S = \int_{t_0}^{t} L\, dt = \int \sum_k p_k dq_k - \int_{t_0}^{t} H\, dt \qquad (2.90)$$

($\dot{q}_k dt = dq_k$) As shown in Figure 2.11, the Δ variation varies the path between fixed end points (x_0, y_0) and (x_1, y_1), but the end point times t_0, t_1 need not be fixed. Furthermore, assume $H = E$ and that all paths have the same energy. This means that the possible varied paths are more restricted than before. The first integral on the right of (2.90) no longer involves the time, and the end points do not vary when we compute ΔS (but the path does vary). Since we assume $H = E =$ constant, the right-most integral is $-E(t - t_0)$. So ΔS

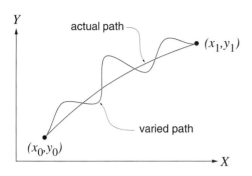

FIGURE 2.11
Δ variation: fixed endpoints.

contains two types of variations:

$$\Delta S = \Delta \underbrace{\int \sum_k p_k dq_k}_{\text{varies due to path only}} - \underbrace{E \Delta t}_{\text{variation of } S \text{ due to end point } \Delta t \neq 0}. \quad (2.91)$$

This is unlike the δS variation discussed previously. Here one term comes from the variation of the end point time t, the other from variations along the path in the integral

$$S_0 \equiv \int \sum_k p_k dq_k. \quad (2.92)$$

Since *any* variations in the action due to the choice of path must vanish by Hamilton's Principle, for this type of variation

$$\Delta S_0 = 0. \quad (2.93)$$

This is known as Maupertuis' Principle. S_0 is sometimes called the "extended action" or, confusingly, the action. The proof here is a subtle one and depends on separating ΔS into two types of variation, one due to the path and the other due to the end points. The part depending only on the path variation must vanish, as a special case of the more general form of Hamilton's Principle. We give a more detailed proof below.

We emphasize that this principle holds only for a certain type of variation. All paths must have the same energy E. They must start and finish at the same point. The arrival times can vary for different paths, however.

⟹ Example 2-D Example of Maupertuis' Principle

We consider a two-dimensional problem as a simple example. $V(x, y)$ is the potential energy. For fixed total energy E, $p = \sqrt{2m}\sqrt{(E - V(x, y))}$, and $p_x dx + p_y dy = p ds$, with $ds^2 = dx^2 + dy^2$. Putting all this into

$$\Delta S_0 = 0 \quad (2.94)$$

we get

$$\Delta \int \sqrt{2m} \sqrt{(E - V(x, y))} \sqrt{1 + \left(\frac{dy}{dx}\right)^2} \, dx = 0. \tag{2.95}$$

You can apply the calculus of variations with x as the independent variable in the usual way to find a differential equation for the path $y(x)$.

More About Maupertuis' Principle

The Lagrangian for the "cannonball" problem is

$$L = \frac{1}{2}m(\dot{x}^2 + \dot{y}^2) - mgy. \tag{2.96}$$

(We assume the trajectory of the cannonball lies in the XY plane.) A suitable solution to the Euler–Lagrange equations found by using Hamilton's Principle is

$$x(t) = x_0 + v_{0,x}t, \qquad y(t) = y_0 + v_{0,y}t - \frac{g}{2}t^2, \tag{2.97}$$

where $v_{0,x}$, $v_{0,y}$, x_0, y_0 are all constants found from the initial conditions.

We know that the equation for the orbit is a parabola (assume $x_0 = y_0 = 0$ for simplicity):

$$y(x) = \frac{v_{0,y}}{v_{0,x}}x - \frac{g}{2v_{0,x}^2}x^2. \tag{2.98}$$

What we would like to have is a variational principle that would give us this orbit equation directly, eliminating the time from the problem.

Consider plotting the trajectory of a system (like the cannonball) with two degrees of freedom. We can use a three-dimensional graph with (X, Y, T) as the axes. The trajectory then appears as a space curve. The curve we seek is the projection of this curve onto the XY plane. Let AB denote the physical trajectory between $t = t_0$ and $t = t_1$ as shown in Figure 2.12. The action integral is

$$S_{AB} = \int_{t_0}^{t_1} L(x(t), y(t), \dot{x}(t), \dot{y}(t)) \, dt. \tag{2.99}$$

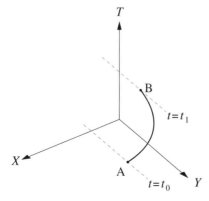

FIGURE 2.12

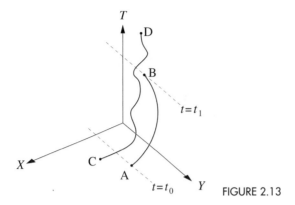

FIGURE 2.13

Our goal is to calculate the *change* in the action if instead we consider another curve CD, not the physical trajectory, "near" to AB.

$$\delta S \equiv S_{\text{CD}} - S_{\text{AB}}. \qquad (2.100)$$

This is illustrated in Figure 2.13. We have expressed Hamilton's Principle previously by saying $\delta S = 0$, but a more precise statement would be

$$\delta S = O[\delta^2]. \qquad (2.101)$$

The right side is a symbolic way of saying the deviations from AB to CD contribute only in second order to the difference in the action integrals. CD is a curve that differs from AB by a δ type variation. This is a necessary condition for the validity of Hamilton's Principle. In order to use Equation (2.101), the definition of a δ variation means we have required that CD must begin and end at the same (x, y) point as AB *and* at the same starting and end times. The two curves must look like Figure 2.14.

Assume the system is holonomic and that $H = E$. Then from (1.65)

$$L = \frac{\partial L}{\partial \dot{x}} \dot{x} + \frac{\partial L}{\partial \dot{y}} \dot{y} - E = p_x \dot{x} + p_y \dot{y} - E. \qquad (2.102)$$

For the δ type variations, the value of the total energy E does not have to be the same for AB and CD.

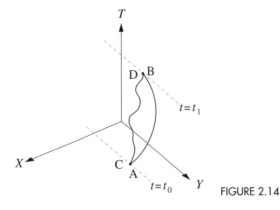

FIGURE 2.14

APPENDIX ABOUT MAUPERTUIS AND WHAT CAME TO BE CALLED "MAUPERTUIS' PRINCIPLE"

Below we shall give a more formal version of our previous proof of the variational principle we have called Maupertuis' Principle. First note that for Maupertuis' Principle, the allowed type of path variations differs from Hamilton's Principle. For these Δ variations, by definition, CD must begin and end at the same points as AB, and the total energy must be the same for both curves. Notice that the departure and arrival times do not have to be the same. This is different from the set of curves we considered for Hamilton's Principle. Using the definition of the Δ type of variation, if

$$\Delta x_1 \equiv x_{CD}(t_1 + \Delta t_1) - x_{AB}(t_1) = 0, \tag{2.103}$$

$$\Delta x_0 \equiv x_{CD}(t_0 + \Delta t_0) - x_{AB}(t_0) = 0 \tag{2.104}$$

and

$$\Delta y_1 \equiv y_{CD}(t_1 + \Delta t_1) - y_{AB}(t_1) = 0, \tag{2.105}$$

$$\Delta y_0 \equiv y_{CD}(t_0 + \Delta t_0) - y_{AB}(t_0) = 0 \tag{2.106}$$

and also

$$E_{CD} = E_{AB} = E, \tag{2.107}$$

then

$$\boxed{\int_{CD} (p_x dx + p_y dy) - \int_{AB} (p_x dx + p_y dy) = O[\Delta^2].} \tag{2.108}$$

Maupertuis' Principle

Here $p_{x,y}$ have their usual meanings as canonically conjugate momenta to x, y.

Proof

This proof relies on the fact that Hamilton's Principle holds for the odd path sketched in Figure 2.15. The path ACDB satisfies all of the necessary conditions, although we note that the $\dot{q}_k$s are discontinuous at C and D as well as at A and B. Call the physical trajectory AB. Start at point A. Now make a varied path that remains at the same x and y positions but moves forwards or backwards in time to $t_0 + \Delta t_0$. (Δt_0 could be negative.) Then travel along the varied path CD until the time $t_1 + \Delta t_1$. Finally, move to the point B by changing only the time. This variation is of the type for which Hamilton's Principle holds. We write this Principle symbolically as

$$\int_{ACDB} L \, dt - \int_{AB} L \, dt = O[\Delta^2]. \tag{2.109}$$

On the pieces AC and DB no coordinate changes, so $\int \sum_k p_k dq_k = 0$. Hence using (2.90) $\int_{AC} L \, dt = -E \Delta t_0$, where E is the common energy of the CD and AB trajectories, and

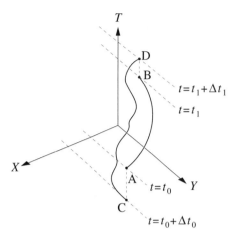

FIGURE 2.15

$\int_{DB} L\, dt = E\Delta t_1$. We have

$$\int_{ACDB} L\, dt = -E\Delta t_0 + \int_{CD} (p_x dx + p_y dy)$$
$$-E(t_1 + \Delta t_1 - t_0 - \Delta t_0) + E\Delta t_1, \quad (2.110)$$

$$\int_{AB} L\, dt = \int_{AB} (p_x dx + p_y dy) - E(t_1 - t_0). \quad (2.111)$$

Taking the difference between Equations (2.110) and (2.111), we get Maupertuis' Principle

$$\int_{CD} \vec{p} \cdot d\vec{r} - \int_{AB} \vec{p} \cdot d\vec{r} = O[\Delta^2]. \quad (2.112)$$

This variational principle involves only the path and not the time. We emphasize that the varied path must not only have the same end points as the physical path, but also must have the same total energy E as the physical path.

It is left as an exercise for the student to show that another form of Maupertuis' Principle is

$$\boxed{\Delta \int \sqrt{T}\, ds = O[\Delta^2]. \qquad (2.113)}$$

Maupertuis' Principle

Here T is the kinetic energy and ds is the element of arc length. Writing $T = E - V$ and treating E as constant in the Euler equation derived from the variational calculus applied to (2.113) guarantees the use of Δ type variations if the end points are also fixed.

QUESTION 9: *Cannonball Orbit* Derive the parabolic form of a cannonball orbit $y(x)$ using Maupertuis' Principle.

CHAPTER THREE

LINEAR OSCILLATORS

OVERVIEW OF CHAPTER 3

An oscillator is a system with periodic motion. In mechanical systems, there is a restoring force that can do both positive and negative work as the system moves. Positive work done by this restoring force changes the kinetic energy into potential energy. Negative work done by the force turns the potential energy back into kinetic energy. If the force is linearly proportional to displacement, the oscillator is a *linear* or *simple harmonic* oscillator. Linear oscillators have many special properties. In particular, linear oscillators have the important property that the oscillation frequency is independent of amplitude. (This is not true if the oscillator is nonlinear.) The importance of linear oscillators in mechanics lies in the fact that, for small vibration amplitudes, we can approximate the dynamics of most mechanical systems as linear oscillators. Not only mechanical systems like a vibrating airplane wing, but, beyond the realm of mechanics, electrical systems and even an electron bound in an atom can be usefully modeled in this way. To understand large-amplitude oscillatory motion, we have to study nonlinear oscillators. The pendulum is an example of an oscillator that is linear at small amplitudes, yet becomes nonlinear at large amplitudes.

To discuss linear oscillators in a physically realistic way, we must depart from our dealings with conservative systems and introduce a special "damping" force which extracts energy from the oscillator. In the case of mechanical oscillators, this occurs due to friction. Such oscillators are called *damped* oscillators.

Oscillators, either damped or undamped, can be driven by external forces. It is possible to calculate the response of a linear oscillator to an arbitrary external force, if that force depends on the time in a known way and is independent of the oscillator amplitude. Such an oscillator is called a *driven* oscillator.

Oscillators of any kind can store energy. They can be driven to large amplitudes by external driving forces, if these forces are nearly synchronous with the oscillator frequency. This phenomenon is called *resonance*. The damping force limits the maximum amplitude of the oscillations at the resonant driving frequency. This means that the ability of the oscillator to store energy is limited by the need to supply energy constantly being removed by the damping force.

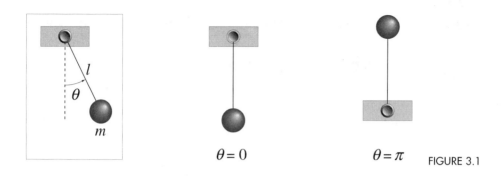

$\theta = 0$ $\theta = \pi$ FIGURE 3.1

3.1 STABLE OR UNSTABLE EQUILIBRIUM?

A mechanical system that remains in its initial state of rest is said to be in *equilibrium*. Such a state of equilibrium holds no matter which set of generalized coordinates we use to describe the motion, since it is a physical condition. Suppose we use the set of coordinates q_k, $k = 1, \ldots N$. (There are N degrees of freedom.) Given the Lagrangian of a system, how could you find the system's equilibrium points? Lagrange discovered the answer to this question: Find the points where all of the generalized forces vanish. For a conservative system, this means you must find all configurations of the system where the potential energy $V(q_1, \ldots, q_N)$ is stationary, that is, where the first derivatives $\frac{\partial V}{\partial q_1}, \ldots, \frac{\partial V}{\partial q_N}$ vanish.

This makes good physical sense. If the body starts from rest, and there is no force, it will remain at rest forever. For example, there are two places where the torque due to gravity acting on a pendulum vanishes, as seen in Figure 3.1. Since the potential $V(\theta) = mgl(1 - \cos\theta)$, the generalized force (which is the torque in this case) is $\mathcal{F}_\theta \equiv -\frac{\partial V}{\partial \theta} = -mgl\sin\theta$. Therefore, $\sin\theta$ vanishes at $\theta = 0, \pi$. The pendulum will remain motionless forever if placed at rest in either of these positions.

These two equilibrium points at $\theta = 0, \pi$ are very different from each other. What if we displace the pendulum very slightly from either position? Will it return to the equilibrium, or will it run away? Everyone will recognize from experience that $\theta = 0$ is a *stable* equilibrium point, whereas $\theta = \pi$ is *unstable*. We are interested in describing the motion for small deviations of θ in both cases. For the stable case, the motion around the equilibrium point is oscillatory, whereas for an upside-down pendulum, the motion runs rapidly away from $\theta = \pi$.

If the constraints change with time it can make a big difference. Try the following experiment: Take a pendulum at rest and shake the support point vertically up and down at a high frequency, gradually increasing the amplitude of the shaking. The configuration with the pendulum upside down ($\theta = \pi$), which is initially an unstable equilibrium point, will become stable at some critical amplitude of shaking, so the pendulum will swing with stable oscillations in the upside-down position! This situation will be discussed quantitatively in Chapter 10.

⊃ Example

We can discuss the pendulum Lagrangian near one of the two equilibrium points by making a Taylor series expansion of L around either $\theta = 0$ or $\theta = \pi$. For θ very close

3.1 STABLE OR UNSTABLE EQUILIBRIUM?

to 0:

$$\cos\theta = 1 - \frac{\theta^2}{2} + \frac{\theta^4}{24} + O[\theta^6], \tag{3.1}$$

$$L(\dot\theta, \theta) = \frac{ml^2}{2}\dot\theta^2 - mgl(1-\cos\theta) = \frac{ml^2}{2}\dot\theta^2 - mgl\frac{\theta^2}{2} + O[\theta^4], \tag{3.2}$$

where ($\dot\theta \equiv \frac{d\theta}{d\tau}$.)

Near to $\theta = 0$ we can drop the θ^4 terms and bring out an overall factor of mgl:

$$L(\dot\theta, \theta) \approx mgl\left(\frac{\left(\sqrt{\frac{l}{g}}\dot\theta\right)^2}{2} - \frac{\theta^2}{2}\right) \tag{3.3}$$

The constant factor mgl doesn't affect the equation of motion and can therefore be omitted. For a further simplification, define a unit of time to be $\sqrt{\frac{l}{g}}$ and change the time variable (and the meaning of $\dot\theta$) to the dimensionless $\tau \equiv t\sqrt{\frac{g}{l}}$. In these new units, the approximated Lagrangian near to the equilibrium point becomes L':

$$L' = \frac{L}{mgl} = \frac{\dot\theta^2}{2} - \frac{\theta^2}{2}, \quad \theta \approx 0. \tag{3.4}$$

Near the unstable equilibrium point θ is very close to π. The Taylor series expansion depends on having a small quantity in order to neglect higher order terms. Define $\theta \equiv \pi + \Delta\theta$. $\Delta\theta$ will be small near the unstable equilibrium point. The Taylor series for $\cos\theta = \cos(\pi + \Delta\theta)$ is $-1 + \frac{\Delta\theta^2}{2} + O[\Delta\theta^4]$. Since π is a constant, $\dot\theta = \dot{\Delta\theta}$. Again, removing the constant factor and scaling the time in the same way as before, the Lagrangian near to the unstable equilibrium point becomes

$$L'' = \frac{L}{mgl} \approx \frac{\dot{\Delta\theta}^2}{2} + \frac{(\Delta\theta)^2}{2}, \quad \theta \approx \pi. \tag{3.5}$$

The crucial difference between stable equilibrium (Equation (3.4)) and unstable equilibrium (Equation (3.5)) is the change of sign in front of the part proportional to the square of the deviation from the equilibrium point. This changes the force from a restoring force to an "antirestoring" force when one calculates the equation of motion.

The Most General Form of a Lagrangian Near Static Equilibrium

Why do most physical systems behave similiarly near their equilibrium points? We will show that this is a consequence of Taylor's theorem applied to the Lagrangian in the vicinity of the equilibrium point.

It is a general property of one-dimensional holonomic systems that these two types of equilibrium, stable or unstable, exist (except for something called "neutral" equilibrium). This is proved by making a Taylor series expansion of the Lagrangian of the system under consideration about an equilibrium point. The equilibrium point of a mechanical system is by definition the point q_{eq} where $\ddot{q} = 0$. This means that, if initially $\dot{q}(0) = 0$, the system never moves: $q = q_{eq}$, $\dot{q} = 0$ for all time. It is no loss of generality to define the constant $q_{eq} = 0$ to simplify the algebra. (If you don't believe this, carry out the calculation without this assumption!)

The most general form of the Taylor series expansion of L to second order* about $q_{eq} \equiv 0$ is

$$L_{approx} = A + Bq + C\dot{q} + Dq^2 + Eq\dot{q} + F\dot{q}^2, \tag{3.6}$$

where A, B, C, D, E, and F are all unknown constants that depend on the system being considered and can be found from derivatives of the Lagrangian evaluated at $(q_{eq}, \dot{q}_{eq})$. For example, $D = \frac{1}{2}\frac{\partial^2 L}{\partial q^2}|_{q_{eq},\dot{q}_{eq}}$ and $F = \frac{1}{2}\frac{\partial^2 L}{\partial \dot{q}^2}|_{q_{eq},\dot{q}_{eq}}$. Since $B = \frac{\partial L}{\partial q}|_{q_{eq},\dot{q}_{eq}}$, it must be true by the definition of equilibrium that $B = 0$. The EOM is then

$$\frac{d}{dt}(C + Eq + 2F\dot{q}) = 2Dq + E\dot{q} \tag{3.7}$$

or

$$\ddot{q} - \frac{D}{F}q = 0. \tag{3.8}$$

The motion depends only on the ratio $\frac{D}{F}$. This is the same EOM we get from the equivalent Lagrangian

$$L_{approx} = \dot{q}^2 + \frac{D}{F}q^2. \tag{3.9}$$

The frequency ω_0 of small oscillations is given by the formula

$$\omega_0^2 \equiv -\frac{D}{F}. \tag{3.10}$$

(The formula above (3.10), only makes sense if $\frac{D}{F}$ is negative.)

By the choice of a new unit of time, $t = \beta\tau$, the equation of motion can be written

$$\ddot{q} - \beta^2 \frac{D}{F}q = 0, \tag{3.11}$$

where β must be real, to avoid having an imaginary time unit. Choose

$$\beta = \sqrt{\left|\frac{F}{D}\right|}. \tag{3.12}$$

---

* The order of a term in the expansion can be found by multiplying q by a scale factor: $q \to \lambda q$. All terms in the Lagrangian that behave like λ^n are said to be of the nth order. By "second order," we mean all terms of the form q^2, $\dot{q}^2$, $q\dot{q}$.

3.1 STABLE OR UNSTABLE EQUILIBRIUM?

This can always be done if $F \neq 0$. In fact, $F > 0$, because it represents part of the kinetic energy term in the Lagrangian, and kinetic energy is always positive.

Working backwards from the equations of motion after we've scaled the time to make it simpler, we find that, by neglecting small terms, the Lagrangian near equilibrium can be put into one of two forms*:

$$L_{\text{stable}} = \frac{1}{2}(\dot{q}^2 - q^2) \quad (3.13)$$

or

$$L_{\text{unstable}} = \frac{1}{2}(\dot{q}^2 + q^2). \quad (3.14)$$

Comparing (3.13, 3.14) to (3.4, 3.5) we see that almost all holonomic, mechanical systems with one degree of freedom behave like the pendulum example near to stable/unstable equilibrium points.

We define "neglecting small terms" to be the neglect of all terms beyond the second order. This approximation will be made frequently to render problems soluble. The price we pay is that we are restricted to discussing only small amplitudes of the motion. The concept of "small" is left somewhat vague. In any particular case, you should really check to see that third- and higher-order terms are truly negligible.

To simplify the formulas, we have eliminated as many constants as possible in (3.13, 3.14). It is a common practice in physics to choose the units to simplify the formulas, including changing the time scale, so we urge you to get used to this technique. It is not difficult to convert to the normal units when you need to do it. There is some practice in scaling and unscaling units in the problems at the end of the chapter.

Since the Lagrangian is $T - V$, if we have a conservative, holonomic, scleronomic system, then T is a quadratic form and the sign of $\frac{D}{F}$ is wholly determined by the curvature of V (i.e., $\frac{\partial^2 V}{\partial q^2}$) at the equilibrium point. L_{stable}, that is, positive curvature of $V(q)$ near $q = q_{\text{eq}} = 0$ gives the equation of motion

$$\ddot{q} + q = 0 \Leftrightarrow \frac{\partial^2 V}{\partial q^2} > 0, \quad \text{at } q = q_{\text{eq}} = 0. \quad (3.15)$$

L_{unstable} or negative curvature of the potential energy gives instead the equation of motion

$$\ddot{q} - q = 0 \Leftrightarrow \frac{\partial^2 V}{\partial q^2} < 0, \quad \text{at } q = q_{\text{eq}} = 0. \quad (3.16)$$

---

* Really three forms are possible: If the curvature of the potential $\frac{d^2 V}{dq^2} = 0$ at the equilibrium point, we have "neutral" equilibrium. Higher-order terms arising from nonlinear forces would then have to be considered.

FIGURE 3.2
A) Curvature of $V > 0$, stable equilibrium; B) curvature of $V < 0$, unstable equilibrium.

A change of the sign in front of q makes the difference between stable and unstable equilibrium. This sign can be determined by examining the curvature of V. As an example and also as a way to remember this rule, a ball bearing placed in the bottom of a bowl (curvature of $V > 0$) is stable, whereas one placed on top of an inverted bowl (curvature of $V < 0$) is unstable. This is shown in Figure 3.2.

The equation of motion for stable motion is known to physicists as the equation for the simple harmonic oscillator, abbreviated "SHO." This equation occurs so often in physics as a first approximation that it is worthwhile to study it in detail.

For certain systems with an infinite number of degrees of freedom (like a stretched string under tension, such as a violin string), Daniel Bernoulli discovered in 1753 that arbitrary motions of the system can be expressed as addition (superposition) of the "modes" of the system. The shape of the mode can be discovered only when the details of the system are given, but, for sufficiently small amplitudes, the amplitude of the mode behaves in time as a linear oscillator. For example, a stretched string under tension and clamped at both ends has a denumerable infinity of modes, each of which oscillates with a single frequency like a linear SHO. Bernoulli's discovery anticipated Fourier analysis but was not accepted by the mathematicians for 100 years. It has been said that that this discovery marked the beginning of theoretical physics. (Although Newton and Galileo might object, it was the beginning of the mechanics of continuous systems with infinite numbers of degrees of freedom.)

QUESTION 1: *Equilibrium 1* In his 1788 treatise *Méchanique Analytique*, Lagrange provided the first general proof that the solution or solutions to the equation

$$\frac{\partial V}{\partial q} = 0 \quad \text{at equilibrium points} \tag{3.17}$$

were the equilibrium points of the general dynamical system with scleronomic, holonomic constraints. Is (3.17) a necessary condition? Is it sufficient? Is it possible to generalize this equation (3.17) to an arbitrary number of degrees of freedom?

The rule of thumb is thus, for a potential $V(q)$, the minima are stable equilibria and the maxima are unstable. This agrees with our intuition. In more than one dimension, we can have saddle points that are stable in one direction and unstable in another. We will discuss > 1-D systems in Chapter 9, where more than one frequency of oscillation is possible near a stable equilibrium.

> **QUESTION 2:** *Equilibrium 2* Prove by using an approximation for the EOM for the pendulum that a small deviation from an unstable equilibrium point grows exponentially. Find the time constant for this exponential growth in terms of the mass and the second derivative of V evaluated at the unstable equilibrium point.

3.2   SIMPLE HARMONIC OSCILLATOR

We begin our study with the frictionless (undamped) simple harmonic oscillator, or SHO, as we refer to it. Because the equation of motion (3.15) is linear in the displacement and its second time derivative, it is a linear oscillator. *The most important property of the linear oscillator is that the frequency of oscillation does not depend on the amplitude of oscillation.* In his work *Dialogues Concerning Two New Sciences*, completed while he was under house arrest and smuggled out of Italy in 1638, Galileo has the character Sagredo say:

> Thousands of times I have observed vibrations, especially in churches where lamps, suspended by long chords, had been inadvertently set into motion... But I never dreamed that one and the same body, when ... pulled aside through an arc of 90° or even 1° or $\frac{1}{2}°$, would employ the same time in passing through the least as through the largest of these arcs, and indeed, it still strikes me as somewhat unlikely.

To which Salviati replies:

> First of all, one must observe that each pendulum has its own time of vibration so definite and determinate that it is not possible to make it move with any other period than that which nature has given it.*

Galileo was the first to realize that we should consider time as the independent variable in physics. This is arguably the beginning of physics as a quantitative science. Before that time, physics was called "natural philosophy."

Galileo was probably aware that a pendulum has a longer period for large amplitudes than for small amplitudes. This is because a pendulum is a linear oscillator only for small amplitudes near $\theta = 0$, as we have shown in the previous example.

---

* A possibly mythical story about the young Galileo tells of him in the cathedral in Pisa, watching the chandelier swinging back and forth. After timing the swings with his pulse, Galileo found that the period of the swing did not depend on the amplitude of the motion. In the story, this is supposed to describe the dawn of understanding the importance of time in physics and also the invention of the pendulum clock. Actually, whether or not the story is true, Galileo did not use pendulum clocks. In later life he did invent an escapement mechanism that would make such a clock possible. The first pendulum clock was built by Huygens not long after Galileo's death in 1642.

Before choosing a new time scale, but after removing a common factor from both parts of the Lagrangian, the Lagrangian for the simple harmonic oscillator can be put into the general form

$$\frac{\dot{q}^2}{2} - \frac{\omega_0^2 q^2}{2}, \tag{3.18}$$

where ω_0^2 is the coefficient of $\frac{q^2}{2}$ when one factors the Lagrangian to make the coefficient of $\frac{\dot{q}^2}{2}$ unity. This form of the Lagrangian gives the EOM

$$\ddot{q} + \omega_0^2 q = 0. \tag{3.19}$$

If we then choose a new unit of time $\tau \equiv \omega_0 t$, then $\frac{d^2 q}{dt^2} = \omega_0^2 \frac{d^2 q}{d\tau^2}$. Changing the meaning of $\ddot{q}$ to mean $\frac{d^2 q}{d\tau^2}$ gives a new form of the EOM:

$$\boxed{\ddot{q} + q = 0.} \tag{3.20}$$

EOM for free oscillator

This is the generic equation of motion for a simple harmonic oscillator. In these time units, the oscillator has the period 2π. In the new time unit, the frequency of the oscillator is $\omega_0 = 1$ by definition.

The Hamiltonian H for the SHO is easily calculated from the approximate Lagrangian, Equation (3.13). From (1.65) $H \equiv \dot{q} \frac{\partial L}{\partial \dot{q}} - L$. Since the time does not appear explicitly in the Lagrangian, H is constant as time evolves:

$$H = \frac{1}{2}(\dot{q}^2 + q^2). \tag{3.21}$$

The dependence on $\dot{q}$ is quadratic, so H is also the total energy of the oscillator.

Mathematical Properties of Second-Order, Linear, Homogeneous ODEs

ODE stands for ordinary differential equation, which is a differential equation with only one independent variable and one dependent variable.

In mathematical language, the equation of motion for a simple harmonic oscillator (3.20) is a *second-order*, *homogeneous*, *linear* ODE with constant coefficients. It is one of the simplest equations of this type and has solutions of the form $\sin t$, $\cos t$. The EOM are of second order because the highest time derivative of q in the equations is $\ddot{q}$, the second derivative with respect to time.

The properties of linearity and of homogeneity in the EOM set the harmonic oscillator apart from many other dynamical systems. The equation is *linear* because only the first

3.2 SIMPLE HARMONIC OSCILLATOR

power of the dependent variable and its derivatives ($q, \dot{q}, \ddot{q}$, etc.) appear. Because Equation (3.20) is *homogeneous* (has zero on the right side, so contains only the dependent variable and its derivatives), multiplying q by a constant leads to a solution if q is a solution.

Since the SHOs equation of motion is a linear equation in q and $\ddot{q}$, we can "superimpose" or add solutions together to create a new solution to the EOM. The ability to add solutions is called the property of *superposition*. This means that if $q_1(t)$ and $q_2(t)$ are solutions to the ODE, so also is $aq_1(t) + bq_2(t)$ if a, b are arbitrary constants. The possibility of superposition is a characteristic feature of all linear differential equations.

It is shown in mathematics textbooks* that the number of initial conditions needed to specify a solution equals the order of the ODE. The most general solution for a second-order ODE must contain two arbitrary constants, which depend on the initial conditions. The physics demands that the motion is determined for all time by specifying the initial displacement and velocity, so the equation of motion must be of second order. This will be true for any mechanical system.

The general solution to Equation (3.20) can be written in the form

$$q(t) = A \sin(t + \phi) = A \sin\phi \cos t + A \cos\phi \sin t = A' \cos t + B' \sin t, \qquad (3.22)$$

where A and ϕ are arbitrary constants. A is called the *amplitude*, and ϕ is the *phase*. Using the trigonometric identity above, the most general solution for $q(t)$ is a linear combination of the two special cases: $\sin t$ and $\cos t$. The initial conditions can be expressed in terms of these constants:

$$q(0) = A \sin\phi, \qquad \dot{q}(0) = A \cos\phi. \qquad (3.23)$$

Equations (3.23) can be inverted to obtain A, ϕ in terms of the initial position and velocity. Physically, the superposition property means that any amplitude is possible, and all amplitudes oscillate with the same frequency.

Complex Solutions for the SHO

The use of complex solutions ($i \equiv \sqrt{-1}$) simplifies the algebra considerably when we consider externally driven simple harmonic oscillators. The most general complex solution to Equation (3.20) can be written as

$$q_{\text{complex}}(t) = \mathcal{A}_c e^{it} = A(\cos(t + \phi) + i \sin(t + \phi)). \qquad (3.24)$$

$\mathcal{A}_c \equiv Ae^{i\phi}$ is now a complex constant in Equation (3.24). What is the physical interpretation of the real and imaginary parts of $\mathcal{A}_c$? To see this, compute the displacement and the velocity of the oscillator at $t = 0$:

$$q(0) = \text{Real}[\mathcal{A}_c], \qquad \dot{q}(0) = \text{Real}[i\mathcal{A}_c] = -\text{Im}[\mathcal{A}_c]. \qquad (3.25)$$

---

* See, for example, *Advanced Calculus*, third edition, W. Kaplan, Chapter 8.

Therefore, the complex constant $\mathcal{A}_c$ contains information about both of the necessary initial conditions:

$$\mathcal{A}_c = q(0) - i\dot{q}(0). \tag{3.26}$$

A shift of the initial conditions to some other time t_0 means we multiply the complex amplitude $\mathcal{A}_c$ by the complex phase factor e^{-it_0}:

$$\mathcal{A}_c = [q(t_0) - i\dot{q}(t_0)]e^{-it_0}. \tag{3.27}$$

Remember to take the real part at the end of the calculation to get the physical solution. We will usually drop the subscript "complex" on q_{complex}, so you will have to remember whether or not you are dealing with the real physical q or else the complex solution, also called q. The real part is the sum $(q + q^*)/2$. q^* is the complex conjugate of the solution $q(t)$. q^* is also a good solution, because the coefficients in the EOM are real.

3.3 DAMPED SIMPLE HARMONIC OSCILLATOR (DSHO)

The Lagrangian formalism we have derived fails to incorporate nonconservative forces. We have to put the friction force into the EOM ourselves. Again choose the time units so that the frequency for the free undamped oscillator $\omega_0 \equiv 1$. Assume that there is a friction force proportional to the velocity $\dot{q}$:

$$\boxed{F_{\text{friction}} = -\frac{1}{Q}\dot{q}.} \tag{3.28}$$

definition of Q

This Equation (3.28) defines the dimensionless constant Q, which is called the "quality factor" of the oscillator. It is related to the way in which the stored energy in the free oscillator drops as a function of time. We discuss this in more detail below. The "free" EOM (3.20) becomes

$$\boxed{\ddot{q} + \frac{1}{Q}\dot{q} + q = 0.} \tag{3.29}$$

DSHO

Why is the friction force assumed to be proportional to velocity $\dot{q}$? We chose it so that the modified EOM remains linear. But reality could be different.[†] A pendulum immersed

---

[†] See A. B. Pippard's *The Physics of Vibration*, p. 31 for a discussion of the real nature of friction forces.

3.3 DAMPED SIMPLE HARMONIC OSCILLATOR (DSHO)

in a viscous fluid that moves so slowly that the flow around the pendulum is laminar will have a retarding force of the form we assumed. For more rapid motion, there may be turbulence in the fluid. This would give a $\dot{q}^2$ dependence instead, spoiling the linearity of the Equation (3.29). Rubbing the pendulum against a fixed surface would, by contrast, lead to a constant force opposed to the motion, but one independent of velocity. Electrical circuits with resistance obey Equation (3.29) rather well. The mathematical and physical concepts developed in this chapter can be directly applied to solving many problems in electrical and radio frequency engineering.

As a guess for a solution to Equation (3.29), one can try the substitution $q(t) = e^{i\alpha t}$ (where α is a constant) into the equation of motion (3.29) above:

$$\left(-\alpha^2 + i\frac{\alpha}{Q} + 1\right)e^{i\alpha t} = 0. \tag{3.30}$$

The solution for α must be a root of the quadratic equation obtained by requiring the coefficient of the exponential in the formula above to vanish. The differential equation has been converted to an algebraic equation. The roots are found solving for the roots of a quadratic equation. Notice that they can be either real or complex:

$$\alpha = \frac{i}{2Q} \pm \sqrt{1 - \frac{1}{4Q^2}}. \tag{3.31}$$

There are three cases, corresponding to the three types of solutions to (3.31): *underdamped*, *overdamped*, and *critically damped*. Once we have found a complex solution that works, mathematics guarantees that there are no other solutions, since q and q^* are linearly independent and a second-order ODE must have two independent solutions.

UNDERDAMPED: $Q > \frac{1}{2}$

In this case the formula for the solution of the quadratic Equation (3.31) gives the square root of a positive number. The general complex solution is

$$q(t) = A_c e^{-\frac{t}{2Q}} e^{\pm i\omega' t}, \quad \omega' \equiv \sqrt{1 - \frac{1}{4Q^2}}. \tag{3.32}$$

In physics only real quantities appear, so the most general physical solution is

$$q(t) = A\, e^{-\frac{t}{2Q}} \sin(\omega' t + \phi). \tag{3.33}$$

The overall amplitude A and the phase ϕ are arbitrary constants. Notice that the solution is exponentially damped. Energy is extracted from the oscillator, usually in the form of heat. This process is irreversible. The smaller Q is, the more quickly the oscillations damp out. There is also a frequency shift, since $\omega' \neq 1$. The oscillator frequency ω' is lower than the undamped oscillator $\omega_0 \equiv 1$ by a factor $\sqrt{1 - \frac{1}{4Q^2}}$. If $Q \gg 1$, this frequency shift is very small, even negligible.

OVERDAMPED: $Q < \frac{1}{2}$

In this case there are no oscillations at all because ω' is imaginary, making the exponential parts real. An example of overdamping is a pendulum swinging in very thick molasses. If we displace it and let go, the pendulum slowly returns to the vertical position without swinging past it. We obtain two solutions to (3.31), so there are two different real exponentials. Both of the exponentials have negative coefficients of the time in the exponent: Both of these are decreasing with time, with no oscillations occurring. The general overdamped solution has the form

$$q(t) = Ae^{\lambda_+ t} + Be^{\lambda_- t} \tag{3.34}$$

where

$$\lambda_\pm = -\frac{1}{2Q} \pm \sqrt{\frac{1}{4Q^2} - 1} < 0 \tag{3.35}$$

and A and B are arbitrary amplitudes set by the initial conditions:

$$q(0) = A + B, \quad \dot{q}(0) = -(\lambda_+ A + \lambda_- B). \tag{3.36}$$

Prove for yourself that both roots, (3.35), are always negative, regardless of Q, if $Q < \frac{1}{2}$.

CRITICAL DAMPING: $Q = \frac{1}{2}$

A third and final possibility exists. If $Q = \frac{1}{2}$, the roots of the quadratic equation are equal and we get a single real solution:

$$q(t) = e^{\lambda t}, \quad \lambda = -1. \tag{3.37}$$

We know from physical reasoning that two initial conditions, $q(0)$ and $\dot{q}(0)$, must be specified. This means we need to find a second solution to the quadratic equation. This can be done by assuming we have an overdamped oscillator with $Q = \frac{1}{2} - \epsilon$, where $\epsilon > 0$. We can look at the two solutions for the overdamped oscillator (3.34). Assume $B = -A$ and let $\epsilon \to 0$. In that limit, one gets a solution proportional to te^{-t}. (The details of this calculation are left to a homework problem.) So the general critically damped solution is

$$q(t) = Ce^{-t} + Dte^{-t}, \tag{3.38}$$

where C and D are arbitrary constants to be determined by the initial conditions. If we wish to build a sensitive galvanometer or seismograph, and want it to return quickly to equilibrium after an impulse of current or earth tremor is recorded, then we want the instrument to be critically damped.

3.3 DAMPED SIMPLE HARMONIC OSCILLATOR (DSHO)

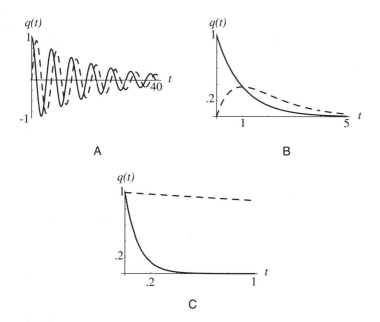

FIGURE 3.3
DSHO responses. The two independent solutions are shown as solid and dashed lines. A) $Q = 10$, underdamped; B) $Q = 0.5$, critically damped; C) $Q = 0.1$, overdamped.

To summarize: A DSHO is *underdamped* if $Q > \frac{1}{2}$, *critically damped* if $Q = \frac{1}{2}$, and *overdamped* if $Q < \frac{1}{2}$. The two independent solutions for each type of oscillator is shown in Figure 3.3. For the underdamped case (Figure 3.3A), the cosinelike (solid curve) and sinelike (dashed curve) solutions are the real and imaginary parts of the complex solution, respectively. Note the difference in time scales for the curves in the figures. The overdamped oscillator usually has two very different decay time constants, as illustrated in Figure 3.3C. If the damping $Q \to \frac{1}{2}$, the two time constants become equal as in Figure 3.3B.

QUESTION 3: *DSHOs* Explain why we need two independent solutions for a complete solution to the second-order differential equation for the DSHO in general (i.e., no matter what Q is). Provide both a mathematical and a physical reason.

What Is the Physical Meaning of Q?

The amplitude of the underdamped oscillator decays exponentially. From Equation (3.32), the amplitude is proportional to $e^{-\frac{t}{2Q}}$ times a function that oscillates with unit magnitude. Since the total energy is proportional to the square of the amplitude, we can calculate the number of periods it takes for the energy to decay to $\frac{1}{e}$ of its original value. If we neglect the small frequency shift (assume Q is large), it will take $\frac{Q}{2\pi}$ complete oscillations to do this.

TABLE 3.1 SOME REPRESENTATIVE Q VALUES (FROM *NEWTONIAN DYNAMICS*, BY R. BEIRLEIN, EXCEPT FOR THE SUPERCONDUCTING MICROWAVE CAVITY)

System	Q
50-gram mass hanging from coil spring	≈ 25
Earth, for oscillations induced by earthquake	≈ 200
FM radio receiver	$\approx 5 \times 10^3$
Tuning fork	$\approx 10^4$
Sodium atom emitting yellow light	5×10^7
Superconducting rf cavity (1,300 MHz)	10^{10}
Iron nucleus (^{57}Fe) gamma ray	3×10^{12}

If we call E the energy stored in the oscillator by the impulse at $t = 0$, then for $Q \gg 1$, the fraction of the stored energy lost per cycle is $\frac{\Delta E}{E} = -\frac{2\pi}{Q}$. Thus

$$\frac{dE}{dt} = -\frac{E}{Q}. \tag{3.39}$$

exponential decay of stored energy

QUESTION 4: *Time Constant* Prove Equation (3.39) and the claim that the time constant for the energy decrease is $\frac{Q}{2\pi}$ oscillation periods. Take the energy of the oscillator to be $E_{\text{osc}} = \frac{1}{2}(\dot{q}^2 + q^2)$.

For real systems we have to deal with an extremely wide range of possible Q values, as seen in Table 3.1.

3.4 AN OSCILLATOR DRIVEN BY AN EXTERNAL FORCE

What has been described above in some detail are the free oscillations of the one-dimensional linear oscillator. Next we will consider the driven, sometimes called forced, SHO/DSHO – an oscillator with an additional external driving force. Whatever supplies the driving force is not considered to be a part of the dynamical system. The driving force is assumed to be a known function of the time, without any dynamics of its own, and is independent of the state of the oscillator. This means that we do not consider any effect of forces the oscillator might exert on the source of the external force. In practice, this is an approximation, of course. The treatment of undamped and damped oscillators is somewhat different.

Assume that an undamped simple harmonic oscillator is driven by an external force $F(t)$, which depends only on the time. We might instead have assumed an arbitrary function

3.4 AN OSCILLATOR DRIVEN BY AN EXTERNAL FORCE

$F(q, t)$ of the displacement and the time, but this would make the mathematics too difficult. The EOM is

$$\ddot{q} + q = F(t). \tag{3.40}$$

SHO with external force

This type of *inhomogeneous* linear equation arises so often in physics, it is worth studying how to solve it in several different ways. Energy can be extracted or stored in the system by means of negative or positive work done by the external force. We would expect that the system energy would no longer be constant. How much energy did the external force F put in or take out of the system? How do we find out?

QUESTION 5: Driven SHO 1 Show that Equation (3.40) can be derived from the Lagrangian $L = \frac{1}{2}(\dot{q}^2 - \omega^2 q^2) + F(t)q$. What is the physical interpretation of the third term (i.e., $F(t)q$)? The energy of the oscillator is $E_{\text{osc}} \equiv \frac{1}{2}(\dot{q}^2 + q^2)$. Find $\frac{dE_{\text{osc}}}{dt}$ and show that this agrees with what you expect from the work done by the external force $F(t)$. Also find $\frac{dH}{dt}$. Does $H = E_{\text{osc}}$ in this case?

There are several ways to find solutions for the behavior of the forced oscillator. Perhaps the most direct method is to solve the equations of motion numerically. Because we are considering linear equations, we have the option of seeking analytic solutions instead of purely numerical ones. (Numerical methods are often the only means available to solve nonlinear EOMs.) Although numerical methods usually work for specific cases, they often do not give sufficient insight into the solution. Thus for linear equations, analytical methods are superior most of the time.

Start by considering the undamped oscillator. Due to the addition of an external driving force, the EOM (3.40) has become *inhomogeneous* (the right side is not zero). It is no longer true that we can create new solutions by arbitrary linear combinations of solutions. Instead, if just one solution to Equation (3.40), called the "steady state" solution $q_{\text{steady state}}$, is known or guessed, the most general solution to this same equation can be found by adding in a solution of the free oscillator, called the "transient" solution, $q_{\text{transient}}$*:

$$q_{\text{most general}} = q_{\text{steady state}} + q_{\text{transient}}. \tag{3.41}$$

This can be seen by substituting Equation (3.41) into Equation (3.40) to prove that $q_{\text{transient}}$ obeys the homogeneous Equation (3.20) if $q_{\text{steady state}}$ obeys Equation (3.40).

---

* In mathematics texts, the steady state solution is known as the "particular" solution, and the transient solution is known as the "free" solution. The terms "steady state" and "transient" are really only appropriate to the damped oscillator, since what we are calling the transient solution will last forever in the absence of damping.

After correctly guessing a steady state solution to the inhomogeneous EOM, one can then add a general solution of the free oscillator (homogeneous) equation. The system can undergo both free and driven oscillations at the same time. Usually these will have different frequencies. The two arbitrary constants in the transient part can be used to make the whole solution satisfy the boundary conditions. For example, to fit specified values of q, $\dot{q}$ at $t = 0$, adjust the amplitude and the phase of the transient oscillations to match these initial conditions. The chief drawback of this method is that we have to guess the steady state solution.

If you cannot or do not want to guess a steady state solution, another method can be used. This method for handling externally driven oscillators introduces the concept of a *Green's function*.[*] It can be used for any linear inhomogeneous equation in any physical problem. The Green's function technique is very common in dealing with electromagnetic fields, for example. There is a very direct way to find the Green's function.[†] It will be seen that it is just the response of the oscillator to a special driving force: an impulse, such as a hammer blow. The advantage of using a Green's function is that the initial conditions are included from the start. One obtains the whole solution, both steady state and transient parts, after doing an integral.

First we will solve (3.40) for a "step function" constant force, turned on at $t = 0$. We will also consider the DSHO response. These examples will illustrate the method of guessing the steady state solution and matching boundary conditions. Then we will solve for the Green's function by solving (3.40) for an impulsive force. We will again consider DSHO responses as well. Next we will show how the Green's function enables us to solve the case of an arbitrary external force $F(t)$. This general case will then be applied to the response of SHOs and DSHOs to a sinusoidal drive force.

3.5 DRIVING FORCE IS A STEP FUNCTION

As a very simple example, consider the pendulum oscillator. Assume that the pendulum is at rest when a horizontal force F_0 is suddenly turned on at $t = 0$. The force is either zero (before $t = 0$) or a constant value F_0 (after $t = 0$) as shown in Figure 3.4. What we wish to know is the motion of the pendulum after the driving force is turned on.

For $t > 0$ the equation of motion (3.40) becomes (using scaled variables)

$$\text{SHO: } \ddot{q} + q = F_0. \tag{3.42}$$

And analogously for the DSHO:

$$\text{DSHO: } \ddot{q} + \frac{\dot{q}}{Q} + q = F_0. \tag{3.43}$$

---

[*] The more correct name is "Green function," but this sounds too awkward, so is rarely used.
[†] Fourier transforms can also be used to find the Green's function, but we will not discuss this here.

3.5 DRIVING FORCE IS A STEP FUNCTION

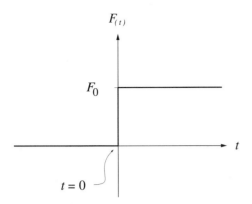

FIGURE 3.4
Constant force turned on at $t = 0$.

Prior to turning on the driving force term, we have the free oscillator equations, $F_0 = 0$, with the solution $q = \dot{q} = 0$.

QUESTION 6: Scaling 1 We chose the unit of time to simplify the EOM. Think about the scaling of the time units in the EOM when an external force is present. Does the force have to be scaled? If so, how?

QUESTION 7: Scaling 2 Explain how to obtain Equations (3.42) and (3.43) above from $F = ma$.

Before solving this problem, use your physical intuition. Think what will happen if a horizontal force is suddenly exerted on a pendulum at rest. Imagine that a magnet is suddenly placed on one side, attracting the pendulum bob. The pendulum will start to oscillate around a new equilibrium point. If the pendulum is damped, the oscillations will eventually die down, and the pendulum will come to rest tilted to the side. When the pendulum is again at rest, the only possible solution is $q = F_0$, since being at rest implies that $\ddot{q} = \dot{q} = 0$. This is the *steady state* solution. The additional oscillations that occur immediately after the force is suddenly applied to the pendulum are the *transient* solution. The mathematics is a little easier in the undamped case, but physically there is always some damping due to friction, and we consider the SHO to be the limit of the DSHO as $Q \to \infty$.

We might as well take $F_0 = 1$. Since the EOM is linear, the solution for an arbitrary magnitude of the driving force is just the solution for $F_0 = 1$ multiplied by F_0. Since the force just after $t = 0$ has a finite jump, there can be no discontinuity in either q or $\dot{q}$. Immediately after the force is applied, q and $\dot{q}$ must remain the same as they were just before the force is turned on. $\dot{q}$ will be continuous, but have a "kink," so that the *second* time derivative, $\ddot{q}$, has a jump of 1 at $t = 0^+$. ($t = 0^+$ stands for a positive time infinitesimally close to $t = 0$.) Only in this way can we maintain consistency with the EOM at all times. A discontinuity in the time derivative $\dot{q}(0)$ would imply an infinite second derivative and

TABLE 3.2*

Transient DSHO solution $q_i(t)$	$q_i(0)$	$\dot{q}_i(0)$
$q_1 \equiv e^{-\frac{t}{2Q}} \cos \omega' t$	1	$-\frac{1}{2Q}$
$q_2 \equiv e^{-\frac{t}{2Q}} \sin \omega' t$	0	ω'

thus, via the EOM, an infinite force at $t = 0$. We will actually consider such a case when we consider an impulsive force.

These dual requirements of continuity give us two equations in two unknowns which completely determine the solution for all positive times. This is called the method of matching boundary conditions.

Use superposition to write the most general solution for $t > 0$ as

$$q(t) = q_{\text{steady state}} + q_{\text{transient}} = 1 + q_{\text{transient}}. \tag{3.44}$$

By substituting Equation (3.44) above into the general equation, either (3.42) or (3.43), we see that the transient solution is a solution to the free oscillator equation.

To know the displacement at all times, we need to know the actual form of the transient solution. For the SHO, the most general solution is $q(t) = 1 + a \cos t + b \sin t$. The constants can be found by using the boundary conditions $q(0^+) = \dot{q}(0^+) = 0$. Thus $q(0) = 1 + a = 0$, $\dot{q}(0) = b = 0$, and we obtain

$$\text{driven SHO: } q(t) = 1 - \cos t. \tag{3.45}$$

For a damped oscillator, the idea is the same, but the formula for the free oscillator solutions is a bit more complicated. The free DSHO has two independent solutions (3.32), which are listed in Table 3.2. Matching the boundary conditions for these transient solutions, we get the equations

$$\begin{aligned} q(0^+) &= 1 + aq_1(0) + bq_2(0) = 0 \quad \text{continuity of } q(0), \\ \dot{q}(0^+) &= a\dot{q}_1(0) + b\dot{q}_2(0) = 0 \quad \text{continuity of } \dot{q}(0). \end{aligned} \tag{3.46}$$

Using the values in Table 3.2, we obtain

$$1 + a = 0, \quad -\frac{1}{2Q}a + \omega' b = 0. \tag{3.47}$$

The result is

$$\text{driven DSHO: } q(t) = 1 - e^{-\frac{t}{2Q}} \left(\cos \omega' t - \frac{1}{2Q\omega'} \sin \omega' t \right). \tag{3.48}$$

In the limit $Q \to \infty$, Equation (3.48) becomes Equation (3.45).

*($\omega' \equiv \sqrt{1 - \frac{1}{4Q^2}}$).

3.6 FINDING THE GREEN'S FUNCTION FOR THE SHO

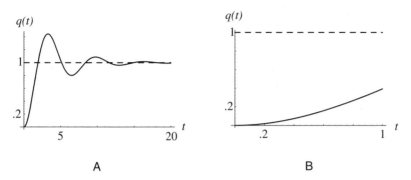

FIGURE 3.5
DSHO response to unit step force ($Q = 2$). Graph B is the same $q(t)$ as graph A but with an expanded time scale to show the behavior near $t = 0$.

A graph of the DSHO oscillator response to a unit step drive force at $t = 0$ appears in Figure 3.5 for $Q = 2$. This agrees with our intuitive solution for the pendulum. A new equilibrium displacement is reached after transient oscillations excited by the force discontinuity damp out. Notice that, for this low Q, the oscillator transient damps out quickly, leaving only the steady state solution (dashed line). The numerical solution also shows that the initial slope and displacement of the oscillator are both zero, because the applied force is finite.

3.6 FINDING THE GREEN'S FUNCTION FOR THE SHO

Imagine that the undamped oscillator receives an impulsive force at $t = t'$. *Impulse* is the time integral of the force. An impulsive force is defined to be an infinite force applied for an infinitesimally short time and having a finite time integral. Again assume that the oscillator was at rest ($q = \dot{q} = 0$) prior to $t = t'$. The Green's function G for this system is the solution for this special kind of impulsive driven force equation. The EOM to be solved is

$$\ddot{G} + G = \delta(t - t'). \qquad (3.49)$$

defining equation for G

Here $\delta(x)$, representing the impulsive force, is called the Dirac delta function, which isn't really a function at all but shorthand for a limit of a function. The defining characteristics of $\delta(x)$ are

$$\delta(x) = 0 \quad \text{for } x \neq 0 \quad \text{and} \quad \int \delta(x)\,dx = 1. \qquad (3.50)$$

This means that we can think of $\delta(x)$ as the limit of a narrow but strongly peaked function with a unit area under it, as we let the width go to zero.

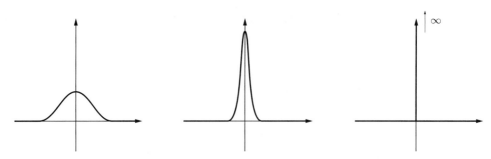

FIGURE 3.6

The δ symbol really stands for doing the calculation with a finite and strongly peaked function, for example a Gaussian. At the end of the calculation, take the limit of the width going to zero, while the area remains one. In that sense, the "delta function" stands for the sequence of operations pictured in Figure 3.6.

Within the (vanishing) width, any well-behaved $f(x)$ is essentially constant and equals $f(0)$ as we let the width of $\delta(x)$ go to zero. Therefore we write the shorthand expression below, for any function $f(x)$,

$$\int f(x)\delta(x)\,dx = f(0). \tag{3.51}$$

Mathematicians can spend considerable time discussing this, but we believe the intuitive meaning is clear; it is an impulsive force exerted on the oscillator. It describes what happens when we hit the oscillator with a short but sharp blow from a hammer, for example. The duration of the hammer blow is so short that the oscillator moves a negligible amount during the time the force is applied.

QUESTION 8: Driven SHO 2 Assume the oscillator is a pendulum. Hit it with an impulsive force at t'. We will prove mathematically that θ is a continuous function of time but $\dot{\theta}$ is discontinuous at $t = t'$. What is the physical reason for this? (Think of the impulsive force as a sharp blow by a hammer.)

Response to Impulse Force

One method for finding the impulse response is to approximate the short impulse by a square pulse of duration $\xi \ll 1$ and height $\frac{1}{\xi}$. As before, define $F_{\text{step}}(x) \equiv 1$ if $x \geq 0$, $F_{\text{step}}(x) \equiv 0$ if $x < 0$. Subtract a solution obtained for the step function force shifted in time by ξ and multiplied by $\frac{1}{\xi}$ from the solution with the step at the origin, also multiplied by $\frac{1}{\xi}$, to obtain

$$F_{\text{square pulse}}(t) = \frac{1}{\xi}[F_{\text{step}}(t) - F_{\text{step}}(t - \xi)]. \tag{3.52}$$

3.6 FINDING THE GREEN'S FUNCTION FOR THE SHO

Likewise,

$$q_{\text{square pulse}}(t) = \frac{1}{\xi}[q_{\text{step}}(t) - q_{\text{step}}(t-\xi)]. \tag{3.53}$$

The time integral $\int F_{\text{square pulse}}(t)\,dt = 1$, even as $\xi \to 0$. The limits of integration must include a finite but arbitrarily small interval around the point $t = 0$. Taking the limit as $\xi \to 0$, we get

$$F_{\text{impulse}}(t) \equiv \delta(t) = \lim_{\xi \to 0} \frac{1}{\xi}[F_{\text{step}}(t) - F_{\text{step}}(t-\xi)], \tag{3.54}$$

$$q_{\text{impulse}}(t) \equiv G(t) = \lim_{\xi \to 0} \frac{1}{\xi}[q_{\text{step}}(t) - q_{\text{step}}(t-\xi)]$$

$$= \frac{d}{dt}(q_{\text{step}}(t)). \tag{3.55}$$

From (3.45),

$$q_{\text{impulse}}(t) \equiv G(t) = \sin t. \tag{3.56}$$

The origin can be shifted to an arbitrary time t' by writing $G(t - t') = \sin(t - t')$. From now on we assume the impulse occurs at a time t'.

Integrating the Discontinuity

There is still another way to find G explicitly without having to guess the solution. Note that, except at $t = t'$, the force applied is zero. This means that G is a solution of an equation we already know how to solve: the free oscillator or *homogeneous* Equation (3.20), which has sines, cosines, or complex exponentials for solutions. The other thing we need to know is that the slope of G has a unit discontinuity at $t = t'$, which we now prove. Integrate the defining Equation (3.49) above for a small interval 2ϵ which encloses t':

$$\int_{t'-\epsilon}^{t'+\epsilon} (\ddot{G} + G)\,dt = \int_{t'-\epsilon}^{t'+\epsilon} \delta(t-t')\,dt = 1. \tag{3.57}$$

Since $\ddot{G}$ is the time derivative of $\dot{G}$, we can do the left integral in Equation (3.57) directly, giving

$$\int_{t'-\epsilon}^{t'+\epsilon} \ddot{G}\,dt = \dot{G}(t'+\epsilon) - \dot{G}(t'-\epsilon). \tag{3.58}$$

Also note that

$$\int_{t'-\epsilon}^{t'+\epsilon} G(t)\,dt \approx 2\epsilon G(t'). \tag{3.59}$$

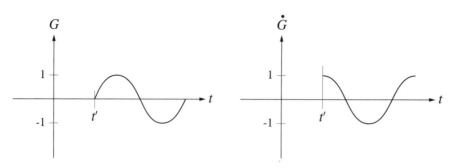

FIGURE 3.7
Note discontinuity in $\dot{G}$ vs. t graph at $t = t'$.

Equation (3.59) holds if G is a continuous function and ϵ is sufficiently small, because then we can write $G(t)$ as a Taylor series near t': $G(t) = G(t') + (t - t')\dot{G}(t') + \cdots$. Neglect all but the first term in the Taylor series if ϵ is sufficiently small. In the limit $\epsilon \to 0$, if G itself is continuous, but has a discontinuous first time derivative, then the left side of Equation (3.57) yields the amount of the jump in $\dot{G}$ at $t = t'$. By the definition of the delta function, the right-hand side of Equation (3.57) is equal to 1. Therefore, the jump at $t = t'$ in $\dot{G}$ is $\dot{G}_{\text{left}} - \dot{G}_{\text{right}} = 1$. For times $t \neq t'$, $G(t)$ must obey the homogeneous differential equation $\ddot{G} + G = 0$.

Since the oscillator was at rest for all times before t', we can use the above information to write down G explicitly:

$$G(t - t') = 0, \quad t - t' \leq 0,$$
$$G(t - t') = \sin(t - t'), \quad t - t' \geq 0. \tag{3.60}$$

This solution for G shown in Figure 3.7 is called the "causal" Green's function because it reflects causality: The oscillator cannot respond before it receives the impulse at $t = t'$.

Summary: How to find G:

1. $G = 0$ for $t < t'$. Solve the free oscillator equation for $t > t'$.
2. Make G continuous at $t = t'$.
3. Make $\dot{G}$ discontinuous with a unit jump at $t = t'$.

QUESTION 9: Green's Functions 1 Is the Green's function given by (3.60) a unique solution to the free simple harmonic oscillator, given an impulsive force at $t = t'$, subject to the boundary conditions that G is continuous at $t = t'$ and $\dot{G}$ is discontinuous with a unit jump at $t = t'$? How about if we include the condition that $G = 0$ for $t < t'$? Why do we chose the "causal" Green's function (i.e., $G(t - t') = 0$ for $t - t' \leq 0$)? (We are looking for a physical reason.)

Green's Function for an Underdamped DSHO

What is the Green's function for this case? For $t - t' > 0$, G is a solution to the free oscillator Equation (3.32). Try

$$G_{\text{DSHO}} = Ae^{-\frac{t-t'}{2Q}} e^{i\omega'(t-t')} + Be^{-\frac{t-t'}{2Q}} e^{-i\omega'(t-t')}. \tag{3.61}$$

For $t - t' < 0$, causality demands that $G = 0$. At $t = t'$, there is a need to satisfy two equations:

$$\begin{aligned} A + B &= 0 \quad \text{continuity of } G \text{ at } t = t', \\ i\omega'(A - B) &= 1 \quad \text{unit discontinuity of } \dot{G} \text{ at } t = t'. \end{aligned} \tag{3.62}$$

The coefficients A, B must be

$$A = \frac{-i}{2\omega'}, \quad B = \frac{i}{2\omega'}. \tag{3.63}$$

Substituting (3.63) into (3.61) we obtain

$$G_{\text{DSHO}}(t - t') = \frac{e^{-\frac{(t-t')}{2Q}}}{2\omega'} [-ie^{i\omega'(t-t')} + ie^{-i\omega'(t-t')}]. \tag{3.64}$$

Taking the real part we get

$$G_{\text{DSHO}}(t - t') = \begin{cases} \dfrac{e^{-\frac{(t-t')}{2Q}}}{\omega'} \sin \omega'(t - t'), & t - t' \geq 0, \\ 0, & t - t' \leq 0. \end{cases} \tag{3.65}$$

You can take it from here and prove that this solution is the derivative of the solution for the unit step in the case of an underdamped oscillator.

3.7 ADDING UP THE DELTA FUNCTIONS – SOLVING THE ARBITRARY FORCE

Now that we have the explicit solution for the "causal" Green's function, we can turn our attention back to solving the more general problem of the response of a DSHO to an arbitrary driving force $F(t)$:

$$\ddot{q} + \frac{\dot{q}}{Q} + q = F(t). \tag{3.66}$$

(To find the SHO solution, take the limit as $Q \to \infty$.)

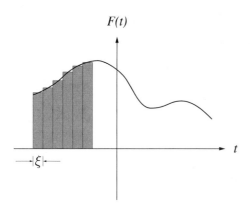

FIGURE 3.8

Just as you did in calculus when calculating the area under a curve, approximate the force $F(t)$ by a set of square pulses as shown in Figure 3.8:

$$F(t) \approx \xi \sum_i F_i \, F_{\text{square pulse}}(t - t_i). \tag{3.67}$$

By definition, the square pulse has the form

$$\begin{aligned} F_{\text{square pulse}}(x) &= \tfrac{1}{\xi}, \quad -\tfrac{\xi}{2} \leq x \leq \tfrac{\xi}{2}, \\ F_{\text{square pulse}}(x) &= 0, \quad \text{otherwise.} \end{aligned} \tag{3.68}$$

The value of the constant F_i is the average value of $F(t)$ in the small interval of width ξ around t_i. The t_is are spaced apart by equal distances of ξ in width. In the limit $\xi \to 0$, the square pulses approach $\delta(t - t_i)$ and the sum (3.67) approaches an integral over the continuous variable t'. We have, in this limit,

$$F(t) = \int_{-\infty}^{+\infty} F(t')\delta(t - t') \, dt'. \tag{3.69}$$

This equation can be crudely interpreted to mean that the delta function is so sharply peaked at $t = t'$ that the whole contribution to the integral is proportional to the value of $F(t')$ at the one point, t.

If we consider Equation (3.69) as expressing the force $F(t)$ as a superposition of impulses, then the superposition principle (extended to an infinite sum) gives the solution to the response as a superposition of Green's functions weighted by $F(t')$. Knowing the Green's function means that we can immediately write the result if we accept the validity of replacing a sum by an integral in the limit:

$$\boxed{q(t) = \int_{-\infty}^{t} F(t')G(t - t') \, dt'.} \tag{3.70}$$

general solution

3.8 DRIVING AN OSCILLATOR IN RESONANCE

The finite upper limit is due to the use of the causal Green's function. For the SHO we have

$$q(t) = \int_{-\infty}^{t} F(t') \sin(t - t') \, dt'. \tag{3.71}$$
simple harmonic oscillator solution

That is all there is to it. The causal Green's function gives us the solution as an integral over all of the past. Green's functions are used throughout physics, wherever the equation we are trying to solve is linear. They are especially useful in solving wave problems encountered with the partial differential equations of acoustics or electromagnetic theory.

QUESTION 10: *Green's Functions 2*   Notice the upper limit of the integral in Equation (3.70) is now t, instead of ∞. Why do we change the limit from Equation (3.70)? Why does this express causality? What specifically do we mean by "cause" and "effect" here? Can "causes" come later than "effects"?

QUESTION 11: *Green's Functions 3*   The causal Green's function is known in electromagnetic radiation theory as the "retarded" Green's function. The opposite of the retarded Green's function is the "advanced" Green's function. In this case "effects" come before "causes." What is the advanced Green's function for the simple harmonic oscillator? How would it change (3.71)? Does this advanced Green's function seem to have any physical purpose?

QUESTION 12: *Green's Functions 4*   Prove, starting from Equation (3.70), that we have achieved what we set out to do: Find the solution $q(t)$ to $\ddot{q} + q = F(t)$.

QUESTION 13: *Green's Functions 5*   Explain in your own words the role of the Green's Function in solving the driven oscillator problem.

3.8 DRIVING AN OSCILLATOR IN RESONANCE

An oscillator can be driven with a sinusoidally varying driving force at a frequency close to the natural oscillator frequency.

In Galileo's time the phenomenon of resonance was recognized and appreciated. We quote from the *Dialogues Concerning Two New Sciences*. Salviati says:

> ... one can confer motion upon even a heavy pendulum which is at rest by simply blowing against it; by repeating these blasts with a frequency which is the same as the pendulum one can impart considerable motion. Suppose by the first puff we have displaced the pendulum from the vertical by, say, half an inch; then if, after the pendulum has returned and is about to begin the second vibration, we add a second puff, we shall

impart additional motion; and so on with other blasts provided they are applied at the right instant, and not when the pendulum is coming toward us since in this case the blast would impede rather than aid the motion. Continuing thus with many impulses we impart to the pendulum such momentum that a greater impulse than that of a single blast will be needed to stop it.

We will consider here the effect of a sinusoidal driving force with a variable frequency ω:

$$F(t) = \sin \omega t. \qquad (3.72)$$

(Remember that the time has been scaled so that the natural frequency of the oscillator is $\omega_0 = 1$. To return to ordinary time units of seconds, replace ω by $\frac{\omega}{\omega_0}$ and replace t by $\omega_0 t$ in all the formulas below.)

First consider the case of an *undamped* SHO. We also simplify to the case where the driving force starts at $t = 0$ and has the *same frequency* as the oscillator, $\omega = \omega_0 = 1$. In this particular case we say that the driving frequency is "in resonance" with the free oscillator frequency:

$$\begin{aligned} F(t) &= 0, \quad t \leq 0, \\ F(t) &= \sin(t), \quad t \geq 0. \end{aligned} \qquad (3.73)$$

Working out the Green's function integral (3.71) explicitly with the help of trigonometric identities, we get

$$q(t) = \int_0^t \underbrace{\sin t'}_{F(t')} \underbrace{\sin(t-t')}_{G(t-t')} dt' = \underbrace{\frac{\sin t}{2}}_{\text{oscillating part}} - \underbrace{\frac{t \cos t}{2}}_{\text{increasing amplitude}}, \qquad (3.74)$$

$$q(t) \to \frac{-t}{2} \cos t \quad \text{as } t \to \infty. \qquad (3.75)$$

Because the amplitude grows without limit as time increases, *an undamped SHO driven at its natural frequency will store an unlimited, ever-increasing amount of energy.* This situation will prevail until either the large amplitude makes the nonlinearities important or else the damping, which must exist at some level for real physical oscillators, becomes important. Real oscillators reach a steady state equilibrium with the driving force after sufficient time elapses. We will explore how the damping affects an oscillator driven at resonance next.

Response of an Underdamped DSHO to a Sinusoidal Driving Force of Variable Frequency ω

Suppose that you have a mass suspended by a spring and that the mass is initially at rest. Take a signal generator that generates a sinusoidally varying current of variable frequency. Turn it on suddenly and have this current pass through a coil which generates a periodic vertical force on the suspended mass. Assume this force is proportional to $\sin \omega t$.

3.8 DRIVING AN OSCILLATOR IN RESONANCE

What is the response of the mass to this force? We can guess the answer. The mass–spring system will start slowly to gain energy, oscillating at two different frequencies, both the free oscillation frequency of the mass–spring combination and the external driving frequency. After a while the system reaches an equilibrium steady state in which the energy supplied by the driving force just balances the losses due to friction. After that, only a single frequency is present, which is the frequency of the signal generator used to provide the driving force. The driven system remains synchronous with the source of the external drive. This means that there will be a constant phase difference between the drive and the oscillator response in the steady state. Only a single frequency is present in the final response of the mass–spring system because the EOM is linear.

Next we formulate what we just said in more mathematical terms. Suppose the oscillator is driven by a force $F(t) = \sin \omega t$ which starts at $t = 0$. We want to discover the response of the oscillator as a function of ω. The "natural frequency" of the free, underdamped oscillator is $\omega' = \sqrt{1 - \frac{1}{4Q^2}}$. (The frequency of the free undamped oscillator (ω_0) and of the free damped oscillator (ω') are extremely close if $Q \gg 1$.)

As before, to satisfy the boundary conditions ($q(0) = 0$, $\dot{q}(0) = 0$), the oscillator response $q(t)$, just after $t = 0$, consists of both a free oscillator and a steady state solution. Due to damping, the free oscillator part of the response, which oscillates at $\omega' \approx 1$, damps out exponentially as time increases. It will be proved that, in about $\frac{Q}{2\pi}$ oscillations, the transient (free) solution damps out, and we are left with the steady state solution, representing the long-term equilibrium state of the oscillator. From the definition of a Green's function (3.70):

$$q(t) = \underbrace{q_{\text{steady state}}(t)}_{\text{frequency}=\omega} + \underbrace{q_{\text{transient}}(t)}_{\text{frequency}=\omega'} = \int_0^t \underbrace{\sin(\omega t')}_{F(t')} G(t-t')\, dt'. \qquad (3.76)$$

We have written G, rather than the explicit form of the Green's function, to avoid confusion with the sinusoidal driving force, which has in general a different frequency from that of the free DSHO. As a homework problem, you can integrate Equation (3.76) explicitly, using the correct form of the causal Green's function for the DSHO (3.65). Integrating the Green's function gives both the transient and the steady state parts of the solution automatically.

Here is a simpler way to carry out the same calculation. We can use a *complex* driving force (i.e., replace $\sin \omega t$ by $e^{i\omega t}$). This is allowed, because we plan to take the real part at the end anyway. *In the steady state, a linear system always responds at the driving frequency*, so $q(t)$ must also be proportional to $e^{i\omega t}$. By substituting $F = e^{i\omega t}$ and $q = q_0\, e^{i\omega t}$ (q_0 is a proportionality constant) into (3.66) we obtain, after solving for q_0,

$$\boxed{q(t) = \frac{e^{i\omega t}}{1 - \omega^2 + \frac{i}{Q}\omega} + q_{\text{transient}}.} \qquad (3.77)$$

complex solution

From the solution to the free DSHO (3.32), $q_{\text{transient}} = \mathcal{A}_c \, e^{-\frac{t}{2Q}} e^{i\omega' t}$, which damps out exponentially. Here the complex notation simplifies everything drastically. The vanishing of the real part of $q, \dot{q}$ at $t = 0$ determines the two arbitrary constants (real and imaginary parts of $\mathcal{A}_c$) in $q_{\text{transient}}$. However, we are really only interested in the *steady state solution* after $q_{\text{transient}}$ has damped out. Physically, both types of oscillations are actually present. If we want the stored energy in the system after a sufficiently long time, we need only to look at $|q(t)|^2_{\text{steady state}}$, since the energy is proportional to the square of the amplitude.

Per unit driving force, this is

$$|q|^2_{\text{steady state}} = \left| \frac{e^{i\omega t}}{1 - \omega^2 + \frac{i}{Q}\omega} \right|^2, \tag{3.78}$$

$$E \sim |q|^2_{\text{steady state}} = \frac{1}{(1-\omega^2)^2 + \frac{\omega^2}{Q^2}}. \tag{3.79}$$

QUESTION 14: *Energy of Driven DSHO* Prove that Equation (3.78) is identical to Equation (3.79).

Any oscillator response (3.79) plotted as a function of driving frequency has a sharp peak, known as the resonant frequency ω_r, as shown in Figure 3.9. We can find this frequency by requiring that $\frac{dE}{d\omega} = 0$ at $\omega = \omega_r$. This gives us

$$\omega_r = \sqrt{1 - \frac{1}{2Q^2}}. \tag{3.80}$$

Notice that ω_r is not equal to ω' (3.32), the frequency of the free DHSO, nor $\omega_0 = 1$, the frequency of the free SHO. If Q is large and $\omega \approx \omega_r \approx 1$, the approximation below is a valid one:

$$1 - \omega^2 = (1-\omega)(1+\omega) \approx 2(1-\omega) \tag{3.81}$$

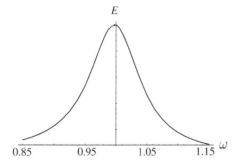

FIGURE 3.9
Oscillator energy vs. driving frequency ω in steady state for a microwave cavity.

3.8 DRIVING AN OSCILLATOR IN RESONANCE

and

$$\frac{\omega^2}{Q^2} \approx \frac{1}{Q^2}. \tag{3.82}$$

The steady state (long time value) of the stored energy E is, in this approximation,

$$\boxed{E \sim \frac{1}{(1-\omega)^2 + \frac{1}{4Q^2}}. \tag{3.83}}$$
stored energy vs. ω

Equation (3.83) is easier to use in practice than Equation (3.79), but it is accurate only if $Q \gg 1$. The formula (3.83) is the famous *Lorentzian* line shape for a driven DSHO. It occurs everywhere in the physics of linear systems with sinusoidal external driving forces. The width $\Delta \omega$ when the energy is at half its maximum value is $\frac{1}{Q}\omega_0$ (including the scaling factor here):

$$\boxed{\frac{\Delta \omega}{\omega_0} = \frac{1}{Q}, \tag{3.84}}$$

where ω_0 is the frequency of the free SHO, which we are using as the resonant frequency since $Q \gg 1$. Equation (3.84) is a very useful one in practice. You can excite the oscillator with an external force, sweep the frequency over the resonance, and measure the frequency difference between the half peak energy points. The relative frequency difference gives $\frac{1}{Q}$. High Qs mean not only slow decay of transient signals, but narrow steady state resonances in the energy (or amplitude) versus frequency plot.

QUESTION 15: *Full Width at Half Max* Prove that the full width ($\frac{\Delta \omega}{\omega_0}$) when the stored energy is at half its maximum value (full width half max) is equal to $\frac{1}{Q}$. Provide both a physical and mathematical explanation.

The graphs in Figure 3.10 plot the oscillator amplitude as a function of the time in dimensionless units. In all three of the cases illustrated above, the driving force (dotted curve) is turned on suddenly at $t = 0$. The transient oscillation beats with a slightly different driving frequency as seen in the Figures 3.10A and C. On a time scale longer than depicted in these graphs, the transient will damp out, leaving only the steady state response at the driving frequency. When driven at its natural frequency, $\omega = 1$, the oscillator builds up steadily to a large steady state amplitude (Figure 3.10B). Notice that the amplitude of

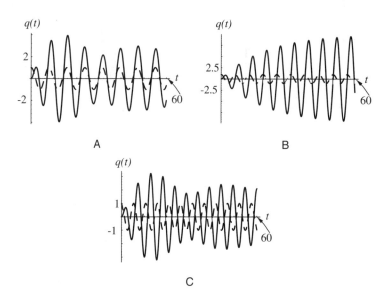

FIGURE 3.10
Damped harmonic oscillator, with $Q = 10$ and driven near resonance with driving force $F(t) = \cos \omega t$. A) $\omega = 0.8$ (below resonance); B) $\omega = 1.0$ (at resonance); C) $\omega = 1.2$ (above resonance).

the oscillator is much larger relative to the driving force for Figure 3.10B than it is for Figures 3.10A and C. It is also possible to see in the figure that, at resonance, the driving force and the response are exactly 90° out of phase. The relative phase shifts rapidly as the driving frequency is varied from below to above the resonance.

3.9 RELATIVE PHASE OF THE DSHO OSCILLATOR WITH SINUSOIDAL DRIVE

What is the phase of the driven oscillator in the steady state? Does the response lead the driving force or lag behind it? As seen in Figure 3.10, the answer depends on whether the driving frequency is lower or higher than the resonance frequency. It is physically apparent that, if the drive frequency is very low compared to the natural oscillator frequency, the response just follows the drive. (Think of a pendulum being driven very slowly.) In this case, we say that the phase between the drive force and the response is zero. It is not so obvious that, in the limit of very high driving frequency, the drive and the response are exactly 180° out of phase. We will now obtain the entire curve of the relative phase as a function of ω, the drive frequency. We will show that the relative phase of the oscillator drops rapidly and dramatically through $-90°$ as the driving frequency is increased through and beyond the resonance frequency.

Suppose the driving force has the form

$$F(t) = \text{Real } [e^{i\omega t}] = \cos \omega t. \tag{3.85}$$

3.9 RELATIVE PHASE OF THE DSHO OSCILLATOR WITH SINUSOIDAL DRIVE

(You could use a sine wave to drive the system – then take the imaginary part rather than the real part. It does not affect the conclusion below, which is a formula for the relative phase.) From (3.77), the steady state response of the system, $q(t)$, must have the form

$$q(t) = \text{Real}[\frac{e^{i\omega t}}{1 - \omega^2 + \frac{i}{Q}\omega}] = \text{Real}[\mathcal{A}_c(\omega) e^{i\omega t}]$$
$$= A(\omega) \cos(\omega t + \phi(\omega)). \quad (3.86)$$

Equation (3.86) defines the complex amplitude $\mathcal{A}_c(\omega)$, which is a strongly varying function of frequency, but a constant with respect to the time. $\mathcal{A}_c = A\, e^{i\phi}$ defines A as the (real) amplitude, and ϕ as the relative phase angle between the oscillator response and the driving force. Equation (3.86) shows that the oscillator response "leads" the drive signal by a time $\Delta t = \frac{\phi}{\omega}$. A negative value of ϕ corresponds to a time lag for the oscillator relative to the drive signal. We say there is a "phase lag" in this case.

To obtain $\phi(\omega)$, rewrite the left side of (3.86) as

$$q(t) = \text{Real}\left[\frac{(1 - \omega^2 - \frac{i\omega}{Q})e^{i\omega t}}{(1 - \omega^2)^2 + \frac{\omega^2}{Q^2}}\right]. \quad (3.87)$$

We will do some purely mathematical manipulations with complex numbers. For any complex number z, with complex conjugate z^*, it is an identity that, if $z = re^{i\theta}$,

$$\frac{1}{z} = \frac{1}{r} e^{-i\theta}. \quad (3.88)$$

Now $r \equiv \sqrt{zz^*}$ is a real number, by definition. Further, $\frac{z^*}{r}$ has a magnitude of unity, so must equal the phase factor $e^{-i\theta}$, that is, $e^{i\phi}$ in our case. Comparing (3.87) with the right side of (3.86) and with (3.88), let $z = 1 - \omega^2 + \frac{i\omega}{Q}$. We see that

$$A = |\mathcal{A}_c| = \frac{1}{\sqrt{(1 - \omega^2)^2 + \frac{\omega^2}{Q^2}}}$$

and $e^{i\phi} = \cos\phi + i\sin\phi = e^{-i\theta} = A(1 - \omega^2 - \frac{i\omega}{Q})$. Thus, taking the ratio of imaginary to real parts of this last expression, we get

$$\boxed{\tan\phi = -\frac{\frac{\omega}{Q}}{1 - \omega^2}. \quad (3.89)}$$

$$\text{relative phase between } q \text{ and } F$$

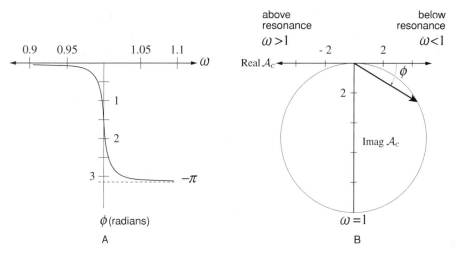

FIGURE 3.11
$Q = 100$ for the left-hand plot; $Q = 10$ for the right-hand plot.

Use the inverse function $\tan^{-1}$ to invert Equation (3.89). A plot of ϕ versus ω appears in Figure 3.11A for ϕ as ω is swept through the resonance. Since $\phi < 0$, the oscillator always lags behind the drive signal. At very low drive frequencies the phase lag is small, but near to the resonance it becomes $-90°$ out of phase with the drive, and at very high frequencies, the phase lag is $-180°$. In Figure 3.11B, the Real$[\mathcal{A}_c]$, Im$[\mathcal{A}_c]$ are plotted on the X and Y axes parametrically versus ω. Here $\mathcal{A}_c$ is the complex amplitude for the oscillator response as defined in Equation (3.86). The trajectory in the complex plane is, to a good approximation if Q is large, a circle of radius $Q/2$, with a center at $(0, -iQ/2)$ for $0 < \omega < \infty$. It is evident that $\phi(\omega \ll 1) \approx 0$, whereas $\phi(\omega \gg 1) \approx -\pi \equiv \pi$. As the driving frequency is slowly increased from a value below the resonant frequency, the complex amplitude $\mathcal{A}_c$ start to traverse the circle in the figure in a clockwise direction. Little happens until the resonance is approached. Then the amplitude swings rapidly clockwise around the circle and returns to the origin as the frequency increases above the resonant frequency.

Figure 3.12 shows an actual measurement of the "out of phase" response (Y axis = imaginary part) versus the "in phase" response (X axis = real part) for a microwave cavity

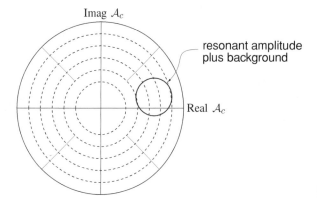

FIGURE 3.12

using a commercial network analyzer. The real part of the amplitude $\mathcal{A}_c$ is the part in phase with the driving signal, while the imaginary part $\mathcal{A}_c$ is the part 90° out of phase. In other words, if $\mathcal{A}_c$ were purely real, the oscillator would move in phase with the driving force. If $\mathcal{A}_c$ were purely imaginary, the phase difference would be 90°. The circle is displaced off center in this particular case because an additional frequency-independent (slowly varying) nonresonant amplitude in the system response was present along with the resonant part. We can see from the figure that the nonresonant part has a complex phase too.

QUESTION 16: *Summary* Describe: 1) the response of underdamped, overdamped, and critically damped free DSHOs; 2) the response of an SHO to a sinusoidal driving force; 3) the response of a DSHO to a sinusoidal driving force.

SUMMARY OF CHAPTER 3

- Static equilibrium points are maxima or minima of the potential energy V for scleronomic systems.
- A minimum of the potential at an equilibrium point means it is stable. A maximum means there is instability. The magnitude of the curvature of V determines the frequency of small-amplitude oscillations for scleronomic systems.
- The frequency of a linear oscillator is independent of amplitude.
- Solutions to the free SHO are of the form $\sin t$, $\cos t$ ($\omega = 1$ scaling). Linear combinations of solutions are also solutions (superposition principle).

Free Damped Simple Harmonic Oscillators

- A frictional force equaling $-\frac{\dot{q}}{Q}$ retards the motion. This defines the quality factor Q.
- Solutions are of the form $e^{i\alpha t}$, $\alpha = \frac{i}{2Q} \pm \sqrt{1 - \frac{i}{4Q^2}}$.
- There are three types of DSHOs: underdamped ($Q > \frac{1}{2}$), critically damped ($Q = \frac{1}{2}$), and overdamped ($Q < \frac{1}{2}$).
- The Q of a DSHO can be found from the exponential decay of the stored energy in the oscillator that drops to $\frac{1}{e}$ of its original value after $\frac{Q}{2\pi}$ oscillations.

Driven (or Forced) SHOs and DSHOs

- Forced linear oscillators do not have a constant total energy. The equation of motion is linear but inhomogeneous. The solution contains both a steady state solution plus a transient solution. The steady state solution has the same frequency as the driving force; the transient solution is a solution to the free oscillator equation.
- There are two ways to solve the driven oscillator: 1) Guess the steady state solution and use the transient solution to match boundary conditions. 2) Use a Green's Function that gives the solution to an oscillator driven by an arbitrary external force $F(t)$:

$$q(t) = \int_{-\infty}^{t} G(t - t') F(t') \, dt'. \tag{3.90}$$

- The causal Green's function for the SHO is

$$G(t) = 0, \quad t \leq 0,$$
$$G(t) = \sin t, \quad t \geq 0. \tag{3.91}$$

- The causal Green's function for the DSHO is

$$G(t) = 0, \quad t \leq 0$$
$$G(t) = \frac{1}{\omega'} e^{-\frac{t}{2Q}} \sin \omega' t, \quad t \geq 0. \tag{3.92}$$

Sinusoidally Driven DSHOs

- The response of a DSHO with $Q \gg 1$ to a sinusoidal $F(t)$ of variable frequency is a bell-shaped curve (Lorentzian) strongly peaked at the resonant frequency of the oscillator ω_r. The higher the Q value, the narrower is the resonance curve. The full width at half maximum of the stored energy versus frequency is $\frac{\Delta\omega}{\omega_0} = \frac{1}{Q}$, which gives a direct method for measuring Q.
- The relative phase between the oscillator and a sinusoidal driving force is an "S-shaped" curve that changes rapidly in the vicinity of the resonance.

PROBLEMS

Equilibrium

Problem 1: *(Particle in a bowl)* A point particle of mass m is confined to the frictionless surface of a spherical bowl. There are two degrees of freedom. Prove that the equilibrium point is the bottom of the bowl. Does the bowl have to be exactly spherical for this to be true? Near to the bottom of the bowl, what is the most general form possible for the shape of the bowl in order to maintain the stability of the equilibrium point at the bottom? In the general case, with a nonspherical bowl of arbitrary shape, no analytic solution is known for the motion of the point mass.

Problem 2: *(Bead on a rotating hoop)* This problem concerns stable and unstable equilibrium in a system with *rheonomic* constraints. Consider a bead of mass m moving without friction on a circular hoop of radius R which rotates at angular frequency Ω about the Z axis as shown in Figure 1.11. There is one degree of freedom, described by the coordinate θ.

a) Find the kinetic energy T and the potential energy V as functions of $\dot\theta, \theta$.
b) From the EOM, show that two cases exist: 1) for $\Omega < \Omega_{\text{critical}}$ there are two possible equilibrium points, $\theta = 0, \pi$; 2) for $\Omega > \Omega_{\text{critical}}$ a third possibility also can exist.

Find the critical angular speed (Ω_{critical}) of the hoop for case 2. Notice that $\mathcal{F}_\theta \neq 0$ at the new equilibrium point in case 2, but $\frac{\partial L}{\partial \theta} = 0$. This proves that Equation (3.17) doesn't necessarily hold for rheonomic constraints.

c) Make a Taylor series expansion of the Lagrangian to second order about the equilibrium points. Find the stable equilibrium points in both cases and the frequency of small oscillations about those points. For example, close to the point at $\theta = 0$, prove that, for small excursions in θ, an equivalent Lagrangian is

$$L = \dot{\theta}^2 + \left[\Omega^2 - \frac{g}{R}\right]\theta^2. \tag{3.93}$$

Is the motion stable or unstable if the bead is put at $\theta = 0$? If it is stable, with what frequency does it oscillate about the equilibrium point? Now do this for the other possible equilibrium points. Be careful to make the expansion about the point you want to consider. It is a little tricky about the third equilibrium point that appears for sufficiently high speed rotation of the hoop, since the location of the point depends on Ω.

d) The bead is placed very near the bottom of the hoop ($\theta = 0$), and the hoop rotation speed is slowly increased from 0 to some value above the critical speed. Describe in words what you would expect to see happen to the bead.

SHOs

Problem 3: *(Relation between H and E)* $H \equiv \sum \dot{q}_k \frac{\partial L}{\partial \dot{q}_k} - L$. Is $H = E$ the total energy for the forced SHO? Also discuss the case where a pendulum's length is changing as a known function of time: $l(t)$.

Problem 4: *(Gas oscillator)* Suppose you have a chamber containing two different gases, divided by a movable partition of mass m and area A. The partition is moved a small distance Δx and then released. Assume the gas pressure for $\Delta x = 0$ is P_0. In general, the pressure P will change with displacement of the partition. Find the frequency of small oscillations if the chamber is a) isothermal with $PV = $ constant and b) adiabatic with $PV^\gamma = $ constant ($\gamma = \frac{5}{3}$). For both cases, use the approximation that the oscillations are small enough to only keep terms of the lowest order. Also, ignore the effect of the kinetic energy of the gases. You can either solve this by finding the generalized equation of motion or by finding the force equation directly.

Problem 5: *(Suspended mass)* A mass m is attached to an elastic string which has an unstretched length L and elastic spring constant k and is under tension T as shown in Figure 3.13.

a) How far down does the mass hang when it is in equilibrium?
b) Now the mass is displaced horizontally from by equilbrium position by a length Δx. Find the frequency of small oscillations for the mass after it is released. If

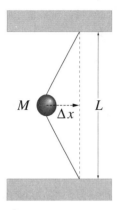

FIGURE 3.13
Elastic string under tension T, with spring constant k.

necessary use the approximation that the oscillations are small enough to allow keeping only terms of second order.

DSHOs

Problem 6: *(Damped pendulum)* A pendulum has a period of 5 seconds. It is damped so that the amplitude falls to one half of its original value in 100 seconds. What is the Q? What would the period be if the damping were not present?

Problem 7: *(Self-resonant frequency of a capacitor)* At high frequencies, simple capacitors behave more like resonant systems than capacitances when placed in a circuit. For a capacitance C, there is also a series inductance L and dielectric loss in the capacitor, which behaves like a shunt resistance R (see Figure 3.14). In unscaled variables, the differential equation for the charge q on the capacitor plates can be written as

$$L\frac{d^2q}{dt^2} + \frac{L}{RC}\frac{dq}{dt} + \frac{q}{C} = V_{\text{external}}(t). \tag{3.94}$$

What is the resonant frequency ω_0 if $C = 10$ picofarads, $L = 20$ nanohenries, and

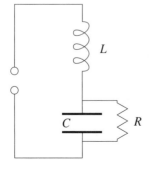

FIGURE 3.14

$R = 10,000$ ohms? What is the Q? At very low frequencies the circuit does behave like a pure capacitance, whereas at very high frequencies it behaves like an inductance, and at its resonant frequency it acts like a pure resistance. Explain and justify these statements.

Problem 8: *(Two masses connected by a spring)* Two masses m_1 and m_2 are connected by a spring. When m_1 is held fixed, m_2 is observed to oscillate with a frequency ω. Find the frequency of linear oscillations when m_2 is held fixed and when both masses are free to move. You can neglect the effect of gravity.

Problem 9: *(Second solution for critical damping)* For the case of critical damping, prove that

$$e^{\lambda_+ t} - e^{\lambda_- t} \to Dt e^{-t}, \tag{3.95}$$

where D is a constant proportional to $\sqrt{\epsilon}$. Assume $Q = \frac{1}{2} - \epsilon$, $\epsilon \to 0^+$. $\lambda_\pm$ are the two time constants for the overdamped oscillator.

Problem 10: *(Critical damping for instruments)* Critical damping is important in the design of instruments. Suppose that, for example, you are designing a seismograph. This is basically a long-period pendulum that responds to an impulsive force caused by earth movements. We want the recorded impulse from an earth tremor to return to the baseline as quickly as possible. Compute the time for a pendulum excited by an impulsive force to return to 1% of its maximum amplitude in the cases of underdamping, overdamping, and critical damping. Choose a Q value to use for each case. Show that critical damping is what you want to use in this application.

Problem 11: *(Numerical simulation)* Experiment with a numerical simulation of a damped SHO on the computer. You can create such a simulation by integrating (3.29) and plotting the solutions $q(t)$ versus t. See how the damping affects the decay of the solution at very long times. Compare critical damping with overdamping and underdamping.

Problem 12: *(Toward a shorter pendulum)* A pendulum consists of a mass M at the end of a massless string of length D. It is free to swing in one direction only, so has one degree of freedom, θ. The frequency of small oscillations is $\omega_0 = \sqrt{\frac{g}{D}}$, with g being the acceleration of gravity. Now suppose the string is *very slowly* shortened by some external agent. The length D varies only a little during the oscillation period. You can assume that $\theta(t) = A \cos \omega t$, where A, ω vary slowly with time but can be treated as constants over times of the order of a single period of the swinging pendulum. $D(t)$ is a given function of the time and is not a dynamical variable. To do this problem, you can treat θ as small, so, for example, $\sin(\theta) \to \theta$, $\cos(\theta) \to 1 - \frac{\theta^2}{2}$.

a) Find the kinetic and potential energies. Find the Lagrangian and prove that the equation of motion is

$$\ddot{\theta} + \frac{2\dot{D}}{D}\dot{\theta} + \omega_0^2 \theta = 0, \tag{3.96}$$

where

$$\omega_0^2 \equiv \frac{g}{D}. \tag{3.97}$$

b) Notice that the total energy of the pendulum $E = T + V$ is no longer constant in time because there is a term proportional to $\dot{\theta}$ in the EOM. The total energy also does *not* equal $H = \dot{\theta}\frac{\partial L}{\partial \dot{\theta}} - L$. Explain why this fact could be deduced from the form of T. Show that $E = H + m\dot{D}^2$.

c) Find $\frac{dE}{dt}$ from $\frac{dH}{dt}$. Hint: Use the fact that L contains the time explicitly.

d) The energy stored in the pendulum oscillations $E_{\text{pend}} = \frac{m}{2}(D^2\dot{\theta}^2 + Dg\theta^2) \approx \frac{m}{2}(gDA^2)$, where A is the amplitude of oscillations and $\dot{\theta} \approx -\omega A \sin \omega t$. $E > E_{\text{pend}}$ because, even if the pendulum doesn't swing, kinetic and potential energy both change. Part of your formula for $\frac{dE}{dt}$ vanishes if $A = 0$. This must be $\frac{dE_{\text{pend}}}{dt}$. Find it, and average it over one oscillation to get $\overline{\frac{dE_{\text{pend}}}{dt}}$. Here the "bar" means an average over one complete period. Prove that

$$\frac{1}{E_{\text{pend}}}\overline{\frac{dE_{\text{pend}}}{dt}} = -\frac{1}{2}\frac{\dot{D}}{D}. \tag{3.98}$$

Use this formula to show that, for small oscillations, the energy stored in the oscillations of the pendulum increases as $\frac{1}{\sqrt{D}}$, no matter how we change D, as long as we do it sufficiently slowly.

e) Show that the amplitude A of oscillations is proportional to $D^{-\frac{3}{4}}$.

The importance of this result is that $\frac{E_{\text{pend}}}{\omega}$ is a constant for any arbitrary variation of the pendulum length $D(t)$, as long as it varies slowly enough so that we can assume A, ω are constant for at least one period. Such a quantity, which remains constant for sufficiently slow variations in a parameter of the problem, is called an *adiabatic invariant*. This result played an important role in the early history of quantum theory. What you have shown applies to any system near its stable equilibrium point and not only to a pendulum.

Driven SHOs

Problem 13: *(Plotting the response of a driven SHO)* An SHO starts out at rest: $q = \dot{q} = 0$. For $t > 0$, a driving force of the form

$$F(t) = \sin at \tag{3.99}$$

is applied for a time of 10 periods of the free oscillator (use scaled time units). Make a plot of the energy stored in the oscillator after 10 periods versus the parameter a, solving the problem by numerical methods by integrating the differential equation of motion. What value of a drives the oscillator to the greatest amplitude? Now assume that the oscillator initially had some energy (i.e., that the initial amplitude was not zero). Find out what happens to the stored energy for driving frequencies both slightly above and below the resonance. Is it possible to extract energy from an oscillator as well as store energy in it by driving it appropriately? Does your answer depend on the phase of the drive signal?

Problem 14: *(Aluminum chime)* A good demonstration that allows us to "listen to the Green's function" can be made with a cylindrical rod constructed from a special low friction aluminum alloy. The rod is supported at two points along its length. The nature of the lowest frequency mode for transverse elastic vibrations is such that the amplitude of oscillation vanishes at these two points for all times. If the center of the bar is struck with a hammer, we hear a piercing note of a single frequency (say ≈ 3.5 kHz), which is damped to a relative amplitude $\frac{1}{e}$ after 10^4 oscillations. What is the time dependence of the amplitude of oscillation at a fixed position along the bar after the bar is struck? Does it obey the equation for the Green's function? Explain.

Problem 15: *(Checking the driven SHO formula)* Prove that the result in Equation (3.74) is correct. Also find the solution in terms of unscaled units. Find the oscillator energy (averaged over one period) as a function of time and show that it is proportional to t^2.

Problem 16*: *(Pulsed driving force)* An undamped simple harmonic oscillator ($\omega = 1, m = 1$) is driven by a time-dependent force

$$F(t) = 1, \quad 0 \leq t \leq T. \tag{3.100}$$

In this situation the oscillator is not excited for $t < 0$.

a) Find $q(t)$ after the force is turned off. Use the Green's function.
b) Use the method of matching boundary conditions for q and $\dot{q}$ to find an explicit form for $q(t)$ at all times. Verify your answer to a). Hint #1: A particular solution to the inhomogeneous equation of motion

$$\ddot{q} + q = 1 \tag{3.101}$$

is $q_{\text{particular}} = 1$. You will have to match the solution you obtain at $t = t_0 = 0$ and $t = t_1 = T$ by using the solution to the homogeneous equation, with $F = 0$. Explain how to do this. Hint #2: For $t \geq T$, it is simplest to use the expression $q(t) = A \sin(t - T) + B \cos(t - T)$. Determine the constants A and B by matching to the solution for $0 \leq t \leq T$.
c) Find a simple expression for the energy E stored in the oscillator ($E = \frac{1}{2}(q^2 + \dot{q}^2)$) as a function of the pulse length T. Plot $E(T)$. Give a physical reason why it behaves the way it does.

d) The change in the Hamiltonian for a system with a Lagrangian that explicitly depends on the time is given by using the formula $\frac{dH}{dt} = -\frac{\partial L}{\partial t}$, which was proved in Chapter 1, Equation (1.68):

$$\Delta H = -\int_{t_0}^{t_1} \frac{\partial L}{\partial t} dt. \tag{3.102}$$

The Lagrangian is $L = \frac{1}{2}(\dot{q}^2 - q^2) + qF(t)$. Find the relation between H and E. Use (3.102) to obtain a simple way to calculate the stored energy as a time integral from $t_0 = 0$ to $t_1 = T$ in the general case of an arbitrary driving force $F(t)$. How is your result related to the work $\int F dq$ done by F on the oscillator? Assume the force is zero outside of the time interval $0 \leq t \leq T$. Apply your result to the case $F = 1$ and check that you get the same result as in c).

Driven DSHOs

Problem 17: *(Two frequencies)* Prove by solving (3.76) explicitly for the DSHO that there are two frequencies present initially, but that the free oscillator frequency is damped out.

Problem 18: *(Resonant amplitude is a circle)* Prove that, with certain approximations valid for high Q, the trajectory of the imaginary (out of phase) part of the solution for the sinusoidally driven DSHO plotted versus the real (in phase) part is a circle in the complex amplitude plane.

Problem 19: *(Superposition)*

a) Prove using superposition that the solution $q(t)$ to the DSHO for a "square" drive pulse of length T and unit magnitude lasting from $-\frac{T}{2}$ to $\frac{T}{2}$ is, for $t > \frac{T}{2}$,

$$\frac{1}{\omega'\left(\omega'^2 + \frac{1}{4Q^2}\right)} e^{-\frac{(t+\frac{T}{2})}{2Q}} \left\{ \omega' \cos\left[\omega'\left(t + \frac{T}{2}\right)\right] + \frac{1}{2Q} \right.$$
$$\left. \times \sin\left[\omega'\left(t + \frac{T}{2}\right)\right] \right\} \text{ minus the same thing with } T \to -T.$$
$$\tag{3.103}$$

b) Find the discontinuity in $\dddot{q}$ at $t = -T/2$? How does it arise?
c) If the force pulse is very short (i.e., $T \to 0$), prove that $q(t) \to TG(t)$.

Problem 20: *(Green's function for a damped oscillator)* Prove that the Green's function for the damped oscillator without scaling the time described by the equation (ω_0 = the oscillation frequency of the free SHO)

$$\ddot{q} + \frac{\omega_0}{Q}\dot{q} + \omega_0^2 q = \delta(t) \tag{3.104}$$

is

$$G(t-t') = \begin{cases} \dfrac{1}{\omega'} e^{-\frac{\omega_0(t-t')}{2Q}} \sin\omega'(t-t'), & t-t' \geq 0, \\ 0, & t-t' \leq 0, \end{cases} \qquad (3.105)$$

where $\omega' \equiv \omega_0\sqrt{1-\frac{1}{4Q^2}}$. Is this the only possible choice for the Green's function? Explain.

Problem 21: *(Driven critically damped oscillator)* A critically damped oscillator has $Q = \frac{1}{2}$. The free oscillator obeys the homogeneous equation of motion $\ddot{q} + 2\dot{q} + q = 0$. The two free oscillator solutions are e^{-t} and te^{-t}.

a) Drive this oscillator with an external driving force that is a discontinuous step at the time $t = 0$, after which $F = 1$. (For $t < 0$, $F = 0$.) Assuming that $q = \dot{q} = 0$ for $t < 0$, find an explicit solution for $t \geq 0$.

b) Consider a different external driving force:

$$F(t) = \begin{cases} 0, & t < 0, \\ \cos t, & t \geq 0. \end{cases} \qquad (3.106)$$

Again, $q = \dot{q} = 0$ for $t < 0$. Find the form of the steady state ($t \gg 1$) solution by first solving for $q(t)$ for the complex driving force $F(t) = e^{it}$, $t > 0$, and then finding the physical displacement of the oscillator $q(t)$ for $F(t) = \cos t$. What is the relative phase between the driving force and the oscillator response in the steady state?

c) To find the exact solution for all positive times you could use a Green's function or you could match boundary conditions at $t = 0$. Use the boundary condition method to find the transient solution. Combine this with the result of part b to find the oscillator's total response to suddenly turning on a $\cos t$ driving force at $t = 0$. Make a sketch of $q(t)$ for $0 \leq t < 4$. For what time is the response maximized?

d) The derivative of a step function is a delta function. From this fact, find the response of this oscillator to a delta function impulse at $t = 0$. Then find the explicit form of the Green's function $G(t - t')$. Write the oscillator response as an integral over t'. What are the limits of integration? Explain. (You don't need to carry out the integral.)

Nonlinear Oscillators:

Problem 22*: *(Nonlinear oscillator with cubic restoring force)* Find a Lagrangian that will lead to this equation of motion:

$$\ddot{q} = -q^3. \qquad (3.107)$$

If the mass $m = 1$, find the kinetic and potential energies. Plot $V(q)$, the potential energy. Find an expression for the total energy in terms of $\dot{q}, q$. Is the total energy constant? Why do we say the oscillator is nonlinear?

Problem 23*: *(Cubic potential)* Suppose that you have one-dimensional motion with $m = 1$. The potential energy $V(q) = -10q + q^3$.

a) Make a graph of this potential for both positive and negative q. Locate the equilibrium points and label which are stable (or neutral) and which are unstable.

b) Find the period of small oscillations around the stable equilibrium point(s).

c) Solve the equation of motion numerically. At some point, the trajectory will become unbounded. Find where this occurs, that is, what initial conditions lead to instability? What energy?

d) What is the total energy in terms of $\dot{q}$ and q? Is it constant? What do the curves of constant energy look like?

CHAPTER FOUR

ONE-DIMENSIONAL SYSTEMS:
CENTRAL FORCES AND THE KEPLER PROBLEM

OVERVIEW OF CHAPTER 4

One of the greatest advances in science was Newton's discovery that the force of gravity is a universal force that not only causes terrestrial objects to fall but also guides the Moon around the Earth and the planets around the Sun. It was not previously understood that the Moon and planets – indeed the universe – obey the same physical laws as terrestial objects. We take this for granted today, but it was a revolution in human understanding, one from which there has been no turning back.

Our goal in this chapter is to show how this problem – the Kepler problem of planetary orbits – can be solved using the powerful analytical techniques of Lagrangian mechanics. We begin by considering the general solution for motion in a one-dimensional potential $V(q)$. Next, we consider a six-dimensional system of two isolated point masses that interact by a mutual force directed along the line between them. This applies to a wide class of physical problems, with results of general significance. By using symmetry properties we can drastically simplify the problem down to a single equation involving only the radial distance between the two masses. At this stage, by introducing the concept of equivalent potential, the problem is reduced to one with only one degree of freedom. Proceeding further, we restrict our consideration to the force of gravity, a force that diminishes according to the inverse of the square of the distance between the attracting bodies. We derive the possible shapes of curves traced out by masses that are attracting each other by gravity. These orbits turn out to be conic sections: ellipses, parabolas, or hyperbolas, a result first shown to follow from the inverse square force law by Isaac Newton in 1687.

4.1 THE MOTION OF A "GENERIC" ONE-DIMENSIONAL SYSTEM

First let's deal with the general one degree of freedom problem that has a Lagrangian $L(q, \dot{q})$. Notice we exclude the case of an explicit time dependence in the Lagrangian. Otherwise our treatment is completely general. Any 1-D problem* can be

---

* The abbreviation "1-D" will stand for "one degree of freedom," "2-D" for "two degrees of freedom," and so on.

solved in principle by expressing the answer as an integral. From this integral, the time is obtained as a function of displacement $t(q)$. It is the inverse of what one usually means by a "solution" (i.e., an explicit form for $q(t)$). Obtaining the inverse solution as an integral is called "solution by quadratures." From this integral for $t(q)$, $q(t)$ cannot always be expressed in analytic terms, however. We will give an example of this for the pendulum.

Qualitative Information About the Solution

Qualitative information about the solution to any such 1-D problem can be deduced by graphing the potential, and then considering what happens as a function of a variable parameter: the total energy E. Suppose the energy E is given by (the coefficient of $\dot{q}^2$ is set to $\frac{1}{2}$ by scaling the time variable appropriately):

$$E = T + V = \frac{1}{2}\dot{q}^2 + V(q). \tag{4.1}$$

Since the Lagrangian doesn't involve the time, the Hamiltonian H is a constant of the motion. For scleronomic constraints or any constraints in which T is a quadratic form in the generalized velocities, $H = E$, energy E is a constant that depends on the initial conditions. Solving Equation (4.1) for the velocity we get

$$\dot{q} = \pm\sqrt{2T} = \pm\sqrt{2(E - V(q))}. \tag{4.2}$$

The kinetic energy is the difference $E - V(q)$ between the total energy and the potential energy. *A physically admissible solution with positive kinetic energy is only possible if the total energy obeys the inequality $E \geq V(q)$. If $T = 0$, the motion momentarily stops and then reverses direction. This is called a "turning point."*

In Figure 4.1, at the turning points C and D, $E - V = 0$; hence $\dot{q} = 0$ there. The point marked A is an unstable equilibrium point, while the point marked B is a stable equilibrium point. Near the stable equilibrium point, the potential is nearly parabolic. For very low energies such as E_0, the motion is very nearly like a simple harmonic oscillator. As the

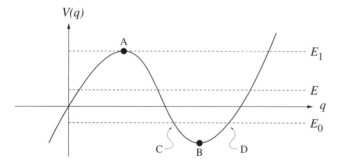

FIGURE 4.1
A) Unstable equilibrium point; B) stable equilibrium point; C), D) turning points for $E = E_0$. To find the kinetic energy use $T = E - V$.

energy increases above E_0, the potential departs from the simple parabolic shape, but the motion will still be oscillatory. The maximum amplitude of the oscillation is determined by the turning points, hence the total energy E. *The period will depend on E only if the potential deviates from a parabola.* For $E = E_1$, one of the turning points is the unstable equilibrium point, so in principle the period becomes infinite – the system gets stuck forever at this point. If $E > E_1$, even slightly, the motion is no longer periodic. As q becomes larger and more negative, the motion becomes unbounded, since the kinetic energy rapidly increases. The system "falls off the edge" of the potential and never returns to the vicinity of the unstable equilibrium point. All this information can be deduced from a graph of $V(q)$ and the total energy of the system.

QUESTION 1: *Potential Graphs* Explain the analogy between a particle on a real hill and the problem of motion in a potential of the same shape as the hill. The particle on a hill is acted upon by gravity. All motion is frictionless. Is the analogy perfect? Especially consider whether the two cases have the same acceleration and whether the mass of the object makes a difference. Now repeat the question for a particle in a real valley versus a potential in the shape of a valley.

Finding the Solution

Since E is constant, a solution by quadratures is obtained by solving for $\dot{q}$:

$$\dot{q} = \pm\sqrt{2(E - V(q))} \quad \text{so} \quad \frac{dq}{\pm\sqrt{2(E - V(q))}} = dt. \tag{4.3}$$

The time t to move from $q = 0$ to $q(t)$ is:

$$\boxed{t = \int_0^q \frac{dq'}{\sqrt{2(E - V(q'))}}.} \tag{4.4}$$

The prime on q' means it is the variable of integration. The expression (4.4) gives the time as a function of the displacement: $t(q)$. In principle, $t(q)$ must be inverted to obtain the solution $q(t)$. This is not always possible, but we consider the problem to be solved. A problem solved as an integral is said to be solved by "quadrature." In the more general case, Equation (4.4) may not be analytically integrable but can be used to get an approximate solution. If the potential is symmetric and the integration is from the stable equilibrium point to a turning point, then $t = \frac{T}{4}$ equals one quarter of the oscillator period. (Why is this true?)

4.2   THE GRANDFATHER'S CLOCK

The "grandfather's clock" was invented by Huygens. It is a nonlinear pendulum, that is, one in which the restoring torque is *not* approximated by θ, but the exact form,

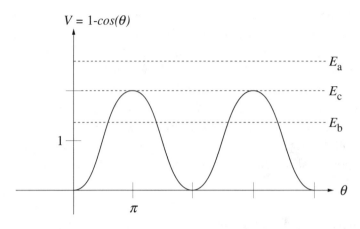

FIGURE 4.2

$\sin\theta$, is used instead. Thus we no longer assume that θ is small. For practical purposes we may wish to know how accurately a grandfather's clock keeps time when the amplitude of the motion changes, since its period does depend on its amplitude.

The pendulum has a potential energy proportional to the height of the pendulum above the point where $V = 0$. If this is taken to be $\theta = 0$, then $V(\theta) = mgl(1 - \cos\theta)$. Making use of the fact that E is a constant of the motion, the equation of motion has the general form of Equation (4.3) above. By choosing appropriate units of time and energy, we can effectively set $mgl = 1$ and work with a scaled Lagrangian $L = \frac{\dot\theta^2}{2} - (1 - \cos\theta)$. The total energy in these units is

$$E = \frac{\dot\theta^2}{2} + (1 - \cos\theta). \tag{4.5}$$

The exact curve of (the scaled) $V(\theta)$ versus θ is shown in Figure 4.2. There are equilibrium points wherever the generalized force (torque in this case) is zero (i.e., at $\theta = 0, \pi, 2\pi, 3\pi, \ldots$ etc.). The points at odd multiples of π are *unstable* equilibrium points; those at even multiples of π are stable equilibrium points. The difference is whether the restoring force for a small displacement acts either to magnify or to decrease the displacement, that is, whether $\frac{d^2V}{d\theta^2}$ is <0 (unstable) or >0 (stable). Draw a horizontal line representing the total energy E across Figure 4.2. Then consider what type of motion is possible for this value of E. There are three distinct types of motion. For E_a, the total energy >2, and the angle θ can increase without limit. This corresponds to a pendulum swinging freely round and round (called "rotation"). For E_b, the total energy <2. Here the motion in θ must be bounded. The pendulum oscillates between the two points where $E = V$ ($T = 0$) (i.e., the turning points of the motion). This periodic motion is called "libration." The regions where $T < 0$ are called *classically forbidden* regions. (In quantum mechanics, a particle can penetrate these forbidden regions to some extent.) Finally, we might have E_c, where $E = 2$, and the pendulum just manages to reach the top of its swing. For this case, which is the boundary between periodic and aperiodic motion, the period becomes infinite. The condition that defines this boundary between oscillation and rotation is determined by the total energy E. If you picture a pendulum, it is certainly intuitively

4.2 THE GRANDFATHER'S CLOCK

clear that whether it swings round and round or just back and forth is determined by the initial kinetic and potential energies. It is only the total energy that separates libration from rotation. The grandfather's clock is a common example of a nonlinear pendulum.

Phase Portrait of the Nonlinear Pendulum: Curves of Constant E

Next we will make a graph that will give valuable qualitative information about the motion in a geometric form. For 1-D motion, $\dot{q}$ is plotted on the Y axis and q is plotted on the X axis. A two-dimensional picture of the motion is the result. This 2-D space is called *phase space*, a concept to be more rigorously defined after we introduce Hamiltonian theory. (Phase space is $2N$-dimensional if there are N degrees of freedom.)

For this oscillator, we will be interested in $\dot{\theta}$ versus θ. First make the approximation of small amplitude: $(\ddot{\theta} + \theta = 0)$. At a certain fixed time, plot a single point on this graph to represent the current position and velocity of the oscillator as shown in Figure 4.3. This point will move as the time changes. This is the phase trajectory. This "path" of the system is traced out as time increases. Because the total energy of the oscillator $\frac{\dot{\theta}^2 + \theta^2}{2}$ is constant, the path in the 2-D phase space is a circle. The phase trajectory moves along the circle in a clockwise direction. Different energies of the SHO correspond to different radii for the circle in phase space. Now drop the restriction to small angles. For larger amplitudes, the phase portrait is more complicated.

Figure 4.4 is a plot of the phase trajectories for the three energies: $E < 2$, $E = 2$, and $E > 2$.

QUESTION 2: *Phase Trajectories* Two phase trajectories can never cross. Prove this statement. An exception may occur at an unstable equilibrium point. Why does this *seem* possible? This is *not* true however. If you look closely at the area around an

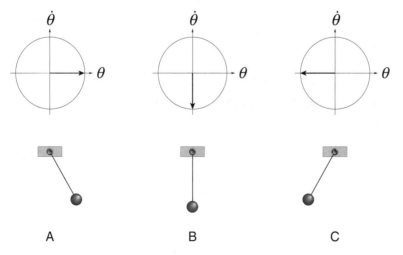

FIGURE 4.3
A) $\theta = 1, \dot{\theta} = 0$; B) $\theta = 0, \dot{\theta} = -1$; C) $\theta = -1, \dot{\theta} = 0$.

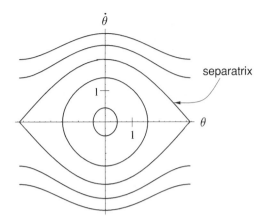

FIGURE 4.4
Phase portrait of the nonlinear pendulum (4.5) for different energies. The *separatrix* corresponds to $E = 2$. It is the boundary between libration and rotation. Inside the separatrix, $E < 2$ and outside $E > 2$.

unstable equilibrium you will see hyperbolas instead of an intersection. Now explain why this is really the case.

How To Calculate the Period for the Grandfather's Clock

Exact analytic solutions to (4.4) can be found for the nonlinear pendulum using elliptic functions. These are not so interesting to us, since they are hard to generalize to other problems. A good approximation to the motion is to assume that the nonlinearity is a small perturbation, which will also reveal how the frequency changes with amplitude. We'll switch the dependent variable to q from θ to emphasize the generality of this approach. For one-dimensional motion (hence 2-D phase space) we have an EOM from (4.1) of

$$\ddot{q} + \frac{dV}{dq} = 0. \qquad (4.6)$$

Since we have scleronomic constraints in which T is a quadratic form in the generalized velocity, the total energy E is a constant.

Let's assume that $V(q) = V(-q)$, that is, that the potential energy is an even function of the displacement. The Taylor series for a general symmetric (even) nonlinear potential must have the form:

$$V(q) = \frac{q^2}{2} + \frac{\epsilon}{4}q^4 + O[q^6]. \qquad (4.7)$$

We can always choose the units such that the coefficient of q^2 is $\frac{1}{2}$. The unknown constant ϵ is a measure of the strength of the nonlinear part of the potential energy versus q. The term proportional to q^4 is an "anharmonic" term, contributing a nonlinear part to the EOM.

4.2 THE GRANDFATHER'S CLOCK

All terms in $V(q)$ of order higher than q^4 will be ignored. We will restrict q to amplitudes where this approximation is valid.

We want to know the variation of the oscillator period with amplitude, first for a generic nonlinear oscillator of this type, then for the specific case of the grandfather's clock. If that is all we want to know, we don't have to resort to full perturbation theory. (If, on the other hand, we also want to know the detailed motion as a function of time, we must use the Lindstedt–Poincaré perturbation theory, which is explained in Chapter 10.) Anharmonic oscillators play a central role in producing chaos, the subject of Chapter 11. We know that

$$E = \frac{1}{2}(\dot{q}^2 + q^2) + \frac{\epsilon}{4}q^4. \tag{4.8}$$

Using (4.4), we integrate from $q = 0$ to $q = q_{max}$ to get the time for $\frac{1}{4}$ of a period. Thus, without any further approximation, if T is the period

$$\frac{T}{4} = \int_0^{q_{max}} \frac{dq}{\sqrt{2E - q^2 - \frac{\epsilon}{2}q^4}}. \tag{4.9}$$

Several tricks can be used to simplify this integral. First of all, note that $E = $ constant, so $E \approx \frac{1}{2}q_{max}^2 + \frac{\epsilon}{4}q_{max}^4$. Substituting this expression into (4.9), and making the substitution to a scaled variable x, $q = q_{max}x$, we get

$$\frac{T}{4} = \int_0^1 \frac{dx}{\sqrt{(1 - x^2) + \frac{\epsilon q_{max}^2}{2}(1 - x^4)}}. \tag{4.10}$$

If we pass over to the limit of a linear system, $\epsilon \to 0$, the period becomes independent of q_{max}, just as we expect for a SHO. Next, factor the term under the square root: $1 - x^4 = (1 - x^2)(1 + x^2)$, and make a power series expansion in the parameter $\epsilon' \equiv \frac{\epsilon q_{max}^2}{2}$:

$$\frac{1}{\sqrt{1 - x^2}} \left(\frac{1}{\sqrt{1 + \epsilon'(1 + x^2)}} \right)$$

$$= \frac{1}{\sqrt{1 - x^2}} \left(1 - \frac{(1 + x^2)\epsilon'}{2} + \frac{3}{8}(1 + x^2)^2\epsilon'^2 + O[\epsilon'^3] \right). \tag{4.11}$$

Finally, to evaluate the integrals, make the substitution $x = \sin u$, $dx = \cos u \, du$:

$$\frac{T}{4} = \int_0^{\frac{\pi}{2}} du \left[1 - \frac{\epsilon'}{2}(1 + \sin^2 u) + \frac{3\epsilon'^2}{8}(1 + \sin^2 u)^2 + \cdots \right]. \tag{4.12}$$

The final result for the period of a nonlinear oscillator with a q^4 term in $V(q)$, after doing the integrals, is

$$T = 2\pi \left(1 - \frac{3}{4}\epsilon' + \frac{57}{64}\epsilon'^2 + \cdots \right). \tag{4.13}$$

Notice that, if $\epsilon' > 0$, the period gets shorter for larger amplitudes. This is what we would expect for a "stiff" spring (i. e., a system that has a restoring force increasing faster than linearly with displacement). The method used here is quite general and could be applied to almost any type of symmetric 1-D motion that deviates only slightly from a simple harmonic oscillator.

Return to the case of the grandfather's clock. If θ is not too large, we can take the next term in the Taylor series beyond our "small angle" approximation, which leads to the harmonic oscillator equation. The Taylor series for $\cos\theta$ is $1 - \frac{\theta^2}{2} + \frac{\theta^4}{24} + \ldots$. Thus $V = mgl(\frac{\theta^2}{2} - \frac{\theta^4}{24} + \cdots)$ for the grandfather's clock. Comparing to (4.13) we see that $\epsilon = -\frac{1}{6}$, so

$$T = 2\pi \left(1 + \frac{\theta_{\max}^2}{16} + \cdots\right). \tag{4.14}$$

The period is longer for large amplitudes because the potential is "soft." We might have guessed this from the physical observation that the restoring force, $\sin\theta$, is less than θ for large angles.

4.3 THE HISTORY OF THE KEPLER PROBLEM

Ptolemy versus Copernicus

The most famous problem ever solved in classical mechanics is the "Kepler Problem" – the detailed description of two bodies interacting by their mutual gravitational attraction. This describes planetary motion around the Sun, the Moon orbiting the Earth, etc. It is probably only a slight exaggeration to say that Newton's mechanics was invented to solve this problem.

To obtain a historical perspective on why the solution of this problem was so important to science, we recall that a geocentric theory of observed motions of the planets with respect to the stars, based on observations between 127 and 151 A.D., was published by Ptolemy of Alexandria in a collection of books called the *Almagest*. Ptolemy invented the idea of epicyles and the equant, as illustrated in Figure 4.5.

It should be emphasized that this theory, which was taught in universities for more than 1,500 years, gave a rather accurate, but somewhat complicated, way to calculate planetary orbits around a *fixed* Earth. Later, the authority of the Church was added, which found justification in the Holy Scripture for the geocentric point of view.

Heliocentric models had been introduced by the Greeks, but these were rejected on the grounds that the Earth was "obviously" at rest, not moving at thousands of miles per hour. The heliocentric theory also could not explain why the Moon would not be left behind in space if the Earth were moving. An objective person might wonder why Nicolaus Copernicus (1473–1543) would invent a new theory in which the planets orbited the Sun rather than the Earth. His heliocentric theory was published in the year of his death. Copernicus worked out the relative distances of the planets from the Sun to surprising accuracy, but he was wrong about absolute distances by more than one order of magnitude.

4.3 THE HISTORY OF THE KEPLER PROBLEM

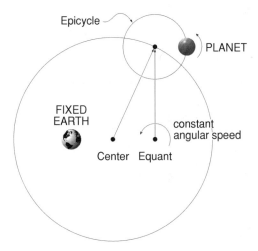

FIGURE 4.5
Ptolemy's theory of planetary orbits.

The usual explanation for why the Copernican theory was preferable to the older Ptolemaic theory was that the orbits, being circles, were much simpler. But, in order to reproduce the observations using circles as his basic solar orbits, Copernicus had to introduce a set of thirty-four circles for the six known planets – a more complex theory than Ptolemy's.

The forward to Copernicus's book carefully explained that this theory could be considered, not as the way things really are, but as a mathematical hypothesis to allow conceptually easier calculations. There was even a dedication to the Pope, since Copernicus himself was a canon of the Church. However, this theory was ridiculed and rejected. There is even a story that Martin Luther himself called Copernicus a fool!*

Kepler and Galileo

This historical background makes it even more astonishing that the work of Kepler took place at all. Following many years of extremely detailed and careful astronomical observations by Tycho Brahe, Johann Kepler established three empirical laws, given in Table 4.1, describing such motion. Kepler, who started in 1601, spent over five years analyzing the motion of the planet Mars. He worked for almost a year to remove a discrepancy of only eight minutes of arc between the observations and the Ptolemaic theory of planetary orbits. The elliptical nature of planetary motion around the Sun was discovered in 1606 and published in 1609. Kepler's third law was published in 1619.

Galileo was a contemporary of Kepler. After being convinced by observation of the phases of Venus and the moons of Jupiter, Galileo, contrary to Copernicus, insisted publically that the heliocentric theory was indeed the actual way the solar system worked. In the climate of the times, with religious wars spreading throughout Europe, this view was

---

* See Alioto, *A History of Western Science*, second edition, p. 184.

> **TABLE 4.1** KEPLER'S LAWS
>
> **First Law**   All planetary orbits are ellipses, with the Sun at the focus.
> **Second Law**   The line joining the planets to the Sun sweeps out equal areas in equal times.
> **Third Law**   The square of the period of a planet is proportional to the cube of its mean distance to the Sun.

considered to be dangerous to established authority. Copernicus's book was placed on the Index (forbidden books) in 1616. Here is a brief excerpt from an account of the infamous heresy trial of Galileo in 1633:

> On 21 June Galileo was examined as to intention; he stated that until the 1616 decree he had considered the two world systems to be freely debatable, but that thereafter he had adhered to the fixed earth and movable sun; in his book he had considered no argument as conclusive and the decision of 'sublime authority' as binding. Asked whether he spoke truly, on pain of torture, he replied: 'I am here to obey, and have not held this opinion after the determination made, as I said.'
> On 22 June the sentence of life imprisonment was read to Galileo at a formal ceremony in the presence of the cardinals of the Inquisition and witnesses, after which he had to abjure on his knees before them.*

A Story About Isaac Newton

Isaac Newton was born in the year of Galileo's death, 1642. The original work on Newton's theory of gravitation was developed about fifty years after Kepler's laws were published, and all of Kepler's purely empirical laws turned out to be consequences of either a central force or the inverse-square force law of gravity. We quote from the book *The Birth of a New Physics* by I. Bernard Cohen:

> ...Thus it came about that Robert Hooke, Edmond Halley, and Sir Christopher Wren, England's foremost architect, met to discuss the question: Under what law of force would a planet follow an elliptical orbit?...

Hooke and the others thought it might be a force falling off as the inverse of the square of the distance from the planet to the Sun. Wren offered Hooke forty shillings if he could prove it within two weeks. Nothing was then heard from Hooke. We continue the quote from the book by Cohen:

> In any event, by January 1684 Halley had concluded that the force acting on planets to keep them in their orbits 'decreased in the proportion of the squares of the distances reciprocally,' but he was not able to deduce from that hypothesis the observed motions of the celestial bodies... In August 1684, Halley decided to go to Cambridge to consult

---

* S. Drake, *Galileo At Work*, p. 351.

Isaac Newton. On his arrival he learned the 'good news' that Newton 'had brought this demonstration to perfection.' Here is DeMoivre's almost contemporaneous account of that visit:

'After they had been some time together, the Dr. [Halley] asked him what he thought the curve would be that would be described by the planets supposing the force of attraction towards the sun to be reciprocal to the square of their distance from it. Sir Isaac replied immediately that it would be an ellipsis. The Doctor, struck with joy, and amazement, asked him how he knew it. Why, saith he, I have calculated it. Whereupon Dr. Halley asked him for his calculation without any further delay. Sir Isaac looked among his papers but could not find it, but he promised him to renew it and then to send it to him. Sir Isaac, in order to make good his promise, fell to work again, but he could not come to that conclusion which he thought he had before examined with care. However, he attempted a new way which, though longer than the first, brought him again to his former conclusion. Then he examined carefully what might be the reason why the calculation he had undertaken before did not prove right, and he found that, having drawn an ellipsis coarsely with his own hand, he had drawn the two axes of the curve, instead of drawing two diameters somewhat inclined to one another, whereby he might have fixed his imagination to any two conjugate diameters, which was requisite he should do. That being perceived, he made both his calculations agree together.'

Spurred on by Halley's visit, Newton resumed work on a subject that had commanded his attention in his twenties when he had laid the foundations of his other great scientific discoveries; the nature of white light and color and the differential and integral calculus...

Newton's theory was published in the 1687 book *Philosophiae Naturalis Principia Mathematica – Mathematical Principles of Natural Philosophy.* However, Newton had actually done the work described above in 1665–66, while Cambridge University was closed due to the plague. More than twenty years passed, while it lay among his notes. He only published the work after much urging from his friend Halley. Writing the *Principia* took two years. His famous three laws of motion appeared for the first time in this book.

4.4 SOLVING THE CENTRAL FORCE PROBLEM

Returning to the present, we are interested in applying the tools we developed earlier in this chapter to the problem of motion due to gravity. The starting point is to consider two bodies interacting via a *central force* as shown in Figure 4.6. A central force is one that depends only on the distance r between the two bodies and is directed along the line between them. It follows that $V = V(r)$ for this type of force.

FIGURE 4.6
Central force.

Systems of Mutually Attracting Bodies

Gravity (and any other inverse-square law force) has the property that a spherically symmetrical mass distribution acts as if all of the mass were concentrated at the center of the body. This greatly simplifies the problem of planetary motion, since the planets can be treated as if they are point particles.

Under what conditions can the motion of N point masses be solved exactly? It turns out that this can be done for $N = 2$ if the force is a central force between the two mass points. For a wide variety of different central potentials, the integral obtained can be expressed in terms of circular or elliptic functions. This is called the "two-body problem."

But if $N \geq 3$, the problem has no solution, even by quadratures. The famous "three-body problem" (e.g., Earth–Moon–Sun) was studied before 1900 by Poincaré, who proved that no analytic solution was possible. This work led Poincaré to a discovery of chaotic dynamics, the importance of which was not fully realized until the late 1960s, almost 300 years after Newton. (See Chapter 11.)

Nevertheless, we can assume that the two-body problem is a good starting point for studying three-dimensional motion. We will specialize to a gravitational interaction later. At the start of solving a problem with more than one degree of freedom, every symmetry of the problem should be exploited fully to reduce the number of degrees of freedom. We will show how the symmetries of the problem allow reduction of the degrees of freedom from the original 6-D down to a 1-D problem, which can then be solved by quadratures. What we mean by "reduction" is that five of the six degrees of freedom equations of motion will become trivial.

Motion of the Center of Mass

We can predict the motion of the center of mass of the fragments of an exploding artillery shell without knowing the internal forces released in the explosion. This is an example of the general theorem that the center of mass of an entire system behaves as a single particle under the action of external gravitational forces.

Use the notation $\vec{r}_1$, M_1 for the position and mass of "particle" number 1 as shown in Figure 4.7. $\vec{r}_2$, M_2 denotes the same for particle number 2. The relative coordinates

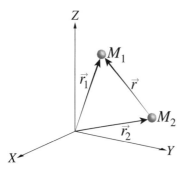

FIGURE 4.7

4.4 SOLVING THE CENTRAL FORCE PROBLEM

$\vec{r} \equiv \vec{r}_1 - \vec{r}_2$. In the most general case, the Lagrangian is

$$\frac{1}{2}M_1\dot{\vec{r}}_1^2 + \frac{1}{2}M_2\dot{\vec{r}}_2^2 - V_{\text{ext},1}(\vec{r}_1) - V_{\text{ext},2}(\vec{r}_2) - V_{12}(\vec{r}_1, \vec{r}_2). \tag{4.15}$$

Any interaction between the two particles V_{12} can depend only on the difference between their positions, $\vec{r}_1 - \vec{r}_2$. This is not true for external forces expressed by V_{ext} above. If the interaction force is a central force, V_{12} can only depend on the magnitude of $\vec{r}_1 - \vec{r}_2$, which we call r.

QUESTION 4: *External versus Interaction Forces 1* Prove that the statements above are true: 1) The interaction forces between two particles depend only on the difference between the particles' positions. 2) External forces can depend on the particles' absolute positions as well.

The Case of No External Forces

Suppose $V_{\text{ext}} = 0$ for both particles, so that the only force is the one between particle 1 and particle 2. Then it *must* be true that the Lagrangian does not change if we arbitrarily translate the origin of our coordinates. This means that we can add a constant vector to both $\vec{r}_1$ and $\vec{r}_2$ without changing the Lagrangian:

$$\begin{aligned}\vec{r}_1 &\to \vec{r}_1 + \vec{a}, \\ \vec{r}_2 &\to \vec{r}_2 + \vec{a}.\end{aligned} \tag{4.16}$$

QUESTION 5: *External versus Interaction Forces 2* Why would Equations (4.16) change the Lagrangian of the system if there were external forces? Give an example.

Since the Lagrangian is a function of the coordinates and the velocities of both particles, and $\vec{a}$ is an arbitrary constant vector, we have

$$\boxed{L = L(\vec{r}_1, \vec{r}_2, \dot{\vec{r}}_1, \dot{\vec{r}}_2) = L(\vec{r}_1 + \vec{a}, \vec{r}_2 + \vec{a}, \dot{\vec{r}}_1, \dot{\vec{r}}_2),} \tag{4.17}$$

definition of translation invariance

Let's consider a special case of translation invariance: an infinitesimal translation. Add a vector $\epsilon\vec{a}$ to the coordinates of both particles. Here ϵ will be made arbitrarily small, while $\vec{a}$ is an arbitrary constant vector. If we expand the translated L in a Taylor series in ϵ, we get

$$L(\vec{r}_1 + \epsilon\vec{a}, \vec{r}_2 + \epsilon\vec{a}, \dot{\vec{r}}_1, \dot{\vec{r}}_2) = L(\vec{r}_1, \vec{r}_2, \dot{\vec{r}}_1, \dot{\vec{r}}_2) + \epsilon\vec{a} \cdot (\vec{\nabla}_1 + \vec{\nabla}_2)L + O[\epsilon^2] \tag{4.18}$$

($\vec{\nabla}_1 \equiv \frac{\partial}{\partial x_1}, \frac{\partial}{\partial y_1}, \frac{\partial}{\partial z_1}$, etc.). Since $\vec{a}$ is arbitrary, the invariance of L (4.17) implies that

$$\vec{\nabla}_1 L + \vec{\nabla}_2 L = 0. \tag{4.19}$$

Taking only the x component of this equation, for example, we get

$$\frac{\partial L}{\partial x_1} + \frac{\partial L}{\partial x_2} = 0. \tag{4.20}$$

The same equation holds for the y and z components. From the Euler–Lagrange equations and the definition of momentum $p \equiv \frac{\partial L}{\partial \dot{q}}$ (1.70), we get

$$\frac{dp_{1x}}{dt} = \frac{\partial L}{\partial x_1}, \quad \frac{dp_{2x}}{dt} = \frac{\partial L}{\partial x_2}. \tag{4.21}$$

Therefore

$$\frac{d\vec{P}_{\text{total}}}{dt} = 0, \quad \vec{P}_{\text{total}} \equiv \vec{p}_1 + \vec{p}_2. \tag{4.22}$$

The significance of this proof is that we have used nothing but the translation invariance to find the constants of the motion $\vec{P}_{\text{total}}$. After you have read Chapter 5, you will recognize that this is a special case of Noether's theorem, which is used to find a constant of the motion from an infinitesimal transformation.

Eliminate the Center of Mass Motion

By the reasoning above, we have conservation of the total system momentum, that is,

$$\frac{d\vec{P}_{\text{total}}}{dt} = \frac{d}{dt}[M_1 \vec{r}_1 + M_2 \vec{r}_2] = 0. \tag{4.23}$$

This means that, using the definition of center of mass,

$$\vec{R}_{\text{cm}} \equiv \frac{M_1 \vec{r}_1 + M_2 \vec{r}_2}{M_1 + M_2} \tag{4.24}$$

moves like a free particle (i. e., with constant velocity).

Change variables from $\vec{r}_1, \vec{r}_2$ to $\vec{R}_{\text{cm}}, \vec{r} \equiv \vec{r}_1 - \vec{r}_2$. Solving for the inverse relation gives r_1, r_2 in terms of $\vec{R}_{\text{cm}}, \vec{r}$:

$$\vec{r}_1 = \vec{R}_{\text{cm}} + \frac{M_2}{M} \vec{r}, \quad \vec{r}_2 = \vec{R}_{\text{cm}} - \frac{M_1}{M} \vec{r}. \tag{4.25}$$

(Note that $M = M_1 + M_2$.) Calculate the kinetic energy:

$$T = \frac{1}{2} M \dot{\vec{R}}_{\text{cm}}^2 + \text{vanishing cross term} + \frac{1}{2} \underbrace{\left(\frac{M_1 M_2}{M}\right)}_{\equiv \mu} \dot{\vec{r}}^2. \tag{4.26}$$

4.4 SOLVING THE CENTRAL FORCE PROBLEM

The *reduced mass* μ is defined above. Note that

$$\frac{1}{\mu} = \frac{1}{M_1} + \frac{1}{M_2}, \tag{4.27}$$

$$L = \frac{1}{2} M \dot{\vec{R}}_{cm}^2 + \frac{1}{2} \mu \dot{\vec{r}}^2 - V(r) \equiv L_{cm} + L_{relative}. \tag{4.28}$$

(Remember $V_{12}(|\vec{r}_1 - \vec{r}_2|) \equiv V(r)$.)

QUESTION 6: *Kinetic Energy* Fill in the steps in the derivation of (4.26) and (4.28).

With no external forces, $\vec{R}_{cm}$ are ignorable coordinates, since they do not appear in L. Hence $\dot{\vec{R}}_{cm}$ is constant. We will mentally transform into the center of mass frame* and adopt our new Lagrangian:

$$\boxed{L \equiv L_{relative} = \frac{1}{2} \mu \dot{\vec{r}}^2 - V(r).} \tag{4.29}$$

From this point of view, we can treat the center of mass as being at rest, thus eliminating half of the six degrees of freedom at a single stroke. The ability to do this rests on the fact that we assumed no external forces. However, some special external forces – for instance, a uniform gravitational field – also allow separation of the center of mass and relative motion into two unrelated pieces. A uniform gravitational force on both particles allows this separation because $V_{external} = m_1 g z_1 + m_2 g z_2 = M g Z_{cm}$.

Angular Momentum Is Conserved for Central Forces

If the potential energy of interaction between the two particles depended on the direction of $\vec{r}_1 - \vec{r}_2$, it would not be rotationally invariant. This could only happen if there were a preferred direction in space. Since we've already assumed a central force, which has no preferred direction by definition, we can assume rotational invariance. As a counter example, if our particles had internal degrees of freedom, like a spin $\vec{S}$ or a magnetic moment, rotational invariance would not follow automatically from the assumption of no external forces. (We could have a term in the Lagrangian like $\vec{S} \cdot \vec{r}$. This is invariant only if we simultaneously rotate both the spin and the coordinates together.) But we do not want to consider such complications here. In this case we conclude

$$\text{no external forces} \Leftrightarrow \text{translational invariance}$$
$$\Leftrightarrow V(\vec{r}_1, \vec{r}_2) = V(\vec{r}_1 - \vec{r}_2) \tag{4.30}$$

---

* The center of mass frame is defined as the reference frame in which the velocity of the center of mass is zero.

plus only central force $\Leftrightarrow$ rotational invariance

$$\Leftrightarrow V(\vec{r}_1 - \vec{r}_2) = V(|\vec{r}_1 - \vec{r}_2|) = V(r). \tag{4.31}$$

The term "central force" thus means that the interaction between the particles doesn't depend on either their absolute spatial position or orientation but only on the distance between them.

In Cartesian coordinates, the square of the velocity is

$$\left(\frac{ds}{dt}\right)^2 = \dot{\vec{r}}^2 = \dot{x}^2 + \dot{y}^2 + \dot{z}^2. \tag{4.32}$$

Going over to spherical polar coordinates (r, θ, ϕ), the element of arc length ("line element") is

$$ds^2 = dx^2 + dy^2 + dz^2 = dr^2 + r^2 d\theta^2 + r^2 \sin^2\theta \, d\phi^2. \tag{4.33}$$

We can get the kinetic energy $T = \frac{1}{2}\mu(\frac{ds}{dt})^2$ directly from Equation (4.33). This in turn enables us to write L (4.29) in the new coordinates without worrying about Jacobians:

$$L = \frac{1}{2}\mu(\dot{r}^2 + r^2\dot{\theta}^2 + r^2\dot{\phi}^2 \sin^2\theta) - V(r). \tag{4.34}$$

spherical polar coordinates

We can see from Equation (4.34) that, due to the central force, ϕ does not appear in the Lagrangian so it is an ignorable coordinate:

$$p_\phi \equiv \frac{\partial L}{\partial \dot{\phi}} = \mu r^2 \dot{\phi} \sin^2\theta = \text{constant} \equiv l_z. \tag{4.35}$$

This constant of the motion, l_z, is the z component of the angular momentum, as can be seen from Figure 4.8. The orientation of the coordinate system is completely arbitrary because of the rotational invariance, so we could reorient the system so that the angular momentum $\vec{l}$ has the x, y, z components $(0, 0, l_z)$. Thus, for this orientation,

$$\vec{l} = l_z \hat{k} = \text{constant}, \tag{4.36}$$

where $\hat{k}$ stands for a unit vector along the z axis. Hence rotational invariance implies that all components of the relative angular momentum in the center of mass system are constants. In Chapter 5 we will use Noether's Theorem to reach the same conclusion from a more general viewpoint, as a special case.

4.4 SOLVING THE CENTRAL FORCE PROBLEM

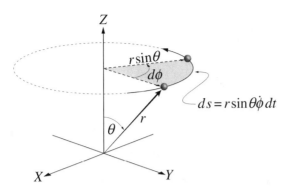

FIGURE 4.8
$\vec{l} = \mu \vec{r} \times \dot{\vec{r}}$. $l_z = \mu r \frac{ds}{dt} \sin\theta = \mu r^2 \sin^2\theta \dot{\phi}$.

QUESTION 7: Constant Angular Momentum 1 Explain how we used the fact that ϕ is ignorable to arrive at the conclusion that the *total* angular momentum is a constant.

QUESTION 8: Constant Angular Momentum 2 An alternate method of proving that the relative angular momentum in the center of mass system is a constant is to show that $\frac{d}{dt}(\vec{r} \times \vec{p})$ is zero, where $\vec{r}$ is the distance between the particles and $\vec{p}$ is defined as $\vec{p} \equiv \mu \dot{\vec{r}}$. Prove that this definition of $\vec{p}$ is consistent with Equation (4.35) and that $\vec{l} = \vec{r} \times \vec{p}$ is constant for the Kepler problem.

There are two consequences of angular momentum conservation:

a) The entire motion takes place in a plane.
b) "Equal areas are swept out in equal times." This is the second of Kepler's Laws. As an empirical observation, the second law proved (with a little hindsight) that the forces between the planets and the Sun are central forces. It has nothing to do with the specific character of gravitational forces, except that they are central forces.

To prove a), we use the following argument: $\vec{l}$ is a constant vector equal to $\vec{r} \times \vec{p}$. This cross product implies that $\vec{p}$ and $\vec{r}$ must be perpendicular to both each other and $\vec{l}$. Thus the motion of $\vec{p}$ and $\vec{r}$ must be in a plane perpendicular to $\vec{l}$, and remain that way for all time, i.e., $\dot{\theta} = 0$. Since l is defined to be along the z direction, this also means that $\theta = \frac{\pi}{2}$.

To prove b), we make use of $\theta = \frac{\pi}{2}$ for all times. As seen in Figure 4.9, the element of area swept out in dt is

$$dA = \frac{1}{2} r^2 \dot{\phi} dt. \tag{4.37}$$

From (4.35) $\mu r^2 \dot{\phi} = l$ is a constant ($\sin\theta = 1$), which proves the assertion b).

Angular momentum conservation has been used to find simple equations for two out of the three remaining degrees of freedom. We have eliminated θ from the problem, and we can find $\phi(t)$ by an integration once $r(t)$ is known. There is still one more equation to find and solve, however.

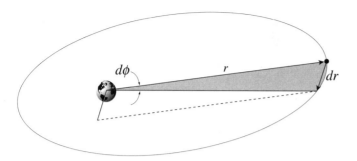

FIGURE 4.9
Note that $dr = r\dot\phi\, dt$. Area of shaded region, $A = \frac{1}{2}$ (area of parallelogram) $= \frac{1}{2} r^2 \dot\phi\, dt$.

The Equivalent 1-D Problem

We wish to use $\vec{l} =$ constant to eliminate all reference to θ, ϕ in our Lagrangian. To summarize, angular momentum conservation has led to the following simplifications:

$$\dot\theta = 0, \qquad \theta = \frac{\pi}{2}, \qquad \dot\phi = \frac{l}{\mu r^2}. \tag{4.38}$$

The order in which you now proceed is important. You can find the energy and eliminate $\dot\phi$ using angular momentum conservation. If instead, you find the Euler–Lagrange equations from the Lagrangian, do not *first* substitute for $\dot\phi$ in the Lagrangian. The result will be a sign error in one of the terms. This is due to the fact that the Euler–Lagrange equations are derived under the assumption of independent variations of each coordinate. They are no longer independent, once Equation (4.38) holds.

Since L (4.34) is a quadratic form in $\dot r, \dot\theta, \dot\phi$ and does not contain explicit reference to the time:

$$H = E = 2T - L = \frac{1}{2}\mu(\dot r^2 + r^2 \dot\theta^2 + r^2 \sin^2\theta\, \dot\phi^2) + V(r). \tag{4.39}$$

Using (4.38), we have

$$\boxed{E = \frac{1}{2}\mu \dot r^2 + \underbrace{\frac{l^2}{2\mu r^2} + V(r)}_{\equiv V_{\text{eff}}}.} \tag{4.40}$$

You may have derived instead an "effective potential" $V(r) - \frac{l^2}{2\mu r^2}$, but this would be due to the sign error caused by the incorrect procedure mentioned above – the minus sign is wrong! The correct procedure will give a plus sign for the "angular momentum barrier" term $\frac{l^2}{2\mu r^2}$. The angular momentum barrier behaves like an additional *repulsive* $\frac{1}{r^3}$ force.

4.5 THE SPECIAL CASE OF GRAVITATIONAL ATTRACTION

We realize from the derivation that the angular momentum barrier comes from the kinetic energy, not from the potential energy. It is responsible for the centrifugal force, which is not a force at all.

From the Lagrangian we can derive the equation of motion for the separation distance r in the final form, where we have eliminated five of the six degrees of freedom we started with:

$$\mu\ddot{r} = \frac{l^2}{\mu r^3} - \frac{dV}{dr}. \qquad (4.41)$$

This is as far as we can go without being more specific about the form of $V(r)$. Once we know $V(r)$, we could always solve this 1-D equation using quadratures (4.4).

QUESTION 9: Kepler Problem Summarize the assumptions needed to obtain Equation (4.41) above.

4.5 THE SPECIAL CASE OF GRAVITATIONAL ATTRACTION

The potential energy between two gravitating bodies is

$$V(r) = -\frac{GM_1M_2}{r}, \qquad (4.42)$$

where G is the gravitational constant. To simplify the notation, define $GM_1M_2 \equiv k$. If we define $\frac{l^2}{2\mu} \equiv \beta$, then

$$V_{\text{eff}}(r) = -\frac{k}{r} + \frac{\beta}{r^2}. \qquad (4.43)$$

We plotted the effective potential versus r in Figure 4.10 for a fixed value of l^2 to get a qualitative picture of the possible types of motion. The effective potential V_{eff} has its minimum value of $-\frac{k^2}{4\beta} = -\frac{k^2\mu}{2l^2}$ at $r_0 = \frac{l^2}{\mu k}$. Five values of the total energy are shown. E_5 is physically impossible, since $T = E - V < 0$. E_4 corresponds to $r = r_0 = $ a constant, so the motion is a circle of this radius.

The actual motion in the plane is determined by both the total energy and the total angular momentum. But knowing the energy alone, we can distinguish qualitatively different types of motion. For E less than E_2 there is bounded motion in r, whereas for $E \geq E_2$, the motion is unbounded, and the two bodies will eventually become infinitely far apart.

QUESTION 10: Effective Potential Why do you think we call the expression in Equation (4.43) V_{eff}? How is it like a potential? How is it not like a potential? Also, when is the attractive term in (4.43) important? When is the repulsive term important?

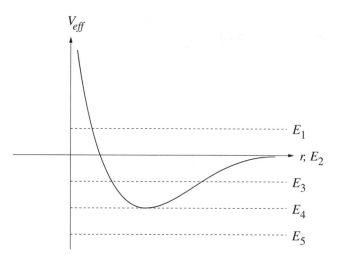

FIGURE 4.10

The equation of motion for gravitational attraction can be derived from the Lagrangian (4.41):

$$\mu \ddot{r} = \frac{l^2}{\mu r^3} - \frac{k}{r^2}. \qquad (4.44)$$

Equation (4.44) says that the inverse-square law attractive force of gravity is opposed by the centrifugal "force," which is repulsive, increasing as the inverse cube of the distance.

The "Magic" $u = \frac{1}{r}$ Transformation

To get a quantitative solution, it is very useful to make the substitution

$$u = \frac{1}{r}. \qquad (4.45)$$

(We could use quadratures instead of the substitution (4.45), but it is more elegant to solve it this way because we can eliminate the time and obtain the equation for the orbit $r(\phi)$ explicitly by solving a simple differential equation.)

Change the dependent variable from r to $u = \frac{1}{r}$. Also convert the independent variable from the time t to ϕ by using (4.38):

$$\mu r^2 d\phi = l\, dt \rightarrow \frac{d}{dt} = \frac{l}{\mu} u^2 \frac{d}{d\phi} \qquad (4.46)$$

and

$$\frac{dr}{dt} = \frac{d(\frac{1}{u})}{dt} = -\frac{1}{u^2}\frac{du}{dt} = -\frac{l}{\mu}\frac{du}{d\phi}. \qquad (4.47)$$

This is not something you would necessarily have thought of doing, but it turns out that solving for the orbit or path of the system is simpler than solving for the time dependence of $r(t)$, $\phi(t)$. Go back to the expression for E (4.40), and express everything in terms of u, $\frac{du}{d\phi}$:

$$E = \frac{l^2}{2\mu}\left[\left(\frac{du}{d\phi}\right)^2 + u^2\right] - ku. \tag{4.48}$$

total energy $E(u, \frac{du}{d\phi})$

Since E is constant, $\frac{dE}{d\phi} = 0$. Carry out the differentiation on the expression (4.48) above. This gives a differential equation for the path in terms of u as the dependent variable and with ϕ as the independent variable. The time is now eliminated:

$$\frac{d^2u}{d\phi^2} + u = \frac{\mu k}{l^2}. \tag{4.49}$$

differential equation for the orbit

Equation (4.49) is that of a simple harmonic oscillator with a constant driving force. Since the coefficient of u on the left side of the equation is 1, the period is 2π, that is, the orbit is a closed one: $u(\phi + 2\pi) = u(\phi)$. Closure of the orbit in space is a very special property of the $\frac{1}{r}$ potential, which is shared by only one other type of force law: a force that increases linearly with r.

4.6 INTERPRETATION OF ORBITS

Equation (4.49) can be solved by inspection. The solution is

$$u = \frac{1}{r} = \frac{\mu k}{l^2} + A\cos\phi. \tag{4.50}$$

(An arbitrary phase could be added to ϕ. This would rotate the orbit in the plane of motion by a fixed amount.) Here A is an arbitrary constant related to the total energy. Let us change to some new parameters and rewrite the solution (4.50) to Equation (4.49):

$$pu = \frac{p}{r} = 1 + \epsilon\cos\phi, \quad p \equiv \frac{l^2}{\mu k}, \quad \epsilon \equiv pA. \tag{4.51}$$

The constant p has the dimensions of length. It determines the actual size of the orbit.

Rewrite Equation (4.51) in Cartesian coordinates:

$$p = r + \epsilon r \cos\phi = \sqrt{x^2 + y^2} + \epsilon x \tag{4.52}$$

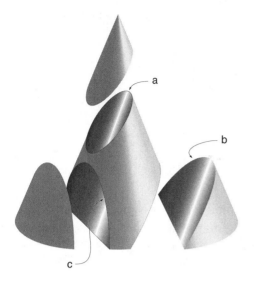

FIGURE 4.11
Conic sections: a) ellipse, b) parabola, c) hyperbola.

or, equivalently,

$$(1 - \epsilon^2)x^2 + 2\epsilon p x + y^2 - p^2 = 0. \tag{4.53}$$

If $\epsilon = 0$, the orbit is a circle of radius p. As we saw in Figure 4.10, the circular orbit has the minimum possible total energy for a system of a given angular momentum. If $0 < \epsilon < 1$, the planet moves in an elliptical orbit around the center of mass (Kepler's First Law). If $\epsilon = 1$, the path is a parabola, whereas if $\epsilon > 1$, the path is a hyperbola, with the two bodies not bound to each other.

For an elliptical orbit, ϵ is the *eccentricity* of the ellipse. In every case, the orbit is described by the curve obtained by the intersection of a surface of a cone with a plane.

QUESTION 11: *Orbits 1* Prove that, if $\epsilon = 1$, the orbit is a parabola. Prove that the planet's motion is not bounded if $\epsilon \geq 1$.

QUESTION 12: *Orbits 2* What is meant by a "closed" orbit? Explain why an orbital period for $r(\phi)$ of 2π means the orbit is closed. What if the period were $2\pi \frac{p}{q}$, with p and q integers? Describe the general features of the orbit if this were the case.

Substituting Equation (4.51) into the expression (4.48) for the energy (and defining $a \equiv \frac{p}{1-\epsilon^2}$), we get

$$E = \frac{\mu k^2}{2l^2}(\epsilon^2 - 1) = \frac{k}{2p}(\epsilon^2 - 1), \tag{4.54}$$

$$E = -\frac{k}{2a}. \tag{4.55}$$

4.6 INTERPRETATION OF ORBITS

TABLE 4.2 SUMMARY OF VARIOUS POSSIBLE TYPES OF MOTION DUE TO GRAVITATIONAL ATTRACTION

Energy E in units of $\frac{\mu k^2}{2l^2}$	Type of Orbit
$E < -1$	impossible
$E = -1$	circular orbit
$-1 < E < 0$	elliptical
$E = 0$	parabolic, not bound
$E > 0$	hyperbolic, not bound

For elliptical orbits, the quantity $a \equiv \frac{p}{1-\epsilon^2}$ is the semimajor axis. (This is discussed in the next section.) The quantity a is defined in the same way for unbounded orbits as well, where it becomes negative. Equation (4.55) is equivalent to (4.54), but it shows explicitly that the total energy depends only on a. Negative total energy corresponds to bounded motion. Positive total energy corresponds to unbounded motion. This makes physical sense, since the planet can travel arbitrarily far from the sun if the motion is unbounded. At a sufficiently great distance, the potential energy becomes negligible, and therefore the total energy, which consists only of kinetic energy, must be positive. For $E \geq 0$, two masses have a "close encounter," and then fly apart, never to meet again.

Elliptical Orbits

Starting from Equation (4.53), assume that $0 \leq \epsilon < 1$. After completing the square, you get the equation of an ellipse:

$$(1 - \epsilon^2)\left(x + \frac{\epsilon p}{1 - \epsilon^2}\right)^2 + y^2 = \frac{p^2}{1 - \epsilon^2}, \quad (4.56)$$

which has the form

$$\frac{(x - x_c)^2}{a^2} + \frac{y^2}{b^2} = 1. \quad (4.57)$$

Equation (4.57) is the standard form for an ellipse, centered at $x_c = -\frac{\epsilon p}{1-\epsilon^2}$, with semimajor axis $a = \frac{p}{1-\epsilon^2}$, and semiminor axis $b = \frac{p}{\sqrt{1-\epsilon^2}}$. The "apsides" or turning points, are defined as the values of r for which $\frac{dr}{dt} = \frac{dr}{d\phi}\frac{d\phi}{dt} = 0$. For motion about the sun, $r_{\max}$ is known as the "aphelion" and $r_{\min}$ is known as the "perihelion."

Period of Elliptical Motion

How long does it take a planet to go around the Sun? We could use our general formula (4.4) to find the period τ of the motion. This would involve an integral between

TABLE 4.3 IMPORTANT QUANTITIES FOR ELLIPTICAL ORBITS*

Quantity	Symbol	Formula
eccentricity	ϵ	$0 \leq \epsilon < 1$
scale factor	p	$\frac{l^2}{\mu k}$
semimajor axis	a	$\frac{p}{1-\epsilon^2}$
semiminor axis	b	$\frac{p}{\sqrt{1-\epsilon^2}}$
ellipse center	x_c	$-\frac{\epsilon p}{1-\epsilon^2}$
apsides (turning points)		
	$r_{\min}$	$\frac{p}{1+\epsilon}$
	$r_{\max}$	$\frac{p}{1-\epsilon}$
total energy	E	$\frac{\mu k^2}{2l^2}(\epsilon^2 - 1) < 0$

*$k = GM_1M_2$, l = total angular momentum, μ = the reduced mass. The numerical value of $G = 6.672(6) \times 10^{-11}$ N m^2/kg^2.

the apsides:

$$\frac{\tau}{2} = \sqrt{\frac{\mu}{2}} \int_{r_{\min}}^{r_{\max}} \frac{dr}{\sqrt{E - V(r) - \frac{l^2}{2\mu r^2}}}. \tag{4.58}$$

This expression can be used to find the period for any central force. In our case, $V(r) = -\frac{k}{r}$. The integral looks rather formidable, though it can be done. However, it can be greatly simplified by a change of variables. (See the discussion in the next section.)

Fortunately, there is an easier way to obtain the period. Since we know the semimajor and semiminor axes of the ellipse, we can calculate the area. This information can be used to find the period directly from Kepler's Second Law, and thus deduce the Third Law. Since $\frac{dA}{dt} = \frac{l}{2\mu}$ = constant, it follows that the period is directly given in terms of the area swept out by the ellipse, πab times $\frac{2\mu}{l}$. Using the expressions for a and b from Table 4.3, we have

$$\tau = \frac{2\mu}{l}\pi ab = \frac{2\mu}{l}\frac{\pi p^2}{(1-\epsilon^2)^{\frac{3}{2}}}. \tag{4.59}$$

Since the semimajor axis $a = \frac{p}{1-\epsilon^2}$,

$$\tau = 2\pi\frac{\mu}{l}\sqrt{pa^3} = 2\pi\sqrt{\frac{\mu}{k}}\sqrt{a^3}. \tag{4.60}$$

Finally, putting in the value of the scale factor $p = \frac{l^2}{\mu k}$, we obtain the period in terms of the semimajor axis and some constants. Notice that the answer does not depend on l. In fact the period depends only on the semimajor axis of the elliptical orbit. Furthermore, since

4.6 INTERPRETATION OF ORBITS

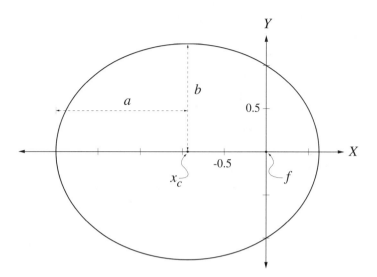

FIGURE 4.12
Planetary orbit with $p = 1$ and $\epsilon = 0.6$. Both the planet and the Sun orbit around the center of mass at the focus f of the ellipse, which is the origin of the coordinates in the figure above. The Sun's orbit is not shown here. The center of the ellipse is at $x_c = -.9375$. The semimajor axis is $a = 1.5625$ and the semiminor axis is $b = 1.25$. The apsides or turning points are at $x = -2.5$ and 0.625.

the Sun is much heavier than any of the planets, μ will be almost the same for every planet:

$$\boxed{\tau = 2\pi\sqrt{\frac{\mu}{k}}\, a^{\frac{3}{2}}.}$$ (4.61)

orbital period $\tau(a)$

Kepler's Third Law says that $\tau^2 \sim a^3$ *with a proportionality constant which is the same for all planets*. We now examine whether this is in fact true. Recall that $k = GM_1 M_2$ and $\mu = \frac{M_1 M_2}{M_1 + M_2}$. Therefore

$$\frac{k}{\mu} = G(M_1 + M_2) \approx GM_2 \quad \text{if } M_2 \gg M_1.$$ (4.62)

For most cases, the mass of the planet, M_1, is negligible compared to the Sun. (The Sun's mass is 3.3×10^5 Earth masses.) But for Jupiter the ratio of masses is more like 1,000:1, so Kepler's Third Law must be corrected by about 0.1% in this case. One wonders what would have happened had Kepler's measurements been accurate enough to detect this. Would he have published his Third Law?

Time Dependence of the Motion

How could we construct the orbit of a comet using only a few observation points of its position as a function of time? When Sir Edmond Halley observed in 1682 what later became known as Halley's comet, he did not have the luxury of waiting 76 years to determine the period. Instead, in 1705, following a graphical method developed earlier by Newton, he reconstructed the complete orbit of this comet from limited data taken over a relatively brief time period. Halley's comet reappeared on Christmas Day, 1758, exactly as he had predicted. Later, in 1801, the young Gauss became famous when he reconstructed the orbit of the "lost" asteroid Ceres from only three observations. Ceres was observed for only a month before it disappeared behind the Sun. It reappeared on New Year's Day, 1802, exactly where Gauss had predicted.

Even after solving the problem of converting earth-bound measurements of the comet/asteroid position against the fixed stars into r and ϕ expressed in the inertial frame of the center of mass, one must then fit the data to the equations for $r(t)$ and $\phi(t)$ as functions of time. This avoids directly solving the equation of motion (4.44) for $r(t)$ – although we will use the method for 1-D systems derived at the beginning of this chapter to solve for T, the elapsed time. First we would like to describe a more geometric way to treat the problem. Consider a body in its elliptical orbit about the Sun and draw the circle that circumscribes the ellipse, as shown in Figure 4.13. The radius of the outer circle is a, the semimajor axis of the ellipse. As indicated in Figure 4.13, drop the perpendicular from the body orbiting the Sun to the point A on the horizontal line along the major axis of the ellipse. The angle $\mathcal{E}$ is called the *eccentric anomaly*. The azimuthal angle ϕ is the *true anomaly*. From the geometry of Figure 4.13, the distance OA is

$$OA = a\cos\mathcal{E}. \tag{4.63}$$

The distance $|x_c|$ from the ellipse center O to the Sun is $a\epsilon$. (Look at Equation (4.57) and remember that the Sun is about at the focus of the ellipse.) Therefore the distance from the point A to the Sun is

$$a(\epsilon - \cos\mathcal{E}) = r\cos(\pi - \phi) = -r\cos\phi. \tag{4.64}$$

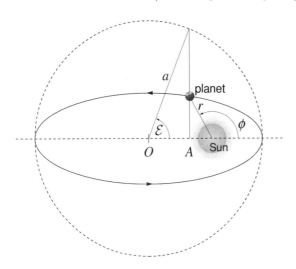

FIGURE 4.13
Construction of the eccentric anomaly.

4.6 INTERPRETATION OF ORBITS

Solve this equation for $\cos\phi$ and put it back into the orbit equation, (4.51). By straightforward algebra we get

$$r = a(1 - \epsilon \cos \mathcal{E}). \tag{4.65}$$

It is a short step from here to the equations for $x(\mathcal{E}) = r\cos\phi$ and $y(\mathcal{E}) = \sqrt{r^2 - x^2}$:

$$\boxed{x = a(\cos\mathcal{E} - \epsilon), \quad y = a\sqrt{1 - \epsilon^2}\sin\mathcal{E}.} \tag{4.66}$$

Next write an expression for the time in terms of an indefinite integral using (4.58):

$$t = \sqrt{\frac{\mu}{2}} \int \frac{dr}{\sqrt{E + \frac{k}{r} - \frac{l^2}{2\mu r^2}}}. \tag{4.67}$$

To simplify the integral over r, note that for elliptic orbits $E = -\frac{k}{2a}$ and make use of the substitutions $\frac{l^2}{\mu k} \equiv p$ and $p = a(1-\epsilon^2)$. By factoring constants out from the integral, obtain

$$\int dt = \sqrt{\frac{\mu}{2k}} \int \frac{r\,dr}{\sqrt{-\frac{r^2}{2a} + r - \frac{1}{2}a(1 - \epsilon^2)}}. \tag{4.68}$$

Finally, substitute for r from Equation (4.65) to put everything in terms of $\mathcal{E}$, the eccentric anomaly. The denominator of the r integral becomes $\sqrt{\frac{a}{2}}\epsilon\sin\mathcal{E}$. Write $dr = a\epsilon\sin\mathcal{E}\,d\mathcal{E}$ by differentiating Equation (4.65) and also substitute for r in terms of $\mathcal{E}$ in the numerator. The result is

$$\int dt = \sqrt{\frac{\mu a^3}{k}} \int d\mathcal{E}\,(1 - \epsilon \cos \mathcal{E}). \tag{4.69}$$

The time integral measures elapsed time. It is customary to choose $t = 0$ when the planet or orbiting body is at its perihelion, so $\mathcal{E} = 0$ there. (See Figure 4.13.) If we call T the elapsed time to any other point on the orbit, we have derived the final result:

$$\boxed{T(\mathcal{E}) = \sqrt{\frac{\mu a^3}{k}}(\mathcal{E} - \epsilon \sin \mathcal{E}).} \tag{4.70}$$

Kepler's equation

Given the observed time interval T from the perihelion, the transcendental Equation (4.70) must be solved to obtain $\mathcal{E}(T)$. From Equation (4.65) the distance $r(\mathcal{E}(T))$ can then be determined. Finally, $\phi(\mathcal{E}(T))$ can be determined from the original orbit Equation (4.51), which is used to give $\cos\phi$ in terms of r. To follow this procedure, one must know the constants a and ϵ.

If the orbit period and eccentricity are unknown, they can be determined by observations of position and time elapsed from the perihelion. From a minimum of two such observations, knowing $r(t_1)$ and also $r(t_2)$, the eccentric anomaly equation for the time, (4.70), and the Equation (4.65) can be used to solve for the two unknown constants: a, the semimajor axis, and ϵ, the eccentricity of the ellipse. After that, everything about the orbit is known. The true anomaly ϕ as well as the eccentric anomaly $\mathcal{E}$ each increase by 2π in one complete period around the ellipse.

Hyperbolic Orbits: "Close Encounters"

If the total energy E is positive, there is still some kinetic energy when the two bodies are infinitely far apart, since at sufficiently large distances the potential energy becomes negligible. The unbounded orbit is one branch of a hyperbola. We could have something like a comet that appears once and is gone. (All known comets are actually on periodic elliptical orbits around the Sun, so this is a purely hypothetical case.)

Positive energy implies that the constant $\epsilon > 1$ (see Equation (4.54)). Equation (4.52) now represents a hyperbolic orbit. Equation (4.53) can again be invoked and rewritten as

$$(\epsilon^2 - 1)x^2 - 2\epsilon px - y^2 + p^2 = 0. \tag{4.71}$$

Completing the square we get

$$(\epsilon^2 - 1)\left(x - \frac{\epsilon p}{\epsilon^2 - 1}\right)^2 - y^2 = \frac{p^2}{\epsilon^2 - 1}. \tag{4.72}$$

The general equation for a hyperbola is

$$\frac{(x - x_c)^2}{a^2} - \frac{y^2}{b^2} = 1. \tag{4.73}$$

Hyperbolic orbits are the only possible solutions for the Kepler problem if the force is repulsive rather than attractive. This situation arises in the scattering of positively charged alpha particles from positively charged heavy nuclei.

In Figure 4.14, the left branch of the hyperbola corresponds to an attractive gravitational force. The curve passes "behind" the center of the force, starting (say) from negative x, y and emerging towards positive y and negative x. The other branch of the hyperbola arises because we have to square the orbital equation (4.52) to get the general form (4.72) for a hyperbola. It is unphysical for an attractive force but is the physical solution for a *repulsive* $\frac{1}{r^2}$ force, as we will see below. It can be seen that the curve is symmetric for positive and

4.7 REPULSIVE $\frac{1}{r^2}$ FORCES

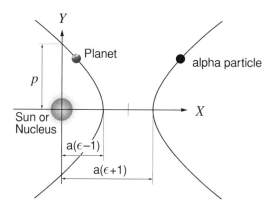

FIGURE 4.14
Two kinds of hyperbolic orbits. The planet and Sun model an attractive force whereas the alpha particle and nucleus model a repulsive force.

negative y. The minimum distance of approach to the origin occurs for $y = 0$, and for the physical branch of the hyperbola is $r_{\min} = \frac{p}{\epsilon+1} = a(\epsilon - 1)$. (For hyperbolic orbits, by definition, $p \equiv \frac{l^2}{\mu|k|}, a \equiv \frac{p}{\epsilon^2-1}$.) As $x \to \infty$, the upper and lower parts of the hyperbola asymptotically approach the lines $y = \pm \frac{b}{a}(x - x_e) = \pm\sqrt{\epsilon^2 - 1}(x - \frac{\epsilon p}{\epsilon^2-1})$.

Since $a \equiv \frac{p}{\epsilon^2-1}$ now, the total energy can be expressed as $E = \frac{|k|}{2a} > 0$. The orbit no longer represents periodic motion, but rather a one time event. We will discuss the question of the time dependence of the motion on hyperbolic orbits after discussing the case of repulsive forces.

4.7 REPULSIVE $\frac{1}{r^2}$ FORCES

The gravitational force is always attractive, but a repulsive $\frac{1}{r^2}$ force can occur between two positively charged nuclei, such as in the Rutherford scattering of alpha particles from gold. Once the incident alpha particle penetrates the cloud of orbital electrons, it "sees" a *repulsive* $\frac{1}{r}$ potential. This experiment established the existence of positive charge concentrated in a small region at the center of the atom (i. e., the existence of the atomic nucleus). In his experiment, Rutherford used the distribution of backscattered alpha particles to establish that there was a point charge of magnitude $+Z|e|$, where Z is the number of atomic electrons and $-|e|$ is the electronic charge. The nearly point charge of the nucleus exerts just the type of force predicted by Coulomb's Law: a $\frac{1}{r^2}$ repulsive force. In the repulsive force case, the orbits are only hyperbolic, since both the potential energy and the kinetic energy are positive.

A repulsive potential of this form is called a "Coulomb potential." The sign of k changes in the expression for the potential. To find the solution, we can use our previous mathematical work, with this change of sign. The equation for $u(\phi)$, Equation (4.49), becomes

$$\frac{d^2u}{d\phi^2} + u = -\frac{1}{p}. \tag{4.74}$$

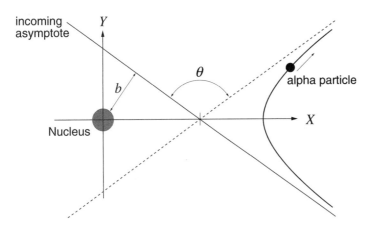

FIGURE 4.15
Rutherford scattering (repulsive force). b is the impact parameter, and θ is the scattering angle.

The definition of the length scale is $p \equiv \frac{l^2}{\mu|k|}$. Expressed in polar coordinates, with $\epsilon > 0$ by definition, the solution to (4.74) is

$$\frac{p}{r} = -1 + \epsilon \cos \phi. \tag{4.75}$$

Since both p and r are positive, no solutions exist unless $\epsilon > 1$, which gives positive total energy. In Cartesian coordinates, the equation to be solved for the orbit resembles (4.52) but the sign of the square root is different:

$$p = -\sqrt{x^2 + y^2} + \epsilon x. \tag{4.76}$$

The physical hyperbola is now the right-hand branch in Figure 4.14. After squaring, we obtain the same equation as (4.71) in the previous section. We can take over our conclusions wholesale if we select the opposite branch of the hyperbola to the one selected in the case of an attractive force. The distance of closest approach can be found by solving for the root of (4.71) with $y = 0$, and then checking which is the right root by looking at (4.76). This procedure reveals that the distance of closest approach to the force center is

$$r_{\min} = \frac{p}{\epsilon - 1} = a(\epsilon + 1). \tag{4.77}$$

If θ is the "scattering angle" (see Figure 4.15), $\tan(\frac{\pi - \theta}{2}) = \sqrt{\epsilon^2 - 1}$, or with the help of trigonometric identities:

$$\sin \frac{\theta}{2} = \frac{1}{\epsilon}. \tag{4.78}$$

Time Dependence of Hyperbolic Orbits: Attractive Case

We would like to discover the time dependence of the hyperbolic orbits, the analog to Equations (4.66, 4.70), which were derived only for elliptical orbits. To do this, we can make a transformation of the equation of motion (4.44). This transformation can then be used to generate new solutions of the equation of motion by transforming the previously obtained solutions. This is a common and very useful technique in theoretical physics. All that is needed is to substitute it' ($i = \sqrt{-1}$) for the time in the equations of motion (4.44). The angular momentum, $\mu r^2 \dot{\phi}$, also undergoes the transformation $l \to -il$, so l^2 reverses sign as does $\ddot{r}$. If the additional transformation $k \to -k$ is made, we return to the original equation (4.44) with an attractive force again. The transformation $t \to it'$, $k \to -k'$ leaves the equation of motion invariant but transforms elliptical orbits into imaginary hyperbolic ones. The length scale $p \equiv \frac{l^2}{\mu k}$ remains positive and is unchanged, but $\frac{\mu k^2}{2l^2}$ does change sign, so we will have transformed negative energy elliptical orbits into positive energy hyperbolic orbits.

This transformation is then used to parametrize the time dependence of the hyperbolic orbits. The result of transforming Kepler's equation (4.70) is given below:*

$$T(\mathcal{E}) = \sqrt{\frac{\mu a^3}{k}}(\epsilon \sinh \mathcal{E} - \mathcal{E}), \tag{4.79}$$

$$r = a(\epsilon \cosh \mathcal{E} - 1), \tag{4.80}$$

$$x = a(\epsilon - \cosh \mathcal{E}), \quad y = a\sqrt{\epsilon^2 - 1} \sinh \mathcal{E}. \tag{4.81}$$

As was previously done for hyperbolic orbits, $a \equiv \frac{p}{\epsilon^2 - 1}$, the constant $\epsilon > 1$, and the energy (4.54) is positive, $E = \frac{k}{2a}$. $\mathcal{E}$ is now a parameter that equals zero at the point of closest approach. This is the standard way to describe the coordinates of the attractive force hyperbolic orbit parametrically as implicit functions of time.

Time Dependence of the Repulsive Hyperbolic Orbit

We merely quote the results here, and leave it to the dedicated reader to verify the equations below. The definitions of the constants a, ϵ remain the same as for the attractive hyperbolic case. The equations are

$$T = \sqrt{\frac{\mu a^3}{k}}(\epsilon \sinh \mathcal{E} + \mathcal{E}), \tag{4.82}$$

$$r = a(\epsilon \cosh \mathcal{E} + 1), \tag{4.83}$$

$$x = a(\cosh \mathcal{E} + \epsilon), \quad y = a\sqrt{\epsilon^2 - 1} \sinh \mathcal{E}. \tag{4.84}$$

---
* Here we have used the mathematical identity $\sin i\alpha = i \sinh \alpha$ for any real α.

The transformation $a \to -a$, $\epsilon \to -\epsilon$ will convert (4.79, 4.80, 4.81) into (4.82, 4.83, 4.84), but we attach no special significance to this.

Precessing Orbits

Suppose the central force gives a potential that is not proportional to $\frac{1}{r}$. It could be $\frac{1}{r^n}$, $n \neq 1$, for example. *Bertrand's Theorem* shows that the *only* values of n giving orbits that close on themselves in space, so maintain the path indefinitely, are $n = 1$ (Kepler problem) and $n = -2$ (Hooke's Law for a space oscillator). Any other form of attractive force law results in an orbit that never repeats itself exactly, that is, it precesses in space. This precession is one of the most sensitive tests of the radial dependence of the force between two gravitating bodies. For example, precession of the perihelion of Mercury is one of the major tests of Einstein's General Theory of Relativity. A proof of Bertrand's Theorem can be found in the Appendix A, of Goldstein's *Classical Mechanics*.

Dark Matter

There is a major puzzle in physics. Spiral galaxies consist of billions of stars rotating about a center. Astronomers are able to study the motion of the stars in a small part of one of these galaxies. They can determine the velocities of these stars by studying the Doppler shifts of the spectral lines in the light emitted. Using experimental data, they can construct a curve of $v(r)$, where v is the star velocity and r the distance from the galactic center. The problem is that the motion is not what one would expect to find if the mass distribution is the same as that of the visible matter. Although the true situation is actually rather complicated, it will not change the basic mystery if we make the simplifying assumption that the galaxy is a continuous distribution of matter, with a density $\rho(r)$. This density, one could assume, would be large only where the luminous matter of the galaxy is located, and negligibly small outside the galaxy. Let the radius r_0 be the outer radius of this matter. We will also assume circular star orbits about the gravitating mass between the star and the galactic center. Remember that it is a property of the inverse-square force law, assuming spherical symmetry of the mass distribution, that only the mass inside the sphere from the center out to the star contributes to the force on the star. By our assumptions, the mass $M(r)$ inside the radius r is

$$M(r) = 4\pi \int_0^r \rho(r')r'^2 \, dr' = \frac{4\pi r^3}{3}\rho \quad \text{(only if } \rho \text{ is constant)}. \tag{4.85}$$

The force divided by the mass of the star equals the radial acceleration, so

$$\frac{v^2}{r} = \frac{GM(r)}{r^2} = \frac{4\pi G\rho r}{3}. \tag{4.86}$$

For radii less than r_0, $v(r) \sim r$. Outside the radius r_0, the mass becomes a constant, $M(r_0)$, so that $v(r)$ for stars in the galactic "halo" should fall like $\frac{1}{\sqrt{r}}$.

4.7 REPULSIVE $\frac{1}{r^2}$ FORCES

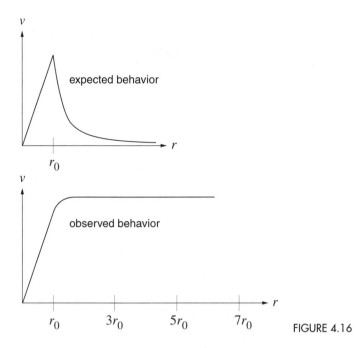

FIGURE 4.16

For constant density, we expect v to rise linearly with r up to the radius of the galaxy, r_0. This rise is observed experimentally, and r_0 found in this way agrees pretty well with r_0 found by just measuring the distribution of luminous (i. e., visible) matter. If $\rho \approx 0$ for $r > r_0$, we would expect $v(r)$ to *fall* like $\frac{1}{\sqrt{r}}$. Now comes the surprise: Nothing like that is observed. Instead, $v(r)$ remains rather constant out to 8 or 9 galactic radii r_0, as shown in Figure 4.16. Many galaxies have been measured, and they all show the same effect.

The estimated additional matter needed to account for this turns out to be 90% of the total matter in the universe! Only 10% of the matter is visible. What could be the source of the additional matter? There are many hypotheses, none of which is completely satisfactory. One hypothesis is that there are many small black holes, by their nature invisible, but contributing strong gravitational fields. A different idea is that there are "brown dwarfs" – Jupiter-sized burnt-out stars,* which are invisible, due to the great distance between us and these other galaxies. A more exotic theory is that we live in a "sea" of neutrinos. This must to some extent be true, since a neutrino once emitted in a radioactive decay cannot easily be reabsorbed. The Sun, for example, is known to be a source of neutrinos. But the gravitational effects of this primordial neutrino sea won't be great enough, unless the neutrino has a small mass. A more radical idea is that dark matter is a sea of "axions" – a so far mythical particle with very low mass, which was suggested for unrelated theoretical reasons. No axions have ever been detected, despite vigorous experimental attempts to discover them, attempts that are still being made. The nature of the "dark matter" remains a mystery.

A recent development has been the discovery by the Hubble telescope of an unexpectedly large black hole with a mass of three billion suns at the center of galaxy M87. Black

---

* This is a special case of MACHOS: Massive Compact Halo Objects. Some recent evidence for MACHOS may have been discovered using gravitational lensing.

holes may in fact exist at the center of most galaxies, including our own. Although of great importance to astrophysics, black holes at the center of galaxies would not resolve the problem of dark matter, because the radial distribution of stellar velocities is consistent only with distributed dark matter.

For further reading on this subject, one reference is *Modern Cosmology and the Dark Matter Problem*, by D. W. Sciama, Cambridge Univ. Press, 1993. In the first part of the book, the author summarizes the evidence for dark matter. In the last part of the book he proposes a speculative theory that involves decay of heavy neutrinos, with a mass about .006% of the electron mass. There is no experimental evidence for the existence of neutrinos of such a mass, but that does not diminish the value of the discussion in the first part of this book, which can serve as an introduction to those who may become seriously interested in the dark matter problem.

SUMMARY OF CHAPTER 4

- For a one-dimensional, scleronomic, conservative problem with Lagrangian $L(q, \dot{q})$, graph the potential $V(q)$ versus q. Plot E as a horizontal line and look for turning points as well as stable and unstable equilibrium points. This gives a qualitative picture of the motion. To get a quantitative solution, calculate the time $t(q)$ as a function of q and invert the function to get $q(t)$, if possible.

> General solution of 1-D potential, with total energy E:
> $$t(q) = \int_0^q \frac{dq'}{\sqrt{2(E - V(q'))}}. \tag{4.87}$$

- Phase space ($\dot{q}$ versus q) portraits can give good qualitative information about the 1-D motion. Plot the curves of constant total energy.
- For a central force $V = V(r)$ between two bodies, the center of mass moves like a free particle with mass $M = M_1 + M_2$. This is a consequence of translation invariance.
- See Table 4.4 for a review of the two-body problem.

PROBLEMS

Phase Portraits

Problem 1*: *(Sketching and interpreting various phase portraits)* Trajectories in the phase space $(q, \dot{q})$ are the contours of constant E.

a) Draw a phase portrait for unstable motion:
$$\ddot{q} - q = 0. \tag{4.88}$$

What quantity is a constant of the motion in this case? Is energy still conserved?

PROBLEMS

TABLE 4.4 REVIEW OF STEPS TO SOLVE THE TWO-BODY PROBLEM

Assumption	Result
1. No external forces exist	1. Center of mass moves freely, 6 DOF → 3 DOF
2. Force is a central force: $V = V(r)$	2. Angular momentum is constant, ⇒ motion lies in a plane, 3 DOF → 1 DOF
3. Force is gravity: $V = -\frac{k}{r}$	3. Can solve for elliptical, parabolic, or hyperbolic orbits
4. Orbit is stationary in space	4. ⇔ Bertrand's Theorem: Force is either $\frac{1}{r^2}$ or r

b) Look at the nonlinear potential $V(q)$ in Figure 4.1. Indicate the region of oscillatory (bounded) motion on a sketch of this potential. Draw the phase portrait for this potential and indicate the direction of "flow" along the trajectories in phase space. Also indicate:
  1) The stable and unstable equilibrium points.
  2) The direction of motion along one of the phase space trajectories.

Problem 2: *(Double-well potential)* Discuss what types of motion are possible for the "double-well" potential depicted in Figure 4.17. Draw a "phase portrait" for this potential.

Problem 3*: *(SHO and physical pendulum)*

a) Why can we say that the simple harmonic oscillator is always bounded but that the physical pendulum is not (for the latter, $V(q) = (1 - \cos\theta)$)? What is the difference? Use the phase portraits to show what the difference is.

b) Make a "phase portrait" of the physical (nonlinear) pendulum for the three cases $E < 2$, $E = 2$, and $E > 2$. In the vicinity of $E = 2$ show how these curves change

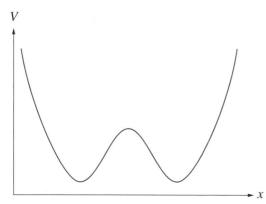

FIGURE 4.17

near the unstable equilibrium point. Explain how you can use these curves to discuss motion near the equilibrium at $\theta = \pi, \dot\theta = 0$ for $E \geq 2$.

Problem 4*: *(Duffing and other nonlinear oscillators)* A Duffing oscillator (with EOM $\ddot{q} + q + \epsilon q^3 = 0$) is a nonlinear oscillator with a "hard" potential, that is, $V(q)$ increases more rapidly than a parabola. Approximate the potential by

$$V(q) = \frac{q^2}{2} + \frac{\epsilon}{4} q^4, \tag{4.89}$$

where $\epsilon > 0$ is a positive constant.

a) For the Duffing oscillator, use a phase portrait to prove that the motion is always bounded. Why can't we use the same argument to reach the false conclusion that physical pendulum motion is always bounded?

b) Suppose instead that $\epsilon < 0$. Discuss the equilibrium points and the phase portrait for this "soft" potential.

c) Now answer the same questions for $V(q) = -\frac{q^2}{2} + \frac{\epsilon}{4} q^4$.

Generic 1-D Problems

Problem 5: *(Grandfather's clock)* A grandfather's clock is adjusted to keep perfect time if the amplitude is $2°$. How many seconds does it lose or gain in a year if instead the amplitude is $10°$?

Problem 6: *(Period of Duffing oscillator)* Use an approximate form of the quadrature solution to find the period of a Duffing oscillator with potential (4.89) as a function of its maximum amplitude q_{max}. Prove that this is given by the formula (dropping second-order terms in ϵ)

$$T = 2\pi \left(1 - \frac{3\epsilon}{32} q_{max}^2\right). \tag{4.90}$$

Why would it be inconsistent to keep terms $O[\epsilon^2]$? For the Duffing oscillator, or any nonlinear oscillator, we could use measurements of the period versus amplitude to determine the value of ϵ. The actual motion will be more complicated than a pure sine wave (i. e., it will contain higher harmonics as well).

Problem 7*: *(Restoring force of q^n)* A certain mechanical system with one degree of freedom has the equation of motion

$$\ddot{q} = -q^3. \tag{4.91}$$

Suppose that the maximum amplitude of the oscillator is known to be q_0. Find an expression for the time it takes to go from $q = 0$ to $q = q_0$ and show that this time is inversely proportional to q_0. Find the coefficient A in the formula $T = \frac{A}{q_0}$ by numerical or other means. Finally, generalize this result to a restoring force of the form q^n (EOM: $\ddot{q} = -q^n$).

Gravity

Problem 8*: *(Potential energy of a spherical shell)* Consider a spherical shell of mass M and radius R. Calculate the gravitational potential for a point mass m_0 in the gravitational field of the spherical shell. The point mass is located at $\vec{r}$. Find the potential energy by using spherical polar coordinates and adding up the contributions to $V(\vec{r})$ from infinitesimally small pieces of the spherical shell. Consider two cases (from the same integral):

a) Suppose $r < R$. The point test mass is *inside* the shell. Show that

$$V = -G\frac{M}{R} = \text{constant.} \quad (4.92)$$

What is the force of gravity on the test mass in this case?

b) Instead, suppose $r > R$. Show that

$$V(r) = -G\frac{M}{r}. \quad (4.93)$$

Again, what is the force on the test mass?

c) What happens when $r = R$? Plot the potential energy vesus r for all r. Plot the force. Is there a discontinuity in the force? In the potential energy?

d) There is an analogy here to Gauss's Law in electrostatics. What is it?

Problem 9*: *(Potential energy of the Earth)*

a) A satellite orbits a spherically symmetric Earth at a radius $r = R_e$ (i.e., just above the tree tops). (R_e is the Earth's radius.) Use elementary Newtonian physics arguments to show that the period of this satellite is the same as that of a pendulum of length R_e undergoing small oscillations in a uniform gravitational field of acceleration g.

b) The gravitational potential energy of a particle of mass m in a hole inside a spherically symmetric Earth of uniform density is given by the formula

$$V(r) = \frac{GM_e m}{2R_e}\left\{\left(\frac{r}{R_e}\right)^2 - 3\right\}, \quad (4.94)$$

where r is the distance to the center of the Earth ($r \leq R_e$), G is the gravitational constant, and M_e is the total mass of the Earth. Prove that this formula for $V(r)$ gives the correct force on a particle inside the Earth at $r \leq R_e$. Sketch the potential energy as a function of r from $r = 0$ out to, say, $r = 2R_e$. Is $V(r)$ continuous at the Earth's surface? Does it have to be continuous? Why or why not?

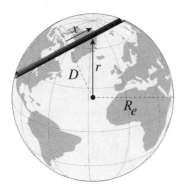

FIGURE 4.18

c) A tunnel is bored through the Earth in a straight line that passes within a distance D from the center. Let x be the distance along this tunnel from the tunnel center as shown in Figure 4.18. When $x = 0, r = D$. Using Lagrangian mechanics (*not vectors!*), with x as the generalized coordinate, find the equation of motion for a particle of mass m dropped into the tunnel. The particle moves under gravity without friction or air resistance. Show that the motion is simple harmonic motion and that the period is the same as that of the satellite in part a) above. If the particle is dropped into the tunnel from the surface at the moment the satellite passes overhead, it will return to the starting point just as the satellite completes one orbit of the Earth.

d) A sloppy physicist known by the initials S. P. makes the following incorrect derivation of $V(r)$ in part b) above: "Let $M(r)$ stand for the part of the Earth's mass inside a sphere of radius $r \leq R_e$. Then $M(r) = M_e(\frac{r}{R_e})^3$, so, for $r \leq R_e$,

$$V_{\text{SP}}(r) = -\frac{GmM(r)}{r}, \tag{4.95}$$

since the matter outside a sphere of radius r does not contribute to the gravitational potential." What equation of motion will S. P. get for part b), and what is wrong with this reasoning?

Center of Mass Motion

Problem 10: *(Two massive bodies in a constant field)* Two massive bodies move in a constant external gravitational field. Show that their motion can be reduced to an equivalent one-body problem, just as we did for the Kepler problem.

Problem 11: *(N-body system)* Prove that, for an N-body system, if the external force on the ith body is $\vec{F}_i$, the center of mass moves like a single particle of mass $M = M_1 + \ldots M_N$ acted upon by a force $\vec{F} = \vec{F}_1 + \vec{F}_2 + \ldots \vec{F}_N$. Use Lagrangian methods to prove this and to prove that we can ignore internal forces as far as center of mass motion is concerned.

PROBLEMS

Problem 12*: *(Explosion of projectile)* A projectile in outer space subjected to no external forces suddenly explodes into three pieces, which have Cartesian coordinates $\vec{r}_1, \vec{r}_2, \vec{r}_3$ with respect to an inertial reference frame. The three masses are m_1, m_2, and m_3. Assume that the forces during the explosion are not known, but it is believed that they can be derived from potentials depending only on the distances between pairs of the particles:

$$V = V_A(|\vec{r}_1 - \vec{r}_2|) + V_B(|\vec{r}_2 - \vec{r}_3|) + V_C(|\vec{r}_3 - \vec{r}_1|). \tag{4.96}$$

a) Show that the center of mass moves at the same constant velocity it had before the explosion.

b) Show that, in the center of mass reference frame, the three fragments lie in a plane after the explosion. Hint: Prove that the total momentum in this reference frame is zero.

c) Derive what happens to the center of mass after the explosion if instead of being "external force-free," the system also had a constant gravitational force on it. Does the result depend on the force being a constant? Is the total momentum of the fragments still zero in the noninertial center of mass reference frame?

Central Force Problems

Problem 13: *(Massive particle moving on a cone)* A massive particle moves under the acceleration of gravity, g, and without friction on the surface of a cone of revolution with half angle α. Find the Lagrangian in plane polar coordinates. Also find the equation of motion for r and the effective potential $V_{\text{eff}}(r)$. If the particle is launched horizontally with velocity v_0 at a height z_0, prove that the condition for circular motion is $v_0^2 = gz_0$.

Problem 14*: *(Two connected masses)* Two masses m_1 and m_2 are connected by a weightless string of fixed total length l_0. Mass m_1 rests on a frictionless table, which has a small hole cut into it. Mass m_2 hangs down vertically from this hole. Assume that m_2 can only move in the vertical direction, so the problem has two degrees of freedom.

a) Assuming that the acceleration of gravity is g, find the Lagrangian and the equations of motion for this system. (Use plane polar coordinates.)

b) The total energy $E = T + V$ is a constant of the motion. How can you see this by inspection of the Lagrangian? There is a second constant of the motion. Explain how to find it, and prove that it is indeed constant. (Call this constant l). What is the physical interpretation of l?

c) Is there a case where the motion of the mass m_1 is a circle of constant radius r_0 from the hole in the table? Find the radius of this circle in terms of l, m_1, m_2, and g. Let $E(r_0)$ be the total energy in this case. Prove that $E(r_0) = \frac{3}{2}m_2 g r_0$. Why is $E(r_0)$ the minimum possible total energy E?

d) For $E > E(r_0)$, solve the radial equation of motion. Put the solution in terms of E and l, the constants of the motion. Express the solution as an integral ("solution by quadratures") that gives the time as a function of r. Is this sufficient to specify the solution completely? How would you find the turning points of the motion? The period?

e) Suppose you treated the EOM for the radial equation as generated by a fictitious 1-D potential? What would be this potential? Find the effective one-dimensional potential $V_{\text{eff}}(r)$ and draw a graph of V_{eff} versus r.

Problem 15: *(Arbitrary central force)* Suppose you have an arbitrary central force potential $V(r)$. Make the $r = \frac{1}{u}$ transformation and find the differential equation for $u(\phi)$. Work out the explicit form of the differential equation in u if $V(r) = -\frac{k}{r^\beta}$ with k, β constants.

Problem 16: *(Using Maupertuis' Principle instead of $u = \frac{1}{r}$ transformation)* Maupertuis' Principle states that $\Delta \int \sqrt{T} \, ds = 0$, where Δ is a variation between fixed end points that leaves the total energy E constant, T is the kinetic energy, and ds is the element of arc length. Recall that, for 2-D motion in the plane, $ds^2 = dr^2 + r^2 d\phi^2$. In the Euler–Lagrange equation it doesn't matter what the independent variable is, so use ϕ. Prove that Maupertuis' Principle gives the same equation for the orbit we obtained by the $u = \frac{1}{r}$ transformation. The potential is an arbitrary central potential $V(r)$.

Problem 17*: *(Tether ball)* A mass m is attached to a weightless string which initially has the length s_0. The other end of the string is attached to a post of radius a. Neglecting the effect of gravity, suppose the string is set into motion, with an initial velocity tangential to the string of v_0. Find the Lagrangian and the equation of motion for the length $s(t)$ of the string. Prove that the time it takes for the string to wind up on the post so that $s = 0$ is $t_{\text{wrap}} = \frac{s_0}{|v_0|}$. Notice that t_{wrap} does not depend on the post radius.

Gravity and Planetary Orbits

Problem 18: *(Elliptic orbits)* For elliptic orbits, prove that the distance from the ellipse center to the focus of the ellipse (position of the Earth–Sun center of mass) is $a\epsilon$, where a is the semimajor axis and ϵ is the eccentricity.

Problem 19*: *(Weighing the Sun, Earth, and Moon)* Kepler's Third Law in its exact form (4.61) allows you to "weigh" the Sun but not the Earth.

a) Determine the solar mass M_S from the length of the year and the mean radius of the Earth's orbit, neglecting the small eccentricity. Use $\bar{R} = 1.49 \times 10^8$ km, and the gravitational constant $G = 6.67 \times 10^{-11}$ N m^2 kg^{-2}.

b) You can find the mass of the Earth. Evaluate approximately the ratio of masses of the Sun to that of the Earth, using only the lengths of the year and of the lunar month (27.3 days) and the mean radii of the Earth's orbit and of the Moon's orbit (3.8×10^5 km). What is the mass of the Earth? What is the Earth's mean density, given that the mean radius of the Earth is 6,378 km? Notice that you can't also "weigh" the Moon by this method. How is the mass of the Moon determined?

Problem 20*: *(Modifying $V(r)$ for the Kepler problem)* How would you modify the equation (4.51) for the orbit in the Kepler problem if, instead of the gravitational potential $V(r) = -\frac{k}{r}$, you had something of the form $V(r) = -\frac{k}{r} + \beta r$? The Sun's gravitational attraction to the Moon makes a small perturbation of the form βr, where β is very small and is itself of the form $\gamma \cos \phi$. What would you expect for the effect of the perturbation? Is it oscillatory or is there a cumulative effect? A potential of this form could also arise in the Earth–Sun system (with a constant β), if there were a uniform density of gravitating matter within the solar system. How could you establish limits on β?

Problem 21: *(Tides)*

a) If the tidal force is due to the gravitational pull of the Earth on the Moon, then why do we have two maximum tides per day? (Once seems to make sense since a body of water is closest to the Moon once per day.)

b) Also, the time of maximum tide actually lags behind the time when the body of water is closest and furthest from the Moon. Can you account for this?

c) Note that in reality the Sun also has an effect on the tides. Calculate the ratio of the tidal force due to the Sun over the tidal force due to the Moon. Hint: The gravitational force on the Earth from the Sun is about 175 times larger than that of the Moon. Does this mean that the Sun dominates the tides?

Problem 22: *(Hyperbolic kick)*

a) Why does an object in a hyperbolic orbit passing close to a planet (which is in orbit about another large object like the Sun) get a velocity "kick" from it?

b) Why does it not work for a stationary planet?

(This is known as gravity assist and is often used by spaceships such as *Voyager 2*.)

Problem 23: *(Comets)* The comet Hyakutake, which appeared in March–May 1996, has the following data associated with it: Its eccentricity ϵ is .999846, and its perhelion is 0.230123 AU (from *Sky and Telescope*, May 1996). Using this information, and assuming the comet is small enough so that its mass is negligible compared to the Sun, calculate when we can predict to see the comet again. Also calculate the aphelion and compare it to Pluto's aphelion. You will need to use the table of astrophysical data the appendix at the end of this chapter. Then repeat the calculation for the comet Hale–Bopp, which was very bright in March–May 1997. Hale–Bopp's ϵ is .995075, and its perhelion is 0.913959 AU (using the March 13, 1997 epoch, since Hale–Bopp's eccentricity is somewhat time dependent).

Problem 24*: *(Precessing ellipses)* Discuss the motion of a particle in a central force potential

$$V(r) = -\frac{k}{r} + \frac{\beta}{r^2}. \tag{4.97}$$

In particular, show that the equation of the orbit has an exact solution that can be put in the form

$$\frac{p}{r} = 1 + \epsilon \cos \alpha \phi. \tag{4.98}$$

This is an ellipse for $\alpha = 1$, but it is a *precessing* ellipse if α is not equal to 1. The precessing motion may be described in terms of the rate of precession of the apsides (turning points). Derive an approximate expression for the rate of precession when α is close to unity. If β is increased to the point where it is no longer small compared to the centrifugal term, how does this affect the orbit?

Problem 25: *(Parametrization of parabolic orbits)* For a parabolic orbit, the total energy E is zero and the eccentricity $\epsilon = 1$. If we consider the parabolic orbit as the limit of an elliptic orbit as $\epsilon \to 1$, we note that $r_{\min} \to \frac{p}{2}$, whereas $r_{\max} \to \infty$ (see Table 4.3). The formulas (4.65) and (4.70) suggest that $\mathcal{E} \to 0$ as $\epsilon \to 1$. After some experimenting, we arrived at the idea of defining a new parameter η by the equation $\mathcal{E} \equiv \eta \sqrt{1 - \epsilon^2}$ and taking the limit as $\epsilon \to 1$. Prove that the limit $\epsilon \to 1$ gives the equations for r, t, x, y in terms of the parameter η:

$$r = \frac{p}{2}(1 + \eta^2), \quad t = \sqrt{\frac{\mu p^3}{k}} \frac{\eta}{2} \left(1 + \frac{\eta^2}{3}\right),$$
$$x = \frac{p}{2}(1 - \eta^2), \quad y = p\eta. \tag{4.99}$$

This is the parametrization for a parabolic orbit. Notice that obtaining $\eta(t)$ involves solving a cubic equation. Since this can be done in terms of algebraic expressions, the time dependence of parabolic orbits could in principle be expressed algebraically, in contrast to the time dependence of elliptical or hyperbolic orbits.

Problem 26: *(Parametrization of hyperbolic orbits)* You may want to derive Equations (4.79)–(4.81) without using the complex time transformation. Prove that these are the correct parametric equations, by substituting Equations (4.81) into Equation (4.52) for the attractive branch of the hyperbola. To check Equation (4.79), prove from

$dt = \frac{\mu}{l} r^2 d\phi$, (4.80) and the identity $\cosh^2 \mathcal{E} - \sinh^2 \mathcal{E} = 1$ that

$$\frac{d\phi}{d\mathcal{E}} = \frac{d}{d\mathcal{E}} \arctan \frac{y}{x} = \frac{\sqrt{\epsilon^2 - 1}}{1 + \epsilon \cosh \mathcal{E}},$$

$$\frac{dt}{d\mathcal{E}} = \frac{\mu}{l} r^2 \frac{d\phi}{d\mathcal{E}}.$$

(4.100)

Then prove that integrating this last equation gives

$$T = \frac{\mu}{l} a \sqrt{\epsilon^2 - 1} (\epsilon \sinh \mathcal{E} - \mathcal{E}).$$

(4.101)

Finally, prove this last equation is equivalent to Equation (4.79).

Rutherford Scattering

Problem 27: *(Asymptotes and impact parameter)*

a) The asymptotes of the hyperbolic orbits (4.52, 4.76) for both attractive and repulsive forces are straight lines. Prove that the equations for these lines in Cartesian coordinates for a repulsive force will have the general form

$$y = \pm(\sqrt{\epsilon^2 - 1})x - \frac{\epsilon p}{\sqrt{\epsilon^2 - 1}}.$$

(4.102)

The asymptotes for an attractive inverse-square force obey a similar equation to (4.102), but with $\epsilon \to -\epsilon$.

b) In Figure 4.15, the *impact parameter b* is defined as the distance of closest approach to the origin along the incoming asymptote. Prove that the angular momentum l is given in terms of the reduced mass and the center of mass relative velocity v_∞ infinitely far away from the origin by the formula $l = \mu v_\infty b$. Then show that the impact parameter can be written as a function of ϵ for constant energy:

$$b(\epsilon) = \frac{k}{2E} \sqrt{\epsilon^2 - 1}.$$

(4.103)

Here E is the total energy in the center of mass, k is defined by $V(r) = \pm \frac{k}{r}$, which includes both the attractive and repulsive inverse-square law cases. By convention k is always positive.

Problem 28: *(Cross section)* The scattering of alpha particles (helium nuclei) from a positively charged nucleus is known as *Rutherford Scattering*. The essence of the experiment is that a parallel beam of energetic alpha particles is sent towards a single gold nucleus.* There is a distribution of angular momenta. For a given alpha particle,

---

* The experimental results were first reported in 1911. In reality there are many gold nuclei in a foil, but they scatter independently, so we consider only one nucleus at a time.

the total angular momentum l is given by $l = \mu v_\infty b$, where μ is the reduced mass, v_∞ is the initial (and final) magnitude of the relative velocity of the alpha particle and the nucleus, and b is the impact parameter as defined in the previous problem. All alpha particles incident at the same impact parameter $b(\epsilon)$ (4.103) scatter through an angle $\theta(\epsilon)$ (4.78). Experimentally, one measures a distribution of scattering angles θ for a known incident energy E. The experimental result is expressed in terms of a "differential scattering cross section $d\sigma$." Imagine each gold nucleus puts up a screen of area $d\sigma$ cm^2. Knowing the flux of alpha particles and the number of gold nuclei per cm^2 in the foil, we can calculate how many α particles will hit this screen per second. Consider the infinitesimal "ring" of area $d\sigma = 2\pi b \, db$. All of the alpha particles hitting this area will be scattered into angles between θ and $\theta + d\theta$ within a solid angle of $d\Omega \equiv 2\pi \sin\theta d\theta$. The differential cross section for scattering by a single gold nucleus into $d\Omega$ per solid angle $d\Omega$ is thus given by

$$\frac{d\sigma}{d\Omega} = \frac{2\pi b \, db}{2\pi \sin\theta d\theta} = \frac{b \, db}{\sin\theta d\theta}. \qquad (4.104)$$

(Since the sign of $\frac{db}{d\theta}$ is negative, we should really use $|\frac{db}{d\theta}|$ to avoid negative cross sections.)

a) Start with the repulsive Coulomb potential:

$$V(r) = \frac{Z_{Au} Z_\alpha e^2}{r}, \qquad (4.105)$$

where $Z_{Au} = 79$ is the number of positive charges in the gold nucleus, $Z_\alpha = 2$ is the number of positive charges in the alpha particle, and $e = $ the electronic charge, which is also the charge on the proton. Prove this gives the famous *Rutherford differential cross section*

$$\frac{d\sigma}{d\Omega} = \left(\frac{Z_{Au} Z_\alpha e^2}{4E}\right)^2 \frac{1}{\sin^4 \frac{\theta}{2}}, \qquad (4.106)$$

where E is the total energy in the center of mass system, which is close to, but not quite equal to, the kinetic energy of the incident alpha particles.

b) For numerical calculations, you may find it useful that $\frac{e^2}{m_e c^2}$ is a length called the classical radius r_c of the electron. ($r_c = 2.817 \times 10^{-13}$ cm, $c = $ velocity of light, $m_e c^2 = 0.511$ MeV $\approx 8.2 \times 10^{-14}$ joules, $m_\alpha c^2 = 3728$ MeV, and $m_{Au} c^2 = 183{,}471$ MeV.) The radius of a gold nucleus is approximately 7.5 fermis (1 fermi $\equiv 1 \times 10^{-13}$ cm). How close (minimum distance of approach) will an alpha particle with 5.3 MeV kinetic energy come to the gold nucleus? (Hint: First assume that the gold nucleus is infinitely massive so that the lab frame is also the center of mass frame, and the total energy in the center of mass frame is the alpha particle's kinetic energy in the lab frame.) Generalize your result to arbitrary masses. Would the scattering cross section be different if the potential were attractive instead of repulsive? Qualitatively, what would you expect will happen to the angular distribution of scattering if the alpha particles are sufficiently energetic to penetrate inside the nucleus?

Bertrand's Theorem

Problem 29: ($\frac{1}{r^n}$ *potential*) Write a computer program to find the orbit (by numerical integration) for an attractive potential of the form $\frac{1}{r^n}$. Verify Bertrand's theorem "experimentally" on the computer for yourself. The best way to do this is to use the $u = \frac{1}{r}$ transformation, and numerically integrate the differential equation for $u(\phi)$. Take units in which $\mu = 1$. Vary l, the angular momentum, and graph the orbit $x(\phi)$, $y(\phi)$ for different n values, including $n = 1$ and $n = -2$, integrating $0 < \phi < 4\pi$ or 6π, for example. Be sure to choose the constant in front of the potential so that the force is attractive. Try $n = 4$ to see if you can get the orbit through the origin described in the next problem.

Problem 30: *(Central force $F = -\frac{k}{r^n}$)* A particle moves in a central force $F(r) = -\frac{k}{r^n}$, where k is a positive constant. If the particle's orbit is circular and passes through the center of the force, prove that $n = 5$. Find the total energy E. For Bertrand's theorem to apply, it is necessary that a circular orbit of constant r exist, which is certainly the case here, although this is not an orbit that passes through the origin. However, Bertrand's theorem determines the n values for which *all* orbits are closed orbits. For other potentials, it is still possible to choose initial conditions so that some exceptional orbits are closed. Notice that it is the force in this problem that varies as r^{-n}, not the potential as in the previous problem. There is a difference of 1 in the n value.

APPENDIX

TABLES OF ASTROPHYSICAL DATA*

TABLE 4.5 PHYSICAL CONSTANTS AND UNITS

Name	Symbol	Value	Units
Gravitational constant	G	$6.672(6) \times 10^{-8}$	N m² kg^{-1}
Speed of light (definition)	c	2.99792458×10^8	m s^{-1}

* From the *Vade Mecum* of the American Institute of Physics and *The Astronomy and Astrophysics Encyclopedia*, ed. S. P. Maran, van Nostrand Reinhold, Cambridge Univ. Press, 1992.

TABLE 4.6 ASTRONOMICAL CONSTANTS AND UNITS

Name	Symbol	Value	Units
Astronomical unit	AU	1.4959780×10^{13}	cm
Parsec	pc	3.085678×10^{18}	cm
		$\frac{648,000}{\pi}$	AU
		3.261633	light years
Light year		9.460530×10^{17}	cm
Sidereal year (1900)	yr	3.1557×10^{7}	s
Hubble constant ($0.5 \leq h \leq 1$)	H_0	$100h$	km s^{-1} Mpc^{-1}
Hubble time	H_0^{-1}	$9.8h^{-1} \times 10^{9}$	yr
Hubble distance	$\frac{c}{H_0}$	$3000h^{-1}$	Mpc
Density of galactic matter		2×10^{-31}	g cm^{-3}
		1×10^{-7}	atom cm^{-3}
Space density of galaxies		0.02	Mpc^{-3}

TABLE 4.7 PARAMETERS OF THE GALAXY

Name	Value	Units
Diameter	25	kpc
Width of disk	2	kpc
Number of stars	10^{11}	
Distance of Sun from center	8.5(5)	kpc
Height of Sun above disk	8	pc
Rotational velocity of Sun	220(10)	km s^{-1}
Period of rotation	2.5×10^{8}	yr

TABLE 4.8 SOLAR DATA

Name	Symbol	Value	Units
Mass	$M_\odot$	$1.989(2) \times 10^{33}$	g
Radius	$R_\odot$	$6.9599(7) \times 10^{10}$	cm
Gravity at surface	$g_\odot$	2.74×10^{4}	cm s^{-2}

TABLE 4.9 PLANETARY DATA

Parameter	Mercury	Venus	Earth	Mars	Jupiter	Saturn	Uranus	Neptune	Pluto
Semimajor axis (AU)	0.387	0.723	1 (def)	1.524	5.203	9.523	19.164	29.987	39.37
Eccentricity	.206	.007	.017	.093	.049	.053	.046	.012	.249
Period	88.0d	224.7d	1.000y	1.88y	11.86y	29.46y	84.01y	164.1y	247y
Mass (10^{27} g)	0.3302	4.8690	5.9742	0.6419	1898.8	568.41	86.987	102.85	0.014
Density (g/cm^3)	5.43	5.24	5.515	3.93	1.33	0.70	1.3	1.8	~0.4
Equatorial radius (km)	2439	6052	6378.14	3393.4	71398	60000	25400	24300	2000
Oblateness	0	0	.00335	.00519	.06481	.10762	.030	.0259	0
Surface gravity (cm/s^2)	370	890	978	371	2288	905	830	1115	~20
Rotation period	176d	116.7d	24h (def)	$24^h 39^m 35^s$	$9^h 55^m 33^s$	$10^h 39^m 24^s$	~16^h	~18.5^h	$6^d 9^h 18^m$
Satellites	0	0	1	2	16	17	15	8	1

CHAPTER FIVE

NOETHER'S THEOREM AND HAMILTONIAN DYNAMICS

OVERVIEW OF CHAPTER 5

Functions of the dynamical variables and their time derivatives that remain constant during the motion are called *conserved quantities* or *constants of the motion*. Noether's Theorem reveals how the symmetries of the Lagrangian can be used to construct constants of the motion from the Lagrangian. These constants of the motion can then be used to reduce the number of variables in the differential equations of motion. For example, angular momentum will no longer appear as a fortuitous combination of coordinates and momenta, but instead will be the answer to the question: "What remains constant if the physical problem exhibits symmetry under three-dimensional rotations?"

There are two distinct forms of analytical mechanics: the Lagrangian formalism and the form developed by Hamilton called *Hamiltonian dynamics*, which we introduce in this chapter. Not only is the Hamiltonian form used to obtain an elegant geometric picture of dynamical motion in phase space, but it also serves as the starting point for most advanced theoretical physics, most notably quantum mechanics. Practical problems are often more easily solved by the Lagrangian formalism, whereas theoretical questions are almost always best stated in terms of the Hamiltonian and the new variable introduced therein, the *canonically conjugate momentum*. The variables q and $\dot{q}$ are not functionally independent, since $\dot{q} = \frac{dq}{dt}$. However, the Lagrangian of each dynamical system will tell us the "other variable," the dynamically independent partner of q. This is $p(q, \dot{q}) \equiv \frac{\partial L}{\partial \dot{q}}$, the momentum canonically conjugate to the variable q. Use of q and p rather than q and $\dot{q}$ leads to equations with a symmetric form in the two variables. This in turn has significant consequences in the theoretical development of mechanics.

5.1 DISCOVERING ANGULAR MOMENTUM CONSERVATION FROM ROTATIONAL INVARIANCE

Start by forgetting everything you have learned previously about angular momentum. We want to be led to discover angular momentum from general principles instead of assuming it. Suppose that the Lagrangian of a particle of mass m contains an arbitrary

5.1 DISCOVERING ANGULAR MOMENTUM CONSERVATION FROM ROTATIONAL INVARIANCE

central force $V(r)$. The kinetic energy is $\frac{1}{2}m(v_x^2 + v_y^2 + v_z^2)$, which depends only on the magnitude and not the direction of the coordinates and velocity. Hence the coordinate system can be freely rotated without changing the Lagrangian (as long as the rotation is time independent). What function of coordinates and velocity is conserved as a consequence of this symmetry?

Coordinate transformations come in two kinds: discrete and continuous. A mirror reflection, $x \to -x$, $y \to -y$, $z \to -z$, is an example of a discrete transformation. Rotation and translations are both examples of continuous transformations, since you can rotate and translate by *any* amount.

Consider rotations about the Z axis. The rotational symmetry of our Lagrangian allows us to rename the coordinates or rotate them so that the axis we want is called the Z axis. If the force were not a central force, different directions would be physically different from each other and this renaming procedure would not be allowed. For any value of θ, a point at (x, y) is transformed to a point (x', y') by the formulas

$$x' = x \cos \theta - y \sin \theta,$$
$$y' = y \cos \theta + x \sin \theta. \tag{5.1}$$

If $\theta = 0$, this is the *identity* transformation. (This is just mathematical jargon for not rotating at all.) It is important to notice that velocities transform the same way as the coordinates do (velocity is, after all, a vector). To "discover" angular momentum, it will be sufficient to consider only an infinitesimal value of $\theta = \delta\theta$. Make $\delta\theta$ arbitrarily small, and thus neglect $\delta\theta^2$ and higher powers. As a limiting case of Equations (5.1), the transformation equation for an infinitesimal rotation becomes

$$x' = x - y\,\delta\theta,$$
$$y' = y + x\,\delta\theta. \tag{5.2}$$

Prior to the infinitesimal rotation, the Lagrangian L had the general form

$$L = L(x, y, z, \dot{x}, \dot{y}, \dot{z}, t). \tag{5.3}$$

After the infinitesimal rotation, the Lagrangian is transformed into L':

$$L' \equiv L(x - y\,\delta\theta, y + x\,\delta\theta, z, \dot{x} - \dot{y}\,\delta\theta, \dot{y} + \dot{x}\,\delta\theta, \dot{z}, t). \tag{5.4}$$

Because $\delta\theta$ is an infinitesimal, we can make a Taylor series expansion of L' in $\delta\theta$ and keep only the first two terms:

$$L' = L + \delta\theta\left[\left(x\frac{\partial L}{\partial y} - y\frac{\partial L}{\partial x}\right) + \left(\dot{x}\frac{\partial L}{\partial \dot{y}} - \dot{y}\frac{\partial L}{\partial \dot{x}}\right)\right] + O[\delta\theta^2]. \tag{5.5}$$

By definition, the invariance of the Lagrangian under these transformations implies that

$$L' = L. \tag{5.6}$$

From Equation (5.6), it must be true that the term multiplied by $\delta\theta$ vanishes, since $\delta\theta$ can take on arbitrary (infinitesimal) values. We use the Euler–Lagrange equations (1.60) to replace $\frac{\partial L}{\partial y}$ by $\frac{d\frac{\partial L}{\partial \dot{y}}}{dt}$ and $\frac{\partial L}{\partial x}$ by $\frac{d\frac{\partial L}{\partial \dot{x}}}{dt}$. The coefficient of $\delta\theta$ in (5.5) is the total time derivative of $xp_y - yp_x$. The standard definition of 3-D canonical momentum is $p_x \equiv \frac{\partial L}{\partial \dot{x}}$, $p_y \equiv \frac{\partial L}{\partial \dot{y}}$, $p_z \equiv \frac{\partial L}{\partial \dot{z}}$. Thus the vanishing coefficient of $\delta\theta$ in Equation (5.5) is seen to be the total time derivative of a constant function of coordinates and momenta:

$$\frac{dl_z}{dt} = 0, \quad l_z \equiv xp_y - yp_x, \tag{5.7}$$

where l_z is the z component of the angular momentum. You should prove to yourself that Equation (5.7) follows from Equations (5.5, 5.6). The same derivation above (Equations (5.1)–(5.7)) can be repeated for rotations around the X and the Y axes. Equation (5.7) will turn out be true for any direction in space; hence

$$\frac{d\vec{l}}{dt} = 0 \Leftrightarrow \text{rotational invariance of } L. \tag{5.8}$$

The notation for angular momentum will be $\vec{l}$. (It is often written in other texts as $\vec{L}$. This should not be confused with L, the Lagrangian, which is a scalar.)

QUESTION 1: *Rotational Invariance* Rotational invariance about the Z axis leads to constant total angular momentum l_z. Prove that this statement is true for a Lagrangian with an arbitrary number of degrees of freedom, using a method similar to the book's method for rotation about the Z axis for one degree of freedom.

QUESTION 2: *Translational Invariance* Translation invariance in the x direction leads to constant total momentum p_x. Prove that this statement is true for a Lagrangian with an arbitrary number of degrees of freedom, using a method similar to the book's method for rotation about the Z axis in one degree of freedom.

5.2 NOETHER'S THEOREM

In some cases, the choice of the best coordinate system to use isn't obvious. *But if a symmetry exists, Noether's Theorem* states that a corresponding constant of the motion exists.* The theorem is a direct generalization of the method we have used in the previous section for rotational invariance. Each conserved quantity can be used to eliminate one degree of freedom, bringing us one step closer to a solution of the problem.

To use the theorem, we need to find a continuous family of transformations (like the example of rotations) for the coordinates of the system. The transformation should depend

---

* Emmy Noether (1882–1935) was a mathematician. The theorem bearing her name was proved in 1918.

5.2 NOETHER'S THEOREM

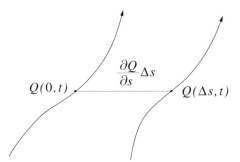

FIGURE 5.1

on one or more continuously variable parameters (such as angles of rotation), which when zero, give the identity transformation (i.e., no change in coordinates). If the Lagrangian is invariant under these transformations (i.e., does not change), Noether's theorem gives a guaranteed way to find constants of the motion. There will be as many constants as there are parameters in the symmetry transformation. For example, rotationally invariant Lagrangians, no matter how complicated the system, must have three constants of the motion, which are the three components of the total angular momentum.

Let the parameter characterizing a general transformation of coordinates be defined as s. (This is the generalization of θ in the previous example.) If $s = 0$, the coordinates are not transformed; it is the identity transformation. If $q(t)$ is a solution of the original EOM for the Lagrangian, we use the notation $Q(s, t)$ for a solution of the Euler–Lagrange equations for any value of s, with $Q(0, t) \equiv q(t)$. (We suppress all indices for multiple degrees of freedom, but the whole derivation can be easily applied to any number of degrees of freedom.) Figure 5.1 shows two possible trajectories, one for $s = 0$, the other for some value $s \neq 0$.

QUESTION 3: *Free Particle* Consider the Lagrangian of a free particle in three-dimensional space. This Lagrangian is invariant under rotations around the Z axis. Explain what is meant by s, $q(t)$, and $Q(s, t)$ for this case.

$L' \equiv L(Q(s, t), \dot{Q}(s, t), t) = L(q, \dot{q}, t)$ is the definition of invariance. If the Lagrangian is invariant, L' must not depend on s (dropping the time dependence for simplicity):

$$\frac{d}{ds} L(Q(s, t), \dot{Q}(s, t)) = 0. \tag{5.9}$$

According to the chain rule

$$\frac{dL}{ds} = \frac{\partial L}{\partial Q} \frac{dQ}{ds} + \frac{\partial L}{\partial \dot{Q}} \frac{d\dot{Q}}{ds}. \tag{5.10}$$

From the Euler–Lagrange equations for the transformed solutions Q (1.60),

$$\frac{\partial L}{\partial Q} = \frac{d}{dt} \frac{\partial L}{\partial \dot{Q}}. \tag{5.11}$$

Thus

$$\frac{dL}{ds} = \frac{d}{dt}\left[\frac{\partial L}{\partial \dot{Q}} \frac{dQ}{ds}\right] = 0. \qquad (5.12)$$

This means that

$$\boxed{I(q,\dot{q}) \equiv p\frac{dQ}{ds}\bigg|_{s=0} = \text{a constant},} \qquad (5.13)$$

where $p \equiv \frac{\partial L}{\partial \dot{q}}$ and dQ/ds is evaluated at $s = 0$ for convenience.

Suppose the invariance transformation is described by more than one parameter. For example, the rotations are specified by three parameters. We need an index subscript on s to distinguish them: s_j, $j = 1, 2, 3 \ldots$. For each parameter s_j, we can repeat the previous derivation to show there is a conserved I_j associated with it. In the formula below, we also explicitly indicate that there are N degrees of freedom.

$$\boxed{I_j(q_1, q_2, \ldots, q_N, \dot{q}_1, \dot{q}_2, \ldots, \dot{q}_N) \equiv \sum_{k=1}^{N} p_k \frac{dQ_k}{ds_j}\bigg|_{\text{all } s=0} = \text{a constant},} \qquad (5.14)$$

where ($p_k \equiv \frac{\partial L}{\partial \dot{q}_k}$). For the space rotations, I_1, I_2, I_3 are the components of the total angular momentum $\vec{l}$.

Equation (5.14) is Noether's Theorem: *If the Lagrangian is invariant under a continuous symmetry transformation, there are conserved quantities associated with that symmetry, one for each parameter of the transformation. These can be found by differentiating each coordinate with respect to the parameters of the transformation in the immediate neighborhood of the identity transformation, multiplying by the conjugate momentum, and summing over the degrees of freedom.*

QUESTION 4: *Transformations* What is a continuous transformation? Explain. Give an example of a continuous transformation and a discrete transformation (not already given). Why do we want only continuous transformations for Noether's Theorem? Also why, in Equation (5.13) above, do we use $p \equiv \frac{\partial L}{\partial \dot{q}}$ and not $\frac{\partial L}{\partial \dot{Q}}$?

Often students first encounter this idea of associating a symmetry of the system with a conserved quantity in connection with quantum mechanics. But this concept has its origin in classical mechanics, although it continues to play an important role in quantum mechanics.

We could speculate on whether the converse of Noether's Theorem holds. For example, the Laplace–Runge–Lenz vector in the Kepler problem* is a conserved quantity. Does this imply a continuous symmetry heretofore unknown to us? It does: It is called the "$O(4)$ symmetry." Like the Lenz vector example, does the existence of conservation laws necessarily imply hidden symmetries of the Lagrangian?

5.3 HAMILTONIAN DYNAMICS

Up to now, we have used a phase space in which we track the development of $q_k(t)$ and $\dot{q}_k(t)$ in time. It will be convenient to use a more general definition of phase space, one that contains the coordinates $q_k(t)$ and the *canonically conjugate momenta* $p_k(t)$, which are defined for the general case of N degrees of freedom by:

$$p_k \equiv \frac{\partial L}{\partial \dot{q}_k}$$
$$(k = 1, \ldots, N).$$
(5.15)

Thus q_k and p_k will become the basic dynamical variables instead of q_k and $\dot{q}_k$. According to the definition (5.15), p_k is a function of $q_k, \dot{q}_k$, which are themselves functions of the time via the equations of motion.

The Lagrangian $L(q, \dot{q}, t)$ is replaced by the *Hamiltonian* $H(q, p, t)$. The Euler–Lagrange equations which determine the motion of the system are replaced by *Hamilton's equations*. Not only does this procedure lead to symmetric equations involving the dynamical variables q_k and p_k, but a whole new approach to classical mechanics is introduced, one that leads to the most powerful and sophisticated tools of theoretical physics. The concept of canonical momentum is the key concept in Hamilton's theory. You must be warned that momentum can lose its familiar definition: $\vec{p} = m\vec{v}$. It will turn out that this is still true in many simple cases, but it is often not true when generalized coordinates are used for convenience in solving a problem.

Although we defined canonical momentum in Chapter 1, let us begin again from the beginning. Our goal is to find a quantity $p(q, \dot{q})$ that is dynamically independent of the generalized coordinate q. We will explain more precisely what is meant by "dynamical independence" later. First we have to do some preliminary mathematical spadework to understand how to eliminate $\dot{q}$ and replace it with the canonical momentum p for Hamilton's theory.

5.4 THE LEGENDRE TRANSFORMATION

We begin with a purely mathematical exercise. The Legendre transformation is a recipe for starting with a function of a variable and generating a new function of a new

---

* The Laplace–Runge–Lenz vector is defined in Problem 2 at the end of this chapter.

variable. If the transformation is repeated, it restores the old function of the old variable. Legendre transformations are used in mathematical treatments of partial differential equations and also are used very extensively in thermodynamics to change from one set of variables to another. To focus on the mathematical content, we will use a notation that does not specifically refer to mechanics. Consider the independent mathematical variables: a passive* variable x and an active variable y. Assume a function $A(x, y)$ of these variables is known explicitly. Now introduce a third variable z and define the function of these three initially independent variables $B(x, y, z) \equiv yz - A(x, y)$. (The minus sign is not essential but will be convenient.) Small changes dx, dy, dz in x, y, z cause a change dB in the function B:

$$dB = z\,dy + y\,dz - \left.\frac{\partial A}{\partial x}\right|_y dx - \left.\frac{\partial A}{\partial y}\right|_x dy. \tag{5.16}$$

Regrouping the terms in Equation (5.16) we get

$$dB = \left(z - \left.\frac{\partial A}{\partial y}\right|_x\right) dy + y\,dz - \left.\frac{\partial A}{\partial x}\right|_y dx. \tag{5.17}$$

So far, z has been an arbitrary independent variable. We now define z to be a function of x and y by the equation

$$z = z(x, y) \equiv \left.\frac{\partial A}{\partial y}\right|_x. \tag{5.18}$$

The coefficient of the term proportional to dy in Equation (5.17) vanishes. The other partial derivatives of B, which is now only a function of x, z can be computed from Equation (5.17):

$$\left.\frac{\partial B}{\partial z}\right|_x = y, \quad \left.\frac{\partial B}{\partial x}\right|_z = -\left.\frac{\partial A}{\partial x}\right|_y. \tag{5.19}$$

To compute B explicitly, we have to invert the relation for z (5.18), solving for $y = y(x, z)$ and then substitute into $B(x, y(x, z), z)$. With the Legendre transformation, $y(x, z)$ is also obtained from the partial derivative $y = y(x, z) = \frac{\partial B}{\partial z}|_x$. This means that, given $B(x, z)$, the transformation can be inverted. For a Legendre transformation, it is possible to work either with $B(x, z)$ or with $A(x, y)$ to find the "passive" partial derivative, since $\frac{\partial B}{\partial x}|_z = -\frac{\partial A}{\partial x}|_y$.

It is often said that the advantage of the Legendre transformation is that it creates a function of x, z alone. This is true. But this could also be done by substituting an arbitrary functional relationship $y = y(x, z)$ into $B(x, y(x, z), z) = B(x, z)$. However, all information about $y(x, z)$ may be lost after the substitution, since the simple relations between

---

* The meaning of "passive" and "active" will become clear from the context.

5.4 THE LEGENDRE TRANSFORMATION

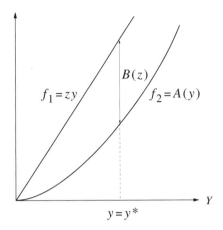

FIGURE 5.2
Graphical representation of the Legendre transformation showing the construction of z and $B(z)$ from y and $A(y)$. At the maximum separation, $\frac{d}{dy}(zy - A) = 0$, so $z = \frac{dA}{dy}|_{y=y^*}$.

partial derivatives ((5.18), (5.19)) would not be valid. A concrete example may help to make this clear.

▷ **Example**

This is really an exercise. Define $A(x, y) \equiv (1 + x^2)y^2$. Prove that $B(x, z) = \frac{z^2}{4(1+x^2)}$ for the Legendre transformation. Show that you can invert the transformation using the partial derivative relations (5.19) and an "inverse" Legendre transformation to find $y(x, z)$ and $A(x, y)$ from $B(x, z)$. Now try the arbitrary substitution $y = z$ and show that the form of $y(x, z)$ cannot be recovered from knowing $B(x, z) \equiv yz - A(x, y) = -x^2 z^2$.

Two different, but completely equivalent, geometric interpretations of the Legendre transformation may help the reader to visualize what the transformation means. In the first way, the distance between a line of variable slope z: $f_1(y) = zy$ and a function $f_2(y) = A(y)$ is maximized to find $y^*(z)$. (We are suppressing the passive variable x.) This shows that only convex functions $A(y)$ can be used for the Legendre transformation, since otherwise the maximum might not exist. To find the maximum distance you must solve the equation $\frac{d}{dy}(zy - A(y)) = 0$, which is the same equation as (5.18). Figure 5.2 shows this construction. The maximum distance is the function $B(z)$.

A second construction, Figure 5.3, shows the dual nature of the Legendre transformation. If z is the slope of the tangent to the curve $A(y)$ then $B(z)$ is the intercept of the line tangent to A at the point y^*. The same is true if instead we start from z and the convex function $B(z)$ and in the same way build y and $A(y)$.

QUESTION 5: *Convex versus Concave* Why does $A(y)$ have to be a convex function? What happens if it is not? (Try $A(y) = y$, for example.) Is $B(z)$ convex?

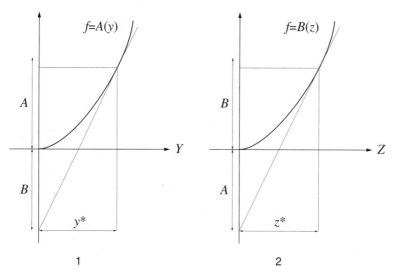

FIGURE 5.3
Dual nature of the Legendre transformation. 1) Construction of z and $B(z)$ from y and $A(y)$. Since the slope of $A(y)$ at $y = y^*$ is equal to $z(y^*)$, then $z = \frac{A(y^*) + B(z)}{y^*}$.
2) Construction of y and $A(y)$ from z and $B(z)$. Since the slope of $B(z)$ at $z = z^*$ is equal to $y(z^*)$, then $y = \frac{A(y) + B(z^*)}{z^*}$.

Why Transform?

In mechanics, start with the Lagrangian $L(q, \dot{q})$ (the possibility of explicit time dependence in the Lagrangian will be temporarily set aside, just to simplify the notation). The active variable is $\dot{q}$, and the passive variable q. By making the Legendre transformation as described above, we pass to the variable p and the Hamiltonian $H(q, p)$:

$$H \equiv p\dot{q} - L(q, \dot{q}), \quad p \equiv \frac{\partial L}{\partial \dot{q}}\bigg|_{\text{constant } q}. \tag{5.20}$$

The transformation is invertible as noted above.

Since the Legendre transformation can be made equally well in either direction, why do we prefer the variable p and the Hamiltonian $H(q, p)$ to the choice of $\dot{q}$ and $L(q, \dot{q})$? The key feature of using the canonical momentum p, which is the tangent to the Lagrangian, instead of $\dot{q}$, is that Hamilton's Principle holds for *independent* variations of q and p. The arbitrary variations δp and δq are truly independent at each point in time, unlike the variations δq and $\delta \dot{q}$. To see this, recall Hamilton's Principle

$$\delta S(\text{action}) = \int \delta L \, dt = 0. \tag{5.21}$$

Calculate δL in terms of the variations of q and p from (5.20):

$$\delta L = \dot{q}\delta p + p\delta \dot{q} - \delta H. \tag{5.22}$$

5.4 THE LEGENDRE TRANSFORMATION

The chain rule for partial derivatives tells us (remember that δp, δq are arbitrary infinitesimal functions of the time) that

$$\delta H = \frac{\partial H}{\partial q}\delta q + \frac{\partial H}{\partial p}\delta p. \tag{5.23}$$

Inserting Equation (5.23) into Equation (5.22) and collecting together the coefficients of δq and δp, we have

$$\delta L = \left(\dot{q} - \frac{\partial H}{\partial p}\right)\delta p - \left(\dot{p} + \frac{\partial H}{\partial q}\right)\delta q + \frac{d}{dt}(p\delta q). \tag{5.24}$$

When Equation (5.24) is integrated with respect to time to compute the variation in the action, the total time derivative on the right-hand side of the equation will contribute nothing if $\delta q = 0$ at the end points. This has always been a requirement of Hamilton's Principle. Thus (5.24) indicates that Hamilton's Principle will only work for independent arbitrary variations of p and q if the coefficients of δp, δq vanish. (The dual transformation already shows us that the coefficient of δp vanishes automatically.)

To prove that the coefficients *do* vanish, start again from the basic defining Equation (5.20) and vary all the variables, $p, q, \dot{q}$ (this does not imply that they are independent), to obtain

$$dH = \dot{q}\,dp + p\,d\dot{q} - \left.\frac{\partial L}{\partial q}\right|_{\dot{q}} dq - \left.\frac{\partial L}{\partial \dot{q}}\right|_{q} d\dot{q}. \tag{5.25}$$

The coefficient of $d\dot{q}$ vanishes due to the definition of p. Varying q for constant p and p for constant q yields the equations

$$\dot{q} = \left.\frac{\partial H}{\partial p}\right|_{q}, \quad \left.\frac{\partial H}{\partial q}\right|_{p} = -\left.\frac{\partial L}{\partial q}\right|_{\dot{q}} = -\frac{d}{dt}\left.\frac{\partial L}{\partial \dot{q}}\right|_{q} = -\dot{p}. \tag{5.26}$$

In the last step, on the right, we have made use of the passive nature of q in the Legendre transformation and the Euler–Lagrange equations of motion.

We have not only derived Hamilton's canonical equations of motion but have proved at the same time that independent infinitesimal variations in δq and δp *from the physical path in phase space* do not change the action, Equation (5.21). We can summarize our result in a single equation representing any change in H due to changes in the arguments of the function $H(q, p)$:

$$dH = \dot{q}\,dp - \dot{p}\,dq. \tag{5.27}$$

The symmetry between q and p is evident here. It is a consequence of the Legendre transformation combined with the Euler–Lagrange equation.

Before, in Chapter 2, we plotted $q(t)$ on the Y axis and time t on the X axis, making small variations from the actual graph of the physical coordinate versus time, and proving

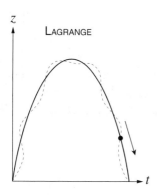 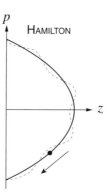

FIGURE 5.4
Two views of the dynamics of a falling body. In the Lagrange case, the height z versus the time t is plotted. In the Hamilton case, momentum p versus height z is plotted. Dashed curves are variations from the physical paths.

that the equations of motion follow from making the action integral an extremum on the physical path. Now we are plotting the phase trajectory of the moving point $q(t)$, $p(t)$ in phase space as shown in Figure 5.4. Varying this trajectory by arbitrary infinitesimal variations in q and p also leaves the action unchanged. Since we take Hamilton's Principle to be the basic law of mechanics, the trajectory in phase space with this extremum property is the solution for the motion. In this view, the time appears as a parameter that we vary in order to trace out the trajectory. There are rewards for this shift in viewpoint, which we will discuss below.

5.5 HAMILTON'S EQUATIONS OF MOTION

For N degrees of freedom, the $2N$-dimensional phase space becomes $\{q_k, p_k\}$ and the Hamiltonian H is

$$H \equiv \sum_{k=1}^{N} p_k \dot{q}_k - L, \qquad (5.28)$$
$$H = H(q_1, q_2, \ldots, q_N, p_1, p_2, \ldots, p_N, t).$$

To consider the possibility that the time might appear explicitly in the Hamiltonian, add a term for this:

$$dH = \sum_{k=1}^{N} \dot{q}_k \, dp_k - \sum_{k=1}^{N} \dot{p}_k \, dq_k + \frac{\partial H}{\partial t} dt. \qquad (5.29)$$

Since the time is also a passive variable in the Legendre transformation we know that

$$\left.\frac{\partial H}{\partial t}\right|_{q_1,\ldots,p_1,\ldots} = -\left.\frac{\partial L}{\partial t}\right|_{q_1,\ldots,\dot{q}_1,\ldots} \qquad (5.30)$$

5.5 HAMILTON'S EQUATIONS OF MOTION

The total time derivative of H can be computed:

$$\frac{dH}{dt} = \underbrace{\dot{q}\dot{p} - \dot{p}\dot{q}}_{=0} + \frac{\partial H}{\partial t} = -\frac{\partial L}{\partial t}. \tag{5.31}$$

If there is no explicit time dependence in L, H will be a constant of the motion. If the kinetic energy is a quadratic form in the $\dot{q}_k$s, H is also the total energy $E = T + V$.

The final result is

$$\boxed{\dot{q}_k = \frac{\partial H}{\partial p_k} \quad \dot{p}_k = -\frac{\partial H}{\partial q_k} \quad \frac{dH}{dt} = -\frac{\partial L}{\partial t}.} \tag{5.32}$$

Hamilton's equations of motion

These are the fundamental equations of Hamiltonian dynamics.

In Lagrangian dynamics N second-order differential equations must be solved. In Hamiltonian dynamics there are $2N$ first-order equations instead. This often makes very little difference in the difficulty of finding explicit solutions. The fact that q and p are treated (almost) symmetrically allows for the discovery of some important theorems: Liouville's Theorem, which we will discuss later in this chapter, and the Poincaré Recurrence Theorem, which is discussed in Appendix B. It also makes possible the development of sophisticated analytical tools such as *canonical transformations*, as we shall see in Chapter 6.

We can get (q, p) at time $t + dt$ from the knowledge of (q, p) at time t by using Hamilton's equations. Thus a step by step time integration can be performed. This is what is actually done when the equations of motion are numerically integrated on the computer.

We now summarize what you must do in order to start from a Lagrangian and convert to the use of Hamilton's dynamics:

1. Define the momentum canonically conjugate to q by the "tangent" to the Lagrangian:

$$p_k \equiv \left.\frac{\partial L}{\partial \dot{q}_k}\right|_{\text{constant } q_k}. \tag{5.33}$$

   Do this for each degree of freedom, holding the coordinates and velocities for the other degrees of freedom constant.
2. Define the Hamiltonian H as in Equation (5.28) above. This is now a mixed function of all the q_k, $\dot{q}_k$ and p_ks. This is still not the final form, since the Hamiltonian must be expressed as a function only of the q_ks and p_ks.
3. Invert the function(s) you obtained in Equation (5.33) to get $\dot{q}_k(q_1, q_2, \ldots, p_1, p_2, \ldots)$.
4. Eliminate the generalized velocities in the temporary form of the Hamiltonian from Equation (5.28). You should now have the Hamiltonian as a function of the p_ks and q_ks only. The time will appear explicitly in the Hamiltonian only if it was explicitly present (due to time-dependent constraints) in the Lagrangian.
5. Solve the $2N$ first-order Equations (5.32).

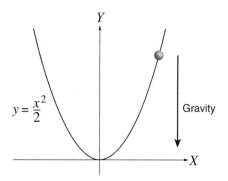

FIGURE 5.5
Mass falling on a parabolic wire.

Examples of Hamiltonian Dynamics

Example 1: Simple Harmonic Oscillator

Consider the simple harmonic oscillator (SHO), which has the Lagrangian* $L = \frac{m}{2}\dot{x}^2 - \frac{k}{2}x^2$. The conjugate momentum is $p = \frac{\partial L}{\partial \dot{x}} = m\dot{x}$. Now write down the Hamiltonian H:

$$H = p\dot{x} - L = m\dot{x}^2 - \frac{m}{2}\dot{x}^2 + \frac{k}{2}x^2 = \frac{p^2}{2m} + \frac{k}{2}x^2. \tag{5.34}$$

(We could have guessed that $H = E = T + V$ here. Why?) Hamilton's equations for $\dot{q}$ and $\dot{p}$ become

$$\dot{x} = \frac{\partial H}{\partial p} = \frac{p}{m}, \quad \dot{p} = -\frac{\partial H}{\partial x} = -kx. \tag{5.35}$$

The right-hand equation is Hooke's Law. If we like, we can differentiate the left-hand equation with respect to the time and eliminate $\dot{p}$ to get $m\ddot{x} + kx = 0$ as expected. No obvious practical advantage is gained by making the transition from Lagrangian to Hamiltonian dynamics in this example, but there are theoretical advantages that will emerge later.

Example 2: Particle Sliding on a Parabolic Wire

In Figure 5.5, a particle of mass m slides under the action of gravity and without friction on a wire shaped into a parabola. We choose x to be the generalized coordinate. The parabola has the shape $y = \frac{x^2}{2}$. The Lagrangian is $L = \frac{m}{2}(\dot{x}^2 + \dot{y}^2) - mgy = \frac{m}{2}(1 + x^2)\dot{x}^2 - \frac{mg}{2}x^2$. Because the kinetic energy is a quadratic form in $\dot{x}^2$, we know immediately that the Hamiltonian is $H = T + V = E$. The canonically conjugate momentum is $\frac{\partial L}{\partial \dot{x}} = m(1 + x^2)\dot{x} = p$. Substituting for $\dot{x}$ in the kinetic energy gives us

$$H = \frac{p^2}{2m(1 + x^2)} + \frac{mg}{2}x^2. \tag{5.36}$$

---

* Here we do not follow our usual practice of scaling all of the variables to get rid of the constants.

5.5 HAMILTON'S EQUATIONS OF MOTION

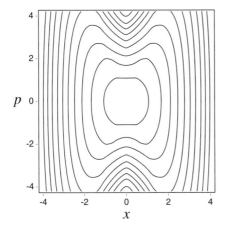

FIGURE 5.6
Contours of constant $H = E$ for Equation (5.36), where we set $m = g = 1$ for the calculation.

From this point we can use Hamilton's equations to find the EOM, which are

$$\dot{x} = \frac{p}{m(1+x^2)}, \quad \dot{p} = -x\left(\frac{p^2}{m(1+x^2)^2} + mg\right). \tag{5.37}$$

Notice that $p \neq m\dot{x}$ in this example.

Since the Hamiltonian is constant in this example, the phase trajectories can be found as the contours of constant H. Even though the EOM are not linear and cannot be (easily) integrated analytically, the trajectories in phase space are plotted for several different values of $H = E$ in Figure 5.6.

⊃ Example 3: Spherical Pendulum

A pendulum is connected at the top to the center of a sphere as shown in Figure 5.7. The length of the pendulum is R; the mass (at the end) is m. The pendulum is free to pivot in any direction about the center of the sphere, while it is acted upon by a constant gravitational field in the z direction. The gravitational acceleration is g.

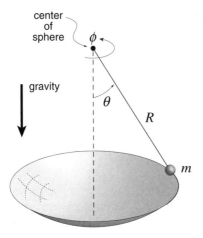

FIGURE 5.7

The pendulum can "swing," changing its polar angle θ, with kinetic energy $\frac{1}{2}m(R\dot\theta)^2$. It can also "orbit," changing the azimuthal angle ϕ, with kinetic energy $\frac{1}{2}m(R\sin\theta\dot\phi)^2$ (see (4.33)). The total kinetic energy T is

$$T = \frac{1}{2}mR^2\dot\theta^2 + \frac{1}{2}mR^2\sin^2\theta\dot\phi^2. \tag{5.38}$$

The potential energy is

$$V = mgR(1-\cos\theta). \tag{5.39}$$

and we can drop the constant term mgR. ϕ is missing from $L = T - V$, so we expect $p_\phi = $ constant.

a) The conjugate momenta are

$$\begin{aligned}p_\theta &= \frac{\partial L}{\partial \dot\theta} = mR^2\dot\theta \equiv l_\theta, \\ p_\phi &= \frac{\partial L}{\partial \dot\phi} = mR^2\sin^2\theta\dot\phi \equiv l_\phi.\end{aligned} \tag{5.40}$$

b) The Hamiltonian is

$$H = T + V = \frac{l_\theta^2}{2mR^2} + \frac{l_\phi^2}{2mR^2\sin^2\theta} - mgR\cos\theta. \tag{5.41}$$

(How do we know immediately that $H = E = T + V$?)

c) From the equations of motion, we find that, indeed, $l_\phi = $ constant. The equations for $\dot\theta, \dot\phi, \dot l_\theta$ are

$$\dot\theta = \frac{\partial H}{\partial l_\theta} = \frac{l_\theta}{mR^2}, \tag{5.42}$$

$$\dot\phi = \frac{\partial H}{\partial l_\phi} = \frac{l_\phi}{mR^2\sin^2\theta}, \tag{5.43}$$

$$\dot l_\theta = -\frac{\partial H}{\partial \theta} = \frac{l_\phi^2 \cos\theta}{mR^2\sin^3\theta} - mgR\sin\theta. \tag{5.44}$$

Differentiating Equation (5.42) and substituting it into Equation (5.44) gives

$$\ddot\theta = \frac{l_\phi^2 \cos\theta}{m^2 R^4 \sin^3\theta} - \frac{g}{R}\sin\theta. \tag{5.45}$$

The first term is complicated and would be hard to obtain by other methods. If $l_\phi = 0$, we have the same equation of motion as for a planar pendulum. If, at the same time, θ is very small, $\omega = \sqrt{\frac{g}{R}}$. For other values of l_ϕ, more complicated types of motion are possible.

5.6 LIOUVILLE'S THEOREM

Consider a phase space distribution of identical dynamical systems governed by identical Hamiltonians. These systems differ only by different initial values of coordinates

5.6 LIOUVILLE'S THEOREM

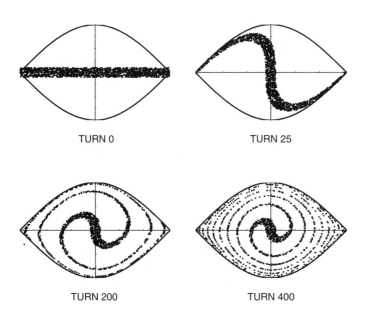

FIGURE 5.8
Capture of debunched beam with fast turn-on.

and momenta. The state of each of the dynamical systems corresponds to a single point which traces out a phase trajectory as time evolves.

We consider a large number of such systems, all located within some bounded area in phase space. Figure 5.8 is an example. It is taken from an accelerator physics computer simulation concerned with the capture of electrons into a circular accelerator. The acceleration is accomplished by the longitudinal electric field in a radiofrequency (rf) resonant cavity. To accelerate the particles to higher energy, the particles must first be "captured" in an "rf bucket." What is shown is a numerical simulation of this capture process. Each point represents the arrival time at the cavity and the energy of a particle in the accelerator.

The dynamics of electron motion are exactly the same as for the large-amplitude pendulum, each different point representing different initial conditions. Deviation from a reference energy (Y axis) is equivalent to p, and injection time or rf phase (X axis) is equivalent to q. The result of the computer simulation is shown after 0, 25, 200, and 400 turns around the machine in Figure 5.8. At turn zero we see electrons injected at all different times within one rf period, but with a limited energy spread. The distribution of energies around the reference energy is otherwise uniform. As the number of turns increases, each particle traces out a trajectory in phase space. We have plotted the separatrix* on the plots in Figure 5.8 to show which electrons will be captured into oscillating (periodic) orbits. These stable orbits will be successfully accelerated in a real accelerator. Particles outside the separatrix will be lost.

Notice in Figure 5.8 that, although the phase space gets progressively more "twisted up," the total area filled with particles remains constant. This is Liouville's Theorem in

---

* See Chapter 4 for the definition of separatrix.

action. It says that, whatever the dynamical equations are, we can't compress a phase space distribution into a smaller area. If we have dissipative forces, then Liouville's Theorem can be violated. But then we can't describe the motion using Hamiltonian dynamics. Hamilton's equations imply Liouville's Theorem: *The area of any small patch in phase space is preserved as time progresses.* The way in which the phase space distribution becomes rather complicated and stringy is called *filamentation*.

If the points in phase space are quite numerous, we can describe them by talking about the density of points in phase space, just as we talk about the density of molecules in a gas. The concept of gas density doesn't make sense for distances on the molecular level, since the density fluctuates wildly. If we chose volumes on a human scale instead, they are likely to contain $\approx 10^{21}$ gas molecules. In the latter scale, it is a good approximation to ignore fluctuations and talk about the "density" of a gas as a continuous function of position, time, etc. The density concept arose before the existence of molecules was discovered, but it remains useful at times to define density $\rho(x, y, z, t)$ as if there actually were such a continuous function to describe the number of gas molecules per unit volume.

The *phase space density* is the limit

$$\rho \equiv \lim_{\Delta A \to 0} \frac{\mathcal{N}(\Delta A)}{\Delta A}, \tag{5.46}$$

where $\mathcal{N}$ = the number of systems in "phase space area" $\Delta A \equiv \Delta q \, \Delta p$. There is a restriction that ΔA never gets so small that density fluctuations become important. In this sense only will we treat $\rho(q, p, t)$ as a continuous function of the phase space variables and perhaps the time. The total number of particles (dynamical systems) contained in a patch of area A is then

$$\mathcal{N}_{\text{particles}} = \iint_A \rho(q, p) \, dq \, dp. \tag{5.47}$$

In the most general case with more than one degree of freedom, the phase space has $2N$ dimensions. Hence ρ is then a function of $2N$ qs and ps.* The phase space area generalizes to a volume ΔV_N in a $2N$-dimensional space:

$$\Delta V_N = \Delta q_1 \Delta p_1 \ldots \Delta q_N \Delta p_N. \tag{5.48}$$

In that case

$$\mathcal{N}_{\text{particles}} = \iint \ldots \int_{\Delta V_N} \rho \, dq_1 \, dq_2 \, \ldots \, dq_N \, dp_1 \, dp_2 \, \ldots \, dp_N. \tag{5.49}$$

We'll stick with one degree of freedom, $N = 1$, for our proof of the theorem. An extension of the proof to $N > 1$ is straightforward if the Jacobian method is used as explained in Appendix A.

---

* The density ρ could also be an explicit function of the time.

5.6 LIOUVILLE'S THEOREM

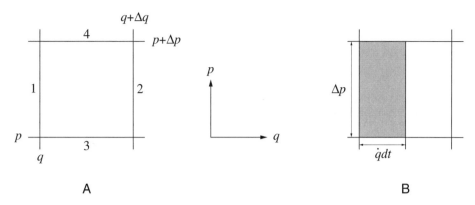

FIGURE 5.9

Liouville's Theorem states that ρ flows over phase space like an incompressible fluid. In this respect it behaves more like a liquid, say a bucket of water, than a gas. We can dump a bucket of water on the floor, but since water is (almost) incompressible, the total volume of water doesn't change. The same is true of the phase "fluid" ρ. It may change its shape in phase space, but if we follow a point q^*, p^* which moves according to Hamilton's equations and measure $\rho(q^*, p^*)$ at that point, we find that the density doesn't change. Of course, if we sit still, at a fixed point, the density can change. All this can be seen in Figure 5.8. What we are describing is just the difference between a total time derivative $\frac{d\rho}{dt}$, where we include changes due to changes in q and p, and a partial time derivative $\frac{\partial \rho}{\partial t}$, where instead we sit at a fixed point with q and p held constant. To prove Liouville's Theorem, we need to prove that $\frac{d\rho}{dt} = 0$ is a consequence of Hamilton's equations of motion. This is the mathematical way to state that the "fluid" of particles in phase space is incompressible.

Start by dividing the phase space into arbitrarily small rectangles of size Δq, Δp, with area $\Delta q \Delta p$. At the end of the calculation, we will take the limit as the rectangle dimensions become infinitesimally small. The number of "particles" (i.e., dynamical systems) inside the little rectangle shown in Figure 5.9A is

$$d\mathcal{N} = \rho \Delta q \Delta p. \tag{5.50}$$

Consider what happens after a time dt. Particles enter or leave at the boundaries marked 1, 2, 3, 4. The number entering at boundary 1 is the number in the little rectangle of area $\Delta p \, \dot{q} \, dt$ in Figure 5.9:

$$d\mathcal{N}_1 = (\rho \dot{q})_q \, dt \, \Delta p. \tag{5.51}$$

(The subscript q means evaluate this quantity in parentheses at q. Strictly speaking, we are averaging the same quantity over Δp, but remember that Δp is very small.) The number leaving the rectangle through boundary 2 is

$$d\mathcal{N}_2 = -(\rho \dot{q})_{q+\Delta q} \, dt \, \Delta p. \tag{5.52}$$

The net change due to both vertical boundaries is

$$d\mathcal{N}_{12} = \{(\rho \dot{q})_q - (\rho \dot{q})_{q+\Delta q}\} \, dt \, \Delta p. \tag{5.53}$$

Because Δq is also small, we can use a Taylor series expansion of the quantity in curly brackets, keeping only the first term:

$$d\mathcal{N}_{12} = -\frac{\partial(\rho\dot{q})}{\partial q} dt\, \Delta q\, \Delta p. \tag{5.54}$$

By identical reasoning applied to the horizontal boundaries 3 and 4, we see that

$$d\mathcal{N}_{34} = -\frac{\partial(\rho\dot{p})}{\partial p} dt\, \Delta q\, \Delta p. \tag{5.55}$$

Inside the rectangle, the number must vary if ρ depends explicitly on the time:

$$d\mathcal{N}_{\text{inside}} = \frac{\partial \rho}{\partial t} dt\, \Delta q\, \Delta p. \tag{5.56}$$

Since the total number of systems entering must equal the local increase in the number of systems in the rectangle, we have a differential equation:

$$\boxed{\frac{\partial \rho}{\partial t} + \frac{\partial(\rho\dot{q})}{\partial q} + \frac{\partial(\rho\dot{p})}{\partial p} = 0. \qquad \text{continuity equation}} \tag{5.57}$$

This is called the *continuity equation*. It occurs in many branches of physics. For example, in electromagnetic theory, there is a similar equation that expresses the fact that charge can neither be created nor destroyed:

$$\frac{\partial \rho}{\partial t} + \vec{\nabla} \cdot \vec{j} = 0. \tag{5.58}$$

Here ρ is the charge density, and $\vec{j}$ is the current density. Equation (5.58) expresses the conservation of charge as a differential equation. Continuity equations are really bookkeeping – nothing more. We have found a differential Equation (5.57) that guarantees that the total number of "particles" (systems) remains constant – a very reasonable thing to require. Classical mechanics hasn't been used yet.

Now make use of the fact that our collection of identical systems obeys Hamilton's equations of motion. Carry out the partial differentiation of the products in (5.57) and regroup the result to get

$$\frac{\partial \rho}{\partial t} + \rho \left\{ \frac{\partial \dot{q}}{\partial q} + \frac{\partial \dot{p}}{\partial p} \right\} + \dot{q}\frac{\partial \rho}{\partial q} + \dot{p}\frac{\partial \rho}{\partial p} = 0. \tag{5.59}$$

The term in curly brackets vanishes, due to Hamilton's equations (5.32) (since $\frac{\partial^2 H}{\partial q \partial p} = \frac{\partial^2 H}{\partial p \partial q}$).

5.7 MOMENTUM SPACE

The rest of the terms are the total time derivative of ρ (by the chain rule). We have achieved our proof, which is

$$\boxed{\frac{d\rho}{dt} = 0.} \qquad \text{Liouville's Theorem}$$ (5.60)

Equation (5.60) means that the area of any distribution in phase space must be preserved. It turns out that this result has many practical applications, such as in the theory of statistical mechanics.

QUESTION 6: *Liouville's Theorem 1* The fact that the area of a small patch in phase space is preserved as time progresses is equivalent to the statement that, if you move along with a phase trajectory, the local density must remain constant. Why?

QUESTION 7: *Liouville's Theorem 2* Explain in detail how the phase space diagrams of the accelerator capture process in Figure 5.8 show the aspects of Liouville's Theorem.

Finally, we mention what happens if a system *does not* obey Hamilton's equations. A damped simple harmonic oscillator is an example of dissipative flow. The equations of motion do not conserve phase space area, and a small patch will shrink to a point at the origin. We can rewrite (5.59) as

$$\frac{d\rho}{dt} + \rho \left\{ \frac{\partial \dot{q}}{\partial q} + \frac{\partial \dot{p}}{\partial p} \right\} = 0.$$ (5.61)

The solution of this equation is

$$\rho(q(t), p(t)) = \rho(q(0), p(0)) \exp\left(- \int_0^t \left[\frac{\partial \dot{q}}{\partial q} + \frac{\partial \dot{p}}{\partial p} \right] dt \right).$$ (5.62)

We see that the sign of $\frac{\partial \dot{q}}{\partial q} + \frac{\partial \dot{p}}{\partial p}$ determines whether the phase volume expands or contracts in the case of non-Hamiltonian motion.

5.7 MOMENTUM SPACE

In quantum mechanics much is made of the fact that working in momentum space and coordinate space are completely equivalent, but, different, ways to formulate a problem. Sometimes it is easiest to solve the problem in momentum space rather than coordinate space. In classical mechanics there is an analogous transformation. Starting with the

Lagrangian $L(q, \dot{q}, t)$, make a "double" Legendre transformation to remove both q and $\dot{q}$, substituting instead p, $\dot{p}$:

$$K(p, \dot{p}, t) \equiv L(q, \dot{q}, t) - p\dot{q} - q\dot{p}. \quad (5.63)$$

The function K is called the *momentum space Lagrangian*. The two kinds of Lagrangian, K and L, are dynamically equivalent, since they differ only by a total time derivative: $\frac{d(pq)}{dt}$. Any problem in classical mechanics can be reformulated in momentum space. See the homework problems for some examples.

5.8 HAMILTONIAN DYNAMICS IN ACCELERATED SYSTEMS

The Hamiltonian formalism lends itself nicely to solving problems in accelerated systems. The form of the Hamiltonian in a noninertial frame is simply related to the Hamiltonian form in an inertial frame. We will demonstrate this with two examples: a rotating system and a linearly accelerated system.

Physics for a Bug Crawling on a Phonograph Turntable

There are two types of motion for the system in Figure 5.10: the rotation of the turntable and a possible motion of the bug on the surface of the turntable. A physicist at rest on the moving turntable wishes to calculate the equations of motion for the bug. The physicist may have no way to refer to the outside world and observes only the motion of the bug and the force on the bug. This dynamics is affected by the fact that the bug moves in a noninertial rotating coordinate system. Use x and y as Cartesian coordinates of the bug on the surface of the turntable as measured in the physicist's (rotating) coordinate frame. Recall, however, that although any coordinates may be used in the Lagrangian, T and V must be evaluated in an inertial frame. The Lagrangian L written in the inertial frame of the lab is

$$L = \frac{m}{2} v_{\text{lab}}^2 - V(x, y) \quad (5.64)$$

Here $V(x, y)$ is the inertial frame potential energy, which we assume can be expressed in terms of x and y in the rotating coordinate system, and $\vec{v}_{\text{lab}}$ is the bug's velocity in

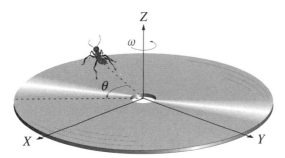

FIGURE 5.10

5.8 HAMILTONIAN DYNAMICS IN ACCELERATED SYSTEMS

the lab frame, which needs to be reexpressed in terms of the physicist's eye view. The transformation equations relating turntable-based bug coordinates (x, y) to lab coordinates $(x_{\text{lab}}, y_{\text{lab}})$ are

$$x_{\text{lab}} = x \cos\theta - y \sin\theta, \quad y_{\text{lab}} = y \cos\theta + x \sin\theta, \tag{5.65}$$

where θ is defined as shown in Figure 5.10. If the turntable rotates counterclockwise at a constant angular speed ω, then $\theta = \omega t$. To get the components of the lab velocity in terms of rotating system coordinates, you must differentiate the transformation equations with respect to the time:

$$\begin{aligned} v_{x,\text{lab}} &= v_x \cos\theta - v_y \sin\theta - \omega(x \sin\theta + y \cos\theta), \\ v_{y,\text{lab}} &= v_y \cos\theta + v_x \sin\theta + \omega(x \cos\theta - y \sin\theta). \end{aligned} \tag{5.66}$$

The time dependence occurs both because the bug could be moving relative to the turntable with velocity (v_x, v_y), and even if the bug rests, the turntable moves with angular velocity ω. The bug's velocity v_{lab}^2, after some algebra, is

$$v_{\text{lab}}^2 = v_x^2 + v_y^2 + 2\omega(v_y x - v_x y) + \omega^2 r^2 \tag{5.67}$$

$(r^2 = x^2 + y^2)$. Substitute this expression into the Lagrangian above (5.64). Calculate the canonical momenta *in the rotating system*:

$$\begin{aligned} p_x &= \frac{\partial L}{\partial v_x} = m(v_x - \omega y), \\ p_y &= \frac{\partial L}{\partial v_y} = m(v_y + \omega x). \end{aligned} \tag{5.68}$$

Finally, construct the Hamiltonian in the usual way:

$$H \equiv p_x v_x + p_y v_y - L. \tag{5.69}$$

Express it as a function of p_x, p_y and get

$$H = \frac{(p_x^2 + p_y^2)}{2m} + \omega(yp_x - xp_y) + V(x, y), \tag{5.70}$$

$$\boxed{H_{\omega \neq 0} = H_{\omega=0} - \omega l_z,} \tag{5.71}$$

where l_z is the z component of the angular momentum as seen in the rotating frame. Equation (5.71) says that to find the Hamiltonian from the physicist's perspective, we only need to subtract ωl_z from the Hamiltonian we would have had if the rotation of the coordinates were not present. For any force on the bug, we can calculate the EOM in the rotating frame without any reference to the outside world. The noninertial frame effects are all contained in the subtraction of ωl_z, regardless of the actual form of the potential energy. In principle, experiments performed on the turntable without any reference to the outside world can detect the presence of such a term and measure the value of ω.

The noninertial effects of rotation would lead to a considerably more complicated form using Lagrangian instead of Hamiltonian dynamics. The simplicity of the Hamiltonian is not an accident, nor is it accidental that the dynamical quantity l_z appears in it.

⊃ **Example: Charged Particle(s) Moving in a Magnetic Field: Larmor's Theorem**

There is an important application of what has just been proved. *Larmor's theorem*, simply stated, says that we can remove the effect of a static magnetic field on a moving charged particle by going to a rotating frame. The effects of the magnetic field are removed to first order in the field strength, so the theorem only applies to relatively weak magnetic fields. It has important applications in atomic physics and elsewhere. Only an electric current or the presence of magnetic dipoles in a para- or ferromagnetic material can produce a static magnetic field $\vec{B}(x, y, z)$.

Because magnetic monopoles don't exist, it is shown in a course on electromagnetic theory that

$$\vec{\nabla} \cdot \vec{B} = 0 \tag{5.72}$$

Since the divergence of the field vanishes, then by a general theorem about vector fields, we can represent $\vec{B}$ as the curl of another field, the vector potential $\vec{A}$:

$$\vec{B} = \vec{\nabla} \times \vec{A}. \tag{5.73}$$

We claim without proof (see the homework problem) that for electric and magnetic fields ($\vec{E}$ and $\vec{B}$), the Lagrangian for a charged particle moving in those fields is

$$L = \frac{1}{2}mv^2 - e\Phi + \frac{e}{c}\vec{v} \cdot \vec{A}. \tag{5.74}$$

(The units are cgs Gaussian units, e is the charge, $\Phi(x, y, z)$ is the electrostatic potential, $\vec{v}$ is the particle velocity, and $\vec{A}(x, y, z)$ is the vector potential.) To find the Hamiltonian, use the usual recipe $\vec{p} = (\frac{\partial L}{\partial v_x}, \frac{\partial L}{\partial v_y}, \frac{\partial L}{\partial v_z})$ to obtain

$$\vec{p} = m\vec{v} + \frac{e}{c}\vec{A} \tag{5.75}$$

(not $\vec{p} = m\vec{v}$!), which gives

$$H \equiv \vec{p} \cdot \vec{v} - L = \frac{1}{2}mv^2 + e\Phi = E \text{ (total energy)}. \tag{5.76}$$

Eliminating v yields

$$H = \frac{\left(\vec{p} - \frac{e}{c}\vec{A}\right)^2}{2m} + e\Phi. \tag{5.77}$$

H is a constant only if the fields are static, which we will assume here for simplicity.

5.8 HAMILTONIAN DYNAMICS IN ACCELERATED SYSTEMS

Specializing to the case of a uniform magnetic field, it can be shown that

$$\vec{A} = \frac{1}{2}(\vec{B} \times \vec{r}). \tag{5.78}$$

(You should convince yourself that Equation (5.78) does in fact describe the vector potential of a *uniform* magnetic field by working out the curl of $\vec{A}$.) In particular, a magnetic field in the z direction means $A_x = -\frac{B}{2}y$, $A_y = \frac{B}{2}x$, $A_z = 0$. Putting in this actual form of the vector potential for the case of a uniform magnetic field in the z direction into (5.77) we get

$$H = \frac{p^2}{2m} - \frac{e}{2mc}l_z B + \frac{e^2}{2mc^2}\left(\frac{1}{4}r^2 B^2\right) + e\Phi. \tag{5.79}$$

The value of Φ is arbitrary here, so the particle can be arbitrarily accelerated, but it must remain nonrelativistic for this form of the Lagrangian to be valid.

Now go into the reference frame that rotates with the constant angular frequency ω_L:

$$\boxed{\omega_L = -\frac{eB}{2mc}.} \tag{5.80}$$

Larmor frequency

Equation (5.80) defines the *Larmor frequency*. In the rotating frame, if $\Phi = 0$, using (5.71) we see that the Hamiltonian is identical with the Hamiltonian of a 3-D harmonic oscillator:

$$H_{\text{rotating frame}} = \frac{p^2}{2m} + \frac{e^2 r^2 B^2}{8mc^2}. \tag{5.81}$$

As promised, the linear term in B is absent. Incidentally, one can make a fine lens for electron beams using a magnetic solenoid – in fact many older-model TV sets have such a lens. The focusing comes from oscillations in the rotating frame – a particle starting from the axis returns to the axis in a time which can be easily calculated from the Hamiltonian above. In the lab frame, the particle motion is a helix. If the angles are small, all particles are not only focused at the same time, but also at the same place down the Z axis.

Linearly Accelerated Systems

A physicist is traveling through intergalactic space in a spaceship without windows. He suspects that his spaceship is being accelerated by a mysterious external force. How does he find out whether this is true?

Assume the existence of a test particle inside the spaceship, one whose coordinates x, y, z we can measure with respect to the ship. With reference to an inertial frame, let us say that the coordinates of this particle are X, Y, Z. We know that

$$X = X_0 + x, \quad Y = Y_0 + y, \quad Z = Z_0 + z, \tag{5.82}$$

where X_0, Y_0, Z_0 are the coordinates of the ship with respect to the hypothetical inertial frame. By the definition of an inertial frame according to Isaac Newton, the *absolute acceleration* with respect to any inertial frame is a physically measurable quantity and should not depend on which inertial frame we choose. Singling out one such hypothetical inertial frame, the kinetic energy T of the test particle is

$$T = \frac{1}{2}m(\dot{X}^2 + \dot{Y}^2 + \dot{Z}^2) = \frac{1}{2}m\big((\dot{X}_0 + \dot{x})^2 + (\dot{Y}_0 + \dot{y})^2 + (\dot{Z}_0 + \dot{z})^2\big). \tag{5.83}$$

The Lagrangian is $L = T - V$. If the forces exerted on the test particle come from within the spaceship, it is most convenient to give the potential energy V as a function of x, y, z the spaceship coordinates. The value of the potential energy must refer to an inertial frame, so it may contain an additional time dependence from the coordinate transformation.

To find the Hamiltonian, we follow the standard prescription, with the particle's spaceship coordinates as the dynamical variables:

$$p_x = \frac{\partial L}{\partial \dot{x}} = m(\dot{x} + \dot{X}_0), \tag{5.84}$$

and similarly for y and z. The usual definition of the Hamiltonian gives us the Hamiltonian H_{ship} in the spaceship variables:

$$H_{\text{ship}} \equiv \dot{x}p_x + \dot{y}p_y + \dot{z}p_z - L, \tag{5.85}$$

$$H_{\text{ship}}(\vec{r}, \vec{p}, t) = \frac{p^2}{2m} + V(x, y, z) - \dot{\vec{R}}_0 \cdot \vec{p}, \tag{5.86}$$

with $\vec{r} \equiv (x, y, z)$, $\vec{p} \equiv (p_x, p_y, p_z)$, $\dot{\vec{R}}_0 \equiv (\dot{X}_0, \dot{Y}_0, \dot{Z}_0)$. We can use formula (5.86) for translational motion just as we can use (5.71) for rotational motion.

If there really is a mysterious force on the the spaceship, how do we measure it within the ship? The Hamiltonian (5.86) contains all the information about the dynamics of a test particle. In the simplest case the potential energy $V(x, y, z) = 0$. According to Hamilton's equations $\frac{dH_{\text{ship}}}{dt} = \frac{\partial H_{\text{ship}}}{\partial t} = -\ddot{\vec{R}}_0 \cdot \vec{p}$. The Hamiltonian in the spaceship will not be constant, nor will it be the total energy, since T is not quadratic in $\dot{x}, \dot{y}, \dot{z}$. We can also use Hamilton's equations to show

$$\dot{x} = \frac{p_x}{m} - \dot{X}_0(t), \quad \dot{p}_x = 0. \tag{5.87}$$

The first equation, combined with the second one, shows that the velocity $\dot{x}$ will be a function of time if $\ddot{X}_0 \neq 0$, even though the momentum p_x is a constant. The astronauts in the ship will observe the backwards acceleration of a particle which has no external forces acting on it. Many other effects of not being in an inertial frame could be demonstrated. Continuous apparent acceleration of a free particle and the apparent lack of energy conservation are perhaps the most conspicuous ways to demonstrate the presence of an accelerated reference frame.

SUMMARY OF CHAPTER 5

- Noether's Theorem gives a method for using symmetries of the Lagrangian to construct constants of the motion. It states that if $s_1, \ldots, s_j, \ldots$ are parameters of the symmetry transformation, then $I_1, \ldots, I_j, \ldots$ are constants of the motion:

$$I_j(q, \dot{q}) \equiv \sum_{k=1}^{N} p_k \frac{dQ_k}{ds_j}\bigg|_{\text{all } s=0} = \text{a constant} \quad (5.88)$$

$$\text{where } p_k \equiv \frac{\partial L}{\partial \dot{q}_k}.$$

$Q(s_1, \ldots s_j, \ldots, t)$ is the transformed coordinate and $Q(0, \ldots, 0, t) = q(t)$.

- The Legendre transformation gives a general and invertible method for replacing one variable by another. Start with any convex function $A(y)$. Solve the equation $z = \frac{dA}{dy}$ for $y = y^*(z)$. Then create the new function $B(z) \equiv zy^* - A(y^*)$. B is a unique function of z only.

- Define the *canonically conjugate momentum* to the coordinate q_k as $p_k \equiv \frac{\partial L}{\partial \dot{q}_k}$. The Hamiltonian is defined by a Legendre transformation in N variables that exchanges the $\dot{q}_k$ dependence for p_k, creating a new function of q_k, p_k:

$$H \equiv \sum_{k=1}^{N} p_k \dot{q}_k - L, \quad (5.89)$$

$$H = H(q_1, q_2, \ldots, q_N, p_1, p_2, \ldots, p_N, t).$$

- Hamilton's equations of motion, which are equivalent to the Euler–Lagrange equations, are

$$\dot{q}_k = \frac{\partial H}{\partial p_k}, \quad \dot{p}_k = -\frac{\partial H}{\partial q_k}, \quad \frac{dH}{dt} = -\frac{\partial L}{\partial t}. \quad (5.90)$$

H is a constant of the motion if L doesn't contain the time explicitly.

- Phase space may be defined in terms of the $q_1, \ldots, q_N, p_1, \ldots, p_N$ space. It has $2N$ dimensions. Liouville's theorem is concerned with the motion in phase space of a large number of identical systems, all governed by the same Hamiltonian. Phase space volume is conserved, that is, the phase space density is like an incompressible fluid.
- In a rotating coordinate system, with angular velocity ω, the Hamiltonian has a simple relation with the $\omega = 0$ (nonrotating) Hamiltonian:

$$H_{\omega \neq 0} = H_{\omega=0} - \omega l_z. \qquad (5.91)$$

An application of this is Larmor's theorem, which allows us to transform away the first-order effects of a magnetic field on a charged particle by going into a frame rotating with the Larmor frequency.

PROBLEMS

Constants of the Motion

Problem 1: *(Constant of the motion for more than one degree of freedom)*

a) For N degrees of freedom prove that the constant I defined in Equation (5.13) becomes Equation (5.14) for a one-parameter continuous transformation ($j = 1$) that leaves the Lagrangian invariant:

$$I(q_1, \ldots, \dot{q}_1, \ldots) = \sum_{k=1,\ldots,N} p_k \frac{dQ_k}{ds}\bigg|_{s=0}. \qquad (5.92)$$

b) Consider the example of Z axis rotations and two point masses interacting by a central potential in three dimensions ($N = 6$). Find an explicit form for I in this case.

Problem 2: *(A new constant of the motion for the Kepler problem)*

a) Prove by using the equations of motion for the Kepler problem (4.38, 4.41) that the "Laplace–Runge–Lenz vector" $\vec{A}$, defined as

$$\vec{A} \equiv \vec{p} \times \vec{l} - \mu k \frac{\vec{r}}{r} \qquad (5.93)$$

(where $\vec{l}$ is the angular momentum, $\vec{p}$ the momentum, and μ the reduced mass), is conserved for motion in a $-\frac{k}{r}$ potential.

b) Find a geometric interpretation by evaluating $\vec{A}$ for an elliptical orbit. (Hint: Since $\vec{A}$ is a constant of the motion, it is only necessary to evaluate it at one convenient point on the orbit.) Is there a corresponding symmetry transformation?

Legendre Transformations

Problem 3: *(Routhians are "reduced" Lagrangians)* The coordinate q_N is ignorable if the Lagrangian contains only the time derivative of the Nth coordinate.

$$L = L(q_1, q_2, \ldots, q_{N-1}, \dot{q}_1, \dot{q}_2, \ldots, \dot{q}_{N-1}, \dot{q}_N, t). \tag{5.94}$$

By using a Legendre transformation, we create a new function, the *Routhian R* (1.71).

$$R(q_1, \ldots, q_{N-1}, \dot{q}_1, \ldots, \dot{q}_{N-1}) \equiv L - p_N \dot{q}_N. \tag{5.95}$$

Since q_N is ignorable in the original Lagrangian, $p_N \equiv \frac{\partial L}{\partial \dot{q}_N}$ is a constant. Prove that the problem is reduced to $N - 1$ degrees of freedom by using the Routhian as a new Lagrangian and showing that the Routhian obeys the Euler–Lagrange equations in the $N - 1$ dynamical variables $q_1, \ldots, q_{N-1}$.

Examples of Hamiltonian Dynamics

Problem 4*: *(Motion along a spiral)* A particle of mass m moves in a gravitational field along the spiral $z = k\theta$, $r =$ constant, where k is a constant, and z is the vertical direction. Find the Hamiltonian $H(z, p)$ for the particle motion. Find and solve Hamilton's equations of motion. Show in the limit $r \to 0$, $\ddot{z} = -g$.

Problem 5*: *(Two particles connected by a spring)* Two particles of different masses m_1 and m_2 are connected by a massless spring of spring constant k and equilibrium length d. The system rests on a frictionless table and may both oscillate and rotate. Find Lagrange's equations of motion. Are there any ignorable coordinates? What are the conjugate momenta? Find the Hamiltonian and Hamilton's equations of motion.

Problem 6: *(Changing the independent variable; time as a dependent variable)* In the theory of special relativity, time is treated on the same basis as the space coordinates x, y, z. We no longer regard time as the independent variable, but instead we choose for that role another parameter, which we will call θ here. Then, in a particular reference frame, the trajectory of a particle would be given parametrically as $x(\theta), y(\theta), z(\theta), t(\theta)$. This can also be done in prerelativity mechanics, although there is no compelling reason to do it. Nevertheless it provides some interesting insights.

a) Let the time be an arbitrary function $t(\theta)$. If $L(q, \dot{q}, t)$ is the Lagrangian of a system with one degree of freedom, show that the Lagrangian corresponding to using θ as the independent variable is

$$L_\theta = t' L\left(q, \frac{q'}{t'}, t\right) \tag{5.96}$$

($t' \equiv \frac{dt}{d\theta}$, $q' \equiv \frac{dq}{d\theta}$). Show using Hamilton's Principle that this Lagrangian leads to the (two) Euler–Lagrange equations with θ as the independent variable.

b) The parameter $t(\theta)$ is now to be regarded as a second dynamical variable. Prove that the momentum conjugate to t is

$$p_t = L + t'\frac{\partial L}{\partial t'} = -H, \tag{5.97}$$

where H is the ordinary Hamiltonian. The time has as its conjugate momentum the negative of the Hamiltonian. Phase space has been enlarged to four dimensions by adding time and energy.

c) Show that the momentum conjugate to q is unchanged by the transformation of the independent variable.

d) Find the Hamiltonian and Hamilton's equations of motion assuming that θ is the independent variable.

Problem 7: *(Particle in a 2-D central force)* Find the Lagrangian for a point particle in a 2-D central force. Work in only two dimensions, using plane polar coordinates. Are there any ignorable coordinates? Find the conjugate momenta. Then find the Hamiltonian and Hamilton's equations of motion. Prove that you obtain equations that are equivalent to (4.38, 4.41).

Problem 8: *(Particle on a cylinder)* Imagine a particle confined to an open cylinder of radius R and bound to the origin by a spring with spring constant k, as shown in Figure 5.11.

a) Prove that the Lagrangian is

$$L = \frac{1}{2}m((R\dot\theta)^2 + \dot z^2) - \frac{1}{2}k(R^2 + z^2). \tag{5.98}$$

b) Next find the conjugate momenta, the Hamiltonian, and Hamilton's equations of motion. Based on these equations, what type of motion do you expect for the particle? Will there be oscillatory motion? How about linear motion?

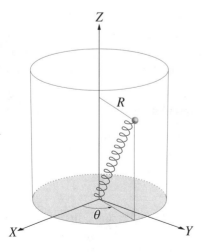

FIGURE 5.11

PROBLEMS

Electrons in a Magnetic Field

Problem 9: *(Electrons focused by a solenoid lens)* An electron is launched from the axis ($x = y = 0$) of a solenoid which has a uniform magnetic field in the z direction of $B = 1$ tesla (MKS units) or $B = 10^4$ gauss (cgs units). The electrons have a kinetic energy of 5 keV (1 eV $= -1.6 \times 10^{-19}$ joules). In cgs units, the charge is 4.8×10^{-10} esu, and the kinetic energy is 8×10^{-9} ergs.

a) Find the ratio of the electron velocity to the velocity of light. Verify that the electrons are not relativistic (i.e., corrections due to special relativity can probably be neglected).

b) Find the Larmor frequency.

c) Show that regardless of v_x or v_y at $t = 0$ (i.e., the "takeoff angle") all electrons will return to the axis (be focused to a point) after a certain distance f. Find a formula for f and calculate the actual distance to this focal length for this case. (Assume the takeoff angle is small, i.e., that most of the velocity is in the z direction.)

Solenoid lenses are often used as focusing elements in electron beams. Most newer-model TV sets use electrostatic lenses however.

Problem 10*: *(Lagrangian for slow charged particles)* The Lagrangian for a nonrelativistic charged particle in an *arbitrary* electric and magnetic field is claimed to be Equation (5.74). (We now include the case of fields changing with time.) $\Phi(x, y, z, t)$ is the electric potential and $\vec{A}(x, y, z, t)$ is the vector potential. The magnetic field is defined as $\vec{B} = \vec{\nabla} \times \vec{A}$. In the general case, the electric field is $\vec{E} = -\frac{1}{c}\frac{\partial \vec{A}}{\partial t} - \vec{\nabla}\Phi$. If the fields are assumed to be static, $\frac{\partial \vec{A}}{\partial t} = 0$, $\Phi = \Phi(x, y, z)$. If the magnetic field is also a uniform field, $\vec{A} = \frac{1}{2}\vec{B} \times \vec{r}$. To do this problem, you can assume that the force on a charged particle is given by the Lorentz force law:

$$\vec{F} = e\left(\vec{E} + \frac{\vec{v}}{c} \times \vec{B}\right). \tag{5.99}$$

a) Prove that if there is a spatially uniform constant magnetic field, the curl of $\frac{1}{2}\vec{B} \times \vec{r}$ gives $\vec{B}$.

b) Assume the magnetic field is constant and uniform, and the electric field is static. Prove that the Lagrangian (5.74) implies equations of motion identical to those obtained from the Lorentz force, Equation (5.99).

c) Remove all restrictions on the nature of the $\vec{E}$ and $\vec{B}$ fields. Prove that Equation (5.99) still follows from the Lagrangian (5.74). This is the most general case.

Hints: The total time derivative of any function $F(x, y, z, t)$ along a particle path $x(t), y(t), z(t)$ is given by $\frac{dF}{dt} = \frac{\partial F}{\partial t} + \vec{v} \cdot \vec{\nabla}F$, where $\vec{v}$ is the particle velocity. To do part c), you will need to use the vector identity for $\vec{a} \times (\vec{b} \times \vec{c})$, where $\vec{a}$ and $\vec{c}$ are vector functions and $\vec{b} = \vec{\nabla}$, the gradient operator, which operates on everything to its right.

Problem 11: *(Relativistic Lagrangian for charged particles of all possible velocities)* You do not need to know the theory of special relativity to do this problem. The Lorentz force equation remains valid for relativistic particles, provided that we recognize that the force is $\dot{\vec{p}}$. If the charged particle moves through electric and magnetic fields at velocities close to the velocity of light, the relativistic form of the Lagrangian is given by

$$L_{\text{relativistic}} = -mc^2\sqrt{1 - \frac{v^2}{c^2}} - e\Phi + \frac{e}{c}\vec{v}\cdot\vec{A}. \qquad (5.100)$$

The reason why this form of the Lagrangian is the correct relativistic Lagrangian will be discussed in Chapter 12.

a) Show that this equation reduces to the nonrelativistic form (5.74) as $v \to 0$. (Gaussian units are used here.)

b) Find the canonical momenta p_x, p_y, p_z and the Hamiltonian H in the relativistic case. Is $H = T + V$? Find Hamilton's equations of motion.

Physics in Noninertial Coordinate Systems

Problem 12*: *(Motion of a particle in an elevator)* A particle of mass m moves in the vertical (z) direction in an elevator. The elevator is accelerated upward with a constant acceleration a ($a > 0$ if upward acceleration). Find the Hamiltonian $H(z, p, t)$ for the particle motion in a uniform vertical gravitational field. Note that p and z are to be measured in the elevator reference frame – a noninertial frame stationary with respect to the elevator. Show that, with the correct choice of a, the particle moves like a free particle.

Problem 13: *(Dynamics inside an orbiting satellite)* A satellite is in a circular orbit of radius R far above the Earth.

a) Neglect gravitational effects due to the finite mass of the satellite. Find the Hamiltonian for motion of a point particle inside the satellite with respect to the center of mass of the satellite under the influence of the Earth's gravity. Assume that the satellite dimensions are very small compared to the Earth's radius so that you can expand the gravitational potential in a Taylor series.

b) Interpret your answer. Particularly consider the application to the possible effects on astronauts. Why do we not see this when we watch films of astronauts floating inside a spacecraft?

c) Would there be any effect due to the gravitational forces between the satellite itself and the point particle? Explain.

d) How would the Hamiltonian of the point particle be affected if the satellite spins on an axis tangential to its motion with constant angular velocity ω? What visible effects would you expect on the motion of the test particle?

Problem 14*: *(Dynamics on a rotating turntable using polar coordinates)* Suppose a bug is crawling on a turntable rotating arbitrarily around an axis perpendicular to

PROBLEMS

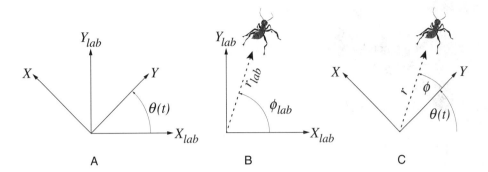

FIGURE 5.12
A) Relation between two coordinate systems. B) Bug's coordinates in lab frame. C) Bug's coordinates in rotating frame.

the plane of the turntable. The bug's position is measured by a physicist standing on the turntable in terms of coordinates x, y, while we (in an inertial frame) measure its coordinates x_{lab}, y_{lab}. The polar coordinates of the bug are r_{lab}, ϕ_{lab} in the lab system and r, ϕ in the rotating system. $r = r_{\text{lab}}$ and $\phi = \phi_{\text{lab}} - \theta(t)$. $\theta(t)$ is the angle between the two coordinate systems, and the rotating frame is moving counterclockwise. The situation and relation between the coordinate systems is pictured in Figure 5.12. The Lagrangian of the bug, in mixed coordinates is

$$L = \frac{m}{2} v_{\text{lab}}^2 - V(r, \phi), \tag{5.101}$$

where v_{lab}^2 is the square of the velocity of the bug in the lab system. The bug (mass m) experiences an arbitrary potential $V(r, \phi)$, which is expressed in terms of its polar coordinates.

a) Why must we use v_{lab}^2 and not v^2 in the Lagrangian?
b) Substitute the rotating coordinates into the expression for the lab kinetic energy in the Lagrangian to find the canonically conjugate momenta (to ϕ, r): p_ϕ and p_r.
c) Calculate the bug's Hamiltonian in terms of r, ϕ, p_r, p_ϕ. Prove that, for arbitrary variation of θ with time,

$$H = H_{\text{lab}} - \dot{\theta} p_\phi, \tag{5.102}$$

where H_{lab} is the bug's Hamiltonian if $\dot{\theta} = 0$.

Momentum Space

Problem 15*: *(Momentum space in classical mechanics)*

a) Consider a harmonic oscillator ($m = 1$) with the Hamiltonian

$$H = \frac{1}{2}(p^2 + \omega^2 q^2). \tag{5.103}$$

Find the explicit form for the harmonic oscillator momentum space Lagrangian K (5.63) as a function of p and $\dot{p}$. Show that K has the same functional form in momentum space as the Lagrangian L has in coordinate space. Find p and $\dot{p}$ using $p = \frac{\partial L}{\partial \dot{q}}$ and $\dot{p} = -\frac{\partial H}{\partial q}$.

b) Justify calling K the "momentum space Lagrangian" by showing that, in general, for any dynamical system the equations of motion are

$$\frac{d}{dt}\left(\frac{\partial K}{\partial \dot{p}}\right) - \frac{\partial K}{\partial p} = 0. \tag{5.104}$$

Hint: $\frac{d(pq)}{dt} = q\dot{p} + \dot{q}p$.

c) Check that the momentum space equations of motion for the harmonic oscillator in part a) derived from (5.104) are correct.

Problem 16*: *(A familiar system expressed in momentum space)* Find the "momentum space Lagrangian" K for the case of a particle falling in a gravitational field:

$$L\left(z, \frac{dz}{dt}\right) = \frac{m}{2}\left(\frac{dz}{dt}\right)^2 - mgz. \tag{5.105}$$

Hint: Add the total time derivative $\frac{1}{2}\frac{d}{dt}z^2 = z\frac{dz}{dt}$ to the Lagrangian L first; then make the double Legendre transformation from ordinary space to momentum space. Don't forget that the definition of p may differ from $m\dot{z}$.

APPENDIX A

A GENERAL PROOF OF LIOUVILLE'S THEOREM USING THE JACOBIAN

It is hard to generalize the proof in Section 5.6 to the case of more than one degree of freedom. We will give an alternate proof of the theorem, still assuming only one degree of freedom, which will be much easier to generalize to the case of $2N > 2$-dimensional phase space.

For a given point in the phase space (q, p) we can regard $(q, p)_{t+dt}$ as functions of $(q, p)_t$. In other words

$$\begin{aligned} p(t+dt) &= p(t) + dt\,\dot{p} \equiv P(p(t), q(t)), \\ q(t+dt) &= q(t) + dt\,\dot{q} \equiv Q(p(t), q(t)). \end{aligned} \tag{5.106}$$

We know that these functions are related to derivatives of the Hamiltonian, but for now treat them as arbitrary functions. Calculate a volume element in the new $(t+dt)$ variables

APPENDIX A   A GENERAL PROOF OF LIOUVILLE'S THEOREM USING THE JACOBIAN   203

in terms of a volume element in the old variables at t:

$$dV_{t+dt} \equiv dp_{t+dt}\, dq_{t+dt} = \det[J]\, dp_t\, dq_t \qquad (5.107)$$

("Volume" is used in a generic sense; here it is just area because the space is two dimensional.) Equation (5.107) is a purely mathematical statement, holding for any transformation of variables. J is the Jacobian matrix* for this particular transformation:

$$J \equiv \frac{\partial(P, Q)}{\partial(p, q)} \equiv \begin{pmatrix} \frac{\partial P}{\partial p} & \frac{\partial P}{\partial q} \\ \frac{\partial Q}{\partial p} & \frac{\partial Q}{\partial q} \end{pmatrix} = \begin{pmatrix} \frac{\partial p_{t+dt}}{\partial p_t} & \frac{\partial p_{t+dt}}{\partial q_t} \\ \frac{\partial q_{t+dt}}{\partial p_t} & \frac{\partial q_{t+dt}}{\partial q_t} \end{pmatrix}; \qquad (5.108)$$

"$\det[J]$" stands for the determinant of the matrix J. The partial derivatives of (5.108) will be computed using Hamilton's equations (5.32). We will then evaluate the determinant, letting $dt \to 0$ while retaining terms of $O[dt]$. The phase volume changes between t and $t + dt$ by a term of $O[dt]$:

$$dV_{t+dt} = dV_t + dt\, \frac{dV}{dt}. \qquad (5.109)$$

We intend to prove that $\frac{dV}{dt} = 0$, so that area in phase space is preserved. If you start with a certain area in phase space, although the shape may be distorted as time passes, the area itself will not change, even after a finite time interval. By comparing Equations (5.107, 5.109), above, it can be seen that $\det[J] = 1$ is the condition for $\frac{dV}{dt} = 0$.

To prove the phase space volume is constant, substitute Hamilton's equations (5.32) into (5.106):

$$p_{t+dt} = p_t - dt\, \frac{\partial H}{\partial q}, \qquad q_{t+dt} = q_t + dt\, \frac{\partial H}{\partial p}. \qquad (5.110)$$

Taking the necessary partial derivatives, you should find that the Jacobian is the matrix

$$J = \begin{pmatrix} 1 - dt\, \frac{\partial^2 H}{\partial p \partial q} & -dt\, \frac{\partial^2 H}{\partial q^2} \\ dt\, \frac{\partial^2 H}{\partial p^2} & 1 + dt\, \frac{\partial^2 H}{\partial q \partial p} \end{pmatrix}. \qquad (5.111)$$

Take the determinant of the 2×2 matrix in Equation (5.111). Since we are going to the limit $dt \to 0$, neglect terms of $O[dt^2]$. The terms of $O[dt]$ *cancel* due to Hamilton's equations. Thus the determinant of J is $1 + O[dt^2]$.

Phase space volume is preserved as a consequence of Hamilton's equations. This is a completely general result, which is not difficult to generalize to any number of dimensions. For N degrees of freedom the volume element in phase space is given by the expression

$$dV = dq_1\, dp_1\, dq_2\, dp_2\, \ldots\, dq_N\, dp_N. \qquad (5.112)$$

---

* See Kaplan, *Advanced Calculus*, 3rd ed., p. 99 for a discussion of Jacobians.

In higher-dimensional spaces ($2N > 2$) the volumes at $t + dt$ and t are still related by a factor of the determinant of J. Define the matrix S by the equation $J = I + dt\, S + O[dt^2]$. It can be shown (usually in a course on linear algebra) that for any matrix J sufficiently close to the identity matrix I:

$$\det[J] = 1 + dt\, \text{Trace}[S] + O[dt^2]. \tag{5.113}$$

(The *trace* of a matrix is the sum of its diagonal elements.) This means that $\text{Trace}[S] = \frac{dV}{dt}$ in the general case. Hamilton's equations imply that $\text{Trace}[S] = 0$, and hence phase space volume remains constant for an arbitrary number of degrees of freedom.

APPENDIX B

POINCARÉ RECURRENCE THEOREM

Liouville's Theorem can be used to prove a very general result about dynamical systems. We will use two examples to illustrate this result. First, consider a mass point moving under the influence of gravity on the arbitrary 2-D surface as pictured in Figure 5.13. The potential energy $V = mgz$. The motion is in three dimensions but is constrained to move on a 2-D surface. Assume that the total energy E is bounded. Since $T > 0$ and $V > 0$, V must also be bounded, which means that x, y, z are bounded, so the 4-D phase space has a finite volume.

Since we can't draw pictures in four dimensions, we will just schematize what happens in Figure 5.14. Consider any arbitrarily small neighborhood Ω_0 with finite phase volume V containing a point $x(0), y(0), p_x(0), p_y(0)$ describing the initial state of the system at $t = 0$. After some time T, this little neighborhood with volume V is mapped into another neighborhood Ω_1 of the point $x(T), y(T), p_x(T), p_y(T)$, which has the same volume V in phase space, with a possibly distorted shape (see Figure 5.14A). Observe the same phase space volume again and again at discrete time intervals T. Eventually the whole phase space must be filled up. There must *not* exist an integer k and a time $t = kT$ such that the

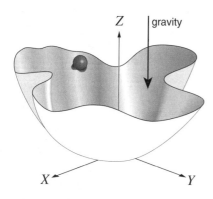

FIGURE 5.13

APPENDIX B POINCARÉ RECURRENCE THEOREM

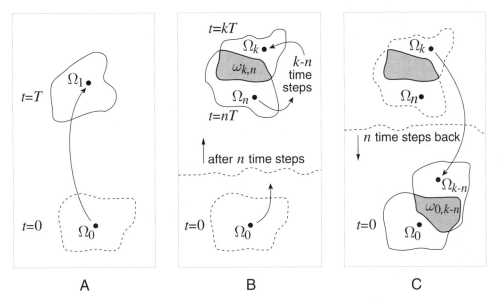

FIGURE 5.14
Poincaré Recurrence Theorem: A) Hamilton's EOM are used to map the initial state of the system at $t = 0$ to the state at time $t = T$. B) After n time steps, we find two regions that overlap. The region of overlap is called $\omega_{k,n}$. C) If we then map n time steps back to return to $t = 0$, we find the recurrence of part of Ω_0 in Ω_{k-n}. This region of overlap is called $\omega_{0,k-n}$.

set of points in some previous Ω_n overlaps a finite portion of Ω_k. If not, then if k becomes sufficiently large, the volume kV of all the points $\{\Omega_0, \Omega_1, \ldots \Omega_k\}$ must become larger than the allowed total phase space volume. Since our phase volume is finite, there must be at least one region of overlap between Ω_k and Ω_n for some $n < k$. Call the common region $\omega_{k,n} \equiv \omega_k = \omega_n$ (see Figure 5.14B). The situation is analogous to filling a circle of one meter radius with dimes. After a finite, but large number of dimes are placed in the circle, it is impossible to avoid overlapping one dime with another previously placed in the circle. The shape of the "dimes" is not constant here, but the area of each dime remains the same.

Next consider the set of points in phase space $\omega_{k,n}$ that belong both to Ω_n and Ω_k. Let's reverse the direction of time through n time steps, and trace back from Ω_k to Ω_{k-n} to locate the subregion $\omega_k \to \omega_{k-n}$. We can also reverse through the same n time steps from Ω_n to Ω_0 to find the subregion $\omega_n \to \omega_0$ in the original neighborhood Ω_0. Since the mapping is invertible and unique, the subspace ω_0 must be identical with ω_{k-n} (see Figure 5.14C) and we will call this region $\omega_{0,k-n}$ ($\omega_0 = \omega_{k-n} \equiv \omega_{0,k-n}$). This means if we discover that two regions Ω_k and Ω_n overlap, then some portion of Ω_{k-n} must overlap with the original region Ω_0. That we will return arbitrarily close to the original starting point in phase space after $k - n$ time steps is the Poincaré Recurrence Theorem.* Recall that the original neighborhood can be arbitrarily small. Of course the smaller the neighborhood of the original point, the more time must elapse to find the recurrence of some portion of it

---

* The discussion here closely follows that in V. I. Arnold's, *Mathematical Methods of Classical Mechanics*.

at a later time. For the mass on the 2-D bowl surface, it means that the initial conditions will be almost exactly repeated eventually. The longer you wait, the closer you will be to matching the initial conditions.

The second and classic example of this theorem is a box with a partition in the middle. On the right side, we have a gas, with N molecules, and on the left a vacuum. There are $3N$ degrees of freedom, and the phase space is $6N$ dimensions. It is important to realize that a *single* point on the multidimensional phase space specifies the positions and velocities of *all* the molecules in the box. If we lift the partition, the gas will fill both sides of the box uniformly. Assuming all collisions with other gas molecules and with the walls are not dissipative, the theorem we've just proved says that, if we wait long enough, at some time the molecules will reassemble themselves in the right side of the box! The catch is that "long enough" could be much longer than the lifetime of the universe, especially with a large number of degrees of freedom. The theorem gives no information about actual recurrence times.

QUESTION 8: *Recurrence Theorem* Is the Poincaré Recurrence Theorem consistent with the Second Law of Thermodynamics? Must entropy, which is a measure of randomness or disorder, always increase in light of the Poincaré Recurrence Theorem? Explain.

CHAPTER SIX

THEORETICAL MECHANICS:
FROM CANONICAL TRANSFORMATIONS TO ACTION–ANGLE VARIABLES

OVERVIEW OF CHAPTER 6

Canonical transformations are transformations from one set of canonically conjugate variables q, p to another conjugate set Q, P. A transformation is said to be canonical if, after the transformation, Hamilton's equations are still the correct dynamical equations for the time development of the new variables. The new Hamiltonian may look quite different from the old one. It may prove easier to solve the EOM in terms of the new variables Q, P. The concept of a *generating function* is introduced, which gives an "automatic" method for producing canonical transformations. There are four types of generating functions for canonical transformations. It will be explained how these different generating functions are connected by Legendre transformations.

Poisson brackets will be introduced, which are invariant under canonical transformations. If Hamilton's dynamics is formulated in terms of Poisson brackets, we have a coordinate-free way to express the equations of motion. The close resemblance of Poisson brackets used in classical mechanics to commutators of operators in quantum mechanics is not an accident, since Poisson brackets played a fundamental role in the invention of quantum mechanics.

We proceed from the general notion of a generating function to the special generating function S, which produces a canonical transformation leading to the *Hamilton–Jacobi equation*. The Hamilton–Jacobi equation leads to a geometric picture of dynamics relating the dynamics to wave motion. There is a close connection between the Hamilton–Jacobi equation in mechanics and the Schrödinger equation in quantum mechanics.

The generating function S turns out to be the time integral of the Lagrangian – the action. The Hamilton–Jacobi equation for time-independent Hamiltonians describing periodic motion leads to the concept of a special set of canonically conjugate variables – *action–angle* variables. The action variables are constants of the motion, and the angle variables increase linearly with time. Thus the time development of the dynamical system takes a simple form when cast in terms of action–angle variables. These variables are also important for further theoretical analysis of dynamical systems.

Conservative systems (systems without damping) come in two types: *integrable* and *nonintegrable*. Only the latter type can exhibit chaos. Integrable systems with N

degrees of freedom by definition have N constants of the motion. Each of these constants confines the motion to a $(2N-1)$-dimensional subspace of the $2N$-dimensional phase space. The intersection of N of these subspaces requires that all the motion takes place on an N-dimensional surface embedded in the $2N$-dimensional phase space. Trajectories on this "surface" can either be periodic or quasiperiodic.

Separable systems are a subset of integrable systems. All analytically soluble mechanics problems are of this type. The Hamilton–Jacobi equation is the most powerful technique for solving separable systems.

6.1 CANONICAL TRANSFORMATIONS

Up to now we have discussed only coordinate transformations (technically known as *point transformations*) between two different sets of space coordinates Q and q:

$$Q = Q(q(t), t). \qquad (6.1)$$
$$\text{point transformation}$$

or, with more degrees of freedom,

$$Q_k = Q_k(q_1, q_2, \ldots q_N, t) \quad (k = 1, \ldots, N). \qquad (6.2)$$

You have probably always referred to point transformations as a change of variables. Notice that this kind of transformation is in the mechanical system's configuration space. It is not the most general mathematical transformation possible in phase space. There could be more general transformations in which the coordinates and the momenta are interdependent. Transformations of this general type are called *contact transformations*:

$$Q = Q(q, p, t), \quad P = P(q, p, t) \qquad (6.3)$$
$$\text{contact transformation}$$

The terminology originated with projective geometry.

Transformations of the type (6.3) are *equivalent* descriptions of the dynamics of a given system if there exists a new Hamiltonian, a function of Q, P and perhaps t, that gives equations of motion in terms of the new variables, which are again Hamilton's equations (5.32). *Canonical transformations will, by definition, take us from one set of coordinates q and canonically conjugate momenta p to another set Q, P, in such a way that the structure of Hamilton's equations for* all *dynamical systems is preserved by this transformation.* The canonically conjugate relation between Q and P will also be preserved. This means that Hamilton's equations will continue to describe the motion for a given specific dynamical

6.1 CANONICAL TRANSFORMATIONS

system, but with a new Hamiltonian which is a function of the new variables Q, P. The new Hamiltonian for any particular system can be derived from the old Hamiltonian by applying a simple rule. A given canonical transformation does not depend on any specific problem or Hamiltonian; it necessarily preserves the form of Hamilton's equations for all dynamical systems to which it is applied. The main application is towards a better understanding of the theoretical structure of mechanics.

Canonical transformations may or may not be of practical use in solving problems. In some cases the equations of motion can be drastically simplified such that the main features of the motion are more clearly revealed. If the contact transformation were *not* canonical, we would sacrifice all of the theoretical advantages that flow out of Hamilton's analytical mechanics, such as Liouville's theorem. It is usually not easy to guess the form of the canonical transformation that will simplify the EOM for a specific dynamical system, but there is a definite mathematical technique that will guarantee to produce a canonical transformation, useful or not.

Recall that two different descriptions of the same physical system are equivalent if their Lagrangians differ by a total time derivative of the form $\frac{dF(q,t)}{dt}$. You may want to review a proof of this before proceeding further. (See Problem 1.6 and/or Question 2.5. Why can't F depend on $\dot{q}$?) Imagine that we have two ways of describing a physical system. Call $\bar{L}(Q, \dot{Q}, t)$ the Lagrangian of the system using the "Q" description, and $L(q, \dot{q}, t)$ the Lagrangian using the "q" description. The two descriptions refer to the same physical system if

$$\bar{L}(Q, \dot{Q}, t) = \lambda L(q, \dot{q}, t) - \frac{dF(q, Q, t)}{dt}. \tag{6.4}$$

Time derivatives of q and/or Q are not allowed to appear in F. (We choose the minus sign in front of F for convenience.) λ is a constant factor. However, by definition, only $\lambda = 1$ can be called a canonical transformation. $\lambda \neq 1$ is associated with a change of units, which is not considered to be a canonical transformation in the most common sense of the term.

We will use Hamilton's Principle to prove that the Euler–Lagrange equations still hold in terms of the new variables for $\bar{L}$ if they hold in terms of the old variables for L. Integrating (6.4), we obtain

$$\int_{t_1}^{t_2} \bar{L} \, dt = \int_{t_1}^{t_2} L \, dt + F(q(t_1), Q(t_1), t_1) - F(q(t_2), Q(t_2), t_2). \tag{6.5}$$

Since Hamilton's Principle holds in the old (q) system, it must also hold in the new (Q) variables. This follows immediately by taking the variation of Equation (6.5) and assuming that arbitrary variations $\delta q(t)$ imply arbitrary variations $\delta Q(t)$. It is necessary to replace our previous assumption that $\delta q = 0$ at the end points of the action integral with the new assumption $\delta F = 0$ at the end points. Then the two descriptions are equivalent, that is, the physics is the same, independent of which coordinate system we use to describe the system.

The function F can be used with any specific Lagrangian to "generate" a new, but equivalent, description of the particular physical system described by this Lagrangian. We think of the canonical transformation as associated with a given form for F rather than with a particular L. There are some restrictions on what you can use for $F(q, Q, t)$. A necessary and sufficient condition for an acceptable F is that $\frac{\partial^2 F}{\partial q \partial Q} \neq 0$. If the mixed second derivative vanishes, it can be shown that the transformation will not be invertible.

F is called a *generating function*. There are four possible types of generating functions, as will be discussed below. The chain rule for the time derivative of $F(q, Q, t)$ is

$$\frac{dF}{dt} = \frac{\partial F}{\partial q}\dot{q} + \frac{\partial F}{\partial Q}\dot{Q} + \frac{\partial F}{\partial t}. \tag{6.6}$$

Since from (6.4) $\dot{q}$ does not appear explicitly in $\bar{L}$,

$$\frac{\partial \bar{L}}{\partial \dot{q}} = \frac{\partial L}{\partial \dot{q}} - \frac{\partial F}{\partial q} = 0, \quad \text{i.e., } p = \frac{\partial F}{\partial q}, \tag{6.7}$$

$$P \equiv \frac{\partial \bar{L}}{\partial \dot{Q}} = -\frac{\partial F}{\partial Q}. \tag{6.8}$$

To summarize:

$$\boxed{P = -\frac{\partial F}{\partial Q}, \quad p = \frac{\partial F}{\partial q}.} \tag{6.9}$$

Equations (6.9) give two equations for the two unknowns, $P(p, q)$, $Q(p, q)$. To find an explicit form for the transformation, solve Equation (6.7) to express $Q = Q(q, p, t)$, and then insert this relation into Equation (6.8) (after taking the partial derivative) to get $P = P(q, p, t)$. In some cases this may be difficult or even impossible to carry out analytically.

> **Example**
>
> As a rather simple example, suppose we take $F = qQ$. Then, according to Equations (6.9), $P = -\frac{\partial F}{\partial Q} = -q$ and $p = \frac{\partial F}{\partial q} = Q$. This particular generating function interchanges the role of coordinate and momentum. To anticipate the result derived below, the new Hamiltonian $\bar{H}(Q, P) = H(-P, Q)$, if $H(q, p)$ is the original Hamiltonian. The reader should check that $\frac{\partial \bar{H}}{\partial P} = \dot{Q}$ and $\frac{\partial \bar{H}}{\partial Q} = -\dot{P}$ if these transformations are made. (Notice that the minus sign in (6.9) is necessary to preserve the form of Hamilton's equations.) We emphasize that Hamilton's equations will always be preserved, since a generating function will automatically generate a canonical transformation.

What if there is more than one degree of freedom? Then F becomes a function of the q_ks and Q_ks ($k = 1, \ldots, N$), and possibly the time. Using $p_k = \frac{\partial F}{\partial q_k}$ and $P_k = -\frac{\partial F}{\partial Q_k}$ we now have the $2N$ equations like (6.9), which give us the transformation rules implicitly, since we have to solve $2N$ equations in $2N$ unknowns Q_k, P_k. With more variables, we

6.1 CANONICAL TRANSFORMATIONS

have to keep track of the indices, but the basic rules for transformation remain the same as for one degree of freedom.

To find the new Hamiltonian $\bar{H}(Q, P)$, we need to return to the definition of how the Hamiltonian is derived from the Lagrangian by a Legendre transformation (5.20):

$$\bar{H}(Q, P, t) \equiv P\dot{Q} - \bar{L} = -\frac{\partial F}{\partial Q}\dot{Q} - L + \frac{\partial F}{\partial q}\dot{q} + \frac{\partial F}{\partial Q}\dot{Q} + \frac{\partial F}{\partial t}. \quad (6.10)$$

(We've used (6.6) for $\frac{dF}{dt}$ above.) Thus

$$\bar{H}(Q, P, t) = p\dot{q} - L + \frac{\partial F}{\partial t} \quad (6.11)$$

and so

$$\boxed{\bar{H}(Q, P, t) = H(q(Q, P), p(Q, P), t) + \frac{\partial F(q(Q, P), Q, t)}{\partial t}.} \quad (6.12)$$

Equation (6.12) says: "*to find the new Hamiltonian, just insert the inverse of the transformation equations expressing P and Q in terms of p and q into the old Hamiltonian H. If F had an explicit time dependence, then add $\frac{\partial F}{\partial t}$ as well.*" This procedure will preserve Hamilton's equations of motion as the new equations of motion, since we know Hamilton's Principle is obeyed for either set of variables. It is only rarely the case that we have an explicitly time-dependent F, so usually $\bar{H}(Q, P) = H(q(Q, P), p(Q, P))$.

In summary, the recipe for a canonical transformation involves these steps:

1. Specify a specific generating function $F(q, Q, t)$.
2. Equations (6.9) give a set of implicit equations for the canonical transformation.
3. Use (6.12) to find $\bar{H}(Q, P, t)$, expressing p and q in terms of P and Q.

There are methods for finding F that we will discuss in Section 6.5, but often you simply make an educated guess.

If we start with a contact transformation in the form of Equations (6.3), how do we find the F which generates it? First, express p, P as functions of q, Q, and t. Then consider Equations (6.9) as partial differential equations to be solved for $F(q, Q, t)$. This may or not be possible to solve, however. Not every possible contact transformation is a canonical transformation. The Hamilton–Jacobi equation is a special case of this "inverse" procedure which we will discuss later.

QUESTION 1: *Canonical Transformation* Follow the recipe for a canonical transformation outlined in the previous section for $F = q + Q$ (use it on your favorite

Hamiltonian). You will find that you do not obtain Hamilton's equations in terms of the new variables. Why does this happen? What is wrong with our generating function?

QUESTION 2: *Change of Scale* A scale change (change of units) clearly does not change the motion of any dynamics. Let's look at what happens to the Hamiltonian and the Lagrangian in this case. 1) Prove that if $H(p, q)$ is the original Hamiltonian, and you make the scale change $Q = \mu q$, $P = \nu p$, then $\bar{H}(Q, P) = \lambda H(\frac{Q}{\mu}, \frac{P}{\nu})$ is the new Hamiltonian in Q, P, where $\lambda = \mu \nu$ where μ, ν are constants. 2) Knowing the form of the Hamiltonian above, prove that the new Lagrangian is of the form of Equation (6.4) with $\lambda = \mu \nu$.

⊃ Example: Harmonic Oscillator Solved by a Canonical Transformation
For this problem (see (5.34))

$$H(q, p) \equiv \frac{1}{2}(p^2 + \omega^2 q^2). \tag{6.13}$$

With 20/20 foresight, choose*

$$F(q, Q) = \frac{1}{2}\omega q^2 \cot 2\pi Q. \tag{6.14}$$

Carry out the canonical transformation:

$$\begin{aligned} p &= \frac{\partial F}{\partial q} = \omega q \cot 2\pi Q, \\ P &= -\frac{\partial F}{\partial Q} = \frac{\pi \omega q^2}{\sin^2 2\pi Q}. \end{aligned} \tag{6.15}$$

Solve the implicit transformation equations for the explicit (inverse) transformation equations:

$$\begin{aligned} p &= \sqrt{\frac{\omega P}{\pi}} \cos 2\pi Q, \\ q &= \sqrt{\frac{P}{\pi \omega}} \sin 2\pi Q. \end{aligned} \tag{6.16}$$

Substitute Equations (6.16) into H:

$$\bar{H} = \frac{\omega}{2\pi} P. \tag{6.17}$$

---

* Not the most obvious function to choose for F, we admit.

Since Q is an ignorable coordinate in $\bar{H}$, P is a constant of the motion. From Hamilton's equations (5.32):

$$\dot{Q} = \frac{\omega}{2\pi}. \tag{6.18}$$

The trivial integration of this equation gives

$$Q = \frac{\omega}{2\pi} t. \tag{6.19}$$

Finally, the familiar solution (in terms of the constant P) is:*

$$p = \sqrt{\frac{\omega P}{\pi}} \cos \omega t,$$
$$q = \sqrt{\frac{P}{\pi \omega}} \sin \omega t. \tag{6.20}$$

(If we wish to be completely general, we could shift the origin of time arbitrarily as well.)

QUESTION 3: *Harmonic Oscillator* Derive the transformation Equations (6.16) from the generating function by the procedure outlined in the previous section. Guess another generating function $F(q, Q)$ and try the same procedure as done in the previous section. What do you get for $\bar{H}$? Is this new equation a valid description of the system?

6.2 DISCOVERING THREE NEW FORMS OF THE GENERATING FUNCTION

All generating functions, of any type, contain one of the old coordinates q or p and one of the new coordinates Q or P. Thus there are four possibilities for each degree of freedom. However, the transformation equations will depend on which of these combinations we are using. We have to derive them for each type. A Legendre transformation is used to replace the dependence on one variable with another, so as to convert one type of generating function to one of the three others.

All four forms of the generating functions found in this way, by successive Legendre transformations, will represent the same *canonical transformation.*

The basic type of generating function is $F(q, Q, t)$. We do not need to know this F explicitly. We are still assuming that $\delta F = 0$ at $t = t_1$ and $t = t_2$, the end points of the time integral for the action. This will always be satisfied, since we are describing the same transformation in different ways. Since $\delta F = p\, \delta q - P\, \delta Q$, if the end point condition is satisfied for one generating function, it will be satisfied for all of them.

---

* H. Goldstein calls this example "cracking a peanut with a sledge hammer"!

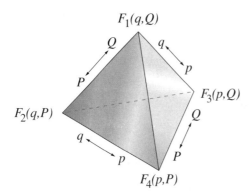

FIGURE 6.1

QUESTION 4: End Points Prove using (6.9) that $\delta F = p\,\delta q - P\,\delta Q$.

We have previously emphasized that generating functions do not depend on the specific problem. The generating function used for the harmonic oscillator (6.14) could also be used for some other problem. The transformations will all be determined by partial derivatives of F, without reference to any particular Lagrangian or Hamiltonian. Of course, the practical usefulness of a particular canonical transformation does depend very much on the problem being considered.

We'll call the type of generating function we've already discussed F_1. The other types will be defined in terms of the independent variables (using a standard notation): $F_2(q, P, t)$, $F_3(p, Q, t)$, $F_4(p, P, t)$. The relations are presented graphically in Figure 6.1.

To get $F_3(p, Q, t)$, make a Legendre transformation on F_1 to eliminate q and substitute p instead, assuming that $p = \frac{\partial F_1}{\partial q}$ as in (6.9):

$$F_3(p, Q, t) = F_1(q, Q, t) - qp. \tag{6.21}$$

Since F_1 is not an explicit function of p, it follows that $\frac{\partial F_1}{\partial p} = 0$, so from Equation (6.21):

$$q = -\frac{\partial F_3}{\partial p}. \tag{6.22}$$

It is still true from Equation (6.9) that

$$P = -\frac{\partial F_1}{\partial Q} = -\frac{\partial F_3}{\partial Q}. \tag{6.23}$$

To derive Equations (6.22) and (6.23), we assumed that q, p, Q, P were all independent variables and not implicit functions of each other. The transformation that makes Q, P functions of q, p is defined by then inverting Equations (6.22) and (6.23). It does take some thought to convince yourself that this procedure of treating the variables as independent until after the partial derivatives are calculated is perfectly valid.

QUESTION 5: Generating Functions 1 Comment on the validity of the procedure used in deriving Equations (6.22, 6.23). In other words, if $Q = Q(q(t), p(t))$ and $P =$

$P(q(t), p(t))$, explain how to derive Equations (6.22, 6.23). What is being held constant and when does it become a function of the other variables?

⊃ **Example**

Start with $F_1(q, Q) \equiv \frac{1}{2}\omega q^2 \cot 2\pi Q$ for the harmonic oscilator and define $F_3(p, Q) \equiv F_1 - qp$. We know that $p = \frac{\partial F_1}{\partial q}$. Use this relation to solve for $q(p, Q) = \frac{p}{\omega} \tan 2\pi Q$. This implies $F_3 = -\frac{p^2}{2\omega} \tan 2\pi Q$. Equation (6.22) gives $q(p, Q)$ above. Equation (6.23) gives $P = \pi \frac{p^2}{\omega} \sec^2 2\pi Q = \pi \omega \frac{q^2}{\sin^2 2\pi Q}$ as in Equation (6.15). We see explicitly that F_3 and F_1 represent the same transformation.

F_4 depends on p and P (and perhaps t). Eliminate P from F_4 to get F_3:

$$F_4(p, P, t) = F_3(p, Q, t) + PQ. \tag{6.24}$$

We know this makes a valid F_3, which depends only on p, Q if F_4 depends only on p and P. Taking partial derivatives to find the transformation rules again, we have

$$q = -\frac{\partial F_4}{\partial p}, \quad Q = \frac{\partial F_4}{\partial P}. \tag{6.25}$$

Finally, define

$$F_2(q, P, t) = F_1(q, Q, t) + QP, \tag{6.26}$$

$$p = \frac{\partial F_2}{\partial q}, \quad Q = \frac{\partial F_2}{\partial P}. \tag{6.27}$$

QUESTION 6: Generating Functions 2 Fill in the steps in the proof of the relations (6.24)–(6.27) for F_4 and F_2. Why do you think we changed the sign of the Legendre transformation in Equations (6.24) and (6.26)? Try the opposite sign. What happens to $\bar{H}$?

⊃ **Example: Famous Examples of Canonical Transformations**

Choose $F_2 = qP$. By using (6.27), we see that we have the identity transformation

$$Q = q, \quad P = p. \tag{6.28}$$

We already tried another simple one: $F_1 = qQ$. For the transformation rules, we got

$$Q = p, \quad P = -q. \tag{6.29}$$

Except for the minus sign, the roles of coordinates and momenta were exchanged. The distinct identity of coordinates and momenta lost its meaning with the canonical transformation. In a general case, the phase space can be transformed at will, possibly leading to a simple solution to whatever problem we want to consider, as demonstrated with the harmonic oscillator.

If there are more degrees of freedom, the examples above can be generalized to become

$$F_2 = \sum_{k=1}^{N} q_k P_k \qquad (6.30)$$

and

$$F_1 = \sum_{k=1}^{N} q_k Q_k. \qquad (6.31)$$

Still another example serves to prove that any point transformation is canonical. Choose

$$F_2 = f(q)P. \qquad (6.32)$$

Here $f(q)$ is an arbitrary function of q. We get the transformation

$$Q = f(q), \quad p = \frac{df}{dq} P. \qquad (6.33)$$

QUESTION 7: *Point Transformations of Many Degrees of Freedom* Try generalizing Equation (6.33) to allow more than one degree of freedom. Also prove that any point transformation is canonical for more than one degree of freedom.

⊃ Example: An *Infinitesimal* Canonical Transformation

Choose $F_2 = qP + \epsilon G(q, P)$. This is very close to the identity transformation if ϵ is infinitesimally small. So far G is an arbitrary function. Carry out the transformation equations for an F_2-type generating function (6.27):

$$\begin{aligned} p &= P + \epsilon \frac{\partial G}{\partial q}, \\ Q &= q + \epsilon \frac{\partial G}{\partial P}. \end{aligned} \qquad (6.34)$$

Now let $\epsilon = dt$, a small time interval. Rearrange the equations above to get P, Q:

$$\begin{aligned} Q &= q + dt \frac{\partial G}{\partial P}, \\ P &= p - dt \frac{\partial G}{\partial q}. \end{aligned} \qquad (6.35)$$

If $G(q, P) = H(q, P) \approx H(q, p)$ as $dt \to 0$, we see that we have reproduced Hamilton's equations. *The Hamiltonian can be viewed as generating a series of infinitesimal canonical transformations which transform us from $(q, p)_t$ to $(q, p)_{t+dt}$.* This point of view on the time development of Hamilton's equations as the iteration of a sequence of canonical transformations proves to be very fruitful for the advanced theory of classical mechanics.

TABLE 6.1 LOOK-UP TABLE OF ALL POSSIBLE GENERATING FUNCTIONS FOR CANONICAL TRANSFORMATIONS

Generating function	Implicit transformation equations
$F_1(q, Q, t)$	$p = \frac{\partial F_1}{\partial q}$; $P = -\frac{\partial F_1}{\partial Q}$
$F_2(q, P, t)$	$p = \frac{\partial F_2}{\partial q}$; $Q = \frac{\partial F_2}{\partial P}$
$F_3(p, Q, t)$	$q = -\frac{\partial F_3}{\partial p}$; $P = -\frac{\partial F_3}{\partial Q}$
$F_4(p, P, t)$	$q = -\frac{\partial F_4}{\partial p}$; $Q = \frac{\partial F_4}{\partial P}$

6.3 POISSON BRACKETS

A certain combination of partial derivatives of two arbitrary functions F and G with respect to canonically conjugate variables $\{q_k\}, \{p_k\}$ is called a "Poisson bracket" $[F, G]_{q,p}$ and is defined as

$$[F, G]_{q,p} \equiv \sum_{k=1}^{N} \left(\frac{\partial F}{\partial q_k} \frac{\partial G}{\partial p_k} - \frac{\partial F}{\partial p_k} \frac{\partial G}{\partial q_k} \right). \tag{6.36}$$

If there is only one degree of freedom, Equation (6.36) becomes

$$[F, G]_{q,p} \equiv \frac{\partial F}{\partial q} \frac{\partial G}{\partial p} - \frac{\partial F}{\partial p} \frac{\partial G}{\partial q}. \tag{6.37}$$

We will restrict the following discussion to this case of only one degree of freedom, but the generalization to many degrees of freedom is completely straightforward.

The value of the Poisson brackets does not depend on which set of conjugate variables are used in the partial derivatives. If Q, P are conjugate variables related by a canonical transformation to another set q, p it is true that

$$[\bar{F}, \bar{G}]_{Q,P} = [F, G]_{q,p}, \tag{6.38}$$

where $\bar{F}$ and $\bar{G}$ are the transformed functions. For this reason we drop the subscripts and refer only to $[F, G]$. Choose any set of variables related by a canonical transformation, and therefore that are dynamically equivalent, to evaluate this.

The proof of (6.38) is reserved for the homework problems. Here we want to point out just one very important consequence. Let $F = Q$ and $G = P$. Then by the fundamental theorem of Poisson brackets, Equation (6.38),

$$[Q, P]_{Q,P} = [Q(q, p), P(q, p)]_{q,p} = 1. \tag{6.39}$$

Without knowing the form of the generating function for the canonical transformation $q, p \to Q, P$, Equation (6.39) gives to us a test of whether a given relationship is canonical or not.

⊃ Example

Consider the harmonic oscillator again. From (6.16) we have the transformation $Q, P \to q, p$, so we need the partial derivatives for $[q, p]_{Q,P}$:

$$\frac{\partial q}{\partial Q} = 2\pi \sqrt{\frac{P}{\omega \pi}} \cos 2\pi Q, \qquad \frac{\partial p}{\partial P} = \frac{1}{2}\sqrt{\frac{\omega}{\pi P}} \cos 2\pi Q,$$
$$\frac{\partial q}{\partial P} = \frac{1}{2}\sqrt{\frac{1}{\omega \pi P}} \sin 2\pi Q, \qquad \frac{\partial p}{\partial Q} = -2\pi \sqrt{\frac{\omega P}{\pi}} \sin 2\pi Q. \tag{6.40}$$

By taking the appropriate combinations to form the Poisson brackets, $[q, p] = 1$, as was inevitable since the transformation was canonical.

In mathematical language, the property (6.39) is sufficient as well as necessary. Not only do canonical transformations yield conjugate pairs of variables with the property, but if this property holds, the transformation must be canonical. Had we doubted that the relations (6.16) could be a canonical transformation, the Poisson bracket test would be conclusive proof. (We reserve the proof of this for a homework problem.)

The canonical relationship between old and new coordinates and momenta (6.39) can be shown to be equivalent to requiring that the Jacobian $\frac{\partial(Q,P)}{\partial(q,p)}$ is a symplectic matrix. (See the appendix at the end of the chapter for the definition of "symplectic" and further discussion.) For one degree of freedom, this means the change of variables preserves phase volume (i.e., $dQ\,dP = dq\,dp$), which is proved in the homework problems.

The close connection between Poisson brackets and commutators of operators in quantum mechanics is beyond the scope of this book.

6.4 HAMILTON–JACOBI EQUATION

Up to now, the solution to a problem with N degrees of freedom involves solving Hamilton's equations of motion; a set of $2N$ first-order ordinary differential equations (ODE) with $2N$ dependent variables and one independent variable, the time. We will show that the same mechanical problem can be solved by finding a solution $S(q_1, \ldots, q_N, t)$ to a single first-order partial differential equation (PDE) containing only the $N + 1$ first-order partial derivatives with respect to the q_ks, the q_ks, and the time. The function S is a special generating function producing a remarkable canonical transformation, as we will explain.

Assume there exists a canonical transformation for which the transformed Hamiltonian equals zero. By the general rule for transforming a Hamiltonian using any type of generating

TABLE 6.2 COMPARISON OF HAMILTION'S EQUATIONS TO THE HAMILTON–JACOBI EQUATION

	Hamilton's equations	Hamilton–Jacobi equation
# of independent variables	1: (time t)	$N + 1$: ($\{q_k\}, t$)
# of dependent variables	$2N$: (q_k, p_k)	1: (the action S)

6.4 HAMILTON–JACOBI EQUATION

function F (6.12):

$$\bar{H}(Q_1, \ldots, Q_N, P_1, \ldots, P_N, t) = 0 = H(q_1, \ldots, q_N, p_1, \ldots, p_N, t) + \frac{\partial F}{\partial t}. \quad (6.41)$$

All the Ps and Qs will be constants, since Hamilton's equations for $\bar{H}$ would give $\dot{P}_k = \dot{Q}_k = 0$, $k = 1, \ldots, N$. Since the new variables are constants, the information about the time development of the system's motion is contained in the canonical transformation. Further, assume there exists a generating function* $S \equiv F_2(q_1, \ldots, q_N, P_1, \ldots, P_N, t)$ that produces a canonical transformation with the desired property that $\bar{H} = 0$. Then $p_k = \frac{\partial S}{\partial q_k}$ and $Q_k = \frac{\partial S}{\partial P_k}$, as we have seen from Equation (6.27). The Equation (6.41) we have to solve for S becomes

$$\boxed{H\left(q_1, \ldots, q_N, \frac{\partial S}{\partial q_1}, \ldots, \frac{\partial S}{\partial q_N}, t\right) + \frac{\partial S}{\partial t} = 0.} \quad (6.42)$$

Hamilton–Jacobi equation

The solution S to Equation (6.42) is called *Hamilton's Principal Function*. Equation (6.42) is a first-order differential equation in $N + 1$ independent variables. By the general theory of differential equations, there will be $N + 1$ arbitrary constants. There could be one arbitrary additive constant in S, since $S +$ constant is a solution for (6.42) if S is a solution. The other N constants are defined as the P_k, $k = 1, \ldots, N$. These constants are fixed by the initial conditions, as are also the constant Q_ks, which equal $\frac{\partial S}{\partial P_k}$. However, there is a great deal of freedom in how these constants P_k are defined. It is good to realize that the Hamilton–Jacobi equation is more of a theoretical framework than a specific algorithm for solving problems. The art, in fact, lies in how the constant new momenta are defined, if we wish to solve a given problem. As a tool for theoretical understanding, it isn't necessary to be specific, however.

To remind ourselves that the P_ks in $S(q_1, \ldots, q_k, P_1, \ldots, P_k, t)$ are constants, let's rename the P_ks to be α_ks. (This notation is used here because it seems to be a standard convention with other authors, notably Goldstein in *Classical Mechanics*, second edition, and Landau and Lifschitz in *Mechanics*, third edition.) Also, the Q_ks $= \frac{\partial S}{\partial P_k}$ comprise N additional constants, which we rename to be β_ks:

$$Q_k \equiv \beta_k = \frac{\partial S(q_1, \ldots, q_N, \alpha_1, \ldots, \alpha_N, t)}{\partial \alpha_k} = \text{constants}. \quad (6.43)$$

If we solve (6.43) for the q_ks, we have $q_k(\alpha_1, \ldots, \alpha_N, \beta_1, \ldots, \beta_N, t)$. The actual values of the αs and βs are determined by the initial conditions. Solving the Hamilton–Jacobi

---

* S must have this form to be an F_2-type generating function. Since the P_k are constants, we can consider S to be a function of only the q_ks, if it is convenient to do so.

equation is equivalent to solving the original mechanical problem, because if you can solve for S, then (in principle) you have the canonical transformation that gives the full solution in terms of the initial conditions. To see this explicitly, suppose that you know $q_k(0)$, $p_k(0)$, the coordinates and momenta at $t = 0$. Solve for the α_k, β_k constants of the motion by solving the $2N$ implicit equations:

$$p_k(0) = \left.\frac{\partial S}{\partial q_k}\right|_{t=0}, \quad \beta_k = \left.\frac{\partial S}{\partial \alpha_k}\right|_{t=0}. \tag{6.44}$$

As a practical matter, solving (6.42) and Equations (6.43, 6.44) may or may not actually be possible for a given problem. It may be possible in one coordinate system but not in another. As it is a purely theoretical concept so far, we do not consider such questions at this stage. However, much better understanding can be gained by following the worked examples below and by doing the homework problems. The theory at this point is rather abstract.

Consider the total time derivative of S:

$$\frac{dS}{dt} = \sum_k \frac{\partial S}{\partial q_k} \dot{q}_k + \frac{\partial S}{\partial t}. \tag{6.45}$$

From Equations (6.27, 6.42) it follows that

$$\frac{dS}{dt} = \sum_k p_k \dot{q}_k - H = L. \tag{6.46}$$

The generating function S is the time integral of the Lagrangian (which was defined as the action!):

$$\boxed{S = \int L\, dt.} \tag{6.47}$$

This is an elegant theoretical result, but one soon realizes that the entire problem must first be solved for $q_1(t), \ldots$ in terms of S in order to know $L(q_1(t), \ldots, \dot{q}_1(t), \ldots, t)$. However, the result (6.47) does show that "least action" (Hamilton's Principle) is a principle that could lead to the solution of partial differential equations.

QUESTION 8: Hamilton–Jacobi Equation Explain why choosing S to satisfy the equation $\bar{H} = 0$ is equivalent to solving the differential equation (6.42) with 1 dependent variable and $N + 1$ independent variables.

⊃ Example: A Particle of Mass m in a Gravitational Field

The three-dimensional Hamiltonian for a particle of mass m in a gravitational field in the z direction is

$$H = \frac{p_x^2 + p_y^2 + p_z^2}{2m} + mgz, \tag{6.48}$$

6.4 HAMILTON–JACOBI EQUATION

where p_k is the conjugate momentum for the particle in the kth direction. From (6.42), the Hamilton–Jacobi equation for the generating function S is

$$\frac{1}{2m}\left(\left(\frac{\partial S}{\partial x}\right)^2 + \left(\frac{\partial S}{\partial y}\right)^2 + \left(\frac{\partial S}{\partial z}\right)^2\right) + mzg + \frac{\partial S}{\partial t} = 0. \tag{6.49}$$

We can "separate," this equation, that is, write S as the sum of separate terms, each depending on a single coordinate or the time:

$$S = W_1(x) + W_2(y) + W_3(z) - Et, \tag{6.50}$$

where E is a constant. Writing S in this way is called the method of *separation of variables*. This form will be justified by showing that it leads to a solution of (6.49). If the Hamiltonian contains the time explicitly we cannot separate the time dependence of S in this way. Substituting (6.50) into (6.49) yields

$$\frac{1}{2m}\left(\left(\frac{dW_1(x)}{dx}\right)^2 + \left(\frac{dW_2(y)}{dy}\right)^2 + \left(\frac{dW_3(z)}{dz}\right)^2\right) + mzg = E. \tag{6.51}$$

Since the constant E is the sum of terms which each depend on a separate independent variable, the individual terms must themselves be constants: $\alpha_1 + \alpha_2 + \alpha_3 \equiv E$. Notice that we have defined the meaning of the new momenta $\alpha_1, \alpha_2, \alpha_3$ to be what are called *separation constants*. From this argument we obtain a set of three ordinary differential equations to solve:

$$\frac{1}{2m}\left(\frac{dW_1(x)}{dx}\right)^2 = \alpha_1, \quad \frac{1}{2m}\left(\frac{dW_2(y)}{dy}\right)^2 = \alpha_2,$$

$$\frac{1}{2m}\left(\frac{dW_3(z)}{dz}\right)^2 + mgz = \alpha_3. \tag{6.52}$$

The solutions are

$$W_1(x) = \pm x\sqrt{2m\alpha_1}, \quad W_2(y) = \pm y\sqrt{2m\alpha_2},$$

$$W_3(z) = \pm\sqrt{\frac{8}{9mg^2}(\alpha_3 - mgz)^{3/2}}. \tag{6.53}$$

(An additional additive constant can always be included in S.) We have found the solution in terms of the necessary three constants.

The crucial step in performing the separation of variables to solve the Hamilton–Jacobi equation is to require the $N = 3$ constants $\alpha_1, \alpha_2, \alpha_3$ created by the separation of variables to be the new "momenta" in a canonical transformation generated by $S = S(x, y, z, \alpha_1, \alpha_2, \alpha_3, t)$. Notice that we *define* the α_i to be the new momenta. We also have the constants β_k defined by (6.43). From the definition of S in (6.50) and

Equations (6.53) above, we find

$$\beta_1 = \frac{\partial S}{\partial \alpha_1} = \pm x\sqrt{\frac{m}{2\alpha_1}} - t, \quad \beta_2 = \frac{\partial S}{\partial \alpha_2} = \pm y\sqrt{\frac{m}{2\alpha_2}} - t,$$

$$\beta_3 = \frac{\partial S}{\partial \alpha_3} = \pm\sqrt{\frac{2(\alpha_3 - mgz)}{mg^2}} - t.$$

(6.54)

Finally, invert (6.54) to solve for x, y, z in terms of the constants and the time:

$$x = \pm\sqrt{\frac{2\alpha_1}{m}}(\beta_1 + t), \quad y = \pm\sqrt{\frac{2\alpha_2}{m}}(\beta_2 + t),$$

$$z = \frac{\alpha_3}{mg} - \frac{g}{2}(\beta_3 + t)^2.$$

(6.55)

This is the right form for the motion of a free particle in a gravitational field. The constants $\alpha_1, \alpha_2, \alpha_3, \beta_1, \beta_2, \beta_3$ depend on initial conditions, the values of x, y, z, and their time derivatives at $t = 0$.

What Can Be Done If the Hamiltonian Does Not Contain the Time Explicitly

We already know that a Hamiltonian that does not contain the time is constant. H is usually also the total energy E. Let S have a special time dependence

$$S(q_1, \ldots, q_N, \alpha_1, \ldots, \alpha_N, t) \equiv W(q_1, \ldots, q_N, \alpha_1, \ldots, \alpha_N) - Et. \quad (6.56)$$

Since, by assumption, this new function W does not contain the time explicitly, the Hamilton–Jacobi Equation (6.42) becomes

$$H\left(q_1, \ldots, q_N, \frac{\partial W}{\partial q_1}, \ldots, \frac{\partial W}{\partial q_N}\right) = E. \quad (6.57)$$

Hamilton–Jacobi equation (second form)

Any solution to Equation (6.57) will define a new function W. Since there are N dependent variables, there will be N independent constants in the solution W. W is called *Hamilton's Characteristic Function*. We can give W a more physical interpretation by noticing that

$$W = S + Et = \int (L + H)\,dt = \int \sum_{k=1}^{N} p_k \dot{q}_k\,dt = \sum_{k=1}^{N} \int p_k\,dq_k. \quad (6.58)$$

The function W can serve in its own right as a new generating function of type F_2. The canonical transformation generated by W is different from that generated by S because

6.4 HAMILTON–JACOBI EQUATION

$\bar{H} = E \neq 0$. A set of equations is obtained in which the time appears only once. This method is valuable for finding constants of the motion and for determining orbit equations for which the time has been eliminated. In the case of S, all the new momenta and coordinates were constants. With W the situation is somewhat different. The new coordinates $Q_k, k = 1, \ldots, N$ are all ignorable in the new Hamiltonian (6.57), which implies that the N new momenta P_k are constants, since $\dot{P}_k = -\frac{\partial H}{\partial Q_k} = 0$.

There is a wide latitude of possible choices available. The simplest choice is to define $P_1 \equiv E$. If we make this choice, then $\dot{Q}_1 = \frac{\partial H}{\partial P_1} = \frac{\partial E}{\partial P_1} = 1$ and thus

$$Q_1 = t + \text{constant} = \frac{\partial W}{\partial E}. \tag{6.59}$$

Q_1 is no longer a constant, as it was with S. It is the time. The time is the coordinate that is canonically conjugate to the energy. The arbitrary constant sets the zero of time. It is a constant of integration and determines the lower limits of the integrals defining W. The $N-1$ momenta $P_2, \ldots, P_N$ are independent of P_1, so differentiating the Hamiltonian with respect to the other momenta gives zero, and hence the $N-1$ new coordinates $Q_2, \ldots, Q_N$ are constants as well. Change the previous definition of the constants α, β to have an index i which runs from 1 to $N-1$: $\alpha_i \equiv P_{i+1}, \beta_i \equiv Q_{i+1}$.

The solution to the Hamilton–Jacobi equation (6.57) can be written as a function:

$$W(q_1, \ldots, q_N, E, \alpha_1, \ldots, \alpha_{N-1}).$$

There are $N - 1$ equations that do not involve the time:

$$\beta_i = \frac{\partial W}{\partial \alpha_i}, \quad i = 1, \ldots, N-1. \tag{6.60}$$

The N equations $p_k = \frac{\partial W}{\partial q_k}$ can be evaluated at $t = 0$ to give $\{E, \alpha_i\}$ and $\{0, \beta_i\}$ for $i = 1, \ldots N-1$ in terms of the $2N$ initial values $q_k(0), p_k(0)$. The $q_k(t)$s can then be found at other times $t \neq 0$ by inverting (6.59, 6.60). This is almost the same as the procedure we followed with the S generating function, except that *one* of the constants is the energy, and the time has been isolated as one of the new coordinates in the equations.

Since the $N - 1$ equations for $\beta_i = \frac{\partial W}{\partial \alpha_i}$ do not contain the time, you can choose any one of the N q_k as an independent variable and solve for the other $N - 1$ q_ks as dependent variables – the orbit. We did this in Chapter 4 for $r(\phi)$ in the Kepler problem.

In practice, solving for either the Principal Function S or the Characteristic Function W can only be done explicitly if the variables can be separated, the general method for which is described below. This does not diminish the more abstract theoretical value of S and W, however. Their great value lies in giving a constructive procedure for finding all the constants of the motion, provided that a solution to the Hamilton–Jacobi equation exists. The existence of solutions was once considered to be obvious, although as we will see, it is by no means always a correct assumption!

The preceding discussion about using W instead of S is rather abstract. It can be put in more concrete terms by using the same example we have already given above.

Example: Parabolic Orbit of a Falling Body Obtained Using W

Consider the previously solved example of the falling body. To carry out this procedure, we must alter the meaning of the αs and βs. The philosophy has changed, since $\bar{H}$ is no longer equal to zero. The actual equations are almost identical with those we solved before with this example. The first new momentum, P_1, is assigned to equal the Hamiltonian, which means that $Q_1 = t$ plus an arbitrary integration constant. The time is separated from the rest of the problem at the beginning. There are now $N - 1 = 2$ remaining Ps and Qs to be found from the solution for the generating function W. After separating the variables, the characteristic function W takes the form

$$W(x, y, z, \alpha_1, \alpha_2) = W_1(x, \alpha_1) + W_2(y, \alpha_2) + W_3(z, E - \alpha_1 - \alpha_2). \quad (6.61)$$

The second form of the Hamilton–Jacobi equation is

$$\frac{1}{2m}\left(\left(\frac{\partial W_1}{\partial x}\right)^2 + \left(\frac{\partial W_2}{\partial y}\right)^2 + \left(\frac{\partial W_3}{\partial z}\right)^2\right) + mgz = E. \quad (6.62)$$

The solution for S found above also gives the solution for W. Define the new momenta $P_2 \equiv \alpha_1$, $P_3 \equiv \alpha_2$:

$$\alpha_1 = \frac{\left(\frac{\partial W_1}{\partial x}\right)^2}{2m},$$

$$\alpha_2 = \frac{\left(\frac{\partial W_2}{\partial y}\right)^2}{2m}, \quad (6.63)$$

$$E - \alpha_1 - \alpha_2 = \frac{\left(\frac{\partial W_3}{\partial y}\right)^2}{2m} + mgz.$$

The solutions to Equations (6.63) are

$$W_1(x, \alpha_1) = \pm x\sqrt{2m\alpha_1},$$
$$W_2(y, \alpha_2) = \pm y\sqrt{2m\alpha_2}, \quad (6.64)$$
$$W_3(z, E - \alpha_1 - \alpha_2) = \pm\sqrt{\frac{8}{9mg^2}}(E - \alpha_1 - \alpha_2 - mgz)^{\frac{3}{2}}.$$

$\beta_{1,2} = \frac{\partial W}{\partial \alpha_{1,2}}$. Notice that $\alpha_{1,2}$ appear now in two places. It is necessary to find the derivative in both places. This is a key difference between using S and using W. We find

$$\beta_1 = \pm x\sqrt{\frac{m}{2\alpha_1}} \mp \sqrt{\frac{2}{mg^2}(E - \alpha_1 - \alpha_2 - mgz)},$$

$$\beta_2 = \pm y\sqrt{\frac{m}{2\alpha_2}} \mp \sqrt{\frac{2}{mg^2}(E - \alpha_1 - \alpha_2 - mgz)}. \quad (6.65)$$

Evaluate the constants $\alpha_{1,2}, \beta_{1,2}$ using (6.65) and $p_x = \frac{\partial W}{\partial x}, p_y = \frac{\partial W}{\partial y}$ at $t = 0$. To simplify the algebra, assume specific values $x(0) = y(0) = z(0) = 0$ and $p_y(0) = 0$ as initial conditions. Equations (6.65), evaluated at $t = 0$, becomes

$$\beta_1 = \mp \frac{1}{g}\sqrt{\frac{2}{m}(E - \alpha_1 - \alpha_2)}, \qquad (6.66)$$

$$\beta_2 = \beta_1.$$

Since it makes no difference, select the upper sign in the equations above. From the equations for $p_{x,y}$, obtain

$$p_x(0) = \sqrt{2m\alpha_1}, \quad \alpha_1 = \frac{p_x^2(0)}{2m},$$
$$0 = \sqrt{2m\alpha_2}, \quad \alpha_2 = 0. \qquad (6.67)$$

Subtract the two equations (6.66). A contradiction is avoided only if $y = 0$ for all times. This makes physical sense as well. Setting $y = 0$, evaluate

$$\beta_1 = \frac{1}{g}\sqrt{\frac{2}{m}\left(E - \frac{p_x^2(0)}{2m}\right)} = \frac{x}{v_x(0)} - \frac{1}{g}\sqrt{\frac{2}{m}\left(E - \frac{p_x^2(0)}{2m} - mgz\right)}, \qquad (6.68)$$

where $v_x(0) = \frac{p_x(0)}{m}$. Equation (6.68) is the promised equation for the orbit, but in rather convoluted form. After squaring and further manipulation of (6.68), and using $E = \frac{p_x^2(0) + p_z^2(0)}{2m}$ and $p_z(0) = m v_z(0)$, we obtain a recognizable equation for the parabolic orbit $z(x)$ of a falling body:

$$z = x\frac{v_z(0)}{v_x(0)} - \frac{x^2}{2g v_x^2(0)}. \qquad (6.69)$$

Although (6.69) could easily have been found without solving the Hamilton–Jacobi equation, it does illustrate how directly constants of the motion can be discovered. They arise automatically with this method.

The time dependence of the three coordinates was obtained directly using S. If the time dependence is desired, use $t = \frac{\partial W}{\partial E}$.

Separability

For systems with $N > 1$ degrees of freedom, if Hamilton's Principal Function can be written as the sum of N terms, each depending on a single coordinate including the time,

$$S(q_1, q_2, \ldots, q_N, \alpha_1, \alpha_2, \ldots, \alpha_N, t) = \sum_{k=1}^{N} W_k(q_k, \alpha_1, \ldots, \alpha_N) - Et, \qquad (6.70)$$

then the Hamilton–Jacobi Equation (6.42) can be separated into N separate equations. (We write the new momentum variables as α_k as before, as a reminder that they are constants.) An even simpler form of separability exists if the Hamiltonian itself is a sum of N independent ("uncoupled") parts:

$$H(p_1, \ldots, p_N, q_1, \ldots, q_N) = \sum_{k=1}^{N} H_k(p_k, q_k). \tag{6.71}$$

The Hamilton–Jacobi equation then splits into N independent equations

$$H_k\left(q_k, \frac{\partial W}{\partial q_k}\right) = \alpha_k, \quad E = \sum_{k=1}^{N} \alpha_k. \tag{6.72}$$

Each α_i term is a constant. Hence the problem dissolves into N separate 1-D problems to be solved by quadratures, as we have already discussed. The previous example demonstrated this method.

Whether or not this can be done depends both on the problem and on the coordinate system we choose. Problems that can be solved by separating the variables include: 1) central force motion, including the Kepler problem (spherical polar coordinates only), 2) the problem of two fixed centers of gravitation (use elliptic coordinates), and 3) a particle attracted to a fixed center of force and also subjected to a uniform gravitational field. Ignorable coordinates can always be separated. For orthogonal coordinate systems, there are certain very general conditions, called the *Staeckel conditions*, that determine which kinds of potentials can be separated.* For example, in spherical polar coordinates, one has a separable problem only for a potential of the form

$$V = V_r(r) + \frac{V_\theta(\theta)}{r^2} + \frac{V_\phi(\phi)}{r^2 \sin^2 \theta}. \tag{6.73}$$

A central force is a special case of this, with $V_\theta = V_\phi = 0$. Separation of the variables in an appropriate coordinate system is considered to be the most powerful analytic technique for solving mechanics problems. But not all problems can be solved in this way. An example is the "Sun–Earth–Moon" problem – the famous (or infamous) three-body problem, for which no analytic solution is possible. This is closely connected with the existence of chaos, which we will discuss in Chapter 11.

> **Example**
>
> Here is an example of a separable system in the sense of (6.70) but not in the very simple sense of (6.71). It is the case of a particle attracted by a fixed gravitating body while also being in a uniform gravitational field oriented along the Z axis. The potential energy ($m = 1$) is
>
> $$V(r, z) = -\frac{k}{r} + gz, \tag{6.74}$$

---

* For a further discussion of this subject, see H. Goldstein, *Classical Mechanics*, second edition, p. 453.

6.4 HAMILTON–JACOBI EQUATION

where r is the distance to the origin in spherical coordinates, z is the height above the XY plane, and $k > 0$. This problem involves a noncentral force, with rotational symmetry around the Z axis.

From the general principles of solving the problem with the second form of the Hamilton–Jacobi equation, we expect to find four constants of the motion, two αs and two βs, since this problem has two non-trivial degrees of freedom.

The first step is to select an appropriate coordinate system, since separability in one coordinate description does not necessarily imply separability in another. We will choose parabolic coordinates ξ, η, which are defined by

$$\xi \equiv r + z, \quad \eta \equiv r - z, \quad r = \sqrt{x^2 + y^2 + z^2}. \tag{6.75}$$

Since $r \geq |z|$, both ξ and η range from 0 to ∞. Surfaces of constant ξ and constant η are paraboloids of revolution. The third coordinate is the azimuthal angle ϕ. In Cartesian coordinates we have

$$x = \sqrt{\xi\eta}\cos\phi, \quad y = \sqrt{\xi\eta}\sin\phi, \quad z = \frac{1}{2}(\xi - \eta).$$

The next step is to find the kinetic energy from the velocity ($\frac{ds}{dt}$):

$$dx = \frac{1}{2}\left(\sqrt{\frac{\eta}{\xi}}\,d\xi + \sqrt{\frac{\xi}{\eta}}\,d\eta\right)\cos\phi - \sqrt{\xi\eta}\sin\phi\,d\phi,$$

$$dy = \frac{1}{2}\left(\sqrt{\frac{\eta}{\xi}}\,d\xi + \sqrt{\frac{\xi}{\eta}}\,d\eta\right)\sin\phi + \sqrt{\xi\eta}\cos\phi\,d\phi, \tag{6.76}$$

$$dz = \frac{1}{2}(d\xi - d\eta),$$

$$ds^2 = dx^2 + dy^2 + dz^2 = \frac{1}{4}\left(\left[1 + \frac{\eta}{\xi}\right]d\xi^2 + \left[1 + \frac{\xi}{\eta}\right]d\eta^2\right) + \xi\eta\,d\phi^2. \tag{6.77}$$

By differentiating the kinetic energy with respect to $\dot\xi, \dot\eta, \dot\phi$, obtain p_ξ, p_η, p_ϕ:

$$p_\xi = \frac{m}{4}\left(1 + \frac{\eta}{\xi}\right)\dot\xi, \quad p_\eta = \frac{m}{4}\left(1 + \frac{\xi}{\eta}\right)\dot\eta, \quad p_\phi = m\xi\eta\dot\phi. \tag{6.78}$$

To get the Hamiltonian, first express the kinetic energy in terms of canonical momenta:

$$T = \frac{1}{2}m\left(\frac{ds}{dt}\right)^2 = \frac{2}{m(\xi+\eta)}\left[\xi p_\xi^2 + \eta p_\eta^2\right] + \frac{p_\phi^2}{2m\xi\eta}. \tag{6.79}$$

In parabolic coordinates, the potential energy is

$$V(\xi, \eta) = -\frac{2k}{(\xi + \eta)} + \frac{g}{2}(\xi - \eta). \tag{6.80}$$

This is an example of a problem where the separation of variables is possible, but the energy is not a sum of separate terms each involving a single coordinate. To separate the variables, set the sum $T+V$ equal to the energy E and multiply through by $\frac{1}{2}(\xi+\eta)$. We get

$$\underbrace{\xi p_\xi^2 + \frac{g}{4}\xi^2 + \frac{p_\phi^2}{4\xi} - \frac{E}{2}\xi}_{\text{depends on } \xi} + \underbrace{\eta p_\eta^2 - \frac{g}{4}\eta^2 + \frac{p_\phi^2}{4\eta} - \frac{E}{2}\eta}_{\text{depends on } \eta} = k. \tag{6.81}$$

Since ϕ is an ignorable coordinate in the Hamiltonian, p_ϕ is a constant of the motion. We have now reduced the problem to one with two degrees of freedom, and so we can expect one α and one β as new constants of the motion. Separating the variables in Equation (6.81) determines a constant of the motion α. Write $W = p_\phi \phi + W_\xi + W_\eta$ and substitute $p_\xi \to \frac{\partial W_\xi}{\partial \xi}$ in the first equation below to get

$$\begin{aligned}
\alpha &\equiv \xi p_\xi^2 + \frac{g}{4}\xi^2 + \frac{p_\phi^2}{4\xi} - \frac{E}{2}\xi, \\
k - \alpha &= \eta p_\eta^2 - \frac{g}{2}\eta^2 + \frac{p_\phi^2}{4\eta} - \frac{E}{2}\eta.
\end{aligned} \tag{6.82}$$

Do the same for η in the second equation. In the nearly identical equation for $W_\eta(\eta, k-\alpha)$, only the sign of g is reversed. The new constant of the motion can be written in more symmetrical form by subtracting the second equation above from the first one:

$$2\alpha = k + \underbrace{\left[\xi p_\xi^2 - \eta p_\eta^2\right] + \frac{g}{2}(\xi^2 + \eta^2) + \frac{p_\phi^2}{4}\left(\frac{1}{\xi} - \frac{1}{\eta}\right) - \frac{E}{2}(\xi - \eta)}_{\text{constant of the motion}} \tag{6.83}$$

(If $g = 0$, this problem reduces to the familiar Kepler problem. The constant above is then proportional to the z component of the Lenz vector, which is a conserved quantity for an inverse-square force law.) The other new constant of the motion is $\beta = \frac{\partial W}{\partial \alpha}$. The solution can be written in terms of definite integrals:

$$W_\xi(\xi, \alpha) = \int p_\xi \, d\xi = \frac{1}{\sqrt{2}} \int_{\xi(0)}^{\xi} d\xi \sqrt{E - \frac{g}{2}\xi + \frac{\alpha}{\xi} - \frac{p_\phi^2}{2\xi^2}}, \tag{6.84}$$

$$W_\eta(\eta, k-\alpha) = \int p_\eta \, d\eta = \frac{1}{\sqrt{2}} \int_{\eta(0)}^{\eta} d\eta \sqrt{E + \frac{g}{2}\eta + \frac{(k-\alpha)}{\eta} - \frac{p_\phi^2}{2\eta^2}}. \tag{6.85}$$

Since our purpose was only to demonstrate the separability of this problem, we will not go any further. Information about the existence of bounded orbits ($E < 0$) and the turning points of the motion can be deduced from the requirement that both quantities under the square roots in (6.84, 6.85) be positive only for bounded regions in ξ and η. One must therefore examine the possible roots of these quantities as ξ and η are varied.

6.4 HAMILTON–JACOBI EQUATION

The Relation of Hamilton–Jacobi Theory to Quantum Mechanics

Classical mechanics is to quantum mechanics as ray (geometric) optics is to wave optics. For classical geometric optics the wavelength of the light is irrelevant. One has only to draw the light rays and propagate them through the medium according to the index of refraction, using Fermat's Principle of Least Time. In simple cases this amounts to using Snell's Law at the boundaries between different media.

The historical motivation for the Hamilton–Jacobi equation was to exploit this analogy between optics and mechanics. The momenta $p_1, \ldots$ are proportional to the gradient of S in N-dimensional *configuration* space. Therefore, surfaces on which S is constant are normal to the possible trajectories $q_1(t), \ldots$ in configuration space. The surfaces of constant phase advance because surfaces on which S is constant move through the configuration space due to the Et part of S. In classical optics these surfaces are surfaces of constant phase (i.e., wave fronts) and the trajectories are light rays, which are always perpendicular to the wave fronts. The way light propagates through a medium with variable index of refraction is described by an equation exactly like (6.42).

However, the invention of quantum theory showed that the correspondence between mechanics and optics was more than just an analogy. In quantum mechanics, S is proportional to the phase of the wave function, so we have a definite mathematical connection between the Schrödinger equation and mechanics. This connection between quantum mechanics and classical mechanics must exist, since in the limit of Planck's constant $\hbar \to 0$, quantum mechanics must reduce to classical mechanics.

Start with the 1-D Schrödinger equation

$$H\Psi = -\frac{\hbar^2}{2m}\frac{\partial^2 \Psi}{\partial x^2} + V(x)\Psi = i\hbar \frac{\partial \Psi}{\partial t}, \qquad (6.86)$$

where Ψ is a complex function

$$\Psi = \sqrt{\rho(x,t)}\, e^{i\frac{S(x,t)}{\hbar}}, \qquad (6.87)$$

and ρ and S are *real* functions of x and t, where ρ is the probability density. Substituting (6.87) into (6.86) above gives

$$-\frac{\hbar^2}{2m}\left\{\frac{\partial^2 \sqrt{\rho}}{\partial x^2} + \frac{2i}{\hbar}\frac{\partial \sqrt{\rho}}{\partial x}\frac{\partial S}{\partial x} - \frac{1}{\hbar^2}\sqrt{\rho}\left(\frac{\partial S}{\partial x}\right)^2 + \frac{i}{\hbar}\sqrt{\rho}\frac{\partial^2 S}{\partial x^2}\right\}e^{i\frac{S}{\hbar}}$$
$$+ V(x)\sqrt{\rho}\, e^{i\frac{S}{\hbar}} = i\hbar\left\{\frac{\partial \sqrt{\rho}}{\partial t} + \frac{i}{\hbar}\sqrt{\rho}\frac{\partial S}{\partial t}\right\}e^{i\frac{S}{\hbar}}. \qquad (6.88)$$

Assume $\hbar|\frac{\partial^2 S}{\partial x^2}| \ll |\frac{\partial S}{\partial x}|^2$. (Here $\hbar$ is regarded as small. We will comment on the physical meaning of this assumption below.) Then collecting terms without $\hbar$ dependence to get the limit of the Schrödinger equation as $\hbar \to 0$, we get

$$\boxed{\frac{1}{2m}\left(\frac{\partial S}{\partial x}\right)^2 + V(x) + \frac{\partial S}{\partial t} = 0.} \qquad (6.89)$$

Quantum mechanics, in the limit of $\hbar \to 0$, gives the Hamilton–Jacobi equation* (6.42) for a 1-D particle in a conservative force. Also, in the $\hbar \to 0$ limit, we may associate the phase of the wave function as the classical action divided by $\hbar$. This forms the basis for the WKB method of quantum mechanics. It can be shown that the limit $\hbar |\frac{\partial^2 S}{\partial x^2}| \ll |\frac{\partial S}{\partial x}|^2$ corresponds to *making the potential essentially constant over many de Broglie wavelengths*. This defines the short wavelength limit or semiclassical WKB limit.

6.5 ACTION–ANGLE VARIABLES FOR 1-D SYSTEMS

Assume that we have a 1-D mechanical system undergoing periodic motion. Suppose that we could make a canonical transformation from variables p, q to a different, but still canonical, set of variables I, ψ. Further suppose that in terms of the new variables the new Hamiltonian lacks any dependence on ψ (i.e., $H = H(I)$), so ψ is then an ignorable coordinate. Because H is constant and depends only on I, I itself must be a constant of the motion. Hamilton's equations in the new variables become quite simple:

$$\dot{I} = -\frac{\partial H}{\partial \psi} = 0, \quad \dot{\psi} = \frac{\partial H}{\partial I} = \text{constant}. \tag{6.90}$$

The variable ψ must increase linearly with the time:

$$\omega(I) \equiv \dot{\psi} = \frac{\partial H}{\partial I}, \quad \psi = \omega[t - t_0]. \tag{6.91}$$

Here I plays the role of a momentum and is known as the *action variable*, while ψ is the coordinate conjugate to I and is called the *angle variable*.

How might we find a transformation that accomplishes this desirable task of transforming us from (p, q) to (I, ψ)? We will use a type-F_1 generating function, $\tilde{W}(q, \psi)$, which is a function of both old and new coordinate variables. Since we have assumed that the motion is periodic in p, q, then the motion must also be periodic in ψ, so $\tilde{W}(q, \psi)$ is a periodic function of ψ. From the rules (6.9) for generating functions of this type

$$d\tilde{W} = p\, dq - I\, d\psi. \tag{6.92}$$

($\tilde{W}(q, \psi)$ is related to Hamilton's Characteristic Function $W(q, I)$ by a Legendre transformation.) If we integrate over a single period of the motion, q returns to its original value, while ψ advances by an amount that we can choose to be 2π per period:

$$\oint d\tilde{W} = 0 = \oint p\, dq - \oint I\, d\psi. \tag{6.93}$$

Because I is a constant, it can be taken out of the integral. The integral $\oint d\psi$ gives 2π, so

$$\boxed{I = \frac{1}{2\pi} \oint p\, dq.} \tag{6.94}$$

---

* Which means that, in this limit, the classical equations of motion are restored.

6.5 ACTION–ANGLE VARIABLES FOR 1-D SYSTEMS

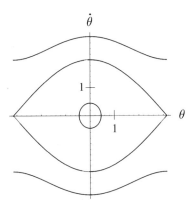

FIGURE 6.2
Pendulum phase plane trajectories: The inner curve is at $E = 0.1$, the middle curve at $E = 2.0$, and the outer curve at $E = 4.0$. Note that we scaled the time here so that $\omega = 1$.

This last equation can be taken as the definition of I, the action variable.* Equation (6.94) applies to an integral around a single period of the motion. Since the area enclosed by a closed curve in phase space will always be finite, the integral (6.94) and thus the action variable will always exist.

Because $d\tilde{W}$ is an exact differential (like the differential of the potential energy), we can find $\tilde{W}(q, \psi)$ by doing an indefinite integral of (6.92). In practice this is often unnecessary.

If Ω is the area of phase space enclosed by the closed phase trajectory ($\mathcal{C}$) of a periodic system, by Stoke's theorem[†]

$$\oint_{\mathcal{C}} p \, dq = \iint_{\Omega} dp \, dq = \text{enclosed area in phase space.} \tag{6.95}$$

⊃ **Example**

The most common example of a periodic 1-D mechanical system is the pendulum.

Phase trajectories for three energies, each of which represents a different type of motion, are plotted in Figure 6.2 for a "scaled" pendulum with $H(p, \theta) = \frac{p^2}{2} + (1 - \cos \theta)$. (Note that we made a similar plot in Chapter 4 – see Figure 4.4.) The inner curve corresponds to an oscillating pendulum, the middle curve is the separatrix, and the outer curve corresponds to rotary motion. The motion of the pendulum is periodic in any of these cases, and the action variable is the area enclosed in phase space divided by 2π. One can imagine the phase space wrapped into a cylinder, since $\theta = \pi$ is the same physical angle as $\theta = -\pi$. We have to make the "closed curve" in the case of rotation to be the upper curve from left to right and the lower curve from right to left. On the cylinder these two physically different curves enclose a finite area. Divided by 2π, this is the generalized action variable for rotation. As the energy grows very large, $E \gg 2$, $I \to 2\sqrt{2E}$.

---

* Often this is shortened to the *action*, but since we want to avoid confusion with the action S, we will always refer to I as the action variable.
[†] See Kaplan, *Advanced Calculus*, 3rd ed., p. 328.

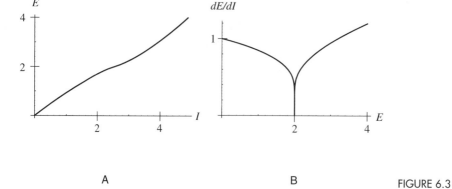

A B FIGURE 6.3

It is important to recognize that the variable ψ canonically conjugate to the action is *not* the physical angle θ of the pendulum, but rather an angle defined in phase space.

We can calculate I analytically for small angles by writing H as

$$H(p, \theta) = E = \frac{p^2}{2} + \omega^2 \frac{\theta^2}{2}. \tag{6.96}$$

Here we use the unscaled Hamiltonian, with $\omega = \sqrt{\frac{g}{l}}$ a constant. Then the phase trajectories are ellipses with the horizontal semiaxis being $\theta_{max} = \frac{\sqrt{2E}}{\omega}$ and the vertical semiaxis being $p_{max} = \sqrt{2E}$. I is simply the area of this ellipse divided by 2π, which gives us $I = \frac{E}{\omega}$. This implies that the frequency of the angle variable is the natural frequency of the pendulum. We could also have obtained this result by doing the line integral (6.94).

We plot the energy as a function of the action variable in Figure 6.3A ($H = E(I)$ here), found by doing the line integral (6.94) for a pendulum without using a small angle approximation. The graph looks rather unremarkable, with nothing visibly strange happening as we cross the separatrix. If, however, we plot the slope of this curve $\frac{dE}{dI} = \omega(I)$ versus the energy, we see clearly in Figure 6.3B that the frequency is zero very near the separatrix. Physically this corresponds the fact that a pendulum released from rest with $\theta = \pi$ has an infinite period, that is, $\omega = \frac{dE}{dI} = 0$ for $E = 2$.

QUESTION 9: Action Find I for Equation (6.96) by doing the line integral (6.94). Prove that you obtain the same result $I = \frac{E}{\omega}$.

QUESTION 10: Rotation Physically, when undergoing rotation for $E = 4$, the pendulum follows only the upper curve or the lower curve in Figure 6.3. Why then do we make the "closed curve" for rotation consist of the upper curve from left to right and the lower curve from right to left? Also prove that as $E \gg 2$, $I \to 2\sqrt{2E}$.

⟳ Example

For the pendulum, we would also like to find the generating function $\tilde{W}(\theta, \psi)$ for the transformation to action–angle variables implicitly defined by Equation (6.92) above. It will be sufficient to restrict ourselves to the case of small angles, where the pendulum oscillates with the Hamiltonian (6.96). From Equation (6.92), the generating function is

$$\tilde{W}(\theta, \psi) = \int d\tilde{W} = \int p\, d\theta - I \int d\psi = \int \sqrt{2E - \omega^2 \theta^2}\, d\theta - I\psi. \quad (6.97)$$

Introducing the dimensionless variable $u \equiv \frac{\omega\theta}{\sqrt{2E}}$, we have the indefinite integral $\int \sqrt{1-u^2}\, du = \frac{1}{2}(\sin^{-1} u + u\sqrt{1-u^2})$. Using $I = \frac{E}{\omega}$, we can write $\tilde{W}$ as

$$\tilde{W} = \frac{E}{\omega}[(\sin^{-1} u - \psi) + u\sqrt{1-u^2}]. \quad (6.98)$$

Now $\tilde{W}$ must be periodic in ψ, since $\oint d\tilde{W} = 0$. This will happen if $u = \sin\psi$, since the nonperiodic term (the one in parenthesis on the right side of the equals sign) in (6.98) will then be zero. This implies that $\sin\psi = \frac{\omega\theta}{\sqrt{2E}}$ or, equivalently, $E = \frac{1}{2}\omega^2\theta^2 \csc^2\psi$. Using this last relationship to express $\tilde{W}$ solely in terms of θ, ψ, we achieve what we are after:

$$\tilde{W}(\theta, \psi) = \frac{1}{2}\omega\theta^2 \cot\psi. \quad (6.99)$$

Originally, this generating function was introduced – see Equation (6.14) – to perform what may have seemed like a rather magical canonical transformation to what we now realize are action–angle variables. Now we see how to derive the appropriate generating function in a more systematic way.

Adiabatic Invariants

The action variable plays another role which has historically been very important in physics. If one of the parameters defining the motion (like the length l or the acceleration of gravity g for a pendulum) is varied slowly, the energy is no longer constant in time. However, the action variable remains almost constant for sufficiently slow variations of the parameters and is thus an *adiabatic invariant*. For the pendulum example we found that $I = \frac{E}{\omega}$ for small angles. If I remains invariant, it is true therefore that the energy is proportional to the frequency, even when the parameters l and/or g are slowly changed. The physical result that the number of quanta remains constant when the parameters of a system are slowly varied was proposed in 1911 by Einstein to be related to this adiabatic invariance of the action variable. The original Bohr–Sommerfeld quantization rules specified integer values for the action variable divided by Planck's constant. The action variables played a fundamental role in the transition from classical to quantum mechanics.

Next we will prove the action variable is an adiabatic invariant in the general case. This proof will also make clear what we mean by "slowly changing." Suppose the Hamiltonian depends not only on p, q but also on a parameter* $\alpha(t)$, which is a function of the time. The motion in phase space will not be a closed curve, but rather a kind of spiral, since

---

* Remember the difference between a dynamical variable and a parameter.

the motion is no longer strictly periodic. If we change the parameter α such that the energy changes by an amount small compared to the energy itself in a single period, and no resonance is produced, then this is called *adiabatic change*. We make the assumption that $\dot\alpha$ is nearly* constant over a period. Previously, the transformation to action–angle variables was generated by $\tilde W(q, \psi)$. $\tilde W$ now becomes a function of α also and hence of the time. Using the rules for time-dependent canonical transformations (6.12), the new Hamiltonian $\bar H$ differs from the old one by

$$\bar H = H(p(I, \psi, \alpha), q(I, \psi, \alpha), \alpha) + \frac{\partial \tilde W}{\partial t}. \tag{6.100}$$

Hamilton's equations in the new variables give

$$-\dot{\bar I} = \frac{\partial \bar H}{\partial \psi} = \underbrace{\frac{\partial H}{\partial p}\frac{\partial p}{\partial \psi} + \frac{\partial H}{\partial q}\frac{\partial q}{\partial \psi}}_{=0} + \frac{\partial^2 \tilde W}{\partial \psi \partial \alpha}\dot\alpha. \tag{6.101}$$

The first two terms on the right side are zero because we assume (6.90) still holds, even when α is slowly varying. The last term arises because the only explicit time dependence in $\tilde W$ occurs through the dependence on α.

Next average this change in the action variable over a period. This involves integrating the remaining term in (6.101) over ψ from 0 to 2π and dividing by 2π:

$$-\langle \dot{\bar I}\rangle_{\text{avg}} = \frac{1}{2\pi}\int_0^{2\pi} d\psi\, \dot\alpha \frac{\partial^2 \tilde W}{\partial \psi \partial \alpha}. \tag{6.102}$$

If we treat $\dot\alpha$ as nearly constant, neglecting second and higher time derivatives of α, we obtain

$$-\langle \dot{\bar I}\rangle_{\text{avg}} \approx \frac{\dot\alpha}{2\pi}\left[\frac{\partial \tilde W}{\partial \alpha}(q, \psi + 2\pi, \alpha(T)) - \frac{\partial \tilde W}{\partial \alpha}(q, \psi, \alpha(0))\right], \tag{6.103}$$

where T is the period of the motion. Finally, remembering that $\tilde W$ is periodic in ψ, we Taylor-expand the term in (6.103) between the brackets [...] and neglect all but the first term in the expansion. This gives the final result:

$$\langle \dot{\bar I}\rangle_{\text{avg}} \approx \frac{T}{2\pi}\dot\alpha^2\frac{\partial^2 \tilde W}{\partial \alpha^2} \approx 0. \tag{6.104}$$

By changing the parameter α slowly enough, we can make the change in the action variable extremely small, since the rate at which the action changes is proportional to the square of the rate at which α changes. Because I is nearly constant, the energy $E(I, \alpha)$ can be changed by an amount determined only by the final and initial value of the parameter α.

---

* That is, we neglect terms of order $\dot\alpha^2$ and higher as well as derivatives of order $\ddot\alpha$ and higher.

QUESTION 11: *Pendulum of Variable Length* A pendulum consists of a mass on a string of variable length. For small oscillations, the amplitude of the motion is proportional to the square root of the total energy. After the pendulum starts to swing, with a total energy E_0 and amplitude A_0, the string is slowly shortened from the original length l_0 to the final length $l_1 < l_0$. Use the adiabatic invariance of the action variable to predict how the final energy and amplitude depend on the length of the string.

6.6 INTEGRABLE SYSTEMS

Analytical methods have gone somewhat out of fashion in favor of geometric and topological concepts, which are augmented and guided by numerical work with the computer. This is due to the limited number of interesting problems that can be solved with analytic techniques. Action–angle variables remain useful in the wider context, however.

A system of N degrees of freedom is said to be *integrable* if N constants of the motion exist. There is a further restriction that these N constants of the motion be in *involution*, a concept to be explained below. It is important to note that these N constants do not have to be known in analytic form; they just need to exist. Since the Hamiltonian is constant (if it does not contain the time), $N = 1$ systems are always integrable.

For integrable systems, the motion in the $2N$-dimensional phase space must be confined to a $2N - N = N$-dimensional subspace of phase space. (The N constants reduce the dimension by N.) If we associate a vector with the tangent to the phase trajectory at each point, we have what is called a *vector field*. We can consider that Hamilton's equations generate this vector field, with the time as a parameter.

Assuming that the motion is periodic and therefore occupies a finite volume in phase space, there is an important theorem that tells us that the N-dimensional subspace in $2N$-dimensional phase space cannot have the topology of an N-sphere, but must instead have the topology of an N-torus (like a doughnut). These topologies are inherently different because one cannot map a torus onto the surface of a sphere without tearing it. (Topologically equivalent curves are curves that can either be deformed into each other or else shrunk to a point.)

If we think of the vector field (for $N = 2$) of phase trajectory tangents as little hairs, there is no way to comb hairs onto the surface of a sphere without having a singularity where the vector field is not defined. The direction of the hair at the pole is not defined, as seen in Figure 6.4A. This is known informally as the "hairy ball" theorem and more correctly as the Poincaré–Hopf theorem. We can see intuitively (for $N = 2$ at least) that the problem is removed if the trajectories move on the surface of a torus instead, as seen in Figure 6.4B. Figure 6.4C is equivalent to the 2-torus. It is a rectangular patch with opposite sides connected together so that the two angle variables ψ_1, ψ_2 obey $\psi_1 + 2\pi = \psi_1$ and $\psi_2 + 2\pi = \psi_2$. The motion is doubly periodic with a possible set of phase space canonical variables being I_1, ψ_1, I_2, ψ_2. The key property of a 2-torus is that two topologically inequivalent curves can be drawn on its surface. Neither can be deformed into the other without tearing the surface. Generalizing to N dimensions, we have an N-torus or, alternatively, an N-dimensional cube with opposite sides joined together.

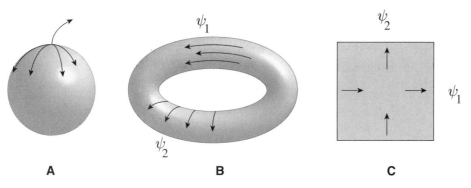

FIGURE 6.4

Even if it cannot be separated, an integrable problem can be reduced to quadratures in principle. However, it is not sufficient that N constants of the motion exist for N degrees of freedom. These constants must be compatible with each other (i.e., they are in *involution* with each other). Involution means that all mutual Poisson brackets vanish identically.

To understand why this must be required, we have to digress into a more geometric picture. Consider only two degrees of freedom and assume that there are two constant functions $F = F(q_1, p_1, q_2, p_2)$ and $G = G(q_1, p_1, q_2, p_2)$, each defining three-dimensional subspaces of the four-dimensional phase space. The gradient operator $grad_{qp} \equiv (\frac{\partial}{\partial q_1}, \frac{\partial}{\partial p_1}, \frac{\partial}{\partial q_2}, \frac{\partial}{\partial p_2})$, when acting on either F or G, defines the local 4-D vector that is orthogonal to the 3-D surface of constant F or G. We can define a *skew gradient* as $sgrad_{qp} \equiv (-\frac{\partial}{\partial p_1}, \frac{\partial}{\partial q_1}, -\frac{\partial}{\partial p_2}, \frac{\partial}{\partial q_2})$. The vector proportional to the skew gradient of F lies in the 3-D surface of constant F – it is constructed to be orthogonal to $grad\, F$.* We can then consider two vector fields, again both defined locally at a point: $sgrad\, F$ and $sgrad\, G$. These could be used to define a local two-dimensional coordinate system as shown in Figure 6.5. To use the vector fields defined by the skew gradients of F and G as a local coordinate system we must use infinitesimal vectors.† Let ϵ be an infinitesimal scalar quantity. Use this to define the infinitesimal 4-vector $\xi_F \equiv \epsilon\, sgrad\, F$. Along ξ_F, the function G should remain constant and vice versa. This condition can be expressed by the equation

$$\xi_F \cdot grad\, G = \Delta G_{(\text{along } sgrad\, F)} = 0. \tag{6.105}$$

If we write out this expression (6.105) in terms of the partial derivatives with respect to q_i, p_i we find only one condition instead of the two we expected ($\Delta F_{(\text{along } sgrad\, G)} = 0$, too):

$$-\frac{\partial F}{\partial p_1}\frac{\partial G}{\partial q_1} + \frac{\partial F}{\partial q_1}\frac{\partial G}{\partial p_1} - \frac{\partial F}{\partial p_2}\frac{\partial G}{\partial q_2} + \frac{\partial F}{\partial q_2}\frac{\partial G}{\partial p_2} = 0. \tag{6.106}$$

---

* There are two more skew vectors that can be constructed to be orthogonal to $grad\, F$, but the one we have mentioned is special in a way we shall see below.

† Taylor's theorem in four dimensions tells us that moving by an infinitesimal vector $\xi \equiv (\xi_1, \xi_2, \xi_3, \xi_4)$ away from the reference point changes F by $\xi \cdot grad\, F$, otherwise known as a directional derivative. In 3-D space, the directional or Lie derivative is $\vec{\xi} \cdot \vec{\nabla}$. See Kaplan, *Advanced Calculus*, 3rd ed., p. 135.

6.7 INVARIANT TORI AND WINDING NUMBERS

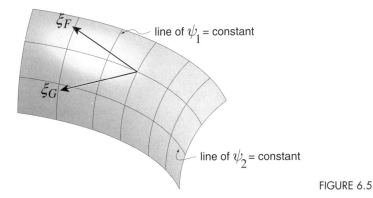

FIGURE 6.5

QUESTION 12: Skew Gradients Prove that you do get the same equation (6.106) when taking $\Delta F_{(\text{along } s\text{grad } G)}$ instead of $\Delta G_{(\text{along } s\text{grad } F)}$.

This is immediately recognized as a condition on the Poisson bracket of F and G (6.36). It is necessary that $[F, G] = 0$ for us to be able to find a two-dimensional surface along which the motion develops with both F and G remaining constant. $[F, G] = 0$ means that F and G are in involution. Since the Poisson bracket is invariant under canonical transformations, the relation $[F, G] = 0$ holds for any set of phase space variables related by a canonical transformation to the original set. One is reminded of the quantum mechanical requirement for compatible observables: They must commute with each other.

It can be shown that if we can find a curvilinear coordinate system in which the components of the two vector fields ($s\text{grad } F$, $s\text{grad } G$) are always constants, the coordinate system *is* the action–angle set of canonical variables. The system is then said to be integrable. Integrable systems are relatively rare, however. Except for separable systems solved by the Hamilton–Jacobi equation, no systematic technique exists for finding constants of the motion that are not directly related to obvious symmetries. (If a symmetry can be identified, Noether's Theorem generates the relevant constant of the motion.)

Chaotic behavior is a characteristic of nonintegrability, which is a property of most nonlinear mechanical systems. For example, the motion of stars in the gravitational field of the galaxy is nonintegrable. It can be regular, following smooth curves in phase space, or irregular and chaotic. For many initial conditions, the chaotic motion can be confined to such small regions of phase space and thus be of little actual importance. It was the study of this stellar galactic motion that first alerted physicists to the importance of chaos in physics.*

6.7 INVARIANT TORI AND WINDING NUMBERS

As a rather trivial example, we choose two uncoupled linear oscillators, with frequencies ω_1 and ω_2. It can be shown (see Chapter 9) that any set of *linear* coupled

---

* According to Gutzwiller: "That discovery was crucial in understanding the observed velocity distribution of stars in our solar neighborhood." (M. C. Gutzwiller, *Chaos in Classical and Quantum Mechanics*, p. 99.)

oscillators can be transformed into uncoupled oscillators. For two degrees of freedom the energy can be written as

$$E = \omega_1 I_1 + \omega_2 I_2. \tag{6.107}$$

The action variables I_k for $k = 1, 2$ can be found by integrating

$$\boxed{I_k = \frac{1}{2\pi} \oint_{\mathcal{C}_k} p_1 \, dq_1 + p_2 \, dq_2.} \tag{6.108}$$

The two closed curves $\mathcal{C}_1, \mathcal{C}_2$ in 4-D phase space must be topologically inequivalent. The angle variables are

$$\psi_k = \omega_k t + \beta_k, \tag{6.109}$$

where the β_k are arbitrary integration constants.

Figure 6.6 shows a slice through two tori, each corresponding to different values of I_2; hence energy. I_1 is the same for both tori. A phase trajectory is a moving point that traces out the curve shown winding around the torus. We can take each of the curves $\mathcal{C}_1, \mathcal{C}_2$ to follow the motion of one of the oscillators, ψ_1 or ψ_2 as indicated, while the other is held constant.

The time derivatives of ψ_1, ψ_2 are ω_1, ω_2. When the angle ψ_1 advances by 2π, the angle ψ_2 will advance by $2\pi \frac{\omega_2}{\omega_1}$. The ratio of the two frequencies is called the *winding number* Ω. The geometric picture is that the trajectory winds $\Omega = \frac{\omega_2}{\omega_1}$ times around the torus for each complete period of ψ_1. There are two distinct cases: Ω is a rational fraction of the form $\frac{p}{q}$, with p, q both integers, or Ω is irrational. For the rational winding number case, the trajectory is periodic, traversing a single curve that winds p times around the torus in the direction of increasing ψ_2 for q complete traversals in the direction of increasing ψ_1. Most of the surface area of the torus is never encountered in this case. By contrast, an irrational Ω means that the motion is *quasiperiodic*, never exactly repeating itself, and filling the area of the surface as time progresses (or at least coming arbitrarily close to every point). This last type of motion is also called *almost periodic*, since every irrational number can be arbitrarily well approximated by a rational number. We will make more use of the concept of winding number in the chapter on chaos, Chapter 11.

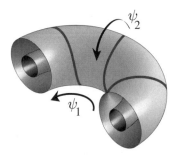

FIGURE 6.6

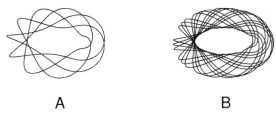

FIGURE 6.7
A) $\Omega = \frac{9}{4}, q = 4$; B) $\Omega = \sqrt{5}, q = 12$.

In Figure 6.7, we contrast the two cases of rational or irrational winding numbers. If ψ_1 increases by $2\pi q$, there are q trips around the torus. In Figure 6.7A, for $\Omega = \frac{p}{q} = \frac{9}{4}$, the curve closes and becomes periodic after four traversals of the torus. In Figure 6.7B, for $\Omega = \sqrt{5}$, the curve never closes exactly on itself. $q = 12$ traversals are shown.

SUMMARY OF CHAPTER 6

- If we have a suitable generating function $F_1(q, Q, t)$ and require that

$$P = -\frac{\partial F_1}{\partial Q}, \quad p = \frac{\partial F_1}{\partial q}, \quad (6.110)$$

the contact transformation obtained by solving (6.110) will be canonical because Hamilton's equations of motion are preserved. The new Hamiltonian $\bar{H}$ in terms of the new variables becomes

$$\bar{H}(P, Q, t) = H(p, q, t) + \frac{\partial F_1}{\partial t}. \quad (6.111)$$

- There are four possible types of generating functions: $F_1(q, Q, t)$, $F_2(q, P, t)$, $F_3(p, Q, t)$, and $F_4(p, P, t)$. F_2, F_3, and F_4 can be created from F_1 by Legendre transformations. The missing variables are proportional to partial derivatives of the generating functions. Hamilton's equations are preserved if we start with a function that is one of these four types and use the rules appropriate to that generating function to obtain the other variables.
- If the variables Q, P are related to the variables q, p by a canonical transformation, then the Poisson bracket $[Q(q, p), P(q, p)]_{q,p} = 1$.
- The Hamilton–Jacobi equation is an alternative approach to analyzing a mechanics problem, in which Hamilton's equations are replaced by Hamilton's Principal Function S. The equation obeyed by S is a partial differential equation in $N + 1$ independent variables. $S = \int L\, dt$, so S is the action.

- If the Hamiltonian is time-independent, $S = W - Et$. W is Hamilton's Characteristic Function. W is often used to find the orbit equations directly without the time dependence. $W(q_1, \ldots, q_N, P_1, \ldots, P_N) = \sum_k \int p_k \, dq_k$.
- If a problem is separable, the S function can be written as a sum of functions of only one q_k. This property depends on the choice of coordinate systems. Separating the Hamilton–Jacobi equation is the most powerful method for finding analytic solutions to mechanics problems.
- If the motion is periodic, canonical angle–action variables can be defined. The action variables are

$$I_k \equiv \frac{1}{2\pi} \oint_{C_k} \sum_{j=1}^{N} p_j \, dq_j, \quad k = 1, \ldots N. \tag{6.112}$$

  The closed curves in $2N$-dimensional phase space, C_k, are topologically independent. The action variables are constants of the motion. The angle variables have the simple time dependence $\psi_k = \omega_k(I_1, I_2, \ldots, I_N) \, t + \beta_k$, with $\omega_k = \frac{\partial E}{\partial I_k}$ and β_k an integration constant.
- More generally, we can classify motion as being either integrable or nonintegrable. If there are N compatible constants of the motion for N degrees of freedom, the problem is integrable, even if there is no analytic solution. In the case of integrability, the motion is confined to an N-dimensional surface in phase space.
- Periodic motion for integrable systems in $2N$-dimensional phase space takes place on an N-torus. Each trajectory in phase space is a curve that winds around the torus. If $N = 2$, the winding number Ω is defined as $\frac{\omega_2}{\omega_1}$. Motion with rational winding numbers gives a single periodic curve on the torus. If the winding number is irrational, the curve never exactly repeats and fills the entire surface of the torus.

PROBLEMS

Generating Functions and Canonical Transformations

Problem 1*: *(Generating function for the harmonic oscillator interpreted)* Consider the generating function $F_1(q, Q)$ for the harmonic oscillator problem with the Hamiltonian

$$H = \frac{1}{2}(p^2 + \omega^2 q^2), \quad m \equiv 1. \tag{6.113}$$

F_1 defines the transformation to the new variables Q, P. It is given by

$$F_1(q, Q) \equiv \frac{1}{2}\omega q^2 \cot 2\pi Q. \tag{6.114}$$

Without referring to the worked example in the text:

a) Find expressions for q, p in terms of Q, P.

b) Find the new Hamiltonian $\bar{H}$ and solve Hamilton's equations of motion.
c) Interpret the new canonically conjugate variables Q, P geometrically in q, p phase space.
d) Check whether the requirement $\delta F_1 = 0$ at the end points of the time integral for the action is equivalent to $p\,\delta q - P\,\delta Q = 0$ at the end points.
e) Find generating functions of the other three types that will generate this same canonical transformation.

Problem 2: *(Generating function produces a canonical transformation)* You are given the generating function $F_3(p, Q) = -(e^Q - 1)^2 \tan p$. Prove that it generates the canonical transformation

$$Q = \log(1 + \sqrt{q} \cos p),$$
$$P = 2(1 + \sqrt{q} \cos p)\sqrt{q} \sin p. \quad (6.115)$$

(From Goldstein, 1980.)

Problem 3: *(Motion in an arbitrary reference frame solved with a canonical transformation)* Let z be the position of a particle (mass m) in an inertial frame, and Z be the position measured from the origin of a possibly noninertial frame displaced by the function $D(t)$ from the origin of the inertial frame. Then $Z = z - D(t)$. In the inertial frame, the Hamiltonian is

$$H(z, p) = \frac{p^2}{2m} + V(z). \quad (6.116)$$

Find an explicit form for the generating function $F_2(z, P, t)$ that generates the transformation from z, p to Z, P. What is the canonical transformation $Z(z, p)$, $P(z, p)$? Find $\bar{H}(Z, P, t)$ and Hamilton's equations of motion in terms of Z, P, t. Show that, in this case, there is a problem with converting F_2 into F_1 by the usual method of a Legendre transformation and explain why this occurs.

(From Percival and Richards.)

Problem 4: *(Solve the freely falling body with a canonical transformation)* The Hamiltonian of a freely falling body is, in one dimension (neglect x, y motion),

$$H = \frac{p^2}{2m} + mgz. \quad (6.117)$$

Find a time-independent generating function $F_4(p, P)$ such that $\bar{H}(Q, P) = P$. Determine the explicit form of the canonical transformation: $Q(z, p)$ and $P(z, p)$. Solve for $z(p, P)$. Prove that Q is the time.

(From Percival and Richards.)

Problem 5: *(Jacobians and canonical transformations)* For the case of 1 degree of freedom, the determinant of the Jacobian of dynamical variables related by a canonical transformation is

$$\boxed{\det\left|\frac{\partial(P, Q)}{\partial(p, q)}\right| = 1.} \tag{6.118}$$

For any functions $F(x, y)$, $G(x, y)$, the Jacobian is defined by

$$\frac{\partial(F, G)}{\partial(x, y)} \equiv \begin{pmatrix} \frac{\partial F}{\partial x} & \frac{\partial F}{\partial y} \\ \frac{\partial G}{\partial x} & \frac{\partial G}{\partial y} \end{pmatrix}. \tag{6.119}$$

Prove that Equation (6.118) holds for any canonical transformation generated by a generating function of the type $F_1(q, Q)$ with q, Q considered to be independent variables. Since we can always produce a generating function of this type by a Legendre transformation or series of transformations from F_2, F_3, F_4, proving Equation (6.118) for F_1-generated transformations is completely general.

One consequence of (6.118) is that a closed curve in p, q space encloses the same area that the image of that curve encloses in P, Q space. An application of (6.118) is used in the proof of the canonical invariance of Poisson brackets in Problem 8.

Hints: The *general chain rule* holds for Jacobian matrices* if we transform from the independent variables x, y to the new independent variables u, v, which are functions $u(x, y), v(x, y)$:

$$\frac{\partial(F, G)}{\partial(x, y)} = \underbrace{\frac{\partial(F, G)}{\partial(u, v)} \frac{\partial(u, v)}{\partial(x, y)}}_{\text{matrix multiplication}}. \tag{6.120}$$

As a special case of the general chain rule:

$$\frac{\partial(u, v)}{\partial(x, y)} \frac{\partial(x, y)}{\partial(u, v)} = \text{identity matrix} \equiv I. \tag{6.121}$$

Problem 6: *(Possible generating functions)* Use (6.118) to prove whether or not the two functions below can be used as generating functions:

$$F_1(q, Q) = qe^Q, \quad F_1(q, Q) = q^2 + Q^4. \tag{6.122}$$

If it is a possible generating function, determine the transformation $q, p \to Q, P$ explicitly.

(Adapted from Percival and Richards.)

---

* *Advanced Calculus*, 3rd ed., by Kaplan, pp. 106ff.

Problem 7*: *(Dynamics on a rotating turntable using polar coordinates)* This is a follow-up to Problem 14 in Chapter 5. Consider a generating function of the $F_2(q, P, t)$ type: $F_2 \equiv r_{\text{lab}} P_r + (\phi_{\text{lab}} - \theta(t)) P_\phi$. The old coordinates p, q are the lab coordinates of the bug, and the new coordinates P, Q are the rotating coordinates of the bug. There are two degrees of freedom, so F_2 contains the sum over these. Using the relations in Table 6.1 and the rules for finding the new Hamiltonian $\bar{H}$, show that the relations between coordinates (r, ϕ) and $(r_{\text{lab}}, \phi_{\text{lab}})$ are correct and $\bar{H}$ is the Hamiltonian (5.102). This demonstrates a link between this type of canonical transformation and a change of coordinate systems.

General Properties of Poisson Brackets

Problem 8: *(Poisson brackets are invariant under canonical transformations)* Prove that any canonical transformation leaves the Poisson brackets $[F, G]$ invariant. That is, if P and Q are obtained from q and p by a canonical transformation, then

$$[F, G]_{Q,P} \equiv \frac{\partial F}{\partial Q}\frac{\partial G}{\partial P} - \frac{\partial F}{\partial P}\frac{\partial G}{\partial Q} = \frac{\partial F}{\partial q}\frac{\partial G}{\partial p} - \frac{\partial F}{\partial p}\frac{\partial G}{\partial q} \equiv [F, G]_{q,p}. \tag{6.123}$$

Hints: For one degree of freedom, the Poisson bracket can be regarded as the determinant of a Jacobian (6.119). The determinant of a product of two matrices is the product of their determinants.

Problem 9*: *(How to test when a transformation is canonical)*

a) First prove that Equation (6.39) is *necessary* for canonical transformations. Sufficiency means that (6.39) is a litmus test for the canonical equivalence of two sets of dynamical variables. To prove this, we have to show that (6.39) implies the existence of a generating function that defines the transformation. This problem also provides a method for finding generating functions, at least in principle.

b) Explain, using (6.123) for one degree of freedom, why the areas enclosed by corresponding closed curves in (q, p) or (Q, P) phase spaces are the same. Also explain why this implies

$$\oint p\,dq = \oint P\,dQ. \tag{6.124}$$

(The symbol $\oint$ stands for an integral around a closed curve.) Hint: Use Stokes' theorem.[†]

c) Now take the point of view that Q, q are the independent variables, and p, P are dependent variables, as with F_1-type generating functions. Then Equation (6.124)

[†] See Kaplan, *Advanced Calculus*, 3rd ed., p. 328.

implies that around any closed curve in (q, Q) space

$$\oint p(q, Q)\,dq - \oint P(q, Q)\,dQ = 0. \qquad (6.125)$$

Imagine a closed curve in (q, Q) space. Explain why (6.125) means that any path integral in this space between fixed end points is independent of the path taken, so can be considered a perfect differential of a function of q, Q. Stated as an equation, this is

$$\int p\,dq - P\,dQ = \int dF_1(q, Q). \qquad (6.126)$$

d) Using (6.126), prove that (6.39) implies the correct relationships for generating functions ($p = \frac{\partial F_1}{\partial q}$ and $P = -\frac{\partial F_1}{\partial Q}$).

Problem 10: *(Finding generating functions)* Suppose you try the contact transformation

$$Q = \log\left(\frac{\sin p}{q}\right), \quad P = q \cot p. \qquad (6.127)$$

a) Find $[Q, P]_{q,p}$. Is (6.127) canonical?
b) Now show that

$$p\,dq - P\,dQ = d(pq + q \cot p). \qquad (6.128)$$

c) Find $F_1(q, Q)$ explicitly. Useful information: $\int \sin^{-1} x\,dx = \sqrt{1-x^2} + x \sin^{-1} x$.

Problem 11: *(Possible canonical transformation)* Is the transformation below canonical?

$$Q = \log(1 + \sqrt{q}\cos p), \quad P = (1 + \sqrt{q}\cos p)\sqrt{q}\sin p \qquad (6.129)$$

Problem 12*: *(Poisson brackets for many degrees of freedom)*

a) Prove, using Hamilton's equations of motion and the definition of the Poisson brackets (6.36), that the total time derivative of any function $D(q_1, \ldots, p_1, \ldots)$ of the ps and qs obeys the equation

$$\frac{dD}{dt} = [D, H] + \frac{\partial D}{\partial t}, \qquad (6.130)$$

where H is the Hamiltonian of the system. Equation (6.130) can be regarded as the most general and canonically invariant way to state Hamilton's equations of motion. Explain.

b) For the I_j defined by (5.14), if the Lagrangian is invariant under transformations that generate I_j, show $[I_j, H] = 0$. Notice that I_j does not contain the time explicitly. This in known as the *Hamiltonian form of Noether's Theorem*.

Problem 13*: *(Landau's proof)* In their book *Mechanics*, Landau and Lifshitz give a proof of the important relation (6.123) as follows (our notation and our formula references have been used below):

> First of all, it may be noticed that the time appears as a parameter in the canonical transformation(s)... It is therefore sufficient to prove (6.123) for quantities which do not depend explicitly on time. Let us now formally regard G as the Hamiltonian of some fictitious system. Then, by formula (6.130), $[F, G]_{p,q} = \frac{dF}{dt}$. The derivative $\frac{dF}{dt}$ can depend only on the properties of the motion of the fictitious system, and not on the particular choice of variables. Hence the Poisson bracket $[F, G]$ is unaltered by the passage from one set of canonical variables to another.

Comment on whether or not you think this is a valid and completely general proof of (6.123).

Problem 14*: *(Poisson brackets of constants of the motion can generate new constants of the motion)* Consider the uniform motion of a free particle of mass m. The Hamiltonian is a constant of the motion and so is the quantity F defined as

$$F(x, p, t) \equiv x - \frac{pt}{m}. \tag{6.131}$$

a) Compare $[H, F]$ with $\frac{\partial F}{\partial t}$. Prove from (6.130) that F is also a constant.
b) Prove that the Poisson bracket of two constants of the motion is itself a constant of the motion, even if the constants $F(x, p, t)$ and $G(x, p, t)$ depend explicitly on the time. (Part a is one example of this.)
c) Show *in general* that if the Hamiltonian and a quantity F are constants of the motion then $\frac{\partial F}{\partial t}$ is a constant of the motion also.

Problem 15: *(Poisson brackets with angular momentum)*

a) Angular momentum is defined as $\vec{l} = \vec{r} \times \vec{p}$. Prove that $[l_x, l_y] = l_z$ for all cyclic permutations of l_x, l_y, l_z.
b) Calculate all the Poisson brackets of the components of $\vec{r}$ and $\vec{p}$ with the components of the angular momentum (for example, $[x, l_z]$, $[p_x, l_z]$, etc.).

Problem 16: *(Poisson brackets and spherical symmetry)* Let $\phi(\vec{r}, \vec{p})$ be any function that is spherically symmetric about the origin (invariant under rotations).

a) ϕ can depend only on the components of $\vec{r}$ and $\vec{p}$ through the combinations r^2, p^2, and $\vec{r} \cdot \vec{p}$. Why is this true?
b) Evaluate the Poisson bracket $[\phi, l_z]$ (l_z is the z component of the angular momentum) and show that it vanishes.

(Adapted from Landau and Lifshitz, 1986.)

Hamilton–Jacobi Equation

Problem 17*: *(Harmonic oscillator example)*

a) Write the Hamilton–Jacobi equation for the 1-D SHO ($m = 1$) and solve for S. Use the substitution $S = W - \alpha t$ with $\alpha =$ constant. $W(q, \alpha)$ does not contain the time explicitly. Leave the expression you get for W in the form of an integral over q.

b) Show, by using the equations for the constant β, that the solution to the physical problem, $q(t)$, is

$$q(t) = \frac{\sqrt{2\alpha}}{\omega} \sin \omega(t + \beta). \qquad (6.132)$$

Hint: $\int \frac{dx}{\sqrt{1-x^2}} = \arcsin[x]$.

c) Recall that α is really P and β is really Q. Show that α is the total energy E.

Action–Angle Variables

Problem 18: *("$\tan^2$" potential)* A particle of unit mass moves in a potential of the form

$$V(q) = U \tan^2 (aq), \qquad (6.133)$$

where U and a are positive constants. Find the turning points of the motion. Prove that the action variable I obeys the relation

$$\frac{aI}{\sqrt{2}} = \sqrt{E + U} - \sqrt{U}, \qquad (6.134)$$

where E is the total energy, and thus prove that the frequency ω has the energy dependence

$$\frac{\omega}{a\sqrt{2}} = \sqrt{E + U}. \qquad (6.135)$$

The increase of the frequency with energy reflects the fact that the restoring force increases with displacement q faster than linearly. Hint: The substitution $x = \tan(aq)$ helps in doing the integral for the action variable.

(From Percival and Richards.)

Problem 19: *(Area in phase space is preserved under a canonical transformation)*

a) In (I, ψ) phase space, the infinitesimal area between I and $I + \delta I$ with ψ ranging from zero to ψ is $\delta I\, \psi$, because the phase trajectories are lines of constant I. (Draw a picture.) By substituting $E(I)$ in $p(q, E)$ to obtain $p(q, I)$, prove that this area is

$$\delta I\, \psi = \delta I \int_0^q \frac{\partial p(q', I)}{\partial I} dq'. \qquad (6.136)$$

Notice that (6.136) is the same result you would get by applying the rules for a type F_2 generating function, so (6.136) tells you how to find a generating function of this type for the transformation to action–angle variables.

b) By using (6.136) and the expression for $E(I)$ in the previous problem (6.134), find the angle variable in terms of the displacement in that case:

$$\sin \psi = \sqrt{1 + \frac{E}{U}} \sin(aq). \tag{6.137}$$

(Adapted from Percival and Richards.)

Problem 20: *(A ball bouncing between two hard walls)*

a) A ball bounces between two walls separated by a distance d. In between the walls, no force acts on the ball. Draw a phase trajectory for one complete cycle. Notice that the momentum changes direction discontinuously when the ball bounces from the wall. In dimensionless units the energy is $E = \frac{p^2}{2}$. Calculate the action variable and the area in phase space for one cycle. Then calculate the frequency by differentiating the energy with respect to the action variable. Does this agree with an elementary calculation of the frequency $\omega = \frac{2\pi}{T}$, with T equal to the period? It is not possible to define a Hamiltonian in this case because the force is discontinuous, but the action variable can still be defined. Is this also true of the angle variable?

b) Notice that this problem can be considered as a limiting case of the potential in Problem 18. To define the limit, take the results for $I(E)$ and $\omega(E)$ for the "$\tan^2$" potential (6.134, 6.135) and go to the appropriate limit for U and a. Check that the answer agrees with what you have calculated here for the bouncing ball.

(Adapted from Percival and Richards.)

Problem 21: *(Particle in combined uniform and $\frac{1}{r^2}$ gravitational force)* Find an expression in Cartesian coordinates for the constant in Equation (6.83). See if you can prove the connection with the Laplace–Runge–Lenz vector. (See Problem 2 in Chapter 5. Equation (5.93) is the definition of this vector.)

Problem 22: *(Hamiltonian in parabolic coordinates; most general form of separable potentials)* Prove that the most general potential separable in parabolic coordinates has the form

$$V(\xi, \eta) = \frac{A(\xi) + B(\eta)}{\xi + \eta} = \frac{A(r+z) + B(r-z)}{2r}, \tag{6.138}$$

where A and B are arbitrary functions.

APPENDIX

WHAT DOES "SYMPLECTIC" MEAN?

One often sees a reference to the *symplectic* property of a general canonical transformation. You can test the Jacobian of an arbitrary transformation of phase space variables for this property. If it holds, then the transformation is canonical, and thus the Hamiltonian form of the equations of motion are preserved.

The symplectic property also implies that the sum of projected areas (action) on the different coordinate–momentum planes in phase space is preserved for any closed cycle in phase space:

$$\sum_{i=1}^{N} \oint P_i \, dQ_i = \sum_{i=1}^{N} \oint p_i \, dq_i, \qquad (6.139)$$

where N is the number of degrees of freedom. The transformation converts $\{q_i, p_i\} \to \{Q_i, P_i\}$.

A few preliminaries will be needed before defining the symplectic property. Label the dimensions in phase space by $(q_1, p_1, q_2, p_2, \ldots, q_N, p_N)$. Consider the generalized phase space gradient vector of the Hamiltonian as a column vector in the $2N$-dimensional space:*

$$\nabla_{\mathbf{qp}}\mathbf{H} \equiv \begin{pmatrix} \frac{\partial H}{\partial q_1} \\ \frac{\partial H}{\partial p_1} \\ \vdots \\ \frac{\partial H}{\partial q_N} \\ \frac{\partial H}{\partial p_N} \end{pmatrix}. \qquad (6.140)$$

Let the label $\mathbf{z}$ stand for the generic coordinate

$$\mathbf{z} \equiv \begin{pmatrix} q_1 \\ p_1 \\ \vdots \\ q_N \\ p_N \end{pmatrix}. \qquad (6.141)$$

---

* We will use boldface notation to denote matrices, including row and column vectors. Multiplication of boldface quantities will then imply matrix multiplication.

APPENDIX WHAT DOES "SYMPLECTIC" MEAN?

Hamilton's equations of motion can be written in matrix form if the Γ matrix is defined first:

$$\Gamma \equiv \begin{pmatrix} 0 & 1 & 0 & 0 & \cdots & \cdots & \cdots \\ -1 & 0 & 0 & 0 & \cdots & \cdots & \cdots \\ 0 & 0 & 0 & 1 & \cdots & \cdots & \cdots \\ 0 & 0 & -1 & 0 & \cdots & 0 & 0 \\ \vdots & \vdots & \vdots & \vdots & \ddots & 0 & 0 \\ \vdots & \vdots & \vdots & 0 & 0 & 0 & 1 \\ \vdots & \vdots & \vdots & 0 & 0 & -1 & 0 \end{pmatrix} \tag{6.141}$$

In other words, Γ consists of a string of little 2×2 submatrices of the form $\begin{pmatrix} 0 & 1 \\ -1 & 0 \end{pmatrix}$ down the diagonal. Some properties of Γ are: $\tilde{\Gamma} = -\Gamma$ (the tilde indicates the transpose), $\Gamma^2 = \tilde{\Gamma}\Gamma = \mathbf{I}$, with $\mathbf{I}$ being the identity matrix, and the determinant $\det[\Gamma] = 1$. The matrix form of Hamilton's equations is

$$\frac{d\mathbf{z}}{dt} = \Gamma \nabla_{\mathbf{qp}} H. \tag{6.142}$$

The matrix Γ acts on the column vector $\nabla_{\mathbf{qp}} H$ in Equation (6.142) to produce the column vector $\frac{d\mathbf{z}}{dt}$. Equation (6.142) can also be written in differential form:

$$d\mathbf{z} = dt\, \Gamma \nabla_{\mathbf{qp}} H. \tag{6.143}$$

Next we must use the Jacobian $\mathbf{J}$ of the transformation to relate the infinitesimal $d\mathbf{Z}$ in the Q, P variables to $d\mathbf{z}$ in the q, p variables. The definition of the Jacobian in this context is*

$$\mathbf{J} \equiv \frac{\partial(Q_1, P_1, \ldots, Q_N, P_N)}{\partial(q_1, p_1, \ldots, q_N, p_N)} = \begin{pmatrix} \frac{\partial Q_1}{\partial q_1} & \frac{\partial Q_1}{\partial p_1} & \cdots & \frac{\partial Q_1}{\partial p_N} \\ \vdots & & & \vdots \\ \frac{\partial P_N}{\partial q_1} & \frac{\partial P_N}{\partial p_1} & \cdots & \frac{\partial P_N}{\partial p_N} \end{pmatrix}. \tag{6.144}$$

An elementary application of the definition of the Jacobian and of the partial derivative gives the relation

$$d\mathbf{Z} = \mathbf{J}\, d\mathbf{z}. \tag{6.145}$$

We can consider H to be a function of the $\{q_i, p_i\}$ or of the $\{Q_i, P_i\}$. In matrix form, the differential dH is

$$dH = \widetilde{\nabla_{\mathbf{qp}} H}\, d\mathbf{z} = \widetilde{\nabla_{\mathbf{QP}} H}\, d\mathbf{Z}. \tag{6.146}$$

---

* W. Kaplan, *Advanced Calculus*, 3rd. ed., Addison-Wesley, 1984, p. 95.

The transpose converts a column vector into a row vector. Using Equation (6.145) gives us the connection between gradients in the qp variables and gradients in the QP variables:

$$\nabla_{qp} H = \tilde{J} \nabla_{QP} H. \tag{6.147}$$

Armed with these relations, we can take (6.143) and transform it into a set of $2N$ equations involving only Q, P instead of q, p:

$$d\mathbf{Z} = dt \, \mathbf{J} \mathbf{\Gamma} \tilde{\mathbf{J}} \nabla_{QP} H. \tag{6.148}$$

If the form of Hamilton's equations is to be preserved under the canonical transformation to the $\{Q_i, P_i\}$ variables, it will be necessary and sufficient that

$$\mathbf{J} \mathbf{\Gamma} \tilde{\mathbf{J}} = \mathbf{\Gamma}. \tag{6.149}$$

(Compare (6.148) to (6.143).) If the condition (6.149) is satisfied, the Jacobian $\mathbf{J}$ is by definition a *symplectic matrix*. By taking the transpose of (6.149) and using the property $\tilde{\mathbf{\Gamma}} = -\mathbf{\Gamma}$, one can show that it is also true that

$$\tilde{\mathbf{J}} \mathbf{\Gamma} \mathbf{J} = \mathbf{\Gamma}. \tag{6.150}$$

We can view (6.150) or (6.149) as a "litmus test" that every canonical transformation must pass. This frees us from the need to use a generating function.

It is straightforward to show that the identity matrix $\mathbf{I}$ is a symplectic matrix and that the product of two symplectic matrices is symplectic. We can also see from (6.149) or (6.150) that the determinant $|\tilde{\mathbf{J}} \mathbf{J}| = 1$, which means that $|\mathbf{J}| = \pm 1$. (Liouville's theorem is equivalent to $|\mathbf{J}| = 1$, as proved in Appendix A of Chapter 5.) It can also be shown that at most $N(2N+1)$ elements of $\mathbf{J}$ can be independent.

It is possible to obtain a geometric picture of the symplectic property. In the phase space of the original variables, consider an infinitesimal parallelogram defined by the two arbitrary (but not parallel) vectors in the q_1, p_1 plane with coordinates $(\delta q_1(1), \delta p_1(1))$ and $(\delta q_1(2), \delta p_1(2))$. The area of a parallelogram formed by these vectors is $\delta a(12) \equiv \delta q_1(1) \delta p_1(2) - \delta q_1(2) \delta p_1(1)$. (In three dimensions, the area is just the magnitude of the cross product of the vectors defining the parallelogram.) This can be written in a more compact form as

$$\delta \mathbf{z}(1) \equiv (\delta q_1(1), \delta p_1(1), 0, 0, \ldots), \quad \delta \mathbf{z}(2) \equiv (\delta q_1(2), \delta p_1(2), 0, 0, \ldots), \tag{6.151}$$

$$\delta a(12) = \widetilde{\delta \mathbf{z}(2)} \, \mathbf{\Gamma} \, \delta \mathbf{z}(1). \tag{6.152}$$

It follows immediately from (6.150) that this area is preserved under a canonical transformation from $\mathbf{z} \to \mathbf{Z}$ variables:

$$\delta A(12) = \delta a(12). \tag{6.153}$$

APPENDIX   WHAT DOES "SYMPLECTIC" MEAN?

In terms of the new variables, the infinitesimal parallelogram is not necessarily oriented completely in any one coordinate–canonical momentum plane, but we have proved that the quadratic form

$$\sum_{i=1}^{N} \delta q_i(1)\delta p_i(2) - \delta q_i(2)\delta p_i(1) \tag{6.154}$$

is preserved under a canonical transformation.

Now consider a closed curve in $2N$-dimensional phase space. The "interior" of the projection of the curve onto the N different q_i, p_i planes can be "tiled" with infinitesimal squares (which become parallelograms after the canonical transformation). The sum of the areas of the projected curve on the different coordinate–canonical momentum planes is invariant by (6.153):

$$\sum_{i=1}^{N} \iint dq_i\, dp_i = \sum_{i=1}^{N} \iint dQ_i\, dP_i. \tag{6.155}$$

Finally, use Stoke's theorem to relate (6.155) to the line integrals around the projections of the closed curve in phase space:

$$\sum_{i=1}^{N} \oint p_i\, dq_i = \sum_{i=1}^{N} \oint P_i\, dQ_i. \tag{6.156}$$

Equation (6.156) is an important result in theoretical mechanics, since these line integrals differ from the action only by a factor of $\frac{1}{2\pi}$. The time development of a system is a special canonical transformation of the phase space variables (6.35), so we have also proven that the sum of the actions is constant in time.

With the matrix notation we can to prove the invariance of the Poisson bracket $[F, G]$ of two arbitrary functions F, G of phase space variables. By the definition of the Poisson bracket (work it out for $N = 1$ or $N = 2$):

$$[F, G]_{q,p} = (\widetilde{\nabla_{\mathbf{qp}}F})\, \boldsymbol{\Gamma}\, \nabla_{\mathbf{qp}}G = (\widetilde{\nabla_{\mathbf{QP}}F})\, \underbrace{\mathbf{J}\boldsymbol{\Gamma}\tilde{\mathbf{J}}}_{\boldsymbol{\Gamma}}\, \nabla_{\mathbf{QP}}G = [F, G]_{Q,P}. \tag{6.157}$$

The invariance of $[F, G]$ under canonical transformations follows immediately from the transformation property of the phase space gradient (6.147) and the symplectic property of $\mathbf{J}$ (6.149).

If the generating function has an explicit time dependence, its partial time derivative must be added to the old Hamiltonian to obtain the new one. However, the symplectic property of the Jacobian is not affected by this. A full discussion of this point is beyond the scope of this book, but a more complete treatment is in the book *Classical Mechanics*, 2d. ed. by H. Goldstein, p. 394ff.

CHAPTER SEVEN

ROTATING COORDINATE SYSTEMS

OVERVIEW OF CHAPTER 7

As it would be viewed by an observer on the Sun, you are racing along at 66,700 miles/hr on an elliptical orbit. A different observer, located at the center of the Earth would see you rotating at 1,038 miles/hr. Yet, in everyday life, we are not normally aware of this. The description of motion depends on the reference frame. Inertial reference frames play a special role.

The Earth we live on is *not* an inertial frame. It is possible for someone on Earth to detect the Earth's rotation by detecting small deviations from Newton's Laws. While he was still an undergraduate, A. H. Compton invented a table-top experiment which not only demonstrated the Earth's rotation, but also measured the latitude of the laboratory. We need to develop a systematic way of translating back and forth between the description of motion in a rotating frame and the description in an inertial frame. This is a purely geometric or "kinematic"* mathematical process, because we assume that the relative motion of the two reference frames is fully specified and is not subject to change by the action of forces, at least within the time period of the experiments we wish to do or during the observations we wish to make.

Motion can take place on a rotating body and be observed either with a reference frame fixed in the body, or from outside (i.e., a coordinate system fixed in "space"). The problem of describing motion in rotating reference frames can be broken into three steps.

First: Consider *fixed* rotations of one frame with respect to another. How do we convert or transform the coordinates of a vector in frame #1 into the coordinates of the same physical vector in frame #2? The answer does not depend on the vector, but only on the relative orientation of the two reference frames. All vectors obey the same transformation rule.

Second: Understand how to transform the coordinates of a point fixed in frame #1 into frame #2 if one frame is *rotating* with respect to another. The point in question is fixed in the first frame, but not in the second.

---

* The dictionary definition of kinematics is "the study of motion exclusive of the influences of mass and force."

Third: Understand how to calculate time derivatives of vectors (examples: velocities or electric fields) in one frame and find the relation to time derivatives of the same vector in the second reference frame. The vector need not be fixed in frame #1. The change in frame #2 comes from two causes: a) the rate of change in frame #1, transformed into the coordinates of frame #2, and b) the changing transformation between the two reference frames.

After this is worked out, we will find that it is often more useful to consider "fictitious forces" and pretend the Earth is an inertial frame than to try to transform in and out of an inertial reference frame directly. The "Tower of Pisa" problem, in which a heavy object is dropped from the Leaning Tower, will constitute an example of how a practical problem is solved using the concept of fictitious forces. We will also show why hurricanes circulate counterclockwise in the Northern hemisphere and clockwise in the Southern hemisphere. The precession of a Foucault pendulum will be explained using a Lagrangian method with coordinates fixed in the Earth. All of these effects occur because the reference system is rotating. They would be absent if we viewed the same motion from an inertial reference system, so we say they are purely kinematic effects not of dynamical origin.

7.1 WHAT IS A VECTOR?

The intuitive definition of a vector is "a quantity that possesses both direction and magnitude." When we wish to refer to a physical vector without specifying its coordinates in any particular reference frame, we put a little arrow above the symbol:* $\vec{r}$.

However, sometimes we need to work with actual coordinates of vectors. Coordinates are numbers that are defined with respect to a specific set of coordinate axes, say the XYZ reference frame. The coordinates of $\vec{a}$ in this frame might be written as (a_x, a_y, a_z) or (a_1, a_2, a_3). Even better, but more abstractly, we can write a_k, where the subscript k is understood to run from 1 to 3 in a three-dimensional space. In a different reference frame, the $X'Y'Z'$ system, we would write a'_k. ($\vec{a}'$ doesn't make sense, since the vector notation doesn't refer to any particular set of coordinates.)

In terms of unit vectors $(\hat{\imath}, \hat{\jmath}, \hat{k})$ along the X, Y, or Z axes, we can also write

$$\vec{a} = a_1 \hat{\imath} + a_2 \hat{\jmath} + a_3 \hat{k}. \tag{7.1}$$

This formula defines the coordinates of $\vec{a}$ in the XYZ frame.

The square of the length of $\vec{a}$ is $a^2 \equiv \vec{a} \cdot \vec{a} = a_1^2 + a_2^2 + a_3^2 = \sum_{k=1}^{3} a_k a_k$. A further simplifying convention, introduced by Einstein, *will be used in the rest of this book*:

$$a^2 = \sum_{k=1}^{3} a_k a_k \equiv a_k a_k. \tag{7.2}$$

The right-hand formula employs the "Einstein summation convention": *Although the summation symbol is dropped for convenience, repeated indices imply the summation is there.*

---

* This vector notation was invented by J. W. Gibbs late in the nineteenth century.

The length of a vector is preserved if the reference frame is rotated:

$$a^2 = a_k a_k = a'_k a'_k. \tag{7.3}$$

The scalar product of two vectors $\vec{a}$ and $\vec{b}$ is written in the Einstein notation as $\vec{a} \cdot \vec{b} = a_k b_k$. The scalar product depends only on the angle between the two vectors and on their magnitudes. It does not change if the coordinates themselves are changed by a rotation of the reference system, so

$$a_k b_k = a'_k b'_k. \tag{7.4}$$

To find out how to transform coordinates explicitly, choose two reference frames with the same origin and differing only by a rotation. There are two ways to write the arbitrary vector $\vec{a}$:

$$\vec{a} = \underbrace{a_1 \hat{\imath} + a_2 \hat{\jmath} + a_3 \hat{k}}_{XYZ \text{ coordinates}} = \underbrace{a'_1 \hat{\imath}' + a'_2 \hat{\jmath}' + a'_3 \hat{k}'}_{X'Y'Z' \text{ coordinates}}. \tag{7.5}$$

The scalar product $\hat{\imath}' \cdot \hat{\imath}$ is the projection of the X' axis onto the X axis. By taking the scalar product of (7.5) with $\hat{\imath}'$, $\hat{\jmath}'$, $\hat{k}'$ one at a time, we obtain three equations for the three quantities a'_1, a'_2, a'_3 in terms of a_1, a_2, a_3 and the various projections of one set of coordinate axes onto the other. For example,

$$\begin{aligned} a'_1 &= \hat{\imath}' \cdot \vec{a} = a_1 \underline{\hat{\imath}' \cdot \hat{\imath}} + a_2 \underline{\hat{\imath}' \cdot \hat{\jmath}} + a_3 \underline{\hat{\imath}' \cdot \hat{k}}, \\ a'_2 &= \hat{\jmath}' \cdot \vec{a} = a_1 \underline{\hat{\jmath}' \cdot \hat{\imath}} + a_2 \underline{\hat{\jmath}' \cdot \hat{\jmath}} + a_3 \underline{\hat{\jmath}' \cdot \hat{k}}, \\ a'_3 &= \hat{k}' \cdot \vec{a} = a_1 \underline{\hat{k}' \cdot \hat{\imath}} + a_2 \underline{\hat{k}' \cdot \hat{\jmath}} + a_3 \underline{\hat{k}' \cdot \hat{k}}. \end{aligned} \tag{7.6}$$

The primed coordinates are given by linear equations in the unprimed coordinates a_k with coefficients ($\hat{\imath}' \cdot \hat{\imath}$, etc.) that do not depend on the particular vector but only on the relationship between the two coordinate frames. (We have underlined the coefficients in the linear equations to draw your attention to the fact that these are considered as fixed coefficients, whereas the a_k change depending on the vector being transformed into the primed frame from the unprimed frame.)

The formulas (7.6) give the transformation rule we were seeking. In fact, we could define a vector as something that transforms according to (7.6). This definition is preferred because it distinguishes vectors from tensors, and it avoids the vague concepts of "direction" and "magnitude," which are hard to express without introducing a specific coordinate system. Tensors will be introduced in the next chapter, so we will say nothing more here. Figure 7.1 summarizes the notation used for vectors.

7.2   REVIEW: INFINITESIMAL ROTATIONS AND ANGULAR VELOCITY

A rigid body can be considered as a collection of point masses held together at fixed distances from each other. In the limit of infinitely many such point masses, we can

7.2 REVIEW: INFINITESIMAL ROTATIONS AND ANGULAR VELOCITY

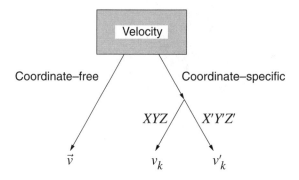

FIGURE 7.1
Notation used for vectors. Note that the components of $v_k \neq v'_k$. However, the square of the length, $v_k v_k = v'_k v'_k$.

consider the rigid body to be a continuous distribution of mass without any internal degrees of freedom. In that case, there are only six degrees of freedom: three translational and three rotational. We will focus here on the mathematics of describing the rotational motion.

Imagine that you have a rigid body rotating about a fixed axis as shown in Figure 7.2. All the points in the body that lie along this axis are not moving. We want to know the motion of an arbitrary fixed point in the body, as seen by a fixed observer located outside of the body. The origin of the fixed observer's coordinate system lies on the axis of rotation. Let the vector between this origin and a point P fixed in the body be called $\vec{r} = \vec{r}(t)$. It will simplify the mathematical description greatly if we initially consider *infinitesimal* rotations instead of finite ones.

How much and in what direction does the point P move in an infinitesimal interval of time dt? Assume that the body rotates in that time interval through an angle $d\phi$ around the (fixed) axis. Let $\vec{r}(t+dt) \equiv \vec{r}(t) + d\vec{r}$ define the infinitesimal vector $d\vec{r}$. From the geometry shown in Figure 7.3, the magnitude of this vector, $|d\vec{r}|$, is an arc of length $r \sin \theta \, d\phi$. The direction of $d\vec{r}$ can be found by noting that $d\vec{r}$ must lie along a vector perpendicular to both the axis of rotation $d\vec{\phi}$ and the vector $\vec{r}$ itself.

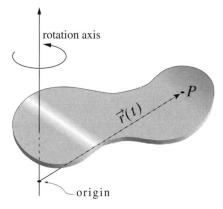

FIGURE 7.2
P = point fixed in body.

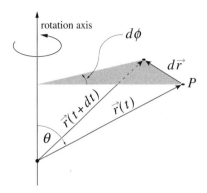

FIGURE 7.3
$|d\vec{r}| = |r|\sin\theta\, d\phi$.

QUESTION 1: Rotations Explain why $|d\vec{r}| = r\sin\theta\, d\phi$, and why $d\vec{r}$ is perpendicular to both the axis of rotation $d\vec{\phi}$ and the vector $\vec{r}$ itself. What is meant by "rotation about an axis"?

Let the direction of the axis of rotation be given by the unit vector $\hat{n}$. Since $\hat{n} \times \vec{r} = |\vec{r}|\sin\theta$ (in a direction perpendicular to $\hat{n}$ and $\vec{r}$), it must be true that

$$d\vec{r} = d\phi\, \hat{n} \times \vec{r}. \tag{7.7}$$

We can define a "vector": $d\vec{\phi} \equiv \hat{n}\, d\phi$. Using this definition, we then write

$$d\vec{r} = d\vec{\phi} \times \vec{r}. \tag{7.8}$$

Notice that $d\vec{r}$ and $d\vec{\phi}$ are infinitesimal quantities, whereas $\hat{n}$ and $\vec{r}$ are not. Formula (7.8) is only valid for infinitesimal rotations.

The velocity of the point P for a continuous rotation in time is $\vec{v}_P = \frac{d\vec{r}}{dt}$. The velocity of any point P in a rotating body in terms of the *angular velocity* $\vec{\omega}$ is, from Equation (7.8) above,

$$\vec{\omega} \equiv \frac{d\vec{\phi}}{dt}, \tag{7.9}$$

$$\vec{v}_P = \frac{d\vec{r}}{dt} = \vec{\omega} \times \vec{r}. \tag{7.10}$$

velocity of point P *fixed in the body*

The relation (7.10) was derived for a fixed time interval t to $t + dt$. By using the infinitesimal dt, we can define the *instantaneous angular velocity* $\vec{\omega}$. In some of the problems we will solve later, the instantaneous axis of rotation is itself varying in both its direction and its magnitude.

7.2 REVIEW: INFINITESIMAL ROTATIONS AND ANGULAR VELOCITY

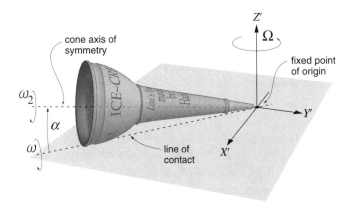

FIGURE 7.4

Why were quotes placed around the word "vector" when defining $d\vec{\phi}$? It is because $d\vec{\phi}$ and $\vec{\omega}$ are not true vectors. An axis of rotation defines a plane in which the rotation takes place. This rotation possesses a helicity or "handedness," which is not the same as a direction to a point located in space. $d\vec{\phi}$ and $\vec{\omega}$ are examples of *pseudovectors*, or as a synonymous term, *axial vectors*. A pseudovector is something that rotates like a vector, but is invariant under reflections,* unlike a vector. For example, vector cross products $\vec{a} \times \vec{b}$ are pseudovectors because under reflection ($\vec{a} \to -\vec{a}, \vec{b} \to -\vec{b}$), the cross product $\vec{a} \times \vec{b}$ remains unchanged. In electromagnetism, the electric field $\vec{E}$ is a true vector, whereas the magnetic field $\vec{B}$ is a pseudovector. We *define* a direction of rotation for a pseudovector by using the "right-hand rule" convention.

Instantaneous Axis of Rotation

What if the instantaneous axis of rotation is not constant? As an example, consider an ice cream cone lying on a flat table as shown in Figure 7.4. Let it roll without slipping in such a way that the apex (point) of the cone remains fixed, while the large end rolls without slipping in a circle on the table.

The velocity of any point on the cone that is momentarily in contact with the table must be zero, since the cone rolls without slipping. This line of contact with the table is the instantaneous axis of rotation and hence gives the direction of $\vec{\omega}$. From the sense of rotation indicated in the figure, we can describe $\vec{\omega}$ in mathematical terms using the unit vectors $\hat{\imath}'$, $\hat{\jmath}'$ in the lab (space) frame along the X', Y' axes, respectively, and magnitude of rotation Ω about the Z' axis (Ω is constant):

$$\vec{\omega} = -|w|(\cos \Omega t \, \hat{\imath}' + \sin \Omega t \, \hat{\jmath}'). \tag{7.11}$$

You should check to make sure that you understand the signs in Equation (7.11). $\Omega > 0$ as shown. A negative value for $\Omega < 0$ would correspond to a cone rolling in the clockwise direction on the plane in the figure.

---

* Definition of reflection: $X \to -X, Y \to -Y, Z \to -Z$.

QUESTION 2: *Rotating Cone 1* Show that $\vec{\omega}$ expressed in a moving coordinate system rotating at an angular frequency Ω around the Z' axis is a constant vector along (say) the X axis in the rotating system. Prove that the central axis of the cone must also lie in a constant direction in this reference frame. Describe the motion of the cone as seen by an observer rotating at angular frequency Ω.

In the example of the rolling ice cream cone, it can be useful to consider $\vec{w}$ as due to the combination of two simultaneous rotations. If the cone rotated with respect to the table at an angular velocity $\vec{\omega}_1 \equiv \Omega \hat{k}'$, it would have to slide at the point of contact with the table. If we add to $\vec{\omega}_1$ an additional simultaneous rotation $\vec{\omega}_2$ around the cone axis of symmetry, the cone's axis will be unaffected by the second rotation and will continue to revolve around the Z' axis at a uniform rate. We can then choose the magnitude ω_2 so that the velocity of any point on the line of contact between the cone and the table is zero, to give us rolling without slipping. This is an intuitive argument that the effect of two angular velocities can be added to equal a third angular velocity. This means that we can replace the effect of two or more simultaneous rotations by a single rotation that has exactly the same effect *if we consider only a short time interval dt*. This equivalence is usually not true for any finite time interval.

Ordinary velocity refers to the motion of a point in space. Angular velocity refers to the motion of a point or collection of points relative to an axis of rotation. The dimensions of velocity are meters second^{-1}, of angular velocity second^{-1}. Nevertheless, it is a remarkable fact that angular velocities add like vectors: $\vec{\omega} = \vec{\omega}_1 + \vec{\omega}_2$. This is proved by noting that, by Taylor's theorem,

$$\underbrace{\vec{r}(d\phi_1, d\phi_2) - \vec{r}(0,0)}_{d\vec{r}} = \underbrace{d\phi_1 \frac{\partial \vec{r}}{\partial \phi_1}}_{d\vec{r}_1} + \underbrace{d\phi_2 \frac{\partial \vec{r}}{\partial \phi_2}}_{d\vec{r}_2} + O\left[d\phi_1^2, d\phi_2^2, d\phi_1 d\phi_2\right]. \quad (7.12)$$

Divide (7.12) above by dt and go to the limit $dt \to 0$. Second- and higher-order terms in the infinitesimal angles $d\phi_1, d\phi_2$ can then be neglected. Using the defining relations for $\vec{\omega}$, (7.8)–(7.10), the vector addition property of angular velocities is proved. Try working out the details of the proof of this vector addition property for yourself.

Figure 7.5 shows the angular velocities from Figure 7.4: $\vec{\omega} = \vec{\omega}_1 + \vec{\omega}_2$, where $\vec{\omega}_1 = \Omega$ is vertical, along Z', $\vec{\omega}_2$ is parallel to the cone axis, and $\vec{\omega}$ is along the line of contact. You

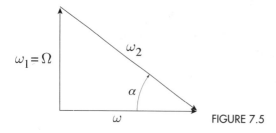

FIGURE 7.5

can determine the directions of the angular velocities by using the right-hand rule. $\vec{\omega}$ and $\vec{\Omega}$ form two sides of a right triangle, with $\vec{\omega}$ parallel to the table (i.e., in the $X'Y'$ plane as shown in Figure 7.5). By the Pythagorean theorem for right triangles, $\omega_2^2 = \omega^2 + \Omega^2$. Knowing the geometry of the cone, you can compute both ω_2 and Ω in terms of ω and the half-angle of the cone.

Remember that we can find the coordinates of $\vec{\omega}$ in *any* coordinate system we want to: the fixed system, the system rotating at frequency Ω, or even a system fixed with respect to the cone itself. Each of these different sets of coordinates describes the same vector in different reference systems.

QUESTION 3: *Rotating Cone 2* Obtain the components of $\vec{\omega}$ in a coordinate system that is fixed to the ice cream cone. Hint: Express the three unit vectors in the frame rotating at $\Omega \hat{k}'$ in terms of unit vectors $\hat{\imath}, \hat{\jmath}, \hat{k}$, which are fixed in the cone.

7.3 FINITE THREE-DIMENSIONAL ROTATIONS

The mathematics of finite rotations in three-dimensional space is quite a bit more complicated than for finite rotations in two-dimensional space, because successive rotations no longer "commute" – a mathematical term meaning that you obtain a different result if you carry out the rotations in a different order. To illustrate this for yourself, do the following experiment. Take a book and lay it flat on a table with the cover right-side up. Imagine a set of coordinate axes with the Z axis vertical, the X axis parallel to the bottom edge of the book, and the Y axis parallel to the binding of the book. Now rotate the book $90°$ about the Z axis (clockwise), followed by a $90°$ clockwise rotation about the original X axis. Finally, make a $90°$ rotation about the original Y axis. Notice the final orientation of the book. Now restore the book to its original position and make the same three rotations, but in the reverse order. If you've done it correctly, the book will be upside down facing you at the end of the first sequence and right-side up facing you at the end of the second. This proves that the order of successive 3-D rotations matters.

A rotation about the Z axis converts a point on the X axis to a point somewhere in the XY plane. This particular rotation is equivalent to a linear transformation of the x, y coordinates. In general a single rotation is a linear transformation of the coordinates in a plane perpendicular to the axis of rotation. A vector directed along the axis of rotation is left unchanged (i.e., is "invariant under this transformation").

7.4 ROTATED REFERENCE FRAMES

Fix a coordinate system in a rigid body, called the "body coordinate system" and align it with a fixed "space coordinate system." Make the origins of the two systems coincide. Then make an arbitrary orientation of the body by rotating it a fixed amount around an axis through the common origin, which is a fixed point in the body as shown in Figure 7.6. (Usually this fixed point is the center of mass for reasons which will become

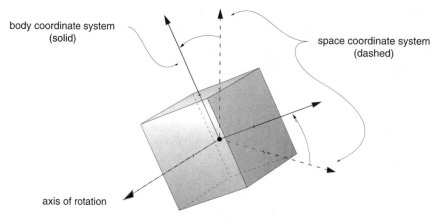

FIGURE 7.6

clear later.)

> Our notation will be:
> SPACE ≡ PRIMED,
> BODY ≡ UNPRIMED.

Definition of the Space Reference Frame K'

Define a *space coordinate system* to be an inertial frame, with coordinate axes labeled by X', Y', Z'. The Cartesian coordinates of the point P (see Figure 7.7) fixed in the body are a set of three numbers which we can arrange in a column:

$$r'_k \equiv \begin{pmatrix} r'_1 \\ r'_2 \\ r'_3 \end{pmatrix}. \tag{7.13}$$

Definition of the Body Reference Frame K

We also define a *rotated* or *body coordinate system* K, which is rotated with respect to K'. Assume that K is not moving with respect to K'. Think of this new coordinate system K as attached to a rigid body, while the space coordinates K' are a fixed inertial reference frame. In Figure 7.7, you can see the orientation of K with respect to K'. The origins of the two coordinates coincide.

In the body system or rotated frame K, the same physical point P has different coordinates:

$$r_k \equiv \begin{pmatrix} r_1 \\ r_2 \\ r_3 \end{pmatrix}. \tag{7.14}$$

In order to completely specify the orientation of K with respect to K', three numbers are required. Call them θ, ψ, ϕ – we'll be more explicit later. This agrees with the fact

7.4 ROTATED REFERENCE FRAMES

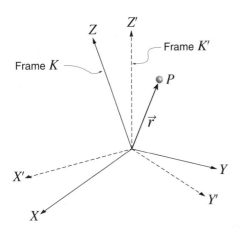

FIGURE 7.7
P = point fixed in body.

that a rigid body with a fixed center of mass has three degrees of freedom. To "make" K from K', first choose an axis (two numbers for the direction), and then rotate by some definite angle around that axis, giving a requirement to specify three parameters in all. Once θ, ψ, ϕ are specified, they will be the same for all possible points P. An arbitrary 3-D rotation can always be described by three parameters: two for the direction of the axis of rotation and one for the amount of the rotation about this axis.

The appropriate math to describe rotations is linear algebra.* We have already found the connection between coordinates of the same physical point P as expressed in two reference frames, K' and K (7.6). This transformation can be represented by a 3 × 3 matrix **U** which describes the linear transformation of any arbitrary point in the space being rotated, assuming that the origin is located at the fixed point. Define **U** as

$$\mathbf{U} \equiv \begin{pmatrix} \hat{i}' \cdot \hat{i} & \hat{i}' \cdot \hat{j} & \hat{i}' \cdot \hat{k} \\ \hat{j}' \cdot \hat{i} & \hat{j}' \cdot \hat{j} & \hat{j}' \cdot \hat{k} \\ \hat{k}' \cdot \hat{i} & \hat{k}' \cdot \hat{j} & \hat{k}' \cdot \hat{k} \end{pmatrix}. \tag{7.15}$$

Since there could be $3^2 = 9$ parameters, when in fact only three independent degrees of freedom exist, there must be 6 equations relating the 9 matrix elements. For the special kind of linear transformation, the rotation described by the matrix **U**, all angles and distances between vectors in the space are preserved. No "stretching" occurs.

From the rule for multiplying a matrix times a column vector, we can rewrite the formulas (7.6) with the elements of **U** (7.15) representing the coefficients of the r_ks:

$$\begin{pmatrix} r_1' \\ r_2' \\ r_3' \end{pmatrix} = \begin{pmatrix} U_{11} & U_{12} & U_{13} \\ U_{21} & U_{22} & U_{23} \\ U_{31} & U_{32} & U_{33} \end{pmatrix} \begin{pmatrix} r_1 \\ r_2 \\ r_3 \end{pmatrix}. \tag{7.16}$$

---

* See, for example, *Linear Algebra with Applications*, 2d ed., by Steven J. Leon.

We can write this either in component form,

$$r'_k = U_{ki} r_i, \quad (7.17)$$

or in the still more efficient form:

$$\mathbf{r}' = \mathbf{U}\mathbf{r}. \quad (7.18)$$

The operation of matrix multiplication is implied in both equations (7.17, 7.18). These are two different ways of writing the same transformation of coordinates from the unprimed to the primed system. In (7.18) we introduce a new notation for the vector $\vec{r}$ from the origin to the point P: the boldface $\mathbf{r}'$ and $\mathbf{r}$. These are understood to be column vectors of the coordinates – a vertical column list of the coordinates. Since it will be very convenient to make use of the properties of matrices to describe rotations, we usually prefer this notation to that of (7.17). The boldface notation will be used only in the context of matrix representations of linear coordinate transformations. Repeated linear transformations can be considered as a single overall linear transformation: If $\mathbf{U}_1$ is the first rotation and $\mathbf{U}_2$ is the second, the net result of these two successive rotations is $\mathbf{U} = \mathbf{U}_2 \mathbf{U}_1$. (Matrix multiplication is implied in the last formula.)

> **QUESTION 4: Linear Transformations** Write out the linear transformation corresponding to rotation about the X axis. Also, how can repeated linear transformations be considered as a single overall linear transformation?

> **QUESTION 5: Rotations** Think of a rotation, and create an explicit matrix $\mathbf{U}$ to describe such a rotation. This rotation can be anything from a simple rotation about the Z axis to something more complex and imaginative. Explore the relation between the rotated frame K and the reference frame K'.

As we explained, all of the matrix elements of the 3×3 matrix $\mathbf{U}$ are not independent, since there are nine matrix elements that must depend on only three parameters. Lengths of vectors are not preserved by arbitrary linear transformations, but they must be preserved by $\mathbf{U}$, because rotations preserve length by definition. Denote the transpose* of $\mathbf{U} = U_{ki}$ by $\tilde{\mathbf{U}} \equiv U_{ik}$. Using (7.13),

$$r_1'^2 + r_2'^2 + r_3'^2 = \begin{pmatrix} r_1' & r_2' & r_3' \end{pmatrix} \begin{pmatrix} r_1' \\ r_2' \\ r_3' \end{pmatrix} = \begin{pmatrix} r_1 & r_2 & r_3 \end{pmatrix} \tilde{\mathbf{U}}\mathbf{U} \begin{pmatrix} r_1 \\ r_2 \\ r_3 \end{pmatrix} = r_1^2 + r_2^2 + r_3^2. \quad (7.19)$$

Since the coordinates are arbitrary, it must be true that $\tilde{\mathbf{U}}$ is the inverse of $\mathbf{U}$ and that the product $\tilde{\mathbf{U}}\mathbf{U}$ is the identity matrix

$$\mathbf{I} \equiv \begin{pmatrix} 1 & 0 & 0 \\ 0 & 1 & 0 \\ 0 & 0 & 1 \end{pmatrix}.$$

---

* In many texts the notation for transpose of $\mathbf{U}$ is $\mathbf{U}^T$.

This being the case, we can write the inverse of Equation (7.18):

$$\mathbf{r} = \tilde{\mathbf{U}}\mathbf{r}'. \tag{7.20}$$

This takes us backwards from coordinates in the space system to coordinates in the body system.

The condition that $\tilde{\mathbf{U}} = \mathbf{U}^{-1}$ means, by definition, that $\mathbf{U}$ is an *orthogonal* matrix. (If the elements of $\mathbf{U}$ were complex instead of real, and if the transpose of the complex conjugate matrix is the inverse, $\mathbf{U}$ is defined as *unitary*.) The condition for $\mathbf{U}$ to be an orthogonal matrix appears to give nine equations. Because $\tilde{\mathbf{U}}\mathbf{U}$ is automatically a symmetric matrix, and a symmetric 3×3 matrix has six independent elements, we can reduce the nine equations to six. To prove this last statement recall that the transpose of a matrix product is the product of the transposes in the reverse order:

$$\widetilde{\mathbf{AB}} = \tilde{\mathbf{B}}\tilde{\mathbf{A}}. \tag{7.21}$$

QUESTION 6: U Matrix Prove by explicitly writing out r'_k in terms of r_k for an arbitrary rotation matrix $\mathbf{U}$ that if the length is to be preserved under rotations then $\tilde{\mathbf{U}}\mathbf{U} = \mathbf{1}$.

7.5 ROTATING REFERENCE FRAMES

Now assume that the body system K may be rotating. This means that $\mathbf{U}$ is a function of time. Also assume that the point we singled out, P, could be moving within K. As a simple example, P could be a bug crawling across a rotating turntable as shown in Figure 4.10. The unprimed coordinates would be the bug's coordinates with respect to the turntable, while the primed coordinates are the bug's coordinates in the fixed reference system K'. The coordinates of the bug at point P are changing in the body system (K) only because the bug moves with respect to the coordinate axes. In the space system (K'), the bug's coordinates may be changing for two reasons: a) the K system is rotating with respect to K' and b) the bug is moving in the body system.

Our goal is to relate the time derivatives of the coordinates of P in the space (fixed) reference system to the time derivatives of P in the body (rotating) system. We will use the coordinate-specific boldface notation introduced above which implies matrix multiplication. Denote by $\mathbf{v}|'_{\text{space}} \equiv \frac{d\mathbf{r}'}{dt}$, the time derivative of the space coordinates of P (space velocity) expressed in the space system K'. Use $\mathbf{v}|_{\text{body}} \equiv \frac{d\mathbf{r}}{dt}$ for the time derivative of the body coordinates of P (body velocity) expressed in the body system. The chain rule for differentiation of the matrix Equation (7.18) with respect to time gives a formula with two terms:

$$\mathbf{v}|'_{\text{space}} = \underbrace{\dot{\mathbf{U}}\mathbf{r}}_{\text{relative rotation of frames}} + \underbrace{\mathbf{U}\mathbf{v}|_{\text{body}}}_{\text{transformation of body velocity into space coordinates}} \tag{7.22}$$

We denote the time derivative of the matrix $\mathbf{U}$ by $\dot{\mathbf{U}}$, the 3×3 matrix that has as its elements the time derivatives of the elements of $\mathbf{U}$. Insert the identity matrix $\mathbf{1} = \tilde{\mathbf{U}}\mathbf{U}$ in between $\mathbf{U}$ and $\mathbf{r}$ on the right side of (7.22):

$$\mathbf{v}|'_{\text{space}} = \dot{\mathbf{U}}\tilde{\mathbf{U}}\mathbf{U}\,\mathbf{r} + \mathbf{U}\mathbf{v}|_{\text{body}}. \tag{7.23}$$

TABLE 7.1 SUMMARY OF THE DIFFERENT NOTATIONS

Quantity	Symbol	Meaning
$\frac{d\mathbf{r}}{dt}$	$\mathbf{v}\vert_{\text{body}}$	body velocity in body coordinates
$\mathbf{U}\frac{d\mathbf{r}}{dt}$	$\mathbf{v}'\vert_{\text{body}}$	body velocity in space coordinates
$\frac{d\mathbf{r}'}{dt}$	$\mathbf{v}'\vert_{\text{space}}$	space velocity in space coordinates
$\tilde{\mathbf{U}}\frac{d\mathbf{r}'}{dt}$	$\mathbf{v}\vert_{\text{space}}$	space velocity in body coordinates

Make use of (7.18):

$$\mathbf{v}'\vert_{\text{space}} = \dot{\mathbf{U}}\tilde{\mathbf{U}}\mathbf{r}' + \mathbf{U}\mathbf{v}\vert_{\text{body}}. \tag{7.24}$$

Now everything in (7.24) is expressed in terms of K' coordinates. $\mathbf{U}\mathbf{v}\vert_{\text{body}}$ is the transformation of the coordinates of $\mathbf{v}\vert_{\text{body}}$ from the K into the K' system. Since the coordinate transformation from the body to the space system for any vector can be made by (matrix) multiplying the vector by $\mathbf{U}$, we can rename $\mathbf{U}\mathbf{v}\vert_{\text{body}} \equiv \mathbf{v}'\vert_{\text{body}}$ to be consistent with our notation. We can also express $\mathbf{v}'\vert_{\text{space}}$ in the body coordinates by multiplying by $\tilde{\mathbf{U}}$: $\tilde{\mathbf{U}}\mathbf{v}'\vert_{\text{space}} \equiv \mathbf{v}\vert_{\text{space}}$.

QUESTION 7: Notation Explain the differences between $\mathbf{v}\vert_{\text{body}}$, $\mathbf{v}'\vert_{\text{body}}$, and $\vec{v}\vert_{\text{body}}$. Does $\mathbf{v}'\vert_{\text{body}} = \mathbf{v}\vert_{\text{space}}$? Also explain why it is meaningless to write $\mathbf{r}'\vert_{\text{body}}$ and $\mathbf{r}\vert_{\text{space}}$.

We will now show that $\dot{\mathbf{U}}\tilde{\mathbf{U}}$ is an antisymmetric matrix, which we will call $\mathbf{A}'$. (The prime is there because we want to work in K' coordinates, and the elements of a matrix do depend on the coordinate system.) The proof of antisymmetry is

$$\mathbf{U}\tilde{\mathbf{U}} = \mathbf{1} \quad \text{so} \quad \dot{\mathbf{U}}\tilde{\mathbf{U}} + \mathbf{U}\dot{\tilde{\mathbf{U}}} = \mathbf{0} \tag{7.25}$$

(chain rule again)

But from (7.21)

$$\widetilde{\dot{\mathbf{U}}\tilde{\mathbf{U}}} = \mathbf{U}\dot{\tilde{\mathbf{U}}}. \tag{7.26}$$

Therefore, the proof is complete: $\tilde{\mathbf{A}}' = -\mathbf{A}'$.

7.6 THE INSTANTANEOUS ANGULAR VELOCITY $\vec{\omega}$

The most general form of an antisymmetric matrix is

$$\mathbf{A}' = \begin{pmatrix} 0 & -\omega'_3 & \omega'_2 \\ \omega'_3 & 0 & -\omega'_1 \\ -\omega'_2 & \omega'_1 & 0 \end{pmatrix} \tag{7.27}$$

7.6 THE INSTANTANEOUS ANGULAR VELOCITY $\vec{\omega}$

$\omega'_1, \omega'_2, \omega'_3$ are numbers that define the matrix $\mathbf{A'}$ in the K' system. If the coordinate system is changed to the unprimed system, the matrix $\mathbf{A}$ will not in general be equal to $\mathbf{A'}$. Instead it can be found by the rule for transforming a matrix from one coordinate system to another $\mathbf{A} = \tilde{\mathbf{U}}\mathbf{A'}\mathbf{U}$. We can interpret this equation as follows: $\mathbf{U}$ converts a vector from its coordinate expression in the unprimed frame into the primed frame; $\mathbf{A'}$ then acts on the (unspecified) vector; and finally $\tilde{\mathbf{U}}$ converts back to the unprimed frame. This rule applies to the transformation of any matrix from one coordinate representation to another.

The special feature of $\mathbf{A'}$ is that it is antisymmetric. It can be shown by direct calculation that (matrix) multiplying $\mathbf{A'}$ into $\mathbf{r'}$ is the same thing as taking $\vec{\omega} \times \vec{r}$, with the coordinates of $\vec{\omega} \equiv (A'_{32}, A'_{13}, A'_{21}) = (\omega'_1, \omega'_2, \omega'_3)$ evaluated in K'. If $\vec{r}$ is parallel to $\vec{\omega}$, we get zero from the cross product, as well as from the operation $\mathbf{A'r'}$ in the matrix notation. This proves that the direction of $\vec{\omega}$ is the axis of the rotation. The magnitude of ω gives us the size of the angular velocity. $\vec{\omega}$ is actually called the "instantaneous angular velocity," since in general $\vec{\omega}$ will not be constant in time.

We can use this result to rewrite Equation (7.24) in vector notation:

$$\vec{v}|_{\text{space}} = \vec{\omega} \times \vec{r} + \vec{v}|_{\text{body}}. \tag{7.28}$$

Compare this to (7.10).

QUESTION 8: *Cross Product* Prove that multiplying $\mathbf{A'}$ into $\mathbf{r'}$ is the same thing as taking the cross product $\vec{\omega} \times \vec{r}$. In other words, $\mathbf{A'}$ is the $\vec{\omega} \times$ operator in matrix notation.

Let $\mathbf{A}$ be the matrix $\mathbf{A'}$ expressed in body coordinates. By the general rules for a change of coordinates induced by $\mathbf{U}$, as mentioned above,

$$\mathbf{A} = \tilde{\mathbf{U}}\mathbf{A'}\mathbf{U} = \tilde{\mathbf{U}}\dot{\mathbf{U}}. \tag{7.29}$$

We can use this relation to compute the components of $\vec{\omega}$ in the body system.

QUESTION 9: *U Matrix* Find the form of $\mathbf{U}$ for a fixed rotation θ about the X axis. (Positive θ means that the Y axis is rotated towards the Z axis.) Now let $\theta = \omega t$. Check that you get the form above for $\mathbf{A'}$, with $\vec{\omega} = (\omega, 0, 0)$. Next find $\vec{\omega}$ in the body coordinates by finding $\mathbf{A}$ using the transformation matrix $\mathbf{U}$.

We recall that the definition of a vector is a set of three numbers that transform like the coordinates of a point in space. The components of any arbitrary vector $\vec{e}$ transform in the same way as $\vec{r}$ by definition:

$$\mathbf{e'} = \mathbf{Ue}. \tag{7.30}$$

Similarly, the time derivative of $\vec{e}$ in the space system can be found in terms of the time derivative of $\vec{e}$ in the body system by applying the operator relation below to $\vec{e}$:

$$\left.\frac{d\vec{e}}{dt}\right|_{\text{space}} = \left[\vec{\omega}\times + \left.\frac{d}{dt}\right|_{\text{body}}\right]\vec{e}. \qquad (7.31)$$

Equation (7.28) is an example of this operator relation being applied to $\vec{r}$. The operator between square brackets [...] can be applied to any vector to connect time derivatives in one frame with those in another. The first term gives the effect of relative motion of the body system with respect to the space system, and the second term arises due to the time derivative of the vector in the body system. Vectors $\vec{e}$ and $\vec{\omega}$ can be expressed in either the body or space coordinates when doing the calculation.

The operator form above is completely equivalent to the matrix form (7.24), but it is often easier to use on $\vec{e}$. In particular, we could use this operator form twice to $\vec{r}$ to calculate the acceleration in the space frame:

$$\vec{a}|_{\text{space}} = \vec{\omega}\times(\vec{\omega}\times\vec{r}) + 2\vec{\omega}\times\vec{v}|_{\text{body}} + \dot{\vec{\omega}}\times\vec{r} + \vec{a}|_{\text{body}}. \qquad (7.32)$$

QUESTION 10: Operator Relation Compare (7.24) to (7.31). Explain how they are similar and how they are different. Convince yourself that the relation above (7.32) is true.

QUESTION 11: Rotating Wheel A wheel of radius r rotates about the Z' axis (see Figure 7.8), at a constant angular velocity ω. Find the velocity $\vec{v}|_{\text{space}}$ and acceleration $\vec{a}|_{\text{space}}$ (relative to the ground) of any point of the rim. Find $\vec{v}|_{\text{space}}$ and acceleration $\vec{a}|_{\text{space}}$ in both the body coordinates and the space coordinates and compare. Provide a physical interpretation for your answers.

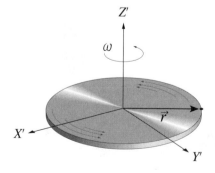

FIGURE 7.8
K frame rotating at speed ω.

7.7 FICTITIOUS FORCES

The force on the mass point is $\vec{F} = m\vec{a}_{\text{space}}$ by Newton's Law of motion. If we pretended that the body system was an inertial system we could define the apparent force as

$$\boxed{\vec{F}_{\text{apparent}} \equiv m\vec{a}_{\text{body}}} \qquad (7.33)$$
$$\text{definition of apparent force}$$

By application of what we have learned from (7.32)

$$\boxed{\vec{F}_{\text{apparent}} = \vec{F} \underbrace{- m\vec{\omega} \times (\vec{\omega} \times \vec{r})}_{\text{centrifugal force}} \underbrace{- 2m\vec{\omega} \times \vec{v}}_{\text{Coriolis force}} \underbrace{- m\dot{\vec{\omega}} \times \vec{r}}_{\text{Euler force}}.} \qquad (7.34)$$

The last three terms on the right side above aren't really forces; they are purely kinematic consequences of the rotation of the body coordinates. We write $\vec{v}$ instead of $\vec{v}_{\text{body}}$ because it is assumed that we are operating in the body coordinate system. For experiments done on the Earth, $\dot{\vec{\omega}} \approx 0$. We should expect to see only the effects of the centrifugal "force" and the Coriolis "force." The prescription for action is thus: "Treat the fictitious forces like real forces, and pretend that you are in an inertial frame." This allows you to work directly in noninertial frames such as the Earth without having to define a separate inertial reference frame.

> **QUESTION 12: Bug on the Earth** Return to the case of a bug crawling, this time on the surface of the Earth. For a bug at the North Pole crawling at a velocity v due south, and a bug at the Equator crawling at a velocity v due east, find the direction and magnitude of all three fictitious forces. For this purpose, assume $\dot{\vec{\omega}}$ cannot be neglected (the Earth's rotation is slowing down slightly). Describe the directions of the forces in words.

7.8 THE TOWER OF PISA PROBLEM

There is a very simple set of experiments we can do to show that the Earth is rotating, even on a cloudy day when we can't see the stars. Drop a plumb bob from the leaning tower of Pisa (height $h = 50$ meters). The gravitational attraction of the Earth pulls the weight on the end of the plumb bob toward the center of the Earth. This defines the vertical direction. In fact, there is no other way to define the vertical! Let the Earth be the body system here. Let us adopt a coordinate system with an origin located at the center of the Earth and with the Z axis pointing outward in a radial direction. Choose the other

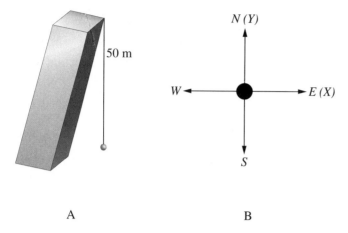

FIGURE 7.9
Z is in the radial direction; X and Y are planar directions.

axes, X and Y, according to Figure 7.9 to make a right-handed coordinate system ($+X$ could point east and $+Y$ could point north, for example). The coordinates of the tower are $(0, 0, R)$, where $R = 6.378 \times 10^6$ m is the radius of the Earth.

What effect will the rotation of the Earth have on the location of the plumb bob? Calculate the two "forces" on the plumb bob from Equation (7.34):

$$\text{gravitational force: } -mg\,\hat{k}, \tag{7.35}$$

$$\text{centrifugal ``force'': } -m\vec{\omega} \times (\vec{\omega} \times \vec{r}). \tag{7.36}$$

Note that the other forces in (7.34) are zero. (In what follows, $\hat{i}, \hat{j}, \hat{k}$ will be unit vectors along the X, Y, and Z axes respectively.) First determine the components of $\vec{\omega}$ in this particular body coordinate system. They are

$$\boxed{\vec{\omega} = |\omega|(0, \cos\lambda, \sin\lambda),} \tag{7.37}$$

$$\text{Earth's angular velocity}$$

where λ is the latitude.

The latitude λ is $0°$ at the equator and $90°$ at the North Pole, as shown in Figure 7.10. For Pisa, $\lambda \approx 45°$ (very approximately). Taking cross products we have

$$\vec{\omega} \times \vec{r} = \omega R \cos\lambda\,\hat{i}, \tag{7.38}$$

$$-\vec{\omega} \times (\vec{\omega} \times \vec{r}) = -\omega^2 R(\cos\lambda \sin\lambda\,\hat{j} - \cos^2\lambda\,\hat{k}). \tag{7.39}$$

We see that the centrifugal force 1) acts to weaken the gravitational force by subtracting $\omega^2 R \cos^2\lambda$ from it, and 2) gives a southward deflection by adding a component along the negative Y direction.

7.8 THE TOWER OF PISA PROBLEM

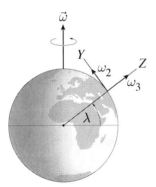

FIGURE 7.10

The tangent of the deflection angle is

$$\tan \theta_{\text{deflection}} = \frac{\omega^2 R \cos \lambda \sin \lambda}{g - \omega^2 R \cos^2 \lambda} \approx \frac{\omega^2 R}{g} \cos \lambda \sin \lambda. \tag{7.40}$$

(Why? How would you prove this last statement?) Substituting numerical values we get

$$\omega = \frac{2\pi}{24 \times 3600} \text{ s}^{-1} = 7.2722 \times 10^{-5} \frac{\text{rad}}{\text{s}}, \tag{7.41}$$

$$\frac{\omega^2 R}{g} = .00344 \quad (g = 9.8 \text{ m s}^{-2}). \tag{7.42}$$

The plumb bob at Pisa is thus deflected by 1.7 milliradians and ends up 8.5 cm south of the coordinate system origin (see Figure 7.11). Unfortunately, there is really no way to measure this deflection, since we can only define the vertical with a plumb bob or a liquid level. All the buildings at this latitude must be tilted by 1.7 milliradians from the direction pointing to the center of the Earth if they were built according to alignment with a plumb bob.

Those people interested in weight reduction should stand at the equator, where the reduction in apparent g is maximal, at about 0.34%. At the North Pole your weight as measured on a bathroom scale will increase by this amount over what it is at the equator. (There are easier ways to find out your latitude!)

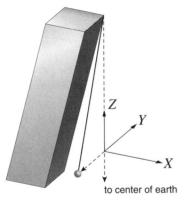

to center of earth   FIGURE 7.11

Dropping a Small BB

The second half of this problem consists of dropping a small BB or lead shot from the tower. We neglect the effects of wind or air resistance, although this might be important in actual practice. The BB will not land on top of the plumb bob, because there is an extra *velocity-dependent* "force" if $\vec{v} \neq 0$: the Coriolis "force." We can move the origin of our coordinates to the point of the plumb bob and make that the new origin. We can calculate the additional deflection due to the Coriolis force using successive approximations, as a power series in the small quantity ωT, where T is the time it takes the BB to fall. (We'll drop the quotes around "force" from now on and pretend these kinematic effects are real.)

Note that the origin here coincides with the origin of a plumb bob (which is not the radial direction as we just discovered).

For the lowest level of approximation, denoted by a superscript (0), we assume that there is no Coriolis force, so the velocity is along the Z axis and equals $\vec{v}^{(0)} = -gt\hat{k}$, where t is the time elapsed since the release of the BB. Then using this as our first approximation, we calculate the magnitude of the Coriolis force and the added deflection it produces:

$$\vec{F}^{(0)}_{\text{Coriolis}} = -2m\vec{\omega} \times \vec{v}^{(0)} = 2m\omega gt \cos\lambda\, \hat{\imath}. \tag{7.43}$$

The direction is determined in Figure 7.12. Integrate the force once with respect to time to get the change in momentum due to the force. Divide by the mass to get the change in velocity:

$$\vec{v}^{(1)} = \vec{v}^{(0)} + \Delta\vec{v}^{(1)}, \tag{7.44}$$

$$\Delta\vec{v}^{(1)} = \omega gt^2 \cos\lambda\, \hat{\imath}. \tag{7.45}$$

The superscript on $\Delta\vec{v}$ reminds us that this is only a first approximation. Integrate again to get the displacement ($\vec{r}^{(0)} = 0$ by definition):

$$\vec{r} = \Delta\vec{r}^{(1)} = \omega g \frac{t^3}{3} \cos\lambda\, \hat{\imath}. \tag{7.46}$$

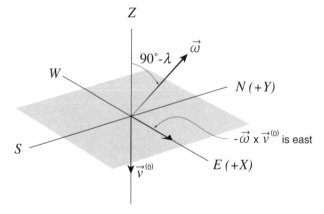

FIGURE 7.12

Call the time from the release of the BB until it hits the ground T. Use the approximate relation $g\frac{T^2}{2} = h$, the tower height, to rewrite the formula above (when the BB hits the ground) as

$$\Delta \vec{r}^{(1)} = \frac{2}{3} h\omega T \cos\lambda\, \hat{\imath}. \tag{7.47}$$

The deflection from the plumb bob is eastward ($+x$), about 5.5 mm when we use the appropriate numbers.

For our second approximation, we use (7.44) as the velocity when calculating the Coriolis force (7.43) and repeat the calculation, obtaining

$$\Delta \vec{v}^{(2)} = -\frac{2}{3}\omega^2 g t^3 \sin\lambda \cos\lambda\, \hat{\jmath},$$
$$\Delta \vec{r}^{(2)} = -\frac{1}{3} h(\omega T)^2 \sin\lambda \cos\lambda\, \hat{\jmath}. \tag{7.48}$$

You should prove Equations (7.48) for yourself. This is an additional southward deflection of about 0.45 microns in this case. An even smaller correction, of order $\omega T \frac{\omega^2 R}{g}$, would result if we had included the centrifugal effects in the calculation.

QUESTION 13: *Tower of Pisa* We just found out by using fictitious forces that if you drop a mass from some height in the Northern hemisphere, it will be deflected eastward. The Earth is rotating from west to east, so, intuitively, you might expect that the Earth rotates while the BB falls, and the BB will therefore land to the west of the plumb bob. Analyze the motion in an inertial frame of reference and show how this simple argument is wrong. The BB follows an elliptical orbit with respect to the center of the Earth. (See the homework problem for a more quantitative analysis.)

7.9 WHY DO HURRICANE WINDS ROTATE?

A hurricane is generated by an extremely low pressure area surrounded by high pressure. If there were no Coriolis forces, you might reasonably expect the wind to blow inward towards the low pressure. The Coriolis force gives an additional $-\vec{\omega} \times \vec{v} = \vec{v} \times \vec{\omega}$ deflection which causes a deflection to the right in the Northern hemisphere and to the left in the Southern hemisphere. This gives rise to a counterclockwise circulation of the wind in the Northern hemisphere (and vice versa for the Southern hemisphere), which is very clearly visible on satellite photographs of hurricanes. Viscosity forces and other effects reduce the Coriolis effect such that the wind maintains a 20–30° angle with respect to the radial direction, as shown in Figure 7.13.

QUESTION 14: *Hurricanes* Explain in detail why hurricanes rotate counterclockwise in the Northern hemisphere and clockwise in the Southern hemisphere. Also, just something to ponder – do you think this is true for other things like bathtubs, toilets, water fountains, etc.?

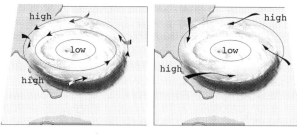

Theoretical   Actual   FIGURE 7.13

7.10 FOUCAULT PENDULUM

One of the most striking experiments that demonstrates the rotation of the Earth is the Foucault pendulum as shown in Figure 7.14. This device is a spherical pendulum, free to swing in both the X and Y directions, and pivoted at the top by a clever bearing that does not restrict the motion. If there were no effect of the Earth's rotation, the x and y motions would be independent and would have the same frequency. For small-amplitude oscillations the frequency would be $\sqrt{\frac{g}{l}}$ in either direction. One possible motion of the pendulum is in a plane, as shown in Figure 7.14 (zero angular momentum), with x and y oscillations in phase; another is a circular or elliptical motion with a phase difference between X and Y projections of the motion. The Earth's rotation causes a slow rotation of the XY plane of the pendulum, with a period that depends on the latitude. If you want to see a Foucault pendulum in action, the National Museum of American History, on the Mall in Washington D.C., has a very long and impressive example. The original demonstration was in Paris by Foucault in 1851. It caused a sensation at the time.

We wish to calculate the additional effects arising from the rotation of the Earth to which the pendulum is attached through the swivel bearing (which decouples the pendulum from the Earth). Angular momentum conservation tells us that if we observe the pendulum in an internal reference frame, the plane of vibration of the pendulum will not change. If

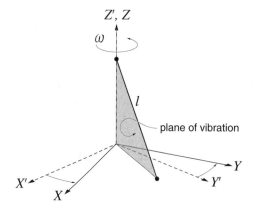

FIGURE 7.14

Foucault pendulum located at the North Pole. K' is space frame. K is Earth frame, rotating at angular velocity ω.

7.10 FOUCAULT PENDULUM

the pendulum were located at the North Pole, then from the viewpoint of an observer at the Pole the pendulum will appear to rotate the plane of vibration once per day, since no torque is transmitted by the bearing. This slow rotation of the pendulum's XY plane of motion is called *precession*.

To obtain the precessional motion in the general case of the pendulum at any latitude, we will resort to Lagrangian methods instead of using the fictitious force approach used before. (It could be done the other way too.) We need to write the Lagrangian in an inertial frame, substitute for the inertial frame coordinates and velocities in terms of Earth frame (noninertial) coordinates and velocities, and find the equations of motion. While doing this, there is a valuable lesson to learn about problem solving. If one tries to proceed without approximation, as is often true in physics, an algebraic morass results. If we *make the approximations in the very beginning*, when writing the Lagrangian, simple equations of motion will result. In principle you could do it by doggedly keeping all terms and approximating at the end, but this turns out to be both hard to do and not very elegant. All the extra work is also a waste of time, because the physics does not lie in a complicated formula. When solving many problems in theoretical physics it is often useful to make your approximations right at the start. Of course, you have to avoid the danger of dropping truly important terms. Experience helps!

We will locate an inertial frame origin to coincide at $t = 0$ with the equilibrium point of the pendulum, that is, the point it hits when it is at rest as shown in Figure 7.14. The Z' direction is vertical; X' and Y' are the horizontal directions. The Lagrangian in the space (inertial) coordinates is

$$T = \frac{1}{2}mv^2|_{\text{space}}, \quad V = mgz', \quad L = T - V. \tag{7.49}$$

The body frame (unprimed) is the same as the frame at rest on the Earth's surface. Use (7.28), and neglect ω^2 terms in the kinetic energy, since $\omega = 7.3 \times 10^{-5}$ s^{-1} is very small. For the pendulum, $z = l(1 - \cos\theta) \approx l\theta^2/2$, and $\theta^2 \approx \frac{(x^2+y^2)}{l^2}$ (small-angle approximation). Also assume $\dot{z}$ is small compared to $\dot{x}, \dot{y}$. Given all these approximations, the Lagrangian, expressed in the Earth (body) frame in terms of unprimed coordinates x, y is

$$L = \frac{m}{2}(\dot{x}^2 + \dot{y}^2 + 2\omega(x\dot{y} - y\dot{x})\sin\lambda + 2\omega z\dot{x}\cos\lambda) - mg\frac{x^2 + y^2}{2l}. \tag{7.50}$$

Finally, we can neglect the term proportional to $z\dot{x}$, because it is small compared to the terms proportional to $x\dot{y}$ and $y\dot{x}$, since most of the motion takes place in the XY plane. After this breathtaking simplification, we proceed in the usual way to find the equations of motion. The effect of the Earth's motion is contained in the term proportional to ω in Equation (7.50) above. The equations are

$$\frac{\partial L}{\partial x} = \frac{m}{2}(2\omega\sin\lambda)\dot{y} - \frac{mg}{l}x, \quad \frac{\partial L}{\partial y} = -\frac{m}{2}(2\omega\sin\lambda)\dot{x} - \frac{mg}{l}y,$$
$$\frac{\partial L}{\partial \dot{x}} = \frac{m}{2}(2\dot{x} - (2\omega\sin\lambda)y), \quad \frac{\partial L}{\partial \dot{y}} = \frac{m}{2}(2\dot{y} + (2\omega\sin\lambda)x). \tag{7.51}$$

From the above, via the Euler–Lagrange equations, we deduce that the EOM are

$$\ddot{x} + \left(\frac{g}{l}\right)x = (2\omega \sin \lambda)\dot{y},$$
$$\ddot{y} + \left(\frac{g}{l}\right)y = -(2\omega \sin \lambda)\dot{x}.$$
(7.52)

The x and y motions are coupled together by a term proportional to the angular velocity of the Earth's rotation in Equations (7.52). If $\omega \to 0$, these equations become uncoupled SHOs. To solve the coupled equations, consider the complex quantity (to make the mathematics easier)

$$\zeta \equiv x + iy.$$
(7.53)

Multiply the second equation of motion by i and add it to the first one. You obtain

$$\ddot{\zeta} + i(2\omega \sin \lambda)\dot{\zeta} + \left(\frac{g}{l}\right)\zeta = 0.$$
(7.54)

This looks just like the EOM for a damped SHO (3.29), except that the damping is complex.

Mathematical solutions can be found by assuming $\zeta = e^{i\alpha t}$ as before. This reduces the differential equation of motion (7.54) to a quadratic equation that must be satisfied by α:

$$-\alpha^2 - 2\omega \sin \lambda\, \alpha + \frac{g}{l} = 0.$$
(7.55)

Since $\omega^2 \ll \frac{g}{l}$, we can approximate the solutions by

$$\alpha \approx -(\omega \sin \lambda) \pm \sqrt{\frac{g}{l}}.$$
(7.56)

So

$$\zeta = x + iy = e^{-i\omega \sin \lambda t}(x + iy)_{\omega=0},$$
(7.57)

where

$$(x + iy)_{\omega=0} = e^{\pm i\sqrt{\frac{g}{l}}t}.$$
(7.58)

(In (7.58), the actual motion for $\omega = 0$ is a linear combination of the positive and negative exponents. The physical sense of rotation for the plane of swinging, if $\omega \neq 0$, does not depend on this choice.) For the physical interpretation, take the real part for the x displacement of the pendulum and the imaginary part for the y displacement. Viewing the motion projected into the XY plane, we find that (7.57) tells us that motion when $\omega \neq 0$ will be seen to precess clockwise at a rate of $\omega \sin \lambda$ radians per second. (Both the sign and the magnitude agree with what we predict at the North Pole, where $\sin \lambda = 1$.)

The period T of this precession depends on the latitude of the pendulum. It is

$$T(\text{precession period}) = \frac{2\pi}{\omega \sin \lambda}. \tag{7.59}$$

In practical units,

$$T = \frac{24}{\sin \lambda} \text{ hours}. \tag{7.60}$$

QUESTION 15: *Foucault Pendulum* If you actually build a Foucault pendulum, it turns out to be very difficult to build a short one. The world record is about 6 inches. Why do you think this is true? The various experimental problems are explained in two articles by Crane in *Physics Teacher* (May 1990, p. 267) and the *American Journal of Physics* (1981, vol. 49, pp. 1004–6).

SUMMARY OF CHAPTER 7

- Three-dimensional rotations do not necessarily commute with each other (i.e., the order of making the rotations matters).
- The most general rotation of a rigid body is described by three parameters. A 3×3 orthogonal matrix $\mathbf{U}$ describes this unique rotation in three-dimensional space.
- We can compute the coordinates of any vector in either the body or the space system. A linear transformation using the matrix $\mathbf{U}$ gives the connection and the inverse between the two sets of coordinates:

$$\mathbf{r}' = \mathbf{U}\mathbf{r}, \quad \mathbf{r} = \tilde{\mathbf{U}}\mathbf{r}'. \tag{7.61}$$

$\mathbf{r} = $ coordinates in body system K;
$\mathbf{r}' = $ coordinates in space system K'

- $\tilde{\mathbf{U}}$ is the transpose of $\mathbf{U}$: $\tilde{U}_{ij} = U_{ji}$. $\mathbf{U}$ is orthogonal: $\tilde{\mathbf{U}} = \mathbf{U}^{-1}$.
- For a system whose orientation is changing with time, define the antisymmetric matrix $\mathbf{A}' \equiv \dot{\mathbf{U}}\tilde{\mathbf{U}}$. The instantaneous angular velocity $\vec{\omega} \equiv (A'_{32}, A'_{13}, A'_{21})$. The axis of rotation is the direction of $\vec{\omega}$, and the magnitude of $\vec{\omega}$ is rate of rotation around the axis.

- The time derivative of an arbitrary vector $\vec{e}$ in the space system can be found in terms of its time derivative in the body system by the relation

$$\left.\frac{d\vec{e}}{dt}\right|_{\text{space}} = \left[\vec{\omega}\times + \left.\frac{d}{dt}\right|_{\text{body}}\right]\vec{e}. \qquad (7.62)$$

- In a noninertial system rotating with angular velocity $\vec{\omega}$ with respect to an inertial system, the fictitious or apparent force is given by

$$\vec{F}_{\text{apparent}} = \vec{F} \underbrace{- m\vec{\omega}\times(\vec{\omega}\times\vec{r})}_{\text{centrifugal force}} \underbrace{- 2m\vec{\omega}\times\vec{v}}_{\text{Coriolis force}} \underbrace{- m\dot{\vec{\omega}}\times\vec{r}}_{\text{Euler force}}. \qquad (7.63)$$

PROBLEMS

Angular Velocity

Problem 1*: *(Locomotive)*

a) A locomotive is rounding a curve of radius 1 km at a speed of 80 km/hr in the counterclockwise direction. The wheels of the locomotive are turning at 1,000 rpm. What are the components of $\vec{\omega}$, the instantaneous axis of rotation, for the wheels in the space system (i.e., for an observer on the ground)?

b) What are the components of $\vec{\omega}$ in the body system of the locomotive (i.e., from the vantage point of the train engineer)? First give an intuitive answer and see if it agrees with a calculation based on the formula for transforming the angular velocity vector components into the body frame (7.29). Does the answer contradict your intuition? How would you explain this result?

c) The components of the angular velocity in the body system can also be found by using the vector transformation property of angular velocity. The chain rule (7.12) implies vector addition of the two angular velocities of the locomotive. Let $\mathbf{U}_1$ be the matrix that transforms from wheel body coordinates to train body coordinates. $\mathbf{U}_2$ will be the matrix transforming from train body coordinates to space (ground-based) coordinates. Then $\mathbf{U} \equiv \mathbf{U}_2\mathbf{U}_1$ transforms from wheel body coordinates to space coordinates. Starting from the definition of the antisymmetric matrix $\mathbf{A}' \equiv \dot{\mathbf{U}}\tilde{\mathbf{U}}$, prove that

$$\mathbf{A}' = \dot{\mathbf{U}}_2\tilde{\mathbf{U}}_2 + \mathbf{U}_2(\dot{\mathbf{U}}_1\tilde{\mathbf{U}}_1)\tilde{\mathbf{U}}_2. \qquad (7.64)$$

Interpret this result and show how it proves that angular velocities add like vectors.

Problem 2: ($\vec{\omega}$ *as pseudovector*) How does the matrix **A** (or, equivalently, the vector $\vec{\omega}$) transform if all coordinates are reversed as under a space reflection? Prove that $\vec{\omega}$ transforms like a pseudovector rather than a true vector. (A vector reverses sign under a spatial reflection; a pseudovector does not.)

Problem 3: (*Rolling cone*) A cone rolls on a flat surface. The instantaneous axis of rotation lies parallel to the point where the cone touches the surface and the angular velocity is $\vec{\omega}$. The motion consists of a motion of the center of mass ($\vec{V}_{cm}$) plus a rotation $\vec{\omega}_{cm}$ about the center of mass. Describe this motion by finding $\vec{V}_{cm}$ and $\vec{\omega}_{cm}$ in the laboratory (space) system.

Problem 4: (*Rolling sphere*) A sphere of radius R rolls without slipping on a flat surface with angular velocity $\vec{\omega}$. Since rolling without slipping means that the velocity of the point of tangency between the sphere and the surface is zero, this gives a relation between $\vec{V}_{cm}$ and $\vec{\omega}$. Find this relation, which is a constraint on the motion. How many degrees of freedom does the sphere have?

Problem 5: (*Charged electron*) In some respects an electron is like a charged spinning top. The electron has internal angular momentum and a magnetic moment, so it behaves like a magnetic dipole oriented along the spin axis. In a magnetic field the equation of motion for the spin angular momentum $\vec{\sigma}$ in a magnetic field $\vec{B}$ is

$$\dot{\vec{\sigma}} = g'(\vec{\sigma} \times \vec{B}), \tag{7.65}$$

where $g' \equiv \frac{e\hbar}{2mc} g$, e is the electronic charge, $\hbar$ Planck's constant divided by 2π, m the electron's mass, and c the velocity of light. The constant g is called the "gyromagnetic ratio." For electrons, $g \approx 2$.

a) Show that in a frame rotating at a certain angular velocity, the effect of the magnetic field can be made to vanish. Find this angular velocity $\vec{\omega}_0$.
b) Suppose the magnetic field has two components: $\vec{B} = B_0 \hat{k} + \vec{B}_1$, where B_0 is a constant, $\hat{k}' = \hat{k}$ is a unit vector in the Z', Z direction, and $\vec{B}_1$ is a rotating magnetic field of constant magnitude: $\vec{B}_1 = B_1(\cos \omega t \, \hat{i}' + \sin \omega t \, \hat{j}')$. Regarding ω as a variable parameter, find the equation of motion in a frame rotating with angular velocity ω and solve it. Describe qualitatively what happens to the spin if a) $\omega = \omega_0$, b) $\omega \neq \omega_0$. (This is the basic equation of NMR – nuclear magnetic resonance – except that the spin is that of an atomic nucleus, not of an electron.)

Orthogonal Matrices

Problem 6: (*The most general form*)

a) Find the most general form of an orthogonal 2×2 matrix. What is the geometric interpretation of such a matrix? What are the complex eigenvalues of such a matrix? What special property do they have? What is the determinant of the general 2-D orthogonal matrix?

b) Write out the 3 × 3 separate matrices for 90° clockwise rotations about the X, Y, and Z axes. Find the products of these rotation matrices about Z first, Y next, and X last. Then find the product for rotating in the reverse order. Interpret the result in terms of the experiment with the book in Section 7.3. (The corresponding questions from part a) for 3 × 3 orthogonal matrices are more difficult to answer. We will develop an explicit form for the 3-D orthogonal matrix **U** in terms of the three Euler angles in the next chapter.)

Problem 7*: *(General properties of orthogonal matrices)* This problem involves proving some general properties of orthogonal matrices in a space of arbitrary dimensions. It will be necessary to know some facts about determinants that hold for any arbitrary $n \times n$ matrix **M**:

$$\det \mathbf{M} = \det \tilde{\mathbf{M}}, \quad \det(-\mathbf{M}) = (-1)^n \det \mathbf{M}, \tag{7.66}$$

$$\det(\mathbf{AB}) = \det(\mathbf{BA}) = \det \mathbf{A} \det \mathbf{B}. \tag{7.67}$$

You may wish to review the derivations of (7.66, 7.67) in a book on linear algebra.[†] Use the above identities to prove the following:

a) If **U** is a real orthogonal $n \times n$ matrix, prove

$$\det \mathbf{U} = \pm 1. \tag{7.68}$$

(If $\det \mathbf{U} = 1$, **U** is a proper rotation; if $\det \mathbf{U} = -1$, **U** is an improper rotation, i.e., a reflection plus a proper rotation.)

b) For a proper rotation of any odd-dimensionality n, prove that the orthogonal matrix **U** has at least one eigenvalue equal to 1; hence there is an "axis" of rotation – a direction that is invariant under the transformation **U**. (Hint: First prove that if there is an eigenvalue equal to 1, $\det(\mathbf{U} - \mathbf{1}) = 0$, where **1** is the identity matrix.)

c) The trace of a matrix is the sum of its diagonal elements. Prove that the trace of any matrix **M** is invariant under an orthogonal transformation: $\mathbf{M}' = \mathbf{U}\mathbf{M}\tilde{\mathbf{U}}$. (Hint: Trace(**AB**) = Trace(**BA**). Prove this first; then prove that Trace(**M**′) = Trace **M**.)

Problem 8*: *(Trace of U)* If **U** is a real orthogonal 3 × 3 matrix, show that the trace of **U** equals

$$\text{Trace } \mathbf{U} = 1 + 2\cos\Phi, \tag{7.69}$$

where Φ is the angle of rotation. (Hint: We proved in Problem 7c that the trace is invariant when the basis vectors are changed to a new set by an orthogonal transformation. Try moving the axis of rotation to the Z axis by such a transformation.)

[†] For example: *Linear Algebra with Applications*, 2d ed., by Steven J. Leon.

Problem 9: *(Inverse operator relation)* Find a formula for $\frac{d}{dt}|_{body}$ that inverts the operator relation, Equation (7.31). What is the inverse relation to the equation (7.32) for the acceleration?

Fictitious Forces

Problem 10*: *(Fictitious forces on a turntable)* A turntable is rotating around the $Z' = Z$ axis with an angular velocity ω. Consider a bug that is free to move without friction on the turntable surface, which is the XY plane as shown in figure 4.10. The bug is not acted upon by any forces parallel to the turntable surface. The bug's Hamiltonian is (5.71)

$$H = \frac{p_x^2 + p_y^2}{2m} - \omega L_z, \tag{7.70}$$

$$L_z \equiv (\vec{r} \times \vec{p})_z.$$

(Everything in (7.70) is expressed in terms of the turntable (body) coordinate system.)

a) Find the equations of motion in terms of x, y for the bug by eliminating p_x and p_y, the canonical momenta, from Hamilton's equations of motion.

b) Show that you get the same answer by considering fictitious centrifugal and Coriolis forces instead. Hint: To check that you have the sign of $\vec{\omega}$ correct, pick a fixed point on the turntable and use the relation (7.28).

c) The bug is initially located at $x = x_0$, $y = 0$, moving with initial velocity $\vec{v}_0$ with respect to the body and initially experiences no acceleration (in either space or body frames). What is $\vec{v}_0$ explicitly in body coordinates? Does the acceleration (both $\vec{a}|_{space}$ and $\vec{a}|_{body}$) remain zero?

Problem 11*: *(WW I Falklands Islands battle)* During a 1915 naval battle between the British and the Germans, which took place in the Falkland Islands (latitude 50° south), the British shells missed by about 100 yards to the left of the German ships. Apparently the British gunsights were intended for battles near Europe (latitude 50° north), and hence they were improperly corrected for the effect of the Coriolis force in Southern latitudes. The spirit of the calculation in this problem is the same as that of the Tower of Pisa problem in Section 7.8, except that the first approximation is constant velocity in a horizontal direction.

a) Derive an approximate formula that gives the amount of deflection for a shell fired horizontally at an opponent a distance D away, due to the Coriolis effect.

b) Explain why the effect observed by the British Navy was twice what you calculated from the Coriolis effect.

c) Assume the velocity of a shell is about 2,000 mph, which is ≈ 1 km/s, or three times the speed of sound. Neglect the effect of gravity in the first approximation – assume the shell travels in a straight line. Estimate how far away the German ships must have been. How does your estimated value for D depend on the direction of firing with respect to north? In the second approximation include

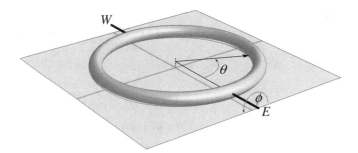

FIGURE 7.15
Compton generator.

the effect of gravity. How does this change the result? Use $\omega_{\text{earth}} = 7.3 \times 10^{-5}$ radians/sec.

Problem 12: *(Deflection of object thrown up into the air)* A heavy object is thrown up into the air. Calculate the deflection of the object when it hits the ground due to the Coriolis force. Compare the results to those of an object dropped at rest from its maximum height.

Problem 13: *(Compton generator)* When he was an undergraduate, the famous physicist A. H. Compton invented a simple way to measure the rotation of the Earth with a table-top experiment. The "Compton generator," as it was called, is a circular hollow glass tube shaped like a doughnut as shown in Figure 7.15. The inside of the tube is filled with water. Imagine that the "doughnut" lies flat on a table and is then turned over by rotating it 180° around a diameter, such that it again lies flat on the table surface, which is horizontal. The result of the experiment is that the water moves with a certain constant drift velocity around the tube after the doughnut has been rotated. If there were no friction with the walls, the water would continue to circulate indefinitely.

a) Prove that the axis about which the doughnut is flipped should be oriented east–west to maximize the drift velocity of the water. What will happen if it is oriented north–south instead?

b) Let θ be defined as the angle between a small volume of the water and the "flip" axis. Calculate the component of the Coriolis force *parallel* to the wall of the circular tube, F_θ, while the circular tube is being flipped 180°. For simplicity, assume it has been flipped through 90° already and is still moving. Draw little arrows for different positions in the tube (different θ values) to show the relative magnitude of the tangential component of the Coriolis force at that point at time. Why don't we have to consider the radial component of this force?

c) Let the angle of rotation about the diameter be ϕ. Then ϕ starts at zero, and when the tube has been flipped over, $\phi = 180°$. From the time integral of the tangential force, calculate the change in the total tangential momentum of the water in the tube. This means integrating over θ around the rim of the tube. Show that the total tangential momentum does not depend on the $\phi(t)$ but only on the total change in

ϕ. Evaluate this total momentum for a change of 180° in ϕ. You can assume the water is equidistant from the center of the circle at some constant radius R. Does the water circulate clockwise or counterclockwise?

d) Since water is incompressible, the water molecules in the tube must all drift with the same velocity after the tube is flipped. With this assumption, you can calculate this drift velocity as a function of the Earth's angular velocity and the latitude. Prove that

$$v_{\text{drift}} = 2\omega R \sin \lambda. \tag{7.71}$$

Compton used small droplets of coal oil mixed in the water to measure the drift velocity under a microscope. The experiment consists of laying the tube flat on a table until the water in it came to equilibrium, then slowly (in about 3 seconds) rotating it about an east–west axis until it had turned 180° and was again lying flat on the table.

e) Compton used this measured drift velocity to determine his latitude. His $\sin \lambda$ was measured to within 3% accuracy, which is pretty good for such a simple device. Assuming $\lambda = \frac{\pi}{4}$ and $R = 1$ meter, what is v_{drift} in mm/s?

Problem 14: *(Force-free motion as seen from a turntable)* Plot the trajectory of the force-free motion of a particle as seen from a frame rotating with a constant angular velocity ω. Assume the particle starts with an initial outward radial velocity v_0, at an initial position halfway towards the rim. If R is the turntable radius, let

$$v_0 = f\omega R. \tag{7.72}$$

Plot this motion for $f = 0, 0.5, 0.7$, and 3. Notice that all trajectories are straight lines in an inertial frame.

Problem 15: *(Hurricanes)* Prove that the steady-state motion of the wind near a low pressure area in the atmosphere is a circle along the lines of constant pressure if air resistance is neglected. Also: Why are there no hurricanes near the equator?

Problem 16: *(Foucault pendulum)* Here is a different method for solving the Foucault pendulum. Assume the motion of the pendulum is given by

$$\begin{aligned} x(t) &= A_x(t)\cos(\omega_P t + \Phi_x(t)), \\ y(t) &= A_y(t)\cos(\omega_P t + \Phi_y(t)), \end{aligned} \tag{7.73}$$

where $\omega_P \equiv \sqrt{\frac{g}{l}}$ is the pendulum frequency. In the absence of Coriolis and centrifugal "forces" $A_{x,y}$, $\Phi_{x,y}$ would be arbitrary constants. If these "forces" are present, they become slowly changing functions of the time. Insert the expressions (7.73) into the equations of motion, (7.52). By averaging over many swings of the pendulum, find the differential equations for $A_{x,y}(t)$ and $\Phi_{x,y}(t)$. Solve them and show that this gives the same solution found in the text.

Problem 17: ($\dot{\vec{\omega}}$) This problem involves the "neglected" term in the transformation rule for acceleration, Equation (7.32). $\dot{\omega} \neq 0$ for the Earth. (We refer here to a change in the magnitude of the Earth's angular velocity. The direction of $\vec{\omega}$ also changes slowly with time. This effect will be discussed in the next chapter.) We know that the Earth's rotation is slowing down because of tidal friction and possibly other dissipative effects. The accumulated effect of this slowing down amounts to 30° of rotation (2 hours) in 1,500 years. What is $\dot{\omega}$? Compare the magnitude of this term in the fictitious force to the centrifugal force term. In principle, how could this very small term be observed experimentally? How much has the day lengthened since the age of the dinosaurs (65 million years ago)?

CHAPTER EIGHT

THE DYNAMICS OF RIGID BODIES

OVERVIEW OF CHAPTER 8

Rigid body motion is an important topic in classical mechanics. E. T. Whittaker once gave a lecture* entitled "Spin in the Universe." We quote:

> Rotation is a universal phenomenon; the earth and all the other members of the solar system rotate on their axes, the satellites revolve around the planets, the planets revolve around the Sun, and the Sun himself is a member of the galaxy or Milky Way system which revolves in a very remarkable way. How did all of these rotatory motions come into being? What secures their permanence or brings about their modification? And what part do they play in the system of the world?

A common simplification for rigid body motion is the separation of the rotational degrees of freedom from those of the center of mass motion, for which the body behaves like a point with all of its mass located at the center of mass. All the information about the mass distribution that we need to solve the rotational dynamics is contained in the *moment of inertia tensor* **I**.[†] This means that two bodies with totally different shapes, but having the same inertia tensor, will have identical rotational dynamics. The inertia tensor relates the instantaneous angular velocity $\vec{\omega}$ to the angular momentum $\vec{L}$. These two vectors are proportional in magnitude but are not always in the same direction. We will see that the inertia tensor determines a natural body coordinate system, called the *principal axis system*. Use of this coordinate system usually greatly eases our calculations.

What is the most general possible motion of a rigid body that has no forces or torques acting upon it? This leads us to Euler's equations and the Poinsot construction. Whenever you throw a football in a forward pass and see the football wobble, you are observing solutions to Euler's equations.[‡] In this chapter, we will also study the

---

* Quoted by S. Chandrasekhar in the preface to his book *Ellipsoidal Figures of Equilibrium*, Dover Press, 1987. Whittaker, a professor of mathematics at Edinburgh University, wrote a famous text on mechanics: *Analytical Dynamics of Particles and Rigid Bodies*, Cambridge University Press, 1904.
† See Appendix A for a general discussion of scalars, vectors, and tensors.
‡ If you travel with the football's center of mass and neglect aerodynamic effects.

rotational dynamics of the Earth, which depends on the Earth's equatorial bulge, a problem first discussed by Isaac Newton.

Finally, we will consider heavy symmetric tops, which are often called *gyroscopes*. These are objects with rotational symmetry about at least one axis, and which both spin rapidly about the symmetry axis and are also under the influence of a torque from the force of gravity. To handle this problem and others, we will introduce the notion of *Euler angles*. The **U** matrix defined in the previous chapter can be obtained as an explicit function of these Euler angles. $\vec{\omega}$ is then obtained as a function of the Euler angles and their time derivatives. We will also find the kinetic energy in a form that can be used for any rigid body problem.

8.1 KINETIC ENERGY OF A RIGID BODY

Consider a fixed "space" coordinate system K' and a collection of mass points m_i, $i = 1, \ldots, M$. These points are said to form a rigid body if, for any two points, the distance remains constant:

$$(\vec{r}_i - \vec{r}_j)^2 = \text{constant}, \quad i, j = 1, \ldots, M. \tag{8.1}$$

Attach a set of body coordinate axes K to this arbitrary rigid body. Within K, we can have relative rotation but no linear velocity since we have a rigid body (i.e., $\vec{v}_i|_{\text{body}} = 0$).

Generalizing slightly from what we have done previously, we can allow the origin of K' (space coordinates) to be located in a different place from the origin of K (body coordinates). Define $\vec{R}$ to be the vector from the origin of K' to the (so far) arbitrary origin of the rigid body coordinate system (see Figure 8.1). Notice that because the origins are allowed to differ, $\vec{r}_i|_{\text{space}}$ is no longer equal to $\vec{r}_i|_{\text{body}}$. The velocity of the ith point in the rigid body as expressed in the space system is

$$\vec{v}_i|_{\text{space}} = \dot{\vec{R}} + \vec{\omega} \times \vec{r}_i|_{\text{body}} \tag{8.2}$$

The vector $\vec{r}_i = 0$ when the point is located at the origin of the body system K (we will drop the specific reference to the body system, since from here on we will *only* work with

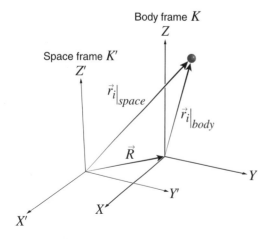

FIGURE 8.1

8.1 KINETIC ENERGY OF A RIGID BODY

$\vec{r}_i$ in the body system). The total kinetic energy to an outside observer in the space system is

$$T = \frac{1}{2}\sum_i m_i v_i^2 \bigg|_{\text{space}} = \frac{1}{2}\left(\sum_i m_i\right)\dot{\vec{R}}^2 + \dot{\vec{R}} \cdot \vec{\omega} \times \left(\sum_i m_i \vec{r}_i\right)$$
$$+ \frac{1}{2}\sum_i m_i (\vec{\omega} \times \vec{r}_i)^2, \tag{8.3}$$

where

$$\sum_i m_i = \text{total mass} = M, \quad \sum_i m_i \vec{r}_i = M\vec{r}_{\text{cm}}. \tag{8.4}$$

A great simplification results if we choose the origin in K to be the center of mass, $\vec{r}_{\text{cm}} \equiv 0$. With this choice, the kinetic energy (8.3) divides neatly into the sum of two separate terms:

$$\boxed{T = T_{\text{translation}} + T_{\text{rotation}}.} \tag{8.5}$$

The kinetic energy associated with rotation is

$$\boxed{T_{\text{rotation}} = \frac{1}{2}\sum m_i (\vec{\omega} \times \vec{r}_i)^2.} \tag{8.6}$$

The translational kinetic energy is the same as if all the mass were concentrated in a point located at the center of mass of the body:

$$\boxed{T_{\text{translation}} = \frac{1}{2}M\dot{\vec{R}}^2.} \tag{8.7}$$

From now on, assume the origin of K is located at the center of mass.

As an example, think about the motion of a baton (or pen) thrown into the air. It follows a rather complicated motion. But, using (8.5) we can separate out the center of mass motion from the rotational motion. The motion of the center of mass of the baton follows a parabola (or is up and down if given no initial horizontal velocity). Since the center of mass motion is just like the motion of a point particle, we want to focus now on the rotational part – the motion of the baton about the center of mass, $\vec{\omega} \neq 0$.

Using vector identities,* we can write

$$(\vec{\omega} \times \vec{r}_i) \cdot (\vec{\omega} \times \vec{r}_i) = \vec{\omega} \cdot (\vec{r}_i \times (\vec{\omega} \times \vec{r}_i)) = \vec{\omega} \cdot \left(r_i^2 \vec{\omega} - (\vec{\omega} \cdot \vec{r}_i)\vec{r}_i\right). \tag{8.8}$$

---

* Kaplan, *Advanced Calculus*, 3rd ed., p. 5 and p. 11.

It will now pay to use explicit vector component indices instead of the vector notation. We will use Greek letters for indices. For example, let $\alpha = 1, 2, 3$ stand for either x, y, or z coordinates respectively (e.g., $\vec{a} \cdot \vec{b} = \sum_\alpha a_\alpha b_\alpha$.) In this new notation we have*

$$T_{\text{rotation}} = \frac{1}{2} \sum_i m_i \left(\sum_{\alpha=1}^{3} r_{i,\alpha}^2 \sum_{\beta=1}^{3} \omega_\beta \omega_\beta - \sum_{\alpha=1}^{3} r_{i,\alpha} \omega_\alpha \sum_{\beta=1}^{3} r_{i,\beta} \omega_\beta \right). \qquad (8.9)$$

Convince yourself that the vector notation in Equation (8.8) and the vector index notation in Equation (8.9) are the same. We can rewrite (8.9) as

$$T_{\text{rotation}} = \frac{1}{2} \sum_{\alpha,\beta} \omega_\alpha \omega_\beta I_{\alpha\beta}, \qquad (8.10)$$

where $\sum_{\alpha,\beta}$ implies $\sum_{\alpha=1}^{3} \sum_{\beta=1}^{3}$, and

$$\boxed{I_{\alpha\beta} \equiv \sum_i m_i \left(r_i^2 \delta_{\alpha\beta} - r_{i,\alpha} r_{i,\beta} \right)} \qquad (8.11)$$

($\delta_{\alpha\beta} \equiv 1$ if $\alpha = \beta$; $\delta_{\alpha\beta} \equiv 0$ otherwise). If we write out the formula above (8.11) explicitly, we get ($\sum \equiv \sum_i$):

$$\boxed{\mathbf{I} = \begin{pmatrix} \sum m_i (y_i^2 + z_i^2) & -\sum m_i x_i y_i & -\sum m_i x_i z_i \\ -\sum m_i x_i y_i & \sum m_i (x_i^2 + z_i^2) & -\sum m_i y_i z_i \\ -\sum m_i x_i z_i & -\sum m_i y_i z_i & \sum m_i (x_i^2 + y_i^2) \end{pmatrix}.} \qquad (8.12)$$

Prove for yourself that (8.12) comes from (8.11).

8.2 THE MOMENT OF INERTIA TENSOR

$I_{\alpha\beta}$ (8.11) is called the *moment of inertia tensor*. If you are unfamiliar with tensors or need a review, see Appendix A. (We will use some results from this appendix here.)

With the notation in (8.10), we have separated the intrinsic properties of the rigid body ($\mathbf{I}$) from the actual motion ($\vec{\omega}$). Because T_{rotation} is a scalar (invariant) under rotations of the space coordinates and $\vec{\omega}$ rotates like a vector, $\mathbf{I}$ must rotate like a second-rank tensor (if you

---

* Note that we have written out the usually implied sums in the equation for clarity.

8.2 THE MOMENT OF INERTIA TENSOR

do not understand this statement, read Appendix A). We can also very conveniently think of **I** as a 3 × 3 symmetric real matrix and the expression (8.10) as matrix multiplication of a row vector · (3 × 3 matrix) · column vector. But remember that **I** is a second-rank tensor, which as explained in Appendix A, is more than just a matrix. In more compact notation, we can write the rotational part of the kinetic energy as*

$$T_{\text{rotation}} = \frac{1}{2}\tilde{\omega}\mathbf{I}\omega, \tag{8.13}$$

where $\tilde{\omega}$ indicates the row rather than column vector ω. Using matrix notation instead of writing out all of the tensor indices is often very useful.

QUESTION 1: *Transformation Law for* I   Since $\vec{\omega}$ rotates like a vector for proper rotations, a fixed change of coordinates (orthogonal transformation **U**) gives new coordinates: $\omega' = \mathbf{U}\omega$ and $\tilde{\omega}' = \tilde{\omega}\tilde{\mathbf{U}}$. Prove that, since T_{rotation} is invariant under a fixed coordinate change, the transformation law for **I** must be

$$\mathbf{I}' = \mathbf{U}\mathbf{I}\tilde{\mathbf{U}}. \tag{8.14}$$

This is the same transformation law found in Question 15 in Appendix A.

Single out a mass point in the sum (8.12) that defines **I** and ask what it contributes to **I**. Suppose the mass is dm and the point where dm is located has the coordinates (x, y, z). Call $d\mathbf{I}$ the contribution to the sum over all of the points:

$$d\mathbf{I} = dm \begin{pmatrix} y^2 + z^2 & -xy & -xz \\ -xy & x^2 + z^2 & -yz \\ -xz & -yz & x^2 + y^2 \end{pmatrix}. \tag{8.15}$$

If the mass distribution is a continuous one, we replace the sum over mass points by an integral over the differential

$$dm = \rho\, dx\, dy\, dz, \tag{8.16}$$

where ρ is the density and $dx\, dy\, dz = dV$ is the element of volume. The I_{33} element would become, for example,

$$I_{33} = \iiint \rho(x^2 + y^2)\, dx\, dy\, dz. \tag{8.17}$$

---

* Since T_{rotation} is the same in both body and space frames you can use either frame to express the coordinates. We use the body frame here.

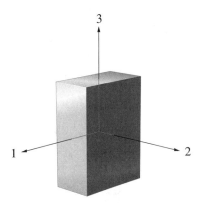

FIGURE 8.2

The inertia tensor is symmetric. If you interchange the indices, the tensor is not changed. A symmetric tensor defined in n-dimensional space has $\frac{n(n+1)}{2}$ independent quantities. An inertia tensor in 3-D space is completely specified by six numbers. Question 1 showed how the form of $\mathbf{I}$ can change under coordinate transformations (8.14). There is a very important theorem for symmetric matrices that uses this fact:

> It is always possible to find an orthogonal transformation $\mathbf{U}$ that diagonalizes $\mathbf{I}$. This is called a *principal axis transformation*.

This will be proved in Appendix B. Making an orthogonal transformation amounts to choosing a particular set of coordinates. Since this can always be done for any rigid body, we can write, with the appropriate choice of a $\mathbf{U}$ matrix,

$$\mathbf{U}\mathbf{I}\tilde{\mathbf{U}} = \begin{pmatrix} I_1 & 0 & 0 \\ 0 & I_2 & 0 \\ 0 & 0 & I_3 \end{pmatrix}. \tag{8.18}$$

Physically, this means that an inertia tensor $\mathbf{I}$ contains within it information about a special set of orthogonal coordinate axes. If we choose these "principal axes" as a body system, calculations involving the moment of inertia tensor will be simplified since $\mathbf{I}$ will be diagonal. Also, I_1, I_2, I_3, which are called the *principal moments*, will all be positive. In matrix language, since all of the eigenvalues of $\mathbf{I}$ are positive, $\mathbf{I}$ is called a *positive definite* matrix.* This guarantees that the kinetic energy will always be positive, no matter what $\vec{\omega}$ is.

In real problems you can often "intuit" the directions of the principal axes from the symmetry of the problem.

If the principal moments of inertia are all different from each other, we say the body is an *asymmetric top*. A good example is a book with different height, width, and breadth as shown in Figure 8.2.

---

* Leon, *Linear Algebra*, 2d ed., Chapter 6.

8.2 THE MOMENT OF INERTIA TENSOR

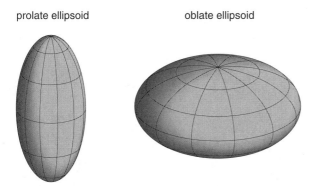

prolate ellipsoid   oblate ellipsoid

FIGURE 8.3
The body 3 axis is in the vertical direction for both cases.

A *symmetric top* has rotational symmetry about one axis. This implies that $I_1 = I_2 \neq I_3$ if we choose the 3 axis as the symmetry axis. There are two different kinds of symmetric tops, as shown in Figure 8.3. A *prolate ellipsoid* (like a football or an egg) must have $I_3 < I_1 = I_2$, with the 3 axis being the long axis of the football. An *oblate ellipsoid*, on the other hand, is like the Earth, which is a "squashed" sphere, being flattened at the pole. In this case, $I_3 > I_1 = I_2$. Of course symmetric tops don't have to be ellipsoids; they must only obey the condition that two of the principal moments are equal. A flat disk of almost zero thickness (like a frisbee) will have $I_3 \approx 2I_1$. The 3 axis is perpendicular to the surface of the disk, passing through the center. A baseball bat with a circular cross section will have its 3 axis along the long dimension of the bat. In this case $I_3 \ll I_1 = I_2$. (Try drawing some pictures and convincing yourself of the truth of these statements.)

An object such as a linear molecule has $I_3 = 0$ (to a good approximation). We call this type of object a *rotator*. It has only two rotational degrees of freedom instead of three.

> **QUESTION 2: *Inertia Tensor*** How do you know that the eigenvalues of the inertia tensor will be positive? Also, explain the meaning of the moment of inertia tensor in detail.

Displaced Axis Theorem

The familiar "parallel axis theorem"* is a special case of what we call the "displaced axis theorem." It states that, if we know the moment of inertia tensor using the center of mass as the origin ($\mathbf{I}_{cm}$), and we wish to know the moment of inertia tensor about an origin displaced by $\vec{a}$, a constant vector, it is given by the formula

$$\mathbf{I}_{\vec{a}} = \mathbf{I}_{cm} + M(a^2 \delta_{\alpha\beta} - a_\alpha a_\beta). \tag{8.19}$$

---

* The moment of inertia, I_a, about an axis parallel to an axis through the center of mass but displaced a distance a from the center of mass is: $I_a = I_{cm} + Ma^2$, where I_{cm} is the moment of inertia about the axis through the center of mass, and M is the mass of the object.

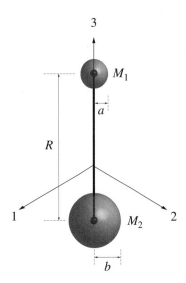

FIGURE 8.4
The origin is at the center of mass.

QUESTION 3: *Displaced Axis Theorem* Prove Equation (8.19) from the definition of **I**.

⊃ Example

As an example, let's calculate the moment of inertia tensor for a dumbbell – two massive spheres M_1, M_2 of different radii a, b respectively, connected by a massless rod of length R as shown in Figure 8.4. We will set our origin at the center of mass.

When calculating **I** it is very important to choose a convenient set of axes. Try to identify the principal axes if you can. Often this involves exploiting the symmetry. If it is not easy to find the principal axis frame, chose the easiest frame you can and then diagonalize the **I** obtained.

In our case, choosing an axis to be along the line connecting the masses (call it the 3 axis) puts us in the principal axis frame. Note that we then have a symmetric top ($I_1 = I_2$). Call $\hat{i}, \hat{j}, \hat{k}$ the unit vectors along the 1, 2, 3 axes respectively. For our coordinate system as shown, M_1 is at the point $\frac{M_2 R}{M_1+M_2} \hat{k}$ and M_2 is at the point $-\frac{M_1 R}{M_1+M_2} \hat{k}$. The simplest way to calculate the components of **I** is to first find the moment of inertia tensor for a sphere and then use the displaced axis theorem.

The moment of inertia tensor for a sphere is most easily calculated in cylindrical coordinates. For the sphere of radius a ($I_1 = I_2 = I_3$):

$$I_{3,(\text{sphere } a)} = \int_{-a}^{a} \int_{0}^{2\pi} \int_{0}^{\sqrt{a^2-z^2}} \rho_1 r^2 \, r \, dr \, d\theta \, dz$$

$$= \frac{8}{15} \rho_1 \pi a^5 = \frac{2}{5} M_1 a^2. \qquad (8.20)$$

(We have used the fact that $M_1 = \frac{4}{3}\pi a^3 \rho_1$). Similarly, for the sphere of radius b, $I_{3,(\text{sphere } b)} = \frac{2}{5} M_2 b^2$.

To find the **I** for the dumbbell, we make use of the displaced axis theorem (8.19):

$$\begin{aligned} I_1 &= I_2 \\ &= \frac{2}{5}M_1 a^2 + M_1 \left(\frac{M_2 R}{M_1 + M_2}\right)^2 + \frac{2}{5}M_2 b^2 + M_2 \left(\frac{M_1 R}{M_1 + M_2}\right)^2, \end{aligned} \qquad (8.21)$$

$$I_3 = \frac{2}{5}M_1 a^2 + \frac{2}{5}M_2 b^2. \qquad (8.22)$$

Once we know **I** for the principal axis frame, we can find it for any other frame by transforming **I** using (8.14). For example, to find **I**′ for a set of axes rotated an angle θ from the 1 axis, we need the **U** matrix:

$$\mathbf{U} = \begin{pmatrix} 1 & 0 & 0 \\ 0 & \cos\theta & \sin\theta \\ 0 & -\sin\theta & \cos\theta \end{pmatrix}. \qquad (8.23)$$

Performing the matrix multiplication yields ($I_1 = I_2$)

$$\mathbf{I}' = \begin{pmatrix} I_1 & 0 & 0 \\ 0 & I_1 \cos^2\theta + I_3 \sin^2\theta & \sin\theta\cos\theta(I_3 - I_1) \\ 0 & \sin\theta\cos\theta(I_3 - I_1) & I_1 \sin^2\theta + I_3 \cos^2\theta \end{pmatrix}. \qquad (8.23)$$

8.3 ANGULAR MOMENTUM OF A RIGID BODY

It is very useful to talk about the "intrinsic" angular momentum of a rigid body, which is the angular momentum it has if the center of mass is at rest. This is usually what we will mean when we talk about the angular momentum. In general, the total angular momentum will depend on our choice of origin, but using the center of mass as the origin simplifies the formulas just as it did for the kinetic energy.*

First we want to prove that it is the inertia tensor that connects intrinsic angular momentum with angular velocity. By the definition of the total angular momentum

$$\vec{L}_{\text{total}} \equiv \sum_i \vec{r}_i \times \vec{p}_i = \sum_i m_i (\vec{r}_i \times \vec{v}_i). \qquad (8.24)$$

Again use body coordinates with the center of mass as the origin (see Figure 8.1):

$$\vec{r}_i |_{\text{space}} = \vec{R} + \vec{r}_i |_{\text{body}}. \qquad (8.25)$$

From (8.2)

$$\vec{v}_i |_{\text{space}} = \vec{V}_{\text{cm}} + \omega \times \vec{r}_i |_{\text{body}}, \quad \vec{V}_{\text{cm}} \equiv \dot{\vec{R}}. \qquad (8.26)$$

---

* We will use $\vec{L}$ to denote the angular momentum. This is not to be confused with the scalar L, the Lagrangian. **L** refers to the column matrix listing the coordinates of $\vec{L}$ in a specific coordinate system.

We calculate (again dropping the reference to the body frame)

$$\vec{L}_{\text{total}} = \vec{R} \times \vec{P} + \sum_i m_i \vec{r}_i \times (\vec{\omega} \times \vec{r}_i) \tag{8.27}$$

($\vec{P} \equiv M\vec{V}_{\text{cm}}$). Since the center of mass is the origin, $\sum_i m_i \vec{r}_i = 0$. Use this fact to prove for yourself that inserting (8.25) and (8.26) into (8.24) gives (8.27). We can write (8.27) as

$$\vec{L}_{\text{total}} = \vec{L}_{\text{cm}} + \vec{L}_{\text{rot}}. \tag{8.28}$$

The second term is the intrinsic angular momentum due to the rotation of the body; the first term ($\vec{L}_{\text{cm}} = \vec{R} \times \vec{P}$) is the angular momentum due to the motion of the center of mass. From now on, the intrinsic (rotational) angular momentum will be called $\vec{L}$. Usually there is no possibility of confusing this with the total angular momentum. Reducing the double cross product with standard vector identities (8.8) we get

$$\vec{L} = \vec{L}_{\text{rot}} = \sum_i m_i \left(r_i^2 \vec{\omega} - (\vec{\omega} \cdot \vec{r}_i)\vec{r}_i \right) = \sum_\beta I_{\alpha\beta} \omega_\beta. \tag{8.29}$$

In the compact matrix notation we have

$$\mathbf{L} = \mathbf{I}\boldsymbol{\omega} \tag{8.30}$$

or

$$\boxed{\vec{L}_{\text{total}} = \underbrace{\vec{R} \times \vec{P}}_{\vec{L}_{\text{cm}}} + \underbrace{\mathbf{I} \cdot \boldsymbol{\omega}}_{\vec{L}_{\text{rot}}}.} \tag{8.31}$$

Thus the magnitude of $\vec{L}$ is proportional to the magnitude of ω. In some special cases $\vec{L}$ and $\vec{\omega}$ are also aligned with each other. For example, a uniformly dense sphere has $I_1 = I_2 = I_3$. Any coordinate orientation can be chosen as a principal axis system due to the degeneracy in the eigenvalues of $\mathbf{I}$. $\mathbf{I}$ will be proportional to the identity matrix, giving perfect alignment between the intrinsic angular momentum and the instantaneous angular velocity.

8.4 THE EULER EQUATIONS FOR FORCE-FREE RIGID BODY MOTION

We will next study the motion of an *arbitrary* rigid body not subjected to any torques,* which means by definition that

$$\frac{d\vec{L}}{dt} = 0. \tag{8.32}$$

The total angular momentum with respect to the center of mass is a constant of the motion. However, this does *not* imply that the angular velocity $\vec{\omega}$ is constant. The easiest way to treat the motion of the angular velocity is to use the principal axis coordinate system, which

---

* There can be forces on it, if they act through the center of mass, but these don't exert torques.

8.5 MOTION OF A TORQUE-FREE SYMMETRIC TOP

is a body coordinate system. Since $\vec{L}$ is a vector, from (7.31)

$$\left.\frac{d\vec{L}}{dt}\right|_{\text{space}} = 0 = \left.\frac{d\vec{L}}{dt}\right|_{\text{body}} + \vec{\omega} \times \vec{L}. \tag{8.33}$$

In terms of the principal axis body coordinate system, using (8.30),

$$\vec{L}_{\text{body}} = \begin{pmatrix} I_1\omega_1 \\ I_2\omega_2 \\ I_3\omega_3 \end{pmatrix}. \tag{8.34}$$

Write out the three equations obtained from the two formulas above ($\frac{d}{dt} \equiv \frac{d}{dt}\big|_{\text{body}}$ here):

$$\boxed{\begin{aligned} I_1 \frac{d\omega_1}{dt} &= \omega_3 L_2 - \omega_2 L_3 = \omega_2 \omega_3 (I_2 - I_3), \\ I_2 \frac{d\omega_2}{dt} &= \omega_1 L_3 - \omega_3 L_1 = \omega_1 \omega_3 (I_3 - I_1), \\ I_3 \frac{d\omega_3}{dt} &= \omega_2 L_1 - \omega_1 L_2 = \omega_1 \omega_2 (I_1 - I_2). \end{aligned}} \tag{8.35}$$

$$\text{Euler's equations}$$

QUESTION 4: Angular Velocity 1 Why aren't $\omega_1, \omega_2, \omega_3 = 0$ in the body frame? That is, in the body frame we are rotating at angular velocity $\vec{\omega}$, so how can we observe it? Explain what exactly is meant by $\omega_1, \omega_2, \omega_3$.

QUESTION 5: Angular Velocity 2 The Euler equations (8.35) are with respect to what frame? Space or body? What is the difference between $\vec{\omega}$ as expressed in the space or body frame anyway? Finally, explain how we obtained (8.35) from (8.33) and (8.34).

Equations (8.35) are known as Euler's equations. In the case of a spherically symmetrical body, Euler's equations imply that $\vec{\omega}$ is also constant. The body can rotate at a constant rotational velocity about a fixed axis. We will discuss a *symmetric top* with $I_1 = I_2 \neq I_3$ next. An *asymmetric top* has none of the moments of inertia I_1, I_2, I_3 equal to each other. The motion can be rather complicated. We will solve this problem in Section 8.9.

8.5 MOTION OF A TORQUE-FREE SYMMETRIC TOP

What is the most general motion possible for a torque-free symmetric top?

Begin by assuming that the top spins* around the 3 axis with a constant angular velocity of Ω radians per second. We also assume the top is tilted at an angle θ from the vertical

---

* In physics, the term "spin" usually refers to angular momentum. Here we use it in a less technical sense to mean rotational angular velocity about a principal axis.

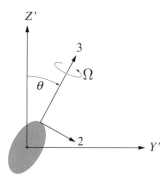

FIGURE 8.5
Body and space coordinate system for the symmetric top.

direction. Define the space Z' axis to be along the angular momentum $\vec{L}$. The magnitude of $\vec{L}$ will be defined by the symbol $L \equiv |\vec{L}|$. Taking a "snapshot" in time, also define the Y' space axis to be in the plane momentarily defined by the 3 axis of the top and the Z' axis (see Figure 8.5).

For a symmetric top, having $I_1 = I_2$ means we can choose the 1 and 2 principal axes to be in any two orthogonal directions in the plane perpendicular to the 3 axis. We will take advantage of this freedom, and rotate the body coordinate system in this instant in time so that the 1 axis points out of the paper and the 2 axis lies in the space $Z'Y'$ plane as shown in Figure 8.5. Then if $\hat{k}'$ is a unit vector along the Z' space axis, and $\hat{j}, \hat{k}$ are unit vectors along the 2 and 3 body axes, momentarily,

$$\hat{k}' = \cos\theta \hat{k} - \sin\theta \hat{j}. \tag{8.36}$$

Because the top has no torques acting on it, T and $\vec{L}$ are fixed constants of the motion. Consistency of the equations we will derive will reveal that θ can be taken as constant as well. This tilt angle of the top, θ, and the instantaneous direction and magnitude of the angular velocity, $\vec{\omega}$, as well as Ω, will be determined from T and $\vec{L}$. Finally, we will give a unified physical picture of the motion.

Part of the contribution to the total angular velocity $\vec{\omega}$ is due to the spin around the body 3 axis, $\Omega \hat{k}$. This part contributes, after matrix multiplication with $\mathbf{I}$, an angular momentum $I_3 \Omega \hat{k}$. At the instant of time pictured in Figure 8.6, $I_3 \Omega \hat{k}$ lies in the $Z'Y'$ plane of the space frame. The vector difference between the total angular momentum $\vec{L} \equiv L\hat{k}'$ and $I_3 \Omega \hat{k}$ must result from matrix multiplying of the inertia tensor by an additional angular velocity, $\vec{\omega}_P$ as shown in Figure 8.6A.

By the definition of $\vec{\omega}_P$ as the "missing piece" of the angular velocity (shown in Figure 8.6B):

$$\vec{\omega} = \Omega \hat{k} + \vec{\omega}_P. \tag{8.37}$$

Since $\vec{\omega}_P$ does not lie along a principal axis, matrix multiplying it by $\mathbf{I}$ to get the missing piece of the angular momentum will give a vector that does not lie in the same direction as

8.5 MOTION OF A TORQUE-FREE SYMMETRIC TOP

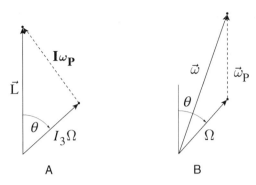

FIGURE 8.6
Comparison of vector diagrams. A) Angular momentum, B) Angular velocity.

$\vec{\omega}_P$. First write $\vec{\omega}_P$ in terms of its body coordinates:

$$\vec{\omega}_P = \omega_{P,1}\,\hat{\imath} + \omega_{P,2}\,\hat{\jmath} + \omega_{P,3}\,\hat{k}. \tag{8.38}$$

Now calculate the missing part of the angular momentum. Since $\mathbf{L} = \mathbf{I}\omega$:

$$\mathbf{I}\omega_P \equiv I_{ij}\omega_{P,j} = I_1(\omega_{P,1}\,\hat{\imath} + \omega_{P,2}\,\hat{\jmath}) + I_3\omega_{P,3}\,\hat{k} = L\,\hat{k}' - I_3\Omega\,\hat{k}, \tag{8.39}$$

where we have used the fact that $I_1 = I_2$. From Figure 8.5, we see that the "missing" angular momentum must lie in the $Z'Y'$ plane, which is also momentarily the 23 plane. Since $\hat{\imath}$ is not in this plane, $\omega_{P,1} = 0$. *We have shown that $\vec{\omega}_P$, and therefore $\vec{\omega}$, lie in the plane defined by the top axis and the vertical direction in space.*

Now we come to the key point of the analysis. Calculate the kinetic energy $T = \frac{1}{2}\tilde{\omega}\mathbf{I}\omega$. By projecting the components of the angular momentum onto the body coordinates, it can be shown that

$$T = I_1\omega_2^2 + I_3\omega_3^2 = \frac{L^2}{2}\left[\frac{\sin^2\theta}{I_1} + \frac{\cos^2\theta}{I_3}\right]. \tag{8.40}$$

Since $I_1 \neq I_3$, the kinetic energy is dependent on θ. Since the rotational kinetic energy is constant, this proves that the top motion cannot change the angle θ between the top 3 axis and the angular momentum. Constant θ implies that $\omega_{P,Y'} = 0$, since otherwise the top would precess about the Y' axis, causing θ to change. There could still be a rotation ("precession") of the top 3 axis about the Z' axis (i.e., $\omega_{P,Z'} \neq 0$). No inconsistency with the requirement that θ be constant arises, because a rotation of the top spin axis about the Z' axis leaves θ unchanged. The direction and magnitude of $\vec{\omega}$ are specified completely by the need to match the total angular momentum, a precondition, and the need to have a constant precession of the top axis around the Z' axis.

Equation (8.40) shows that it is possible to calculate $\theta = \theta(L, T)$ as a function of these constants of the motion. A symmetry axis has no "vector" direction associated with it. This means that we can define $\cos\theta$ to be positive (i.e., $\theta < 90°$), removing the ambiguity in the sign of $\cos\theta$ as determined from Equation (8.40).

We can find the precession rate $|\vec{\omega}_P|$ and the tilt angle θ as functions of $\vec{L}$ and Ω, by separately equating components in (8.39) using the trigonometric relation (8.36):

$$L \cos \theta = I_3(\Omega + \omega_{P,3}),$$
$$-L \sin \theta = I_1 \omega_{P,2}. \tag{8.41}$$

Solving (8.41) for the components of $\vec{\omega}_P$ in the body frame, we obtain

$$\omega_{P,1} = 0, \quad \omega_{P,2} = -\frac{L}{I_1} \sin \theta, \quad \omega_{P,3} = \frac{L}{I_3} \cos \theta - \Omega. \tag{8.42}$$

To gain some insight, let us now express $\vec{\omega}_P$ in the space frame. We again use trigonometric relations between the two coordinate systems:

$$\hat{k} = \cos \theta \hat{k}' + \sin \theta \hat{j}', \quad \hat{j} = \cos \theta \hat{j}' - \sin \theta \hat{k}'. \tag{8.43}$$

Combining Equations (8.38, 8.42) and (8.43) to get an expression for $\vec{\omega}_P$ in space coordinates we get

$$\omega_{P,Z'} = L \left[\frac{\sin^2 \theta}{I_1} + \frac{\cos^2 \theta}{I_3} \right] - \Omega \cos \theta, \tag{8.44}$$

$$\omega_{P,Y'} = 0 = L \sin \theta \cos \theta \left(\frac{1}{I_3} - \frac{1}{I_1} \right) - \Omega \sin \theta. \tag{8.45}$$

As we've said, for θ to remain constant, it must be true that $\omega_{P,Y'} = 0$. Assuming that $\theta \neq 0, \pi$, we obtain from Equation (8.45) an equation determining Ω as a function of L, $\theta(L, T)$ and the constants I_1, I_3:

$$\Omega = L \left(\frac{1}{I_3} - \frac{1}{I_1} \right) \cos \theta. \tag{8.46}$$

The "spin," Ω, is defined to be positive if it lies in the direction of the $+3$ axis, as determined by the right-hand rule. Since $L > 0$ by definition, the sign of Ω is determined from Equation (8.46) by whether $I_3 > I_1$ ($\Omega < 0$) or $I_3 < I_1$ ($\Omega > 0$). This leads to two rather different pictures of the spinning object, depending on whether it is like an oblate or prolate ellipsoid. The next figure illustrates the cases of positive and of negative Ω.

The magnitude of $\vec{\omega}_P$ is now given by Equation (8.44) being just the Z' component of $\vec{\omega}_P$. After combining terms using Equation (8.46) in Equation (8.44), we get the final result for the rate of precession:

$$\omega_P \equiv |\vec{\omega}_P| = \frac{L}{I_1}. \tag{8.47}$$

Equations (8.40, 8.46, 8.47) comprise our main results. They show that the tilt angle θ, the spin Ω, and the precession angular velocity ω_P are constants that are functions of the moments of inertia and the initial conditions: L, T. The "missing piece" of the angular

8.5 MOTION OF A TORQUE-FREE SYMMETRIC TOP

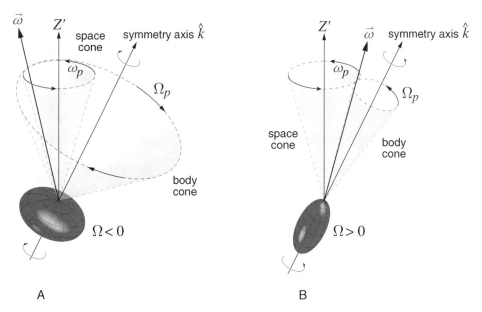

FIGURE 8.7
A) Oblate case: $I_3 > I_1$ ($\Omega < 0$). B) Prolate case: $I_3 < I_1$ ($\Omega > 0$). The axis $\vec{\omega}$ precesses in the space frame at angular velocity ω_p about the Z' axis, and in the body frame at angular velocity Ω_p about the $\hat{k}$ axis.

velocity points straight along the fixed Z' axis. The situation is similar to that of the rolling ice cream cone discussed at the beginning of Chapter 7. The instantaneous axis of rotation is, by definition, $\vec{\omega}$. This motion consists of combining two separate but simultaneous rotations: a "spin" of Ω radians/second around the symmetry axis and a "precession" of ω_P radians/second for the top axis around the Z' space axis. The vector $\vec{\omega}$ must also precess with an angular velocity of ω_P around the Z' axis, since the vectors $\vec{L}$, $\vec{\omega}$, and $\Omega \hat{k}$ must lie in a common plane. If $I_1 \neq I_2$ this coplanarity would not exist. We will deal with the more complicated situation of an asymmetric top (with $I_1 \neq I_2 \neq I_3$) in Section 8.9.

As the top precesses around the Z' axis, $\vec{\omega}$ describes a cone with its axis along the Z' direction. Because $\vec{\omega}$ is an instantaneous axis of rotation, the top "rolls" around $\vec{\omega}$. Thus there are two cones: a "space cone" centered on the fixed space axis Z', which is traced out by $\vec{\omega}$ as time evolves, and a "body cone" centered on the top's 3 axis. This body cone is also being traced by the locus of $\vec{\omega}$. The line where both cones intersect is the direction of $\vec{\omega}$. The moving body cone "rolls" on the fixed space cone. This is shown in Figure 8.7. The precession rate of $\vec{\omega}$ in the body cone is most easily calculated using the Euler equations. We will call this rate Ω_p. It is the rate that keeps the angular velocity vector coplanar with the top axis and with $\vec{L}$.

Calculating Ω_p With the Euler Equations

We can get the same answer using the Euler equations (8.35). However, the viewpoint is different. With the Euler equations we work in body coordinates. The "view from the body" must be physically equivalent to what was done above. The main difference is that we do not take a "snapshot," but rather work with the actual time dependence of $\vec{\omega}$ in body coordinates, $\omega_1, \omega_2, \omega_3$. The third equation in (8.35) says that ω_3, the projection of $\vec{\omega}$ on the 3 axis, is constant since $I_1 = I_2$. Rewriting the first two Euler equations after dividing through by $I_1 = I_2$, we get (defining the constant angular velocity Ω_p)

$$\Omega_p \equiv \omega_3 \left(\frac{I_3}{I_1} - 1 \right), \tag{8.48}$$

$$\frac{d\omega_1}{dt} + \Omega_p \omega_2 = 0, \quad \frac{d\omega_2}{dt} - \Omega_p \omega_1 = 0. \tag{8.49}$$

Combine these equations to obtain uncoupled second-order differential equations

$$\frac{d^2\omega_1}{dt^2} + \Omega_p^2 \omega_1 = 0, \quad \frac{d^2\omega_2}{dt^2} + \Omega_p^2 \omega_2 = 0. \tag{8.50}$$

The solution to these equations is

$$\omega_1 = A \sin(\Omega_p t + \phi), \quad \omega_2 = A \cos(\Omega_p t + \phi), \tag{8.51}$$

where A and ϕ are (so far) arbitrary constants. Since ω_1 and ω_2 are nonzero, uniform precession occurs about the 3 axis. The components of $\vec{\omega}$ perpendicular to the 3 axis precess at a uniform rate Ω_p, as we saw earlier. Making $\omega_1 = 0$ to be consistent with our earlier derivation, if we define $t = 0$ to be the moment of the "snapshot," we must choose $\phi = 0$. As a test of understanding, we will ask the reader to find the value of the constant A in Equation (8.51) from this information.

What we have here is a description of the motion in a moving reference frame (body frame). We've now solved the torque-free symmetric top problem in two quite different ways. Do these solutions agree?

In summary, the motion can be pictured in terms of two cones rolling on each other. It is the most general kind of motion possible if no torque is exerted on the top. The vector $\vec{\omega}$ lies at the common intersection of the body and space cones, which are tangent to each other. From the point of view of an observer fixed in the body, the vector $\vec{\omega}$ is precessing around the body with angular velocity Ω_p. In space, $\vec{\omega}$ precesses around the Z' axis with angular velocity ω_P. For the case $I_3 > I_1$ (Figure 8.7A), the body cone encloses the space cone and rolls around it, whereas if $I_3 < I_1$ (Figure 8.7B), the space cone lies outside the body cone. Visualizing this motion made use of the fact that an angular velocity is at once both a vector and an instantaneous axis of rotation.

A common example of the motion of a torque-free symmetric top is the "wobble" of a spinning football thrown in a forward pass. If the football is thrown so that the angular momentum is slightly off the long axis, the football axis will precess around the fixed direction of the angular momentum at a rate given by Equation (8.47). The center of mass motion is completely independent of the rotational motion and follows a parabolic path just as if the football were a point mass located at the center of mass. Proving this is the subject of a homework problem.

QUESTION 6: *Symmetric Top 1* Explain the motion of the symmetric top in the space frame in your own words. Include how the motion will change it the initial value of θ changes. For example, what will it look like if $\theta = 0$? $\theta = \pi/2$?

QUESTION 7: *Symmetric Top 2* Explain the motion of $\vec{\omega}$ in the body frame. Think about how the motion in the body frame relates to the motion in the space frame. Explain why the two derivations are consistent only for a certain value of the constant A in Equation (8.51) and determine this value.

QUESTION 8: *An Apparent Paradox* For the torque-free symmetric top we discussed above, the angular momentum $\vec{L}$ must be constant. Since the inertia tensor possesses an inverse, $\mathbf{I}^{-1}$, and $\mathbf{L} = \mathbf{I}\boldsymbol{\omega}$, it must be true that $\boldsymbol{\omega} = \mathbf{I}^{-1}\mathbf{L}$. If the inertia tensor (and its inverse) and $\vec{L}$ are constants, $\vec{\omega}$ must also be constant. But we have just finished showing that the angular velocity precesses in space and hence is not a constant vector. Where does the fallacy lie in this argument?

8.6 FORCE-FREE PRECESSION OF THE EARTH: THE "CHANDLER WOBBLE"

The Earth bulges at the equator. It is an oblate ellipsoid, rather than a perfect sphere. This bulge is caused by the rotation of the Earth. It was first estimated by Isaac Newton. His calculation is discussed in more detail in Appendix C at the end of this chapter.

Because it is not a perfect sphere, the Earth has torques exerted on it by the Sun, the Moon, and the other planets. These torques lead to interesting effects on both the Earth's spin axis and on the orbit, which have periods measured in many thousands of years. We will discuss this in Section 8.11. In addition, there is a small precession, which has a much shorter period than these effects. To study this "fast wobble" of the Earth's axis, we can consider the Earth (to a very good approximation) as a symmetric top. If we call the 3 axis the axis through the North Pole (see Figure 8.8), $\omega_3 = 7.3 \times 10^{-5}$ s^{-1}, and

$$\frac{I_1 - I_3}{I_1} = -0.00327 \approx -\frac{1}{305.8}. \tag{8.52}$$

This means that (from (8.48))

$$\Omega_p = \Omega_{\text{earth}} \approx \frac{\omega_3}{305.8}. \tag{8.53}$$

Knowing that the Earth spins on its axis once per day, in 1749 Euler predicted from Equation 8.53 that the precession period due to this effect is 306 days. For 140 years,

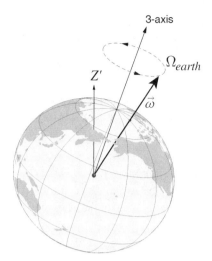

FIGURE 8.8

astronomers tried in vain to observe it. In 1891 an amateur astronomer, S. Chandler, used existing data previously analyzed by others to prove that the spin axis does "wobble," with a maximum excursion of about 10 meters around the Earth's principal axis. The effect had been missed because the period was 427 days, rather than the predicted 306 days. It was then realized that the discrepancy in the period is due to the fact that the Earth is not a perfect rigid body, but instead is deformed due to the tides. This lengthens the predicted period by about 100 days, giving reasonable agreement with observation. Another mystery soon arose, however. The calculated damping period is 10–20 years, but little damping has been observed over what is now a 100-year period of study. This is still not quantitatively understood. The role of earthquakes deep within the Earth seems the best candidate for an explanation at the present time. The recent rather sensational discovery that the liquid iron inner core of the Earth rotates at a different speed than the outer mantle may also lead to speculation that this motion deep inside the Earth drives the "Chandler wobble."

8.7   DEFINITION OF EULER ANGLES

Knowledge of the **U** matrix, (7.15), allows you to convert from body to space coordinates for any physical point:

$$\mathbf{r}'(\text{space coordinates}) = \mathbf{U}\mathbf{r}(\text{body coordinates}) \qquad (8.54)$$

or vice versa

$$\mathbf{r}(\text{body coordinates}) = \tilde{\mathbf{U}}\mathbf{r}'(\text{space coordinates}) \qquad (8.55)$$

($\tilde{\mathbf{U}} = \mathbf{U}^{-1}$). We observed previously that **U** must be a function of three parameters. We now want to make this dependence explicit and find the actual form of **U** in terms of three parameters called the *Euler angles*: θ, ϕ, ψ.

8.7 DEFINITION OF EULER ANGLES

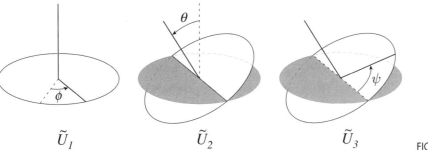

FIGURE 8.9

Having an explicit form for **U** and thus **U̇** will allow you to write down the kinetic energy and Lagrangian for any rigid body in terms of the time derivatives of the Euler angles. This approach will be used to solve the classic problem of the heavy symmetrical top.

Imagine that the space coordinate system, $X'Y'Z'$, is initially aligned with the body coordinate system, XYZ. Imagine a sequence of three rotations acting on the body coordinates which finally convert this coordinate system into one that has an arbitrary orientation with respect to the space coordinates, which remain fixed. Start with the two coordinate systems lined up, so that $K' = K$.

Step 1: Rotate the body system by an angle ϕ about the Z axis as shown in the $\tilde{\mathbf{U}}_1$ part of Figure 8.9. Call the new body coordinates generated in this way ξ, η, ζ. The $\tilde{\mathbf{U}}_1$ matrix is

$$\tilde{\mathbf{U}}_1(\phi) = \begin{pmatrix} \cos\phi & \sin\phi & 0 \\ -\sin\phi & \cos\phi & 0 \\ 0 & 0 & 1 \end{pmatrix}, \tag{8.56}$$

$$\begin{pmatrix} \xi \\ \eta \\ \zeta \end{pmatrix} = \tilde{\mathbf{U}}_1 \begin{pmatrix} r'_1 \\ r'_2 \\ r'_3 \end{pmatrix}. \tag{8.57}$$

For example, the vector $r'_1 = 1, r'_2 = 0, r'_3 = 0$ is $\xi = \cos\phi, \eta = -\sin\phi, \zeta = 0$ in the first body system.

Step 2: Next, rotate this body system through angle θ about the ξ axis (new X axis) as shown in the $\tilde{\mathbf{U}}_2$ part of Figure 8.9:

$$\tilde{\mathbf{U}}_2(\theta) = \begin{pmatrix} 1 & 0 & 0 \\ 0 & \cos\theta & \sin\theta \\ 0 & -\sin\theta & \cos\theta \end{pmatrix}, \tag{8.58}$$

$$\begin{pmatrix} \xi' \\ \eta' \\ \zeta' \end{pmatrix} = \tilde{\mathbf{U}}_2 \begin{pmatrix} \xi \\ \eta \\ \zeta \end{pmatrix}. \tag{8.59}$$

Step 3: Lastly, rotate the body system through an angle ψ about the ζ' axis (new Z axis) as shown in the $\tilde{U}_3$ part of Figure 8.9:

$$\tilde{U}_3(\psi) = \begin{pmatrix} \cos\psi & \sin\psi & 0 \\ -\sin\psi & \cos\psi & 0 \\ 0 & 0 & 1 \end{pmatrix}, \tag{8.60}$$

$$\begin{pmatrix} r_1 \\ r_2 \\ r_3 \end{pmatrix} = \tilde{U}_3 \begin{pmatrix} \xi' \\ \eta' \\ \zeta' \end{pmatrix}. \tag{8.61}$$

The coordinates r_1, r_2, r_3 are the new body system coordinates in K. From (8.54),

$$\tilde{U} = \tilde{U}_3 \tilde{U}_2 \tilde{U}_1, \quad U = U_1 U_2 U_3. \tag{8.62}$$

Multiply the matrices above in the order indicated to obtain

$$U = \begin{pmatrix} \cos\psi\cos\phi - \cos\theta\sin\phi\sin\psi & -\sin\psi\cos\phi - \cos\theta\sin\phi\cos\psi & \sin\theta\sin\phi \\ \cos\psi\sin\phi + \cos\theta\cos\phi\sin\psi & -\sin\phi\sin\psi + \cos\theta\cos\phi\cos\psi & -\sin\theta\cos\phi \\ \sin\theta\sin\psi & \sin\theta\cos\psi & \cos\theta \end{pmatrix}. \tag{8.63}$$

(The complexity of U is somewhat daunting. It is fairly easy to check this form using computer algebra.)

QUESTION 9: *Euler Angles 1* Explain the meaning of Euler Angles (what they are explicitly, what they can be used for, how they are obtained, if the order of operations matter, etc.). Why can you always find Euler angles to describe any kind of rotation?

Figure 8.10 summarizes the operations shown in 8.9.

Physical Interpretation of the Eigenvalues and Eigenvectors of U

Any orthogonal transformation such as U can be represented as a rotation through an angle about an axis which is left unchanged by U (this will be proved in a homework problem). This means that U must have an eigenvalue equal to 1. The eigenvector for the unit eigenvalue gives the direction of the axis of rotation.

Define the angle of rotation about this axis as Φ. How do we find Φ? It is a function of the three Euler angles. If we transform to a new coordinate system where this axis of

8.7 DEFINITION OF EULER ANGLES

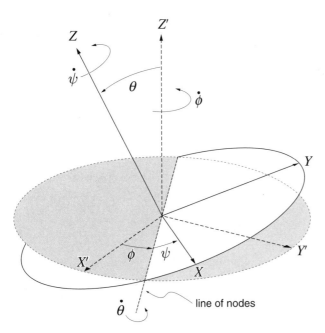

FIGURE 8.10
Definition of Euler angles.

rotation is the new Z axis, the transformed $\mathbf{U}$ will take the form of a rotation Φ about this new Z axis:

$$\mathbf{U}' \equiv \mathbf{C}\mathbf{U}\tilde{\mathbf{C}} = \begin{pmatrix} \cos\Phi & \sin\Phi & 0 \\ -\sin\Phi & \cos\Phi & 0 \\ 0 & 0 & 1 \end{pmatrix}. \quad (8.64)$$

($\mathbf{C}$ is the orthogonal matrix that effects the coordinate change making the rotation axis the "Z" axis.) We can use the fact that the trace of a matrix (sum of the diagonal elements) is invariant under a change of coordinates:

$$\text{Trace}\,[\mathbf{U}'] = \text{Trace}\,[\mathbf{C}\mathbf{U}\tilde{\mathbf{C}}] = \text{Trace}\,[\mathbf{U}]. \quad (8.65)$$

This allows us to find an equation for Φ in terms of θ, ψ, ϕ (using (8.63)):

$$\text{Trace}\,[\mathbf{U}'] = 1 + 2\cos\Phi = \text{Trace}\,[\mathbf{U}] = \cos\theta + \cos(\phi+\psi)(1+\cos\theta). \quad (8.66)$$

A trigonometric identity has been used to simplify the expression for Trace $[\mathbf{U}]$. Further use of trigonometric identities reveals a simple expression for Φ:

$$\boxed{\cos\left(\frac{\Phi}{2}\right) = \cos\left(\frac{\phi+\psi}{2}\right)\cos\left(\frac{\theta}{2}\right).} \quad (8.67)$$

8.8 FINDING THE ANGULAR VELOCITY

It was shown previously that the angular velocity in space coordinates can be read off from the elements of the antisymmetric matrix $\mathbf{A}' = \dot{\mathbf{U}}\tilde{\mathbf{U}}$ (7.27). In body coordinates, we showed that the same angular velocity can be found by looking at $\mathbf{A} = \tilde{\mathbf{U}}\dot{\mathbf{U}}$. It is a completely straightforward calculation, which can be done using computer algebra.

The result is (note that the body coordinates are chosen in the principal axis frame):

$$\text{in body coordinates:}$$
$$\vec{\omega} = \begin{pmatrix} \dot\theta\cos\psi + \dot\phi\sin\psi\sin\theta \\ -\dot\theta\sin\psi + \dot\phi\cos\psi\sin\theta \\ \dot\psi + \dot\phi\cos\theta \end{pmatrix}; \quad (8.68)$$

$$\text{in space coordinates:}$$
$$\vec{\omega} = \begin{pmatrix} \dot\theta\cos\phi + \dot\psi\sin\phi\sin\theta \\ \dot\theta\sin\phi - \dot\psi\cos\phi\sin\theta \\ \dot\phi + \dot\psi\cos\theta \end{pmatrix}. \quad (8.69)$$

There is a less mathematical and more physical way to obtain the components of the angular velocity in terms of the Euler angle time derivatives and the Euler angles. Referring to Figure 8.10, we see that the angular velocity component proportional to $\dot\theta$ lies along the line of nodes. Calling the body axes 1, 2, 3, this contribution to the angular velocity has the components $(\dot\theta\cos\psi, -\dot\theta\sin\psi, 0)$ in the body system. The angular velocity for $\dot\phi$ lies along the Z' axis, so has the body components $(\dot\phi\sin\theta\sin\psi, \dot\phi\sin\theta\cos\psi, \dot\phi\cos\theta)$. (Convince yourself this is true.) Finally, the angular velocity due to $\dot\psi$ lies along the 3 axis so had the components $(0, 0, \dot\psi)$. Putting this all together, we have the components of the most general angular velocity in the body system. Check that this agrees with formula (8.68).

$\vec{\omega}$ Is Not an Integrable Function

Suppose that $\vec{\omega}$ were the total time derivative of a function $\vec{\Lambda}(\theta, \psi, \phi)$. Using the form given in the previous section for the space coordinate form of $\vec{\omega}$ (8.69), the total time derivative of the X' coordinate of $\vec{\omega}$ is, for example,

$$\omega_x = \frac{d\Lambda_x}{dt} = \frac{\partial\Lambda_x}{\partial\theta}\dot\theta + \frac{\partial\Lambda_x}{\partial\psi}\dot\psi + \frac{\partial\Lambda_x}{\partial\phi}\dot\phi = \dot\theta\cos\phi + \dot\psi\sin\phi\sin\theta. \quad (8.70)$$

Therefore

$$\frac{\partial\vec\Lambda_x}{\partial\theta} = \cos\phi,$$
$$\frac{\partial\vec\Lambda_x}{\partial\psi} = \sin\phi\sin\theta. \quad (8.71)$$

8.9 MOTION OF TORQUE-FREE ASYMMETRIC TOPS: POINSOT CONSTRUCTION

If we calculate $\frac{\partial^2 \Lambda_x}{\partial \theta \partial \psi} = \frac{\partial^2 \Lambda_x}{\partial \psi \partial \theta}$ from either formula, we must get the same answer. The top formula gives zero, but the second formula gives something that doesn't always vanish. So it cannot be true that $\vec{\omega}$ is the total time derivative of something. $\vec{\omega}$ is a *nonintegrable* function of the Euler angles and their time derivatives. The nonintegrability is a consequence of the fact that 3-D rotations do not commute.

The Kinetic Energy of Rotation for a Symmetric Top

The results above can be used to solve nontrivial rigid body problems, such as a heavy symmetrical top with one point fixed. First use the expression for $\vec{\omega}$ in body coordinates (8.68) to get the kinetic energy* for the top in terms of Euler angles and their time derivatives:

$$T = \frac{I_1}{2}[\omega_1^2 + \omega_2^2] + \frac{I_3}{2}\omega_3^2. \tag{8.72}$$

(Note that the body coordinates chosen here will be in the principle axis frame, and $I_1 = I_2$ for a symmetric top.) After some cancellations and invoking a few trigonometric identities, we find that

$$\boxed{T = \frac{I_1}{2}(\dot{\theta}^2 + \dot{\phi}^2 \sin^2 \theta) + \frac{I_3}{2}(\dot{\psi} + \dot{\phi} \cos \theta)^2.} \tag{8.73}$$
$$\text{symmetric top}$$

This is a universal expression for the kinetic energy of any symmetric top.

QUESTION 10: *Kinetic Energy* Find the kinetic energy (in the space frame) of an asymmetric top, with $I_1 \neq I_2 \neq I_3$, using body coordinates.

8.9 MOTION OF TORQUE-FREE ASYMMETRIC TOPS: POINSOT CONSTRUCTION

A top's motion is controlled by the fact that both the angular momentum $\vec{L}$ and the kinetic energy T are constants of the motion. We can watch the top move, following the change with time in the orientation of the principal axes for one particular choice of these constants. By changing L and T, we can also observe how the nature of the top's motion depends on these values. L and T can be treated as independent variables within certain limits, to be described below.

---

* In the space (inertial) frame, of course. We can use (8.10) in any coordinate system to get the kinetic energy in the space frame.

Once again we consider a body system attached to the top and aligned with the principal axes. In that system, the kinetic energy takes the form

$$T = \frac{1}{2}\left(I_1\omega_1^2 + I_2\omega_2^2 + I_3\omega_3^2\right). \tag{8.74}$$

It is no restriction to assume $I_1 < I_2 < I_3$. In the same principal axis system of coordinates, the angular velocity $\vec{\omega}$ has the components $\omega \equiv (\omega_1, \omega_2, \omega_3)$.

Although the magnitude of $\vec{L}$ can be specified independently of T, the moment of inertia tensor sets a limited range for L, which depends on T. For a fixed T, the smallest possible L occurs for the top rotating about the 1 axis, which has the smallest principal moment. Conversely, the largest value of L occurs for the top rotating about the 3 axis. We can say then that the physical top can only have L values as indicated by the equation below:

$$L_{\min} \equiv \sqrt{2TI_1} < L < \sqrt{2TI_3} \equiv L_{\max}. \tag{8.75}$$

We are interested in varying the initial condition of L while keeping the initial condition T constant. Then we will look at how the angular velocity evolves as a function of time for each different L using $\mathbf{L} = \mathbf{I}\boldsymbol{\omega}$, given the constraint that the kinetic energy is always constant. This can be done using a geometric method first invented by Poinsot.* By using it we can visualize the rather complicated motion of an asymmetric top without solving the actual equations for the motion.

Angular velocity space is the three-dimensional space of possible values of the components of $\vec{\omega}$. If the kinetic energy is assigned the constant value T, this constraint means that the motion must take place on a surface in the $\vec{\omega}$ space. The surface is an ellipsoid, called the "inertia ellipsoid" or sometimes the "momental ellipsoid." It suffices to prove that the locus of the angular velocity does lie on this type of surface in the body coordinate system, since in the space system we view the same surface from a coordinate system that has only a different orientation.

The equation of an ellipsoid is

$$\frac{\omega_1^2}{a^2} + \frac{\omega_2^2}{b^2} + \frac{\omega_3^2}{c^2} = 1, \tag{8.76}$$

where a, b, and c are constants, taken by convention to be positive. They are the *semi-axes* of the ellipsoid. By comparing Equation (8.74) with Equation (8.76), we see that constant kinetic energy does require that the surface containing the tip of $\vec{\omega}$ be an ellipsoid with semi-axes

$$a = \sqrt{\frac{2T}{I_1}}, \quad b = \sqrt{\frac{2T}{I_2}}, \quad c = \sqrt{\frac{2T}{I_3}}. \tag{8.77}$$

---

* L. Poinsot, *Théorie Nouvelle de la Rotation des Corps*, Paris, Bachelier, 1851.

8.9 MOTION OF TORQUE-FREE ASYMMETRIC TOPS: POINSOT CONSTRUCTION

We emphasize that the ellipsoid is not in real space but in the space of the angular velocity – the instantaneous direction and magnitude of rotation of the top.

Let the angular velocity vector $\vec{\omega}$ point to an arbitrary point on this surface. What is the angular momentum for this direction and magnitude of $\vec{\omega}$? It is the main point of the Poinsot method that *the angular momentum can be found by taking the gradient $\vec{\nabla}_\omega$ of the kinetic energy with respect to $\vec{\omega}$*:

$$\vec{L} = \vec{\nabla}_\omega T. \tag{8.78}$$

In ordinary 3-D space, the gradient of a function $F(x, y, z)$ is a vector, denoted $\vec{\nabla} F$. If the surface $F(x, y, z)$ is a constant, $\vec{\nabla} F$ points along the normal to the surface at the point (x, y, z), because the gradient vector always points along the direction of maximum change of the function F at the point where it is evaluated.* In mechanics, the force is minus the gradient of the potential energy function (1.55). The direction of the force is always orthogonal to the curves of constant potential energy due to this mathematical property of the gradient operator.

In any coordinate system, the gradient in angular velocity space is found in an analogous way to the gradient in ordinary space,

$$\vec{\nabla}_\omega \equiv \left(\frac{\partial}{\partial \omega_x}, \frac{\partial}{\partial \omega_y}, \frac{\partial}{\partial \omega_z} \right). \tag{8.79}$$

We will give a proof of (8.78) that does not make any reference to a specific coordinate system. It makes extensive use of the summation convention for the coordinate indices introduced in Chapter 7. In what appears below, i, j, k all stand for arbitrary vector indices: x, y, z or 1, 2, 3. The symbol δ_{ij} stands for the Kronecker delta. By definition $\delta_{ij} \equiv 1$ if $i = j$ and $\delta_{ij} \equiv 0$ if $i \neq j$. From (8.13),

$$T = \frac{1}{2} \omega_i I_{ij} \omega_j. \tag{8.80}$$

Take the partial derivative of T with respect to ω_k, making use of the identity $\frac{\partial \omega_i}{\partial \omega_k} = \delta_{ik}$:

$$\frac{\partial T}{\partial \omega_k} = \frac{1}{2} \delta_{ik} I_{ij} \omega_j + \frac{1}{2} \omega_i I_{ij} \delta_{jk} = \frac{1}{2} I_{kj} \omega_j + \frac{1}{2} \omega_i I_{ik}. \tag{8.81}$$

Since i, j in Equation (8.81) are dummy indices to be summed over, rename i to j in the last part of Equation (8.81) and make use of the symmetry of the inertia tensor ($I_{jk} = I_{kj}$) to write

$$\frac{\partial T}{\partial \omega_k} = I_{kj} \omega_j = L_k. \tag{8.82}$$

Thus (8.78) is proved.

---

* For a review of the general mathematical properties of the gradient vector, see Kaplan, *Advanced Calculus*, 3rd ed., 1984, p. 183 or an equivalent mathematics textbook.

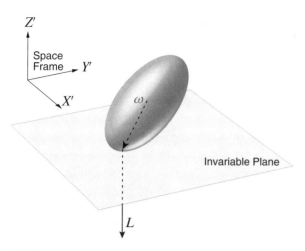

FIGURE 8.11
Poinsot construction.

QUESTION 11: *Proof in Body Coordinates* Prove (8.78), using the principal body system. It is considerably simpler than our proof above. Is your proof using the principal body system valid for any frame? Why?

From (8.78), $\vec{L}$ is normal to the surface of constant T at the point $\vec{\omega}$. The plane normal to $\vec{L}$ is called the *invariable plane*. Since $\vec{L}$ is constant, so is this plane. Figure 8.11 shows $\vec{\omega}$ and $\vec{L}$ at a particular moment in time. The motion of the top must therefore be such that the ellipsoid rolls on the invariable plane, keeping this plane always tangent to the ellipsoid at the point $\vec{\omega}$.

The rolling motion of the top means that the point of contact where the ellipsoid is tangent to the plane traces out two curves – one on the inertia ellipsoid itself and the other on the invariable plane. The curve left on the ellipsoid surface is called the *polhode* and the curve on the planar surface is called the *herpolhode*.

QUESTION 12: *Constant Angular Momentum* We have been emphasizing that the angular momentum is constant as time progresses. But we are now noting that the angular momentum vector traces out the polhode and herpolhode curves. In what sense is the angular momentum constant?

A good description of this motion is by E. T. Whittaker:*

It follows that the body moves as if it were rigidly connected to its momental ellipsoid, and the latter body were to roll about the fixed point on a fixed plane perpendicular to the invariable line, without sliding; the angular velocity being proportional to the radius to the point of contact, so that the component of angular velocity about the invariable line is constant.

---

* E. T. Whittaker, *A Treatise on the Analytical Dynamics of Particles and Rigid Bodies*, 4th ed., Cambridge University Press, 1959, p. 153.

8.9 MOTION OF TORQUE-FREE ASYMMETRIC TOPS: POINSOT CONSTRUCTION

By "invariable line," Whittaker means $\vec{L}$. He asserts that the projection of the angular velocity onto the angular momentum is constant during the motion. This means that the height of the center of the ellipsoid, which is tangent to the plane, is a fixed constant. Let's prove this statement is true and find this height, which we will call h. From (8.13),

$$\vec{\omega} \cdot \vec{L} = \tilde{\omega}\mathbf{I}\omega = 2T. \tag{8.83}$$

Therefore

$$h \equiv \frac{\vec{\omega} \cdot \vec{L}}{L} = \frac{2T}{L}. \tag{8.84}$$

Since T and L are constants, so is h. This completes the qualitative picture of the motion. Everything in the geometric picture can be deduced from knowing h and the shape of the ellipsoid (i.e., the values of a, b, c, the semi-axes of the ellipsoid defined in Equation (8.77)). Where the plane touches the ellipsoid, draw the polhode on the body-oriented ellipsoid and the herpolhode on the invariable space plane in Figure 8.11 as they would appear after an interval of time has elapsed. Notice that the origin remains at a fixed height above the plane due to Equation (8.84), although the magnitude and direction of $\vec{\omega}$ are changing.

Poinsot is said to have been revolting against the insistence of Lagrange on purely analytic solutions. Here we will mount a small counter revolution to obtain analytic forms for the curves traced out by the tip of the axis of revolution – the angular velocity vector. (Obtaining the time dependence is much more difficult. The solution involves elliptic integrals.)

Figure 8.12 assumes $I_1 = 1$, $I_2 = 2$, $I_3 = 3$, $T = 1$, and $L_{min} = \sqrt{2} < L < \sqrt{6} = L_{max}$. Notice that the ellipsoid is longest in the direction of the lowest principal moment, I_1. The first step will be to adopt a set of elliptical polar coordinates in the body system that will

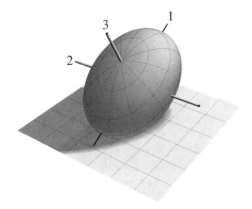

FIGURE 8.12
Poinsot construction: Inertia ellipsoid tangent to invariable plane. The positive end of the principal axes is indicated by the number of the axis near to it. The 3 axis is thickest; the 1 axis, the thinnest.

allow us to roam freely over the surface of the ellipsoid. We call the polar angle in this system u and the azimuthal angle v. The coordinates of an arbitrary point on the surface of the ellipsoid are

$$\omega_1 = a \sin u \cos v, \quad \omega_2 = b \sin u \sin v, \quad \omega_3 = c \cos u. \tag{8.85}$$

You can easily check that this parametrization with u, v takes the tip of $\vec{\omega}(u, v)$ over the ellipsoid surface.

The second step, which was used to construct Figure 8.12 is to use the $\mathbf{U}(\phi, \theta, \psi)$ matrix connecting body and space axes to obtain a relationship between u, v and θ, ψ. In body coordinates, the angular momentum is $(I_1 \omega_1, I_2 \omega_2, I_3 \omega_3)$. In space coordinates it is $(0, 0, -L)$. We can express the relation between body and space frames in the usual way as

$$\mathbf{U} \begin{pmatrix} I_1 \omega_1 \\ I_2 \omega_2 \\ I_3 \omega_3 \end{pmatrix} = \begin{pmatrix} 0 \\ 0 \\ -L \end{pmatrix}. \tag{8.86}$$

The ellipsoid is centered at a distance of $h = \frac{2T}{L}$ above the invariable plane in the space coordinate system. We not only have to make the correct rotation to satisfy (8.86), but we have to move the center up by h. Some thought should convince you that this will achieve the desired tangency for the ellipsoid and the plane, as well as locate the ellipsoid center in the right place.

To find the correct Euler angles in $\mathbf{U}$ (see Equation (8.63)), multiply both sides of Equation (8.86) by $\tilde{\mathbf{U}}$, using (8.85) for $\vec{\omega}$. You then obtain three equations in three unknown angles ϕ, θ, ψ:

$$\sin u \cos v = -\sqrt{\frac{L^2}{2T I_1}} \sin \theta \sin \psi,$$

$$\sin u \sin v = +\sqrt{\frac{L^2}{2T I_2}} \sin \theta \cos \psi, \tag{8.87}$$

$$\cos u = -\sqrt{\frac{L^2}{2T I_3}} \cos \theta.$$

Notice that ϕ does not appear, since we can rotate freely around the Z, Z' axis without changing the tangent condition, and we obtain a connection between u, v and θ, ψ. Because there are three Equations (8.87) and only two variables θ, ψ, there is a functional relationship between θ and ψ which must be satisfied at all times. After squaring and adding the three Equations (8.87), we discover that

$$\frac{2T}{L^2} = \frac{\cos^2 \theta}{I_3} + \left(\frac{\cos^2 \psi}{I_2} + \frac{\sin^2 \psi}{I_1} \right) \sin^2 \theta. \tag{8.88}$$

8.9 MOTION OF TORQUE-FREE ASYMMETRIC TOPS: POINSOT CONSTRUCTION

Solving Equation (8.88) for $\sin^2 \theta$, we obtain

$$\sin^2 \theta(\psi) = \frac{\left(\frac{2T}{L^2}\right) - \frac{1}{I_3}}{\left(\frac{\cos^2 \psi}{I_2} + \frac{\sin^2 \psi}{I_1}\right) - \frac{1}{I_3}}. \tag{8.89}$$

Equation (8.89) is the connection between θ and ψ we seek. Taking ψ as the independent variable, the ellipsoid will roll on the invariable plane as ψ is varied. Notice that, because θ is a polar angle, it only varies between 0 and π, so that $\sin \theta$ is always positive. However, there are two solutions for $\cos \theta = \pm \sqrt{1 - \sin^2 \theta}$. Thus, owing to the reflection symmetry between positive and negative axes of the ellipsoid, two possibilities exist for the polhode. If we fail to remember this, we will encounter rather mysterious results as the angular momentum is varied for a constant value of the kinetic energy.

We can see from (8.89) that θ will be a constant if $I_1 = I_2$, as with a symmetric top. The polhode and the herpolhode will both be circles. This is the circular motion of the angular velocity as viewed from both space and body cones, which we discussed earlier in Figure 8.7. (The polhode is the locus of the tip of $\vec{\omega}$ on the body cone; the herpolhode the locus of the same point on the space cone.)

There may be restrictions on the possible values of ψ that are allowed in order that $0 \leq \sin^2 \theta \leq 1$. We investigate this possibility next. $\sin^2 \theta < 0$ would correspond to $L > \sqrt{2T I_3} = L_{\max}$. This case is already forbidden by our previous limit on L. The other forbidden case of $\sin^2 \theta > 1$ *would* occur if $\left(\frac{\cos^2 \psi}{I_2} + \frac{\sin^2 \psi}{I_1}\right) < \frac{2T}{L^2}$. This *could* happen for certain values of ψ when $L^2 < 2T I_2$. So we must restrict ψ so that this does not occur for $L < L_{\text{crit}}$, where

$$L_{\text{crit}} \equiv \sqrt{2T I_2}. \tag{8.90}$$

If we start the top rotating around its 2 axis, $L^2 = 2T I_2$.

Set (8.89) equal to 1 to find the possible limits on ψ for a given fixed value of L for $L_{\min} < L < L_{\text{crit}}$. Using half-angle trigonometric formulas, the special limiting value of $\psi \equiv \psi_0$ is given by the solution (if one exists) to the equation

$$\cos 2\psi_0 = \frac{\frac{1}{I_1} + \frac{1}{I_2} - \frac{4T}{L^2}}{\frac{1}{I_1} - \frac{1}{I_2}}. \tag{8.91}$$

In the event $L_{\min} < L \leq L_{\text{crit}}$, there are two curves. For one of them, $\psi_0 \leq \psi \leq \pi - \psi_0$. For the other one $\pi + \psi_0 \leq \psi \leq 2\pi - \psi_0$. ψ_0 is a solution to (8.91), while ψ is used as the independent variable to parameterize the two curves, using Equation (8.89) to solve for $\theta(\psi)$ and then Equations (8.87) to actually trace the curves.

The actual surface of the rolling ellipsoid isn't shown in Figure 8.13 because it would be difficult to see the curves (superimpose Figure 8.12 for the more complete picture). We have elected to show both polhodes, but only one of them corresponds to the actual motion for a given set of initial conditions of the top. Only the bottom curve has a tangent to the invariable plane in the figures. The herpolhode is also not shown in the figures to keep them simple.

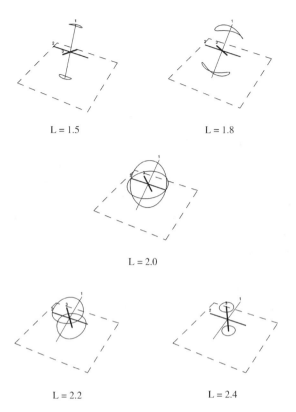

FIGURE 8.13
Polhodes for an asymmetric top of varying L, where $L_{crit} = 2.0$.

What are the limiting cases? For $L \to L_{max}$, the body rotates around the 3 axis. For $L \to L_{min}$, the body rotates around the 1 axis. In both cases, the polhode and the herpolhode shrink to points. These are degenerate cases.

For $L < L_{crit}$ ($L < 2$ in Figure 8.13), there are two disjoint polhode curves, both of which enclose the 1 axis. Of course, the actual motion of the top must select which of the two possible curves it is following. For $L = L_{crit}$ ($L = 2$ in Figure 8.13), the two curves cross on the 2 axis. But physically, an intersection is impossible. This is a singular case. Because of this crossing or "bifurcation," a top rotating about a principal axis corresponding to the intermediate moment of inertia, I_2, is unstable. The top does not "know" which curve to follow from the singular crossing point. It is like an upside-down pendulum being unable to "decide" whether to fall to the left or to the right. For $L > L_{crit}$ ($L > 2$ in Figure 8.13), the two possible polhode curves enclose the 3 axis.

Our goal of a geometric and analytic description of the general motion of a torque-free asymmetric top has thus been achieved, at least for the shape of the orbit followed by the top. Motion about the 1 or the 3 axis is stable. The 2 axis is unstable, because if $\vec{\omega}(t = 0) = (0, \omega_2, 0) + \vec{\epsilon}$, with $\vec{\epsilon}$ having arbitrarily small components in the 1 and 3 directions, the motion will rapidly move away from the 2 axis. The mathematics required to do this is not difficult and is left as a homework exercise.

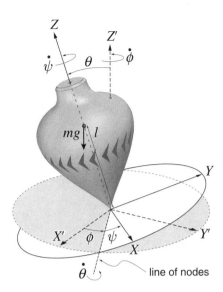

FIGURE 8.14

8.10 THE HEAVY SYMMETRIC TOP

A symmetric top can be mounted so that gravity exerts a torque on it as shown in Figure 8.14. If the fixed point of the top is a distance l from the center of mass, and the top is tilted from the vertical (Z') axis by an angle θ, there will be a torque $mgl \sin\theta$, so the angular momentum need not remain constant. The motion this causes can be rather complex. We will see that uniform precession of the top axis rarely occurs.

The total potential energy is given by summing the potential energy of all of the points in the body:

$$V = g \sum_i m_i z_i = M g z_{\text{cm}} = Mgl \cos\theta. \tag{8.92}$$

Taking the kinetic energy as (8.73), the Lagrangian is

$$L = \frac{I_1}{2}(\dot\theta^2 + \dot\phi^2 \sin^2\theta) + \frac{I_3}{2}(\dot\psi + \dot\phi \cos\theta)^2 - Mgl \cos\theta. \tag{8.93}$$

There are three degrees of freedom. Notice right away that ψ, ϕ are both ignorable coordinates, so it is possible to reduce the problem to one degree of freedom, since p_ψ, p_ϕ will be constants of the motion:

$$p_\psi = \frac{\partial L}{\partial \dot\psi} = I_3(\dot\psi + \dot\phi \cos\theta) = I_3 \omega_3. \tag{8.94}$$

The total angular velocity about the symmetry axis, by definition ω_3 – the "spin" of the top, is thus a constant. (Notice that ω_3 depends on both $\dot\psi$ and $\dot\phi$ from (8.68).) Use the notation

$$a \equiv \frac{I_3 \omega_3}{I_1} \qquad (8.95)$$

to define the constant a. Also,

$$p_\phi = \frac{\partial L}{\partial \dot\phi} = I_1 \dot\phi \sin^2\theta + I_3(\dot\psi + \dot\phi \cos\theta)\cos\theta. \qquad (8.96)$$

Define the constant b by

$$b \equiv \frac{p_\phi}{I_1}. \qquad (8.97)$$

Since the kinetic energy is a quadratic form in $\dot\phi, \dot\theta, \dot\psi$, $H = E$. Thus the total energy $E = T + V$ is constant:

$$E = \frac{I_1}{2}(\dot\theta^2 + \dot\phi^2 \sin^2\theta) + \frac{I_3}{2}\omega_3^2 + Mgl\cos\theta. \qquad (8.98)$$

Using (8.94) and (8.97), we can obtain

$$\dot\phi = \frac{b - a\cos\theta}{\sin^2\theta}. \qquad (8.99)$$

Define another constant:

$$E' \equiv E - \frac{I_3}{2}\omega_3^2. \qquad (8.100)$$

Substituting (8.99) and (8.100) into (8.98) gives a new constant:

$$E' = \frac{I_1}{2}\dot\theta^2 + \underbrace{\frac{I_1}{2}\frac{(b - a\cos\theta)^2}{\sin^2\theta} + Mgl\cos\theta}_{V_{\text{effective}}(\theta)}. \qquad (8.101)$$

Solution of the One-dimensional Equation of Motion

The problem of a heavy symmetric top has been reduced to a problem with only one degree of freedom. This problem can be solved by standard analytic means. The solution comes out in the form of elliptic integrals, which are hard to interpret in physical terms. Nonetheless, we will try to get a feeling for how the solutions work. First transform to the new variable $u \equiv \cos\theta$. Since it is a cosine of an angle, we know that $-1 \le u \le 1$. In terms of u and $\dot u$:

$$E'(1 - u^2) = \frac{I_1}{2}\dot u^2 + \frac{I_1}{2}(b - au)^2 + Mglu(1 - u^2). \qquad (8.102)$$

8.10 THE HEAVY SYMMETRIC TOP

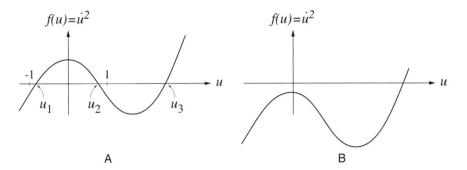

FIGURE 8.15
Possible solutions to $f(u) = \dot{u}^2$: A) three real roots; B) one real and two imaginary roots.

Introduce two new positive constants:

$$\alpha \equiv \frac{2E'}{I_1}, \quad \beta \equiv \frac{2Mgl}{I_1}. \tag{8.103}$$

You should now obtain

$$\dot{u}^2 = (1 - u^2)(\alpha - \beta u) - (b - au)^2 \equiv f(u). \tag{8.104}$$

The function $f(u)$ is a cubic polynomial in u which must be positive for a physically real solution to exist. The problem could be solved now by quadratures, but to get a better physical grasp of what is going on here, we will proceed more qualitatively instead. If $u \to +\infty$, then $f(u) \sim \beta u^3 > 0$, since $\beta > 0$. Conversely, as $u \to -\infty$, $f(u) < 0$, so the shape of the curve resembles those in Figure 8.15. The turning points of the motion are the roots of $f(u)$, which lie between ± 1. A cubic polynomial with real coefficients has either one real root and two complex roots or three real roots. At this point we have two distinct choices. A cubic curve with three real roots is pictured in Figure 8.15A. Note that, for this case $f(0) = \alpha - b^2 > 0$. In Figure 8.15B, we have a curve with one real root and two imaginary roots. For this curve, $f(0) < 0$. It is also possible to have $f(0) > 0$ and only one real root by vertically translating the curve in Figure 8.15B upwards.

Since at the points $u = \pm 1$, $f(u) = -(b \mp a)^2 < 0$, only the case where we have three real roots is physical. Going back to the definition of the constants, this means that $p_\phi < \sqrt{2E'I_1}$. Thus we are requiring that the precession be slower than this upper bound. Since $f(\pm 1) < 0$, one of the three roots must lie outside the physical region, $u_3 > 1$. Call the two physically possible roots u_1, u_2. The situation for sufficiently slow precession (given by the upper bound on p_ϕ) must look like Figure 8.15A. Roots u_1 and u_2 give the minimum and maximum angles respectively for the motion of the top as represented in Figure 8.16.

Between these two limits, what types of motion are possible? Go back to Equation (8.99). Notice that the average value of $\dot{\phi}$ is, in general, nonzero, so there is a net direction of precession. This can go either way. If the sign of $\dot{\phi}$ is always positive, for example, the motion looks like Figure 8.17A. If $\dot{\phi}$ reverses sign as the top's axis moves

FIGURE 8.16
The top's axis (projection onto the unit sphere) must lie between these limits.

between θ_1 and θ_2, we have motion like Figure 8.17B. If $b - a\cos\theta_1 = 0$, the motion looks like Figure 8.17C, in which case the precession in ϕ stops momentarily when the top is at the maximum value of θ.

The motion that changes θ, the "tilt" of the top's axis is called *nutation*. In all there are three frequencies involved in the motion of the top: the spin frequency around the top's axis ($\dot\psi$), the precession frequency of the top's axis ($\dot\phi$), and the nutation frequency of the top's axis ($\dot\theta$).

If initially, at $t = 0$, there is no precession ($\dot\phi = 0$), then it must be true that $b - au_1 = 0$, where $u_1 = \cos(\theta$ at $t = 0)$. If, at the same time, there is also no nutation, then $\dot\theta = 0$, which means that $\alpha - \beta u_1 = 0$ too. In this case, you can write the expression for $f(u) \equiv \dot u^2 = \beta(u_1 - u)[(1 - u^2) - \frac{a^2}{\beta}(u_1 - u)]$. Our cubic equation reduces to a quadratic equation for the turning point u_2. If the top starts with its axis in the vertical position, $u_1 = 1$. The possible roots of the quadratic equation are $u_2 = 1$ or $u_2 = \frac{a^2}{\beta} - 1$. If it happens, by the choice of the initial conditions, that $\frac{a^2}{\beta} > 2$, then the only possible solution is for $u_2 = 1$, (i.e., the top remains with its axis in the vertical position). This is called a "sleeping" top. If you convert this, using the definitions of a and β to a condition on the physical variables, you will have

$$\omega_3^2 > \frac{4I_1 Mgl}{I_3^2} \quad \text{sleeping top condition on } \omega_3. \tag{8.105}$$

For real tops, friction eventually reduces ω_3 so that the top starts to precess and nutate.

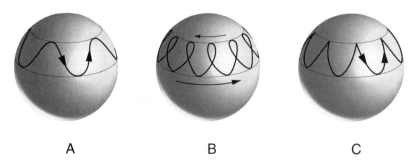

A B C

FIGURE 8.17
Three possible types of motion for the top's axis (projection onto the unit sphere).

QUESTION 13: *Heavy Symmetric Top* Explain how to derive the uniform precession rate for the top when θ is constant and equal to $\frac{\pi}{2}$. Base your work on the methods and answers presented in the text. You should get the result derived in elementary physics courses – look it up and compare. Consider the plot of $f(u)$. For this case what are u_1 and u_2?

There are several features we haven't discussed here: the difference between "fast" and "slow" tops, the applications to gyrocompass guidance systems, etc. There exists a whole well-developed technology based on the rapidly spinning top.*

8.11 PRECESSION OF THE EQUINOXES

The plane of the Earth's orbit around the Sun is called the *ecliptic*. The tilt (θ) of the Earth's axis of rotation with respect to the normal to the ecliptic is 23.445778° (1950). (The technical term for this is the *obliquity*.) It is this tilt that causes the variation in the length of the day between winter and summer and the seasonal change of climate. The Earth's axis has been "nearly" constant in direction relative to an inertial frame. The North Pole currently points in the direction of Polaris, but this has not always been true. Five-thousand years ago, the Southern Cross could be seen from England.[†] There is a retrograde precession ($\dot{\phi}$) of the Earth's axis with a period of 25,730 years. The term "retrograde" means that the axis precesses in a direction opposite to the Earth's direction of travel around the Sun. The obliquity angle also changes slightly ($\dot{\theta} \neq 0$), from 22.1° to 24.5° with a period of 41,000 years. These motions are depicted in Figure 8.18.

Hipparchus of Rhodes observed the longitude (angular height above the ecliptic) of the star Alpha Virginis, comparing his observation with that made 150 years before by the Babylonian astronomers. In that time interval, the longitude had changed by about 2°. Hipparchus correctly interpreted this as a precession of the Earth's axis. The number obtained by Hipparchus for this was 46.8 seconds of arc/year. (The modern value for his era is 48 seconds of arc/year.) The subsequent loss of much of this knowledge is often blamed on Aristotle. Ptolemy (150 A.D.) argued that the heliocentric theory[‡] violated the laws of (Aristotle's) physics, since if the Earth moved, people would fall off. The geocentric Ptolemaic theory was to be unchallenged for almost 1,500 years.

As noted in Section 8.6, the Earth's rotation is responsible for the bulge at the equator. This bulge is the reason that the Sun and the Moon can exert torques on the Earth. That is, the total gravitational potential, integrated over the Earth's volume, will depend on the orientation of the Earth, since the Earth is nonspherical. To calculate this, we will assume that the radius of the Earth is small compared to the distance to the Moon (60.4 Earth radii) and to the Sun. With that assumption, we can expand the gravitational potentials of the

---

* Researchers at Stanford University are working on a gyroscope to be tested on the space shuttle. It will be sensitive enough to detect effects predicted by general relativity, such as frame-dragging. It is known as Gravity Probe B (see *Science*, Nov. 15, 1991, pp. 939–941).
† *The Rotation Of The Earth*, W. H. Munk and G. J. F. Macdonald, Cambridge Univ. Press, 1960.
‡ Aristarchus of Samos (310–230 B.C.) proposed this heliocentric theory.

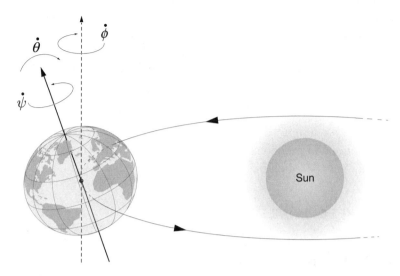

FIGURE 8.18
Period of $\psi = 24$ hours, period of $\phi = 23{,}730$ years, and period of $\theta = 41{,}000$ years.

Sun and the Moon in a Taylor series to second order in the Earth's radius. Fortunately, it is a good approximation that the Moon's orbit lies in the plane of the ecliptic, because this simplifies the calculation considerably. Finally, we note that the periods of the Moon around the Earth and of the Earth around the Sun are much shorter than the time scale of the precession, so we will average over these motions, also assuming for simplicity that the Earth's orbit is circular.

To find the form of the potential energy function for the Earth in the gravitational field of the Sun or the Moon, choose a point $\vec{r}$ within the Earth. We won't assume that the density of the Earth, ρ, is constant, because the core of the Earth is more dense than the outer mantle. We will assume however, that it has spherical symmetry: $\rho = \rho(r)$. We will also assume that the inhomogeneity in density does not contribute to the equatorial bulge. After making these simplifying assumptions, we will obtain an average potential energy that depends only on θ – the angle between the Earth's axis and the normal to the ecliptic. We will then analyze the equations of motion in a way very similar to that used for the heavy symmetrical top. (The potential energy is a different function of θ, however.)

The total gravitational potential energy from either the Sun or the Moon is defined to be

$$V \equiv -MG \iiint \frac{\rho(r)\,dV}{|\vec{R}-\vec{r}|}. \tag{8.106}$$

As shown in Figure 8.19, $\vec{R}$ is a fixed vector from the geometric center of the Earth to the Sun or the Moon, $\vec{r}$ is a variable vector from the center to any point within the Earth's volume, $D(\vec{r})$ is the distance from the variable point to the centers of the Sun or the Moon, dV is an element of volume in the Earth, M is the mass of the Sun or Moon, and $G = 6.6726(5) \times 10^{-11}$ N m^2/kg^2 is the gravitational constant. The next step is to expand

8.11 PRECESSION OF THE EQUINOXES

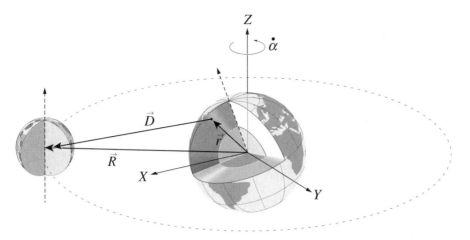

FIGURE 8.19
Moon orbiting Earth. Note: The Earth–Sun figure is similar, but the Earth orbits the Sun instead. This has no effect on V, so you can still use the Earth as the origin as shown for the Earth–Moon case. $\dot\alpha$ is the rotation frequency of the Moon around the Earth or the Earth around the Sun.

the function $\frac{1}{|\vec R - \vec r|}$ as a Taylor series in the three components of $\vec r = (r_1, r_2, r_3)$ around the value at the center, $\vec r = (0, 0, 0)$:

$$\frac{1}{D} \equiv \frac{1}{|\vec R - \vec r|} = \frac{1}{R} + \frac{\tilde{\mathbf{r}}\mathbf{R}}{R^3} + \frac{1}{2}\frac{\tilde{\mathbf{r}}\mathbf{H}\mathbf{r}}{R^5} + \text{terms O}[r^3]. \tag{8.107}$$

We have used the vector and tensor notation introduced in Chapter 7 to avoid writing out all of the components. In the matrix notation, $\tilde{\mathbf{r}}\mathbf{R} \equiv \vec r \cdot \vec R$. We will denote the body frame components of $\vec R$ by (R_1, R_2, R_3). The tensor $\mathbf{H}$ is defined as the collection of second derivatives of the reciprocal of the distance, evaluated at the center:

$$H_{ij} \equiv R^5 \left.\frac{\partial^2 \left(\frac{1}{D}\right)}{\partial r_i \partial r_j}\right|_{r_i, r_j = 0}. \tag{8.108}$$

After a somewhat tedious calculation, it is possible to show that

$$\mathbf{H} = \begin{pmatrix} 3X^2 - R^2 & 3XY & 3XZ \\ 3XY & 3Y^2 - R^2 & 3YZ \\ 3XZ & 3YZ & 3Z^2 - R^2 \end{pmatrix}. \tag{8.109}$$

Equation (8.107) is inserted into (8.106) to obtain V. After factoring out all of the constants, the first integral term gives the mass of the Earth. The second term, proportional to the triple integral $R_i \iiint \vec r_i \rho \, dV$, gives the Earth's mass times the coordinates of the center of mass. If we assume this corresponds with the geometric center, this term vanishes. The third term involves integrals of the type $H_{ij} \iiint r_i r_j \rho \, dV$. In a principal axis body system, with the North Pole along the $+Z$ axis, it can be shown that symmetry of the mass distribution means only the integrals with $i = j$ do not vanish. These can be expressed in terms of the two distinct moments of inertia for the Earth:

$$I_3 \equiv \iiint (x^2 + y^2)\rho\, dV, \quad I_1 = I_2 \equiv \iiint (x^2 + z^2)\rho\, dV, \tag{8.110}$$

giving

$$\iiint \rho\, dV\, \tilde{\mathbf{r}}\mathbf{H}\mathbf{r} = (I_1 - I_3)(3Z^2 - R^2). \tag{8.111}$$

Formula (8.111) makes use of the identity $R^2 = X^2 + Y^2 + Z^2$. To obtain Z we have to remember that we evaluated (8.110) in a body system rotated by the Euler angles ϕ, θ, ψ from the ecliptic plane of the Earth–Sun system (or, approximately, the Earth–Moon system). Assume that the ecliptic plane is the plane $Z' = 0$ in the space coordinate system. Furthermore, assume a circular orbit which has $X' = R\cos\alpha$, $Y' = R\sin\alpha$, $Z' = 0$ at some given moment. We intend to average over the orbital angle α, which varies with time much more rapidly than the effects we are trying to calculate. Using the concept of the $\mathbf{U}$ matrix to take us from body to space frame coordinates, we have the equation

$$\begin{pmatrix} X \\ Y \\ Z \end{pmatrix} = \tilde{\mathbf{U}}(\phi, \theta, \psi) \begin{pmatrix} R\cos\alpha \\ R\sin\alpha \\ 0 \end{pmatrix}, \tag{8.112}$$

$$Z = R(\tilde{U}_{31}\cos\alpha + \tilde{U}_{32}\sin\alpha). \tag{8.113}$$

Squaring, averaging over α, and using the explicit form of the matrix elements (8.63), the average value is

$$\overline{Z^2} = \frac{R^2}{2}\sin^2\theta. \tag{8.114}$$

(Using the exact expression instead of the average would give us a time dependence in the potential, which would lead to short time scale fluctuations for the Euler angles; we ignore these here, because they are very small.) The expression (8.106) becomes, in this approximation,

$$V(\theta) = -\frac{GMM_{\text{earth}}}{R} + \frac{GM}{2R^3}(I_3 - I_1)\left(\frac{1 - 3\cos^2\theta}{2}\right). \tag{8.115}$$

The additional term in the potential energy depends only on the obliquity, θ. It is proportional to $I_3 - I_1$ and hence vanishes for a spherical Earth, as it must. (Had we calculated

8.11 PRECESSION OF THE EQUINOXES

higher-order terms in the Taylor expansion, these would also have to vanish if $I_3 = I_1$. We don't have to worry about including the next term here due to the small coefficient of the term depending on θ.)

The actual potential is the sum of the effect of the Sun and of the Moon. Define

$$\Omega^2 \equiv \underbrace{\frac{GM}{R^3}\bigg|_{\text{sun}}}_{3.964} + \underbrace{\frac{GM}{R^3}\bigg|_{\text{moon}}}_{8.482} = 12.446 \times 10^{-14}\,\text{s}^{-2}. \tag{8.116}$$

We used Kepler's Third Law $\frac{G(M+M_{\text{earth}})}{a^3} = \frac{4\pi^2}{T^2}$ to calculate what we need from the periods of 365.25 days (Sun) and 27.322 days (Moon), correcting the result only by the mass ratio between the Moon or Sun and the Earth. Here a is the semimajor axis for the orbit. We see that the effects of the Sun and the Moon are of comparable size, with the Moon exerting a little more than twice the torque on the Earth than the torque from the Sun. The tidal effects on the Earth are also proportional to $\frac{M}{R^3}$, so the Moon is also twice as effective in creating the ocean tides. The number for Ω^2 should be accurate within our approximations of treating the orbits of the Sun and the Moon as circular and coplanar.

The torque $-\frac{dV}{d\theta}$ is $\frac{\Omega^2}{2}(I_1-I_3)\frac{3}{2}\sin 2\theta$. Since $I_3 > I_1$, if the Earth were not rotating there would be a decrease in θ from this effect until the equatorial plane was parallel to the ecliptic and thus $\theta = 0$. However, the gyroscopic effect of the Earth's rotation causes instead a precession: $\dot{\phi} \neq 0$. The direction of this precession is determined by the sign of the torque.

The next step is to write the Lagrangian for the rotational part of the Earth's motion. Using the formula (8.73) developed previously for the kinetic energy of rotation in terms of the Euler angles and their time derivatives, the Lagrangian is

$$L = \frac{I_1}{2}(\dot{\theta}^2 + \dot{\phi}^2 \sin^2\theta) + \frac{I_3}{2}(\dot{\psi} + \dot{\phi}\cos\theta)^2$$
$$- \frac{\Omega^2}{2}(I_3 - I_1)\left(\frac{1-3\cos^2\theta}{2}\right). \tag{8.117}$$

First let's assume we have no nutation, so $\dot{\theta}$ and $\ddot{\theta}$ are zero. We will seek to find the approximate value of $\dot{\phi}$. The equation of motion for θ is found in the standard way from the Euler–Lagrange equation for this variable and the Lagrangian (8.117). $\ddot{\theta}$ is found to be proportional to $\sin\theta$ times a function of $\cos\theta$, $\dot{\psi}$, and $\dot{\phi}$. Define $\epsilon \equiv \frac{I_3-I_1}{I_1} = .00335281$. Setting $\ddot{\theta} = 0$ then yields the formula (after a small amount of algebra)

$$-\epsilon\dot{\phi}^2 - (1+\epsilon)\frac{\dot{\phi}\dot{\psi}}{\cos\theta} = \frac{3}{2}\Omega^2\epsilon. \tag{8.118}$$

Neglecting the quadratic term in $\dot{\phi}$ because it is so small, we obtain the approximate formula

$$\dot{\phi} \approx -\frac{3}{2}\frac{\epsilon}{1+\epsilon}\Omega^2\frac{\cos\theta}{\dot{\psi}}. \tag{8.119}$$

This last formula (8.119) shows that the precession rate is proportional to the magnitude of $\frac{\text{torque}}{I_1 \sin \theta}$ and inversely proportional to the rotation rate of the Earth on its axis. The number calculated, called $\dot{\phi}_{\text{estimated}}$, is

$$\dot{\phi}_{\text{estimated}} = -7.96 \times 10^{-12} \text{ rad/s}. \tag{8.120}$$

From the 25,730 year precession period,* we obtain

$$\dot{\phi} = -7.73813 \times 10^{-12} \text{ rad/s}. \tag{8.121}$$

The minus sign is due to the clockwise rotation of the Earth's axis compared to the counterclockwise direction of the Earth in its orbit around the Sun, if one takes the $+Z$ direction to lie in the direction of the Northern hemisphere, while the XY plane is the ecliptic.

The value of the precession obtained from (8.120) is within 3% of the actual rate (8.121), an excellent agreement, considering the approximations we made. This calculation was first made by d'Alembert in 1749, although Newton understood the basic reason for the precession and used the connection with the tides to estimate the precession rate within a factor of two.

A precession without change in the obliquity corresponds to minimum energy, where the energy $E' \equiv E - \frac{I_3}{2}\omega_3^2$, as we defined it for the heavy symmetric top (8.100). We can refer back to that calculation to see that there are three relevant constants of the motion:

$$p_\psi \equiv L_3 = I_3(\dot{\psi} + \dot{\phi}\cos\theta), \tag{8.122}$$

$$p_\phi \equiv L_z = I_1 \dot{\phi} \sin^2\theta + I_3(\dot{\psi} + \dot{\phi}\cos\theta)\cos\theta, \tag{8.123}$$

and

$$E' = \frac{I_1}{2}\left(\dot{\theta}^2 + \frac{(L_z - L_3 \cos\theta)^2}{I_1^2 \sin^2\theta}\right) + \frac{\Omega^2}{2}\epsilon I_1 \left(\frac{1 - 3\cos^2\theta}{2}\right). \tag{8.124}$$

E'_{min} can be evaluated by setting $\dot{\theta} = 0$, $\theta = 23.45°$. Any increase in $E' > E'_{\text{min}}$ will lead to a periodic change in θ, with turning points and a period dependent on how much the energy is in excess of the minimum possible energy. Since we know both the approximate turning points and period of the change in obliquity, we can find out if this effect is consistent with a "heavy top nutation" type of effect. The will be done in a homework problem, where we will learn that the gyroscopic motion is *not* the effect causing the 41,000 year period of $\dot{\theta}$. Any true gyroscopic nutation such as we saw with the heavy symmetric top has long since damped out. The 41,000 year period is caused by a slow change in the plane of the ecliptic caused by the perturbations of other planets.

There is also a small forced nutation due to the inclination of the Moon's orbit. This results in an elliptical motion superimposed on the precession, which has a period of 18.6 years, an east–west amplitude of ± 6.87 arcseconds, and a north–south amplitude (semimajor axis) of ± 9.21 arcseconds.

---

* K. Jung, "Figur der Erde," Geophysics I, vol. XLVII, *Encyclopedia of Physics*, Springer-Verlag, 1956.

The Ice Ages

In addition to the retrograde precession ($\dot\phi$) and the change in obliquity ($\dot\theta$), there are periodic changes in the eccentricity e of the Earth's orbit about the Sun that are caused by the effect of the other planets, chiefly Jupiter. (The period of this is 105,000 years.) Right now, $e = .017$, but over time, the eccentricity varies from 0 to .06, at least for the past two million years.

All three of these small effects can cause profound changes in the Earth's climate. Detailed calculations made by Milutin Milankovitch show that the combined effects may have caused the ice ages through variations in the intensity of solar radiation. Data is available for the past 500,000 years from records of the oxygen isotope ratio in deep sea cores. The isotope ratio data is an indirect measurement of average temperature. Since large-scale ice formation lowers the level of the sea, whenever sea level data is available it corroborates the more detailed isotope ratio data. The correlation of this evidence for ancient (and recent) ice ages with the calculations is claimed to be quite good. These questions are discussed in much more detail in the book *Ice Ages; Solving The Mystery* by John and Katherine Palmer Imbrie, Harvard University Press, 1979. More recently, however, a controversy over the explanation of the ice ages has emerged. A group under Walter Alvarez at the University of California, Berkeley, has disputed the interpretation of the isotope data and the ability of the Milankovitch theory alone to explain the ice ages, attributing them instead to cosmic "dust" particles. The passage of time will hopefully clarify the answer to this important geophysical question.

8.12 MACH'S PRINCIPLE

We assume that the laws of Newtonian mechanics will be valid in any inertial frame, but what is the precise definition of an inertial frame? A gyroscope's[*] angular momentum by definition is constant in an inertial frame. It is observed experimentally that a gyroscope appears to rotate with faraway objects such as stars[†] with a high degree of precision (neglecting effects from general relativity), so they must define an inertial frame. However, since our galaxy is rotating, how can the faraway stars in our galaxy determine an inertial frame? The velocity of rotation of the galaxy is almost independent of radius, being about 220 km/s. The Sun is located about $R_{sun} = 8.5$ kparsecs from the galactic center (1 parsec = 3.1×10^{16} m). This galactic rotation corresponds to $\omega \approx 8.4 \times 10^{-16}$ rad/s ≈ 0.5 arcseconds/century at the radial distance of the Sun from the galactic center. (This can be compared with the general relativity prediction of the precession of Mercury of 43 arcseconds/century.) The corresponding acceleration is $\omega^2 R_{sun} \approx 1.9 \times 10^{-14} g$. Such a small acceleration is right at the limit of detection when local inertial frames are compared with the distant stars. It is in this sense that we refer to these faraway stars in our galaxy as defining an inertial frame.

---

[*] See Problem 18 for the definition of a gyroscope.
[†] Sufficiently far away such that there is no parallax due to the Earth's motion.

FIGURE 8.20
Two massive spheres, connected by a spring.

Another definition of a "local" inertial frame comes not from a Foucault pendulum, nor from a gyroscope, but instead from fitting the motions of the innermost planets around the Sun, which has an uncertainty in its rate of rotation of about .4 arcseconds/century.*

Newton defined an inertial frame by declaring that "absolute acceleration" exists and can be distinguished from relative acceleration by experiment. One example he gave was that of two identical massive spheres in empty space. Suppose that these spheres are connected by a string (see Figure 8.20). If a tension exists in the connecting string, the spheres are rotating relative to an inertial frame. This gives an experimental way to detect absolute acceleration (rotation) in a way that is independent of any other reference frame. One can carry the example further and imagine identical small rockets mounted on each sphere which are fired until the tension in the string vanishes. By knowing how much angular momentum was imparted to the spheres by these rockets, we can calculate the product $\mathbf{I}\omega$. ω is the absolute angular velocity.

Ernst Mach believed that such a concept was not philosophically tenable. Instead he proposed that all inertial forces are due to the distributed matter in the universe. The origin of inertial mass (resistance to acceleration) would then be seen to be dynamical. This point of view is known as "Mach's Principle." In an otherwise empty universe the "two-sphere" experiment would not lead to observable tension in the string, because the spheres could not have inertial mass. It would be meaningless to assign a value to ω, since there would be nothing to rotate with respect to. The spheres would not acquire masses until we filled the empty universe with matter. Rotation with respect to this distant matter would generate the observed inertial effect.

From Mach's perspective it is no coincidence that the universe does not rotate – by Mach's Principle the aggregate of matter in the universe is by definition an inertial frame. The plane in which a Foucault pendulum swings appears to rotate with the fixed stars because there is a dynamical force that occurs if there is any relative rotation of the plane of the pendulum and the rest of the matter in the universe.

It is a bit surprising that one could do an experiment to distinguish between Newton's view and Mach's. A massive rotating sphere should lead to an apparent small "centrifugal force" for objects inside the sphere if Mach is right. Newton would predict no such effect. Attempts to measure this force experimentally have not succeeded. One complication is an effect predicted by general relativity called "frame-dragging," which is similar to the effect

---

* See Schiff, L. I., "Observational Basis of Mach's Principle," *Reviews of Modern Physics*, Vol. 36, No. 2, April 1964, pp. 510–511.

predicted by Mach. However, the theory of general relativity does not incorporate Mach's Principle, despite much effort by theorists to include it.* This subject is interesting and of the most fundamental importance to physics. Reinhardt has written an article ("Mach's Principle – A Critical Review," *Z. Naturforsch.* Vol. **28 a.**, pp. 529–537 (1973)) in which he discusses the role of Mach's Principle, including the early history, most notably Newton's rotating water pail experiment. Reinhardt discusses the notion of absolute versus relative acceleration. He ends by concluding that Mach's Principle, "...though an extremely stimulating thought, has *at present* little claim to be a basic physical principle" (italics ours).

Mach's Principle is discussed in a book by D. W. Sciama, which also serves as an introduction to general relativity. The book is *The Physical Foundations of General Relativity* (see the bibliography).

It is perhaps worth mentioning that the famous 3° cosmic black body radiation defines an inertial frame that is independent of the motions of the stars. In fact, recent measurements on the directional dependence of the Doppler-shifted spectrum of this radiation can be used to measure the velocity of our galaxy as it moves through this radiation background that was caused by the Big Bang. Although this means that all inertial reference frames are not strictly equivalent, in practice nothing exists to prevent our making this assumption as far as mechanics is concerned.

SUMMARY OF CHAPTER 8

- The moment of inertia **I** is a second-rank tensor, defined by

$$I_{\alpha\beta} \equiv \sum_i m_i \left(r_i^2 \delta_{\alpha\beta} - r_{i,\alpha} r_{i,\beta} \right). \tag{8.125}$$

- For a rigid body, the kinetic energy is the sum of two terms:

$$T = T_{\text{translation}} + T_{\text{rotation}}. \tag{8.126}$$

These are

$$T_{\text{rotation}} = \frac{1}{2} \tilde{\omega} \cdot \mathbf{I} \cdot \omega, \qquad T_{\text{translation}} = \frac{1}{2} M \vec{V}_{\text{cm}}^2. \tag{8.127}$$

---

* See Brans and Dicke, "Mach's Principle and a Relativistic Theory of Gravitation," *Phys. Rev.*, Vol. 124, No. 3, pp. 925–935.

- A body coordinate system called the principal axis system can always be found. In this system the inertia tensor is diagonal. The three diagonal elements are called the principal moments of inertia.
- The moment of inertia about an axis displaced by a vector $\vec{a}$ from the center of mass is

$$\mathbf{I}_{\vec{a}} = \mathbf{I}_{\text{cm}} + M\left(a^2 \delta_{\alpha\beta} - a_\alpha a_\beta\right). \tag{8.128}$$

- The total angular momentum is

$$\vec{L}_{\text{total}} = \underbrace{\vec{R} \times \vec{P}}_{\vec{L}_{\text{cm}}} + \underbrace{\mathbf{I} \cdot \omega}_{\vec{L}_{\text{rot}}}. \tag{8.129}$$

- The motion of a force-free rigid body, expressed in a principal axis coordinate system, is given by Euler's equations:

$$\begin{aligned} I_1 \frac{d\omega_1}{dt} &= \omega_3 L_2 - \omega_2 L_3 = \omega_2 \omega_3 (I_2 - I_3), \\ I_2 \frac{d\omega_2}{dt} &= \omega_1 L_3 - \omega_3 L_1 = \omega_1 \omega_3 (I_3 - I_1), \\ I_3 \frac{d\omega_3}{dt} &= \omega_2 L_1 - \omega_1 L_2 = \omega_1 \omega_2 (I_1 - I_2). \end{aligned} \tag{8.130}$$

- The Euler angles are defined as a series of three rotations about the body axes, first the Z axis, then the new X axis, and finally the newest Z axis. The angular velocity in either space or body coordinates can be expressed in terms of the Euler angles and their time derivatives (8.68, 8.69).
- The orthogonal matrix $\mathbf{U}$, which allows for a transformation from body to space coordinates, was derived explicitly in terms of Euler angles in formula (8.63).

PROBLEMS

Orthogonal Matrix U

Problem 1: *(Orthogonal linear transformation)* Show that if a linear transformation $\mathbf{U}$ is orthogonal, then it is both necessary and sufficient that the three vectors $\mathbf{Ur}_1$, $\mathbf{Ur}_2$, $\mathbf{Ur}_3$ are mutually orthogonal and normalized to 1 if the basis vectors $\mathbf{r}_1$, $\mathbf{r}_2$, $\mathbf{r}_3$ are orthogonal and normalized to 1.

PROBLEMS

Problem 2: *(Orthogonal transformation represented as rotation)* Prove that any orthogonal transformation **U** can be represented as a rotation though an angle about an axis that is left unchanged by **U**.

Moment of Inertia Tensor

Problem 3: *(Moment of inertia tensor)* Which of the symmetric 3×3 matrices below could represent a physical moment of inertia tensor?

$$I_1 = \begin{pmatrix} 1 & 2 & 1 \\ \ldots & 0 & 2 \\ \ldots & \ldots & 1 \end{pmatrix}, \quad I_2 = \begin{pmatrix} 1.94791 & .0347273 & -.394509 \\ \ldots & 2.42924 & -.823746 \\ \ldots & \ldots & 1.62285 \end{pmatrix} \quad (8.131)$$

Explain. Find the principal axes and principal moments of the ones that are physical.

Problem 4: *(I for a circular hoop)* What is the moment of inertia tensor for a circular hoop of radius R and mass M? What are the principal axes and moments? (Neglect the thickness of the hoop.)

Problem 5: *(I for a thin rod)* Find the inertia tensor, principal axes, and principal moments for a thin rod of length l.

Problem 6: *(I for a circular cylinder)* Find the inertia tensor, principal axes, and principal moments for a circular cylinder of radius R and height h.

Problem 7: *(I for an ellipsoid)* Find the inertia tensor, principal axes, and principal moments for an ellipsoid of semiaxes a, b, c.

Problem 8: *(I for a spherical shell, solid sphere)* Calculate the moment of inertia tensor for a spherical shell of radius R and mass M. Simplify your calculation by using the symmetry to maximum advantage. From this result calculate **I** for a solid sphere of radius R.

Problem 9*: *(I for three mass points)* Three equal mass points are located at $(a, 0, 0)$, $(0, a, 2a)$, and $(0, 2a, a)$. Find the inertia tensor, the principal axes, and the principal moments.

Problem 10*: *(I for a book)* A book of mass M has the dimensions $a = 10\,\text{cm}$ by $b = 20\,\text{cm}$ by $c = 3\,\text{cm}$. Find the principal axes using a symmetry argument. Find the inertia tensor in the principal axis system. Indicate on a diagram the direction of the principal axes and which ones have the least moment of inertia and the greatest moment of inertia.

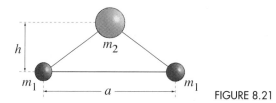

FIGURE 8.21

Problem 11: *(I for a triangle)* Calculate the moment of inertia about the center of mass of an equilateral triangle (see Figure 8.21) with identical masses m_1 at the base located at each end and a mass m_2 located at the top of the triangle. You can use the displaced axis theorem to first calculate the moment of inertia for the rotator formed by the two masses m_1 with an origin located on the triangle base halfway between them, and then use the displacement theorem to move the origin to the actual center of mass. The height of the triangle is assumed to be h, and the base has length a. The principal moments you should find are

$$I_1 = \frac{2m_1 m_2 h^2}{M}, \quad I_2 = \frac{1}{2} m_1 a^2, \tag{8.132}$$
$$I_3 = I_1 + I_2, \quad M = 2m_1 + m_2.$$

Kinetic Energy and Angular Momentum

Problem 12*: *(Rolling cylinder)* We often need to know how to calculate the kinetic energy of a rolling body in a particular geometry. Consider the case of a cylinder rolling on a plane (see Figure 8.22). There is one degree of freedom, denoted by the angle ϕ. If we regard the cylinder as rotating about the point of contact with the plane, it is possible to calculate the kinetic energy knowing only the moment of inertia tensor and the mass M of the cylinder. Assume that one of the principal axes is parallel to the axis of the cylinder (principal moment is **I**) and that the mass is so distributed that the center of mass is off the central axis a distance a as shown in Figure 8.22.

a) Prove that the kinetic energy of the center of mass motion is

$$\frac{1}{2} M(a^2 + R^2 - 2aR\cos\phi)\dot\phi^2. \tag{8.133}$$

b) Find the total kinetic energy $T(\phi, \dot\phi)$.

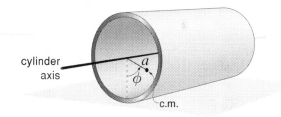

FIGURE 8.22

c) Find the Lagrangian and the frequency of small oscillations about the equilibrium point.
d) If $a \to 0$ the center of mass moves to the axis of the cylinder. What would you expect for the frequency of small oscillations in that case? Does your answer for the general case agree in that limit?

(Landau)

Problem 13: *(L and T for symmetric top)* Analyze the motion of a symmetric top ($I_1 = I_2 \neq I_3$) thrown into the air.

a) First assume the top is thrown with initial angular velocity $\vec{\omega}_0$ along the 3 axis and initial linear velocity $\vec{v}_0$. What is the total kinetic energy? The total angular momentum?
b) Next throw the top into the air with initial angular velocity $\vec{\omega}_0$ along the 1 axis (or 2 axis) and initial linear velocity $\vec{v}_0$ again. What is the total kinetic energy? The total angular momentum?
c) Compare parts a) and b).

Euler's Equations and Euler Angles

Problem 14*: *(Euler equations for components of torque in the body frame)* In this problem we derive the Euler equations for the components of torque in the body frame. We then apply them to the torque-free motion of a symmetric top.

a) Prove that $\frac{d\vec{L}}{dt}$, the rate of change of angular momentum with respect to the center of mass, is given by $\frac{d\vec{L}}{dt} = \vec{\tau} =$ torque $\equiv \sum_i \vec{r}_i \times \vec{f}_i$, where $\vec{f}_i$ is the force on the ith part of the body. In the principal axis frame prove that

$$\tau_1 = I_1 \dot{\omega}_1 + (I_3 - I_2)\omega_2\omega_3 \qquad (8.134)$$

$$+\text{cyclic permutations: } 123 \to 312 \to 231. \qquad (8.135)$$

These equations are "Euler's equations" for a rigid body. Note that they are the projections of the torque onto the *body* set of principal axes.

b) Assume that a symmetrical top ($I_1 = I_2 \neq I_3$) is thrown up in a gravitational field. (One example: a football thrown in a forward pass.) Why is the torque $\vec{\tau} = 0$? Using the Euler equations, determinine the $\vec{\omega}$ in the principal axis frame.
c) Suppose that, instead, $I_1 = I_2 = I_3$. What does this mean physically? What is $\vec{\omega}$ if the torque on the body is zero?

Problem 15*: *(Rotating a rectangular plate)* Using Euler's equations from Problem 14, we will solve for the torque needed to rotate a rectangular plate about a diagonal with a constant angular velocity $\vec{\omega}$, as shown in Figure 8.23.

a) What is $\vec{\omega}$ in the principal axis frame?

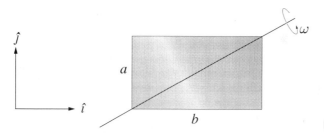

FIGURE 8.23

b) Why is the (body) coordinate system chosen above useful? Why not have it tilted at some angle in the XY plane?

c) Using the fact that $\vec{\omega}$ is constant in the body frame, use the Euler equations to solve for the torque components. Prove that

$$\vec{\tau} = -\frac{Mab\omega^2(b^2 - a^2)}{12(a^2 + b^2)}\hat{k}.$$

(You can assume the plate has zero thickness.)

d) Note that $\vec{\tau} = 0$ if $a = b$. Explain. If $b > a$ draw a vector diagram showing $\vec{L}$ and $\vec{\omega}$ in the body system. Then explain why you need to exert a torque about the plate diagonal axis in order to rotate the plate.

Tops

Problem 16: *(Symmetric top – equating reference frames)* We derived formulas for the symmtetric top using angular velocities in the space frame (8.37, 8.44, 8.45, 8.46) and in the body frame (8.48, 8.51). First find the (time-dependent) linear transformation that connects the two reference frames. Then prove

$$A = -\omega_P \sin\theta, \quad \Omega_P = -\Omega. \tag{8.136}$$

Problem 17: *(Symmetric top – using the Lagrangian)* Use the expression (8.73) for the kinetic energy in the Euler–Lagrange equations to solve for the motion of the force-free symmetric top in the laboratory (space) coordinate system. Show that you get the same answer as we previously obtained.

Problem 18: *(Gyroscope)* A *gyroscope* is a top mounted in a gimbal system such that there is no net torque on the top's center of mass. It is a torque-free symmetric top in the sense we've previously discussed. Consider such a gyroscope in its normal state of uniform precession around the Z axis. Now apply an impulsive torque so that the gyroscope has a small increment of velocity $\delta\dot{\theta}_0$ and $\delta\dot{\phi}_0$ at $t = 0$. Find the subsequent motion of the gyroscope, assuming the equations of motion can be linearized in $\delta\theta(t)$, $\delta\phi(t)$. Is the gyroscope stable with respect to its original motion?

PROBLEMS

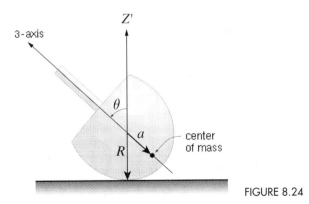

FIGURE 8.24

The answer explains why gyroscopes are used to stabilize ships or video cameras, for example.

Problem 19: *(Routhian for heavy symmetric top)* As an alternative approach to finding the Lagrangian for a heavy symmetric top, use the Routhian technique (5.95) to eliminate two of the three degrees of freedom. Show that this leads to the same answer as (8.93).

Problem 20: *(Tippe top)* This simple device appears to defy the laws of classical mechanics. Start it spinning, and it eventually inverts, raising its center of mass to a higher position, where it spins happily until it runs down. We will examine this strange behavior qualitatively here. The top has a mass M and radius R. The center of mass is a distance a from the center of the sphere as shown in Figure 8.24.

a) First we will neglect friction. Assuming the top is tilted an initial angle θ_0 from the vertical and is given an initial spin ω_0 along the 3 axis, what is the direction of $\vec{\omega}$ and $\vec{L}$ as it spins?

b) Now we will add a frictional force $\vec{F}$ between the top and the surface of the table. In what direction will it point? What is the torque caused by this frictional force? What is the change in $\vec{L}$ due to this torque? How about the change in $\vec{\omega}$? Explain how this change will cause the top to turn over.

c) The top will end up in a higher center of mass configuration than it was initially. Calculate the potential energy required to raise the top's center of mass in this way. Assuming that the energy is supplied by the initial rotational energy of the top, estimate the minimum angular velocity required to make the top stand up.

For an more qualitative analysis of this top, we refer you to an article in the *American Journal of Physics*, Vol. 45, No. 1, January 1977, p. 12ff.

Problem 21*: *(Book as asymmetric top)* A book is an asymmetric top, with no two moments of inertia equal. Assuming that $I_1 < I_2 < I_3$, you can demonstrate experimentally that the motion is stable if the rotation is started about either the 1 or the 3 axis but unstable if started about the 2 axis. To prove this, consider three different situations:

a) $\vec{\omega}(t=0) = (1, 0, 0) + \vec{\delta}_0$. This means that the book is rotating almost around the 1 axis at $t = 0$ but has a small deviation from this, $\vec{\delta}_0$.
b) $\vec{\omega}(t=0) = (0, 1, 0) + \vec{\delta}_0$. This means that the book is rotating almost around the 2 axis at $t = 0$ but has a small deviation from this, $\vec{\delta}_0$.
c) $\vec{\omega}(t=0) = (0, 0, 1) + \vec{\delta}_0$. This means that the book is rotating almost around the 3 axis at $t = 0$ but has a small deviation from this, $\vec{\delta}_0$.

Write the Euler equations in each case, ignoring terms of order δ^2. You should obtain a set of linear first-order equations for $\vec{\delta}, \dot{\vec{\delta}}$. Now assume a solution of the form $\vec{\delta}(t) = \vec{\delta}_0 e^{i\alpha t}$. Setting a certain determinant equal to zero gives you α. Why? If α is real, you have stable solutions; if α is imaginary, you have unstable solutions. Why? Finally, prove that the 2 axis is the unstable axis from these "linearized" Euler equations.

Miscellaneous

Problem 22: *(Equatorial bulge)* Prove that the equation $g_{pole} R_{pole} = g_{equator} R_{equator}(1 - \frac{\omega^2 R}{g})$ derived in Appendix C is valid if the density of the Earth is constant. Remember that the centrifugal reduction in the effective value of g is linear with r.

Problem 23: *(Φ)* Prove Equation (8.67) using half-angle trigonometric identities.

Problem 24*: *(Physical pendulum)* Consider a physical pendulum with pivot points A and B that are a distance L_A and L_B from the center of mass as shown in Figure 8.25.

a) What is the frequency of oscillation about either pivot point? Express your answer in terms of I_{cm}, the moment of inertia of the center of mass, and the distance from the pivot point to the mass center. Hint: Use the displaced axis theorem (8.19).
b) If the frequency of oscillation is adjusted (by varying the lengths L_A and L_B) so the frequency about each pivot point is equal, show that the moment of inertia about the mass center is $I_{cm} = ML_A L_B$, where M is the mass of the pendulum.

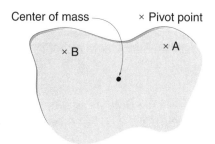

FIGURE 8.25

c) Show that in the special case of b) that the frequency of oscillation is that of a simple pendulum of length $L_A + L_B$.
d) Discuss why this method may be used as a method to determine g to very high accuracy.

Problem 25: *(Is there gyroscopic nutation of the Earth's axis?)* In our discussion of the precession of the equinoxes in Section 8.11, we make the approximation that there was no nutation ($\dot{\theta} = \ddot{\theta} = 0$) in order to derive the rate of precession. It is known that there is a variation in the obliquity of $\pm 1.5°$ with a 41,000 year period. Could this be nutation? In this problem you will show that this effect *cannot* arise from effects due only to the Sun or Moon's gravity.

a) Express the energy E' (see (8.100) for the definition) in units of $I_1 \Omega^2$, the angular momenta L_3, L_z in units of $I_1 \Omega$, and the time in units of $\frac{1}{\Omega}$. Prove that this leads to the simple expression for E':

$$E' = \frac{1}{2}\left(\frac{\dot{u}^2 + (L_z - L_3 u)^2}{1 - u^2}\right) + \epsilon \frac{1 - 3u^2}{2}. \tag{8.137}$$

$\Omega = 1$ in these units, and $\epsilon = \frac{1}{305.8}$.

b) In the same units, show that

$$L_z = 188.655, \quad L_3 = 205.64, \quad E'_{min} = -.00255637. \tag{8.138}$$

c) Prove that $E' \approx 10$ for turning points near $22.24°$ and $24.73°$. This would be the energy needed if the nutation of a heavy top is to explain the variation in obliquity.
d) By doing a numerical integration, prove this leads to a period of approximately 24 hours, rather than 41,000 years. We see that the gyroscopic nutation of the Earth is negligible (although there is a small nutation forced by the Moon). The 41,000 year period cannot be explained by this effect. It is due to perturbations from Jupiter and the other planets.

APPENDIX A

WHAT IS A TENSOR?

The concept of a "tensor" arises as the generalization of "vector." A tensor in 3-D space is a collection of quantities with the same transformation properties under space rotations as products of vector coordinates: 1 (scalar), r_α (vector), $r_\alpha r_\beta$ (second-rank tensor) and so on. More generally, an "nth-rank tensor" in 3-D space is a collection of 3^n quantities that transform in the same way as the products of the components of n

independent vectors. The easiest way to explain the different "ranks" of tensors is to start with examples.

Scalars – Zero-Rank Tensors

A scalar, or zero-rank tensor, is a number. It is invariant under coordinate rotations. Length is an example – it does not matter which coordinate system we are using to measure the distance between two points. It is a geometric quantity that is independent of the specific coordinate system's orientation. The "dot product" of two vectors is a second example: $\vec{a} \cdot \vec{b}$, which is also written in various contexts as $a_\alpha b_\alpha$ (Einstein summation convention) or $\tilde{\mathbf{a}}\,\mathbf{b}$ (matrix notation: "a transpose dot b").

Vectors – First-Rank Tensors

A vector is by definition a first-rank tensor. The components of a vector change when the coordinate system is rotated as we have seen in (7.6). However, the vector still has the same magnitude and direction as it did before the coordinate system was rotated. (The direction *with respect to the coordinate system* has changed, but the direction with respect to any *physical object* is the same.) The dot product of any two vectors gives a scalar. A vector dotted with itself also gives a scalar (7.3, 7.4).

> **QUESTION 14: Vectors** Prove that a two-dimensional vector expressed in coordinates (a_1, a_2) or in coordinates (a'_1, a'_2), which are rotated with respect to (a_1, a_2) by an angle θ, are related such that $(a_1, a_2) \cdot (a_1, a_2) = (a'_1, a'_2) \cdot (a'_1, a'_2)$.

All first-rank tensors (vectors) transform like coordinates under space rotations. Other examples of vectors include velocity, acceleration, electric field, and (under proper rotations) angular velocity and magnetic field.

Second-Rank Tensors

The next step up is a second-rank tensor, which must be multiplied (dot product again) by *two* vectors to obtain a scalar – usually written as: "scalar = vector · second-rank tensor · vector." When a second-rank tensor is "dotted into" two vectors, the result is independent of the orientation of the coordinate system. A second-rank tensor must have two indices: $T_{\alpha\beta}$ with $\alpha, \beta = 1, 2, 3$. There are nine quantities in all. If $T_{\alpha\beta} = T_{\beta\alpha}$ only six of them are independent, and we say the tensor is *symmetric*. In vector notation:

$$S = \sum_{\alpha=1}^{3} \sum_{\beta=1}^{3} a_\alpha T_{\alpha\beta} b_\beta. \qquad (8.139)$$

T is a second-rank tensor with components $T_{\alpha\beta}$, S is a scalar, and $\vec{a}, \vec{b}$ are vectors. Writing this in matrix notation and using the fact that the scalar S is invariant under rotations we

APPENDIX A   WHAT IS A TENSOR?

get

$$S = \begin{pmatrix} a_1 & a_2 & a_3 \end{pmatrix} \begin{pmatrix} T_{11} & T_{12} & T_{13} \\ T_{21} & T_{22} & T_{23} \\ T_{31} & T_{32} & T_{33} \end{pmatrix} \begin{pmatrix} b_1 \\ b_2 \\ b_3 \end{pmatrix}$$

$$= \begin{pmatrix} a'_1 & a'_2 & a'_3 \end{pmatrix} \begin{pmatrix} T'_{11} & T'_{12} & T'_{13} \\ T'_{21} & T'_{22} & T'_{23} \\ T'_{31} & T'_{32} & T'_{33} \end{pmatrix} \begin{pmatrix} b'_1 \\ b'_2 \\ b'_3 \end{pmatrix} \quad (8.140)$$

or, more compactly,

$$S = \tilde{\mathbf{a}} \mathbf{T} \mathbf{b} = \tilde{\mathbf{a}}' \mathbf{T}' \mathbf{b}'. \quad (8.141)$$

Note that (a_1, a_2, a_3), (a'_1, a'_2, a'_3), etc., are row (or column) representations of vectors and therefore obey vector transformation properties under space rotations. The expression (8.140) is the sum of the products of vector and tensor coordinates. Often a second-rank tensor in three dimensions is defined as:

> Alternate definition of a *second-rank tensor*: Any set of nine quantities that transforms under space rotations so that (8.141) is a scalar for arbitrary vectors a_α, b_β.

Since we never refer in this book to higher-rank tensors (of rank >2), we will follow the common practice of dropping the appellation "second-rank" and refer to them as "tensors." It will be clear from the context that we do not mean scalars or vectors. To further distinguish tensors from vectors when we use the matrix notation we will use bold face capital letters (**T**) for tensors and bold face lower case letters (**a**) for vectors.

It is easy to generalize this definition to higher-dimensional spaces. In four-dimensional space, a second-rank tensor will be a set of 16 quantities, for example. (We'll stick to three-dimensional "real" space here.)

QUESTION 15: *Tensors* Starting from the fact that the scalar obtained from each multiplication is invariant, use (8.40) to derive the transformation law for how tensors transform under space rotations (i.e., $\mathbf{T}' = \mathbf{U}\mathbf{T}\tilde{\mathbf{U}}$). Hint: To simplify the algebra, make use of vector transformations in matrix notation such as $\tilde{\mathbf{a}}' = \tilde{\mathbf{a}}\tilde{\mathbf{U}}$ and $\mathbf{a} = \mathbf{U}\mathbf{a}$.

Note that a vector "dotted" into a tensor gives a vector:

$$b_\beta = \sum_{\alpha=1}^{3} T_{\alpha\beta} a_\alpha. \quad (8.142)$$

(Try to prove (8.142). In other words, use the transformation properties of the tensor **T** and the vector **a** to prove that **b** transforms like a vector.)

Besides the inertia tensor other examples of tensors include the stress tensor and the dielectric tensor (used if the displacement vector $\vec{D}$ has a different direction from the electric field vector $\vec{E}$). Tensors are needed to describe the propagation of light in anisotropic media. Remember: Tensors are logically different from *components* of tensors, just like vectors are different from *components* of vectors:

a) The physical tensor itself is independent of coordinates, just like a physical vector (velocity, direction) is independent of coordinates.
b) Tensor components change when you change coordinates: $\mathbf{T}' = \mathbf{U}\mathbf{T}\tilde{\mathbf{U}}$ is the matrix notation for the transformation rule.

APPENDIX B

SYMMETRIC MATRICES CAN ALWAYS BE DIAGONALIZED BY "ROTATING THE COORDINATES"

We will prove that given a real symmetric $n \times n$ matrix $\mathbf{M}$, there exists an orthogonal transformation (rotation) $\mathbf{U}$ of the coordinates that diagonalizes $\mathbf{M}$:

$$\tilde{\mathbf{U}}\mathbf{M}\mathbf{U} = \mathbf{D}, \tag{8.143}$$

$$\mathbf{D} \equiv \begin{pmatrix} d_1 & 0 & 0 & \cdots \\ 0 & d_2 & 0 & \cdots \\ \vdots & \vdots & \ddots & \vdots \\ \cdots & \cdots & 0 & d_n \end{pmatrix}. \tag{8.144}$$

This *fundamental theorem of symmetric real matrices* is the mathematical heart of this chapter. The crucial point is that it is an *orthogonal* transformation that does the diagonalization. In the case of our 3×3 inertia tensor, this means that we can always find a set of principal axes (eigenvalues of $\mathbf{I}$) that form a coordinate system in which $\mathbf{I}$ is diagonal.

To illustrate the method of proof, we will give a specific example. Let

$$\mathbf{I} = \begin{pmatrix} 1 & 1 & 1 \\ 1 & 2 & 1 \\ 1 & 1 & 3 \end{pmatrix},$$

a real symmetric matrix we choose arbitrarily. The first step is to find the eigenvalues λ of $\mathbf{I}$. If $\mathbf{e}$ is an eigenvector of $\mathbf{I}$, by definition this implies that*

$$\mathbf{I}\mathbf{e} = \lambda \mathbf{e}. \tag{8.145}$$

---

* For a review of eigenvalues and eigenvectors see Leon, *Linear Algebra*, 2d ed., Chapter 6.

APPENDIX B SYMMETRIC MATRICES CAN ALWAYS BE DIAGONALIZED

$(\mathbf{I} - \lambda\mathbf{1})\mathbf{e} = 0$ is a set of homogenous equations, and therefore $\det |\mathbf{I} - \lambda\mathbf{1}| = 0$ ($\mathbf{1}$ is the identity matrix). The eigenvalues are the roots of this determinant, which is a polynomial of nth degree, $P(\lambda)$, in the variable λ. After working out the determinant, we find the explicit form of the *characteristic polynomial* in our example:

$$P(\lambda) = 2 - 8\lambda + 6\lambda^2 - \lambda^3. \tag{8.146}$$

The solution to the equation $P(\lambda) = 0$ may be found by numerical methods. We know in advance that the roots will be real numbers. If, in addition, all of these roots are positive, we say that $\mathbf{I}$ is a *positive definite* matrix. If $\mathbf{I}$ were not positive definite, it could not represent a physical inertia tensor (we will prove this later). The roots of Equation (8.146) are $\lambda_3 = 0.324869$, $\lambda_2 = 1.46081$, $\lambda_1 = 4.21432$ with an accuracy of five decimal places.

Select one of these eigenvalues, say λ_1. We need to find the eigenvector $\mathbf{e}^{(1)}$ corresponding to this eigenvalue. Consider the matrix $\mathbf{I} - \lambda_1\mathbf{1}$. This is

$$\begin{pmatrix} 1-\lambda_1 & 1 & 1 \\ 1 & 2-\lambda_1 & 1 \\ 1 & 1 & 3-\lambda_1 \end{pmatrix}. \tag{8.147}$$

We can then think of (8.145) as three simultaneous equations:

$$\begin{pmatrix} 1-\lambda_1 & 1 & 1 \\ 1 & 2-\lambda_1 & 1 \\ 1 & 1 & 3-\lambda_1 \end{pmatrix} \begin{pmatrix} e_1^{(1)} \\ e_2^{(1)} \\ e_3^{(1)} \end{pmatrix} = \begin{pmatrix} 0 \\ 0 \\ 0 \end{pmatrix}. \tag{8.148}$$

These simultaneous equations for the eigenvector must have a nonzero solution because we know that the determinant of the matrix (8.147) vanishes if λ_1 is an eigenvalue. Using "row reduction" (Gauss–Jordan elimination), multiply a row by the appropriate constant and combine it with the other rows until the matrix is in the "reduced echelon form" below:

$$\begin{pmatrix} 1 & 0 & -.525428 \\ 0 & 1 & -.668892 \\ 0 & 0 & 0 \end{pmatrix} \begin{pmatrix} e_1^{(1)} \\ e_2^{(1)} \\ e_3^{(1)} \end{pmatrix} = \begin{pmatrix} 0 \\ 0 \\ 0 \end{pmatrix}. \tag{8.149}$$

The simultaneous equations are now easy to solve, giving $e_1^{(1)}$, $e_2^{(1)}$ in terms of $e_3^{(1)}$. The final step is to determine the value of $e_3^{(1)}$ by normalizing the eigenvector; we obtain $\mathbf{e}^{(1)} = (0.397113, 0.520657, 0.755789)$ for $\lambda_1 = 4.21432$.

For a real symmetric matrix, the eigenvalues will always be real numbers. Because the eigenvalues are real, the coordinates of the eigenvectors are also real, since the Gauss–Jordan process described above deals only with multiplying and subtracting real numbers. The *reality of eigenvalues* is a crucial property of symmetric matrices shared by the more general Hermitian matrices. *Real eigenvectors* only exist in general for symmetric matrices, however.

The proof that the eigenvalues are real for a symmetric real matrix is best made if we write out the indices that are summed over. Let **e** be an eigenvector of **I** with eigenvalue λ. Then

$$I_{ij}e_j = \lambda e_i. \tag{8.150}$$

For each value of i we have an equation in (8.150) above. Take the vector dot product with $\mathbf{e}^*$. (The complex conjugate of the eigenvector **e**):

$$I_{ij}e_i^* e_j = \lambda e_i^* e_i. \tag{8.151}$$

We do not yet assume that the eigenvector has real components. In Equation (8.151) both i and j are summed over. The complex conjugate of this equation (8.151) is (**I** is real)

$$I_{ij}e_i e_j^* = \lambda^* e_i e_i^*. \tag{8.152}$$

Exchanging the names of the summed indices on the left side, we get

$$I_{ji}e_j e_i^* = \lambda^* e_i e_i^*. \tag{8.153}$$

The symmetry of **I** means that if you subtract Equation (8.153) from (8.151,) the left sides are equal, so will be zero after subtraction. Hence $\lambda = \lambda^*$, and thus the eigenvalue is a real number. By using our row reduction technique sketched above, we then see that the components of the eigenvectors are also real.

The eigenvectors corresponding to *different* eigenvalues must be orthogonal. For a proof, assume two eigenvectors with different eigenvalues exist:

$$I_{ij}e_j = \lambda^{(e)}e_i, \quad I_{ij}f_j = \lambda^{(f)}f_i, \tag{8.154}$$

Multiply the first equation by f_i and sum over i. Multiply the second equation by e_i and sum over I. $e_i I_{ij} f_j = f_i I_{ij} e_j$ because **I** is symmetric. Subtract the two equations you get from (8.154) by this operation to obtain $0 = (\lambda^{(e)} - \lambda^{(f)})e_i f_i$. If the **e** and **f** vectors correspond to two different eigenvalues, then $e_i f_i = 0$, which is to say the eigenvectors are orthogonal.

The last step in the proof is to form a matrix **U**, whose columns are the normalized eigenvectors. In the example

$$\mathbf{U} \equiv \begin{pmatrix} 0.397113 & -0.233192 & -0.88765 \\ 0.520657 & -0.739239 & 0.427132 \\ 0.755789 & 0.631781 & 0.172148 \end{pmatrix}. \tag{8.155}$$

The **U** matrix is an orthogonal matrix. $\tilde{\mathbf{U}}\mathbf{U} = \mathbf{1}$. (Prove this by working it out.). Assuming that $e_i e_i = 1$ from the normalization, and that $e_i e_j = 0$ as proved above, this proves that **U** is a real orthogonal matrix.

Work out $\tilde{\mathbf{U}}\mathbf{I}\mathbf{U}$ explicitly in our example and you will find we get a diagonal matrix:

$$\tilde{\mathbf{U}}\mathbf{I}\mathbf{U} = \mathbf{D}. \tag{8.156}$$

We have proved that a real symmetric matrix $\mathbf{I}$ can be diagonalized by a real orthogonal matrix $\mathbf{U}$. This diagonalizing matrix has columns that are the normalized eigenvectors of $\mathbf{I}$. The diagonal matrix $\mathbf{D}$ is just a list of the corresponding eigenvalues along the diagonal, with zeros everywhere else.

If the real symmetric matrix is the 3×3 inertia tensor $\mathbf{I}$, the eigenvectors are the direction cosines of the principal axes, while the eigenvalues are the principal moments I_1, I_2, I_3. The principal axes are tied to the body, so they define a special kind of body frame. For the inertia tensor, the fact that kinetic energy is always positive implies that the principal moments are themselves positive, since $T = \frac{L^2}{2I_i}$ if $\vec{L}$ is directed along the ith principal axis.

If at least two of the eigenvalues of the symmetric tensor are equal, we say there is a *degeneracy*. The proof that eigenvectors are orthogonal breaks down. In our example, we would have discovered that the equations would not have allowed us to express the eigenvector in terms of a single constant $e_3^{(1)}$, but instead it would have been a function of two numbers $e_2^{(1)}, e_3^{(1)}$. We would have to pick two directions in this 2-D subspace arbitrarily. Since any two orthogonal directions in this space would serve as eigenvectors, the choice is not unique.

APPENDIX C

UNDERSTANDING THE EARTH'S EQUATORIAL BULGE*

Isaac Newton made an ingenious calculation predicting an equatorial bulge for the Earth of

$$\epsilon \equiv \frac{R_{\text{equator}} - R_{\text{pole}}}{R_{\text{average}}} = \frac{1}{230}. \tag{8.157}$$

The actual value of $\epsilon = \frac{1}{294}$ corresponds to a difference of about $R_{\text{equator}} - R_{\text{pole}} = 21.7$ km.

At the time, Newton's theoretical prediction was hotly disputed by a leading astronomer, Cassini, who claimed that the Earth bulges at the poles instead (see Figure 8.26). Cassini based this on the best astronomical observations available at the time. The controversy raged for 60 years. It was settled in 1738 in favor of Newton's theory by an expedition to Lapland by Maupertius, who measured the variation in g close to the North Pole.

---

* We follow the works of Newton and Maclaurin as described by S. Chandrasekhar in his book *Ellipsoidal Figures of Equilibrium*, Dover Press, 1987.

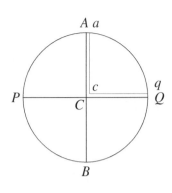
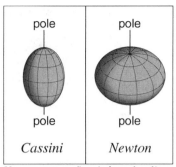

Vous avez confirmé dans les lieux pleins d'ennui. Ce que Newton connut sans sortir de chez lui.

A   B

FIGURE 8.26
A) Illustration from the *Principia* bearing on Newton's arguments for the rotational flattening of the Earth. B) An old time caricature of the controversy between the opposing schools of Newton and Cassini with respect to the figure of the Earth. The quote is by Voltaire. A translation: "you have confirmed, in the desolate wastes, what Newton knew without leaving his place."

Newton suggested the following: Imagine two holes drilled to the center of the Earth, one from the North Pole and one from the equator. Fill both holes with water. The pressure at the center of the Earth must be the same. Since the decrease in g is linear for a spherical Earth, to a good approximation this pressure is $\frac{1}{2} g_{\text{average}} \rho R$ in both cases. (Here g_{average} refers to the radial average of the "effective g," corrected for the centrifugal effect.) Setting the pressures equal we obtain

$$g_{\text{pole}} R_{\text{pole}} = g_{\text{equator}} R_{\text{equator}} \left(1 - \frac{\omega^2 R}{g}\right). \tag{8.158}$$

(We've ignored the variation in R and g for the small centrifugal term.) From the definition of ϵ (8.157): $R_{\text{pole}} = (1 - \frac{\epsilon}{2})R$; $R_{\text{equator}} = (1 + \frac{\epsilon}{2})R$, with R being the average radius. If ϵ is positive, the Earth is an oblate ellipsoid. Solve Equation (8.158) for ϵ, ignoring higher-order terms. You should obtain the equation

$$\epsilon = m + \left(\frac{g_{\text{pole}}}{g_{\text{equator}}} - 1\right) \tag{8.159}$$

with the definition

$$m \equiv \frac{\omega^2 R}{g} = 3.44 \times 10^{-3}. \tag{8.160}$$

Since Newton knew both the radius of the Earth and the value of g, he was able to compute m. We also can make a first estimate of $\epsilon \approx m$. It is purely an accident that this agrees

APPENDIX C UNDERSTANDING THE EARTH'S EQUATORIAL BULGE

better with the experimental value than Newton's prediction (8.157). The discrepancy is due to the fact that the center of the Earth is considerably more dense than the mantle.

To make a second-order estimate we need to determine the dependence of g on ϵ. The centrifugal effect m causes ϵ to be nonzero, which in turn causes $g_{\text{pole}} \neq g_{\text{equator}}$, Newton showed (see below) that

$$\frac{g_{\text{pole}}}{g_{\text{equator}}} = 1 + \frac{\epsilon}{5}, \tag{8.161}$$

which means that

$$\epsilon = \frac{5}{4}m = 4.30 \times 10^{-3}. \tag{8.162}$$

The correct answer is $\epsilon = 3.40 \times 10^{-3}$.

As we said above, we can regard the bulge as being caused by the rotation of the Earth. The gravitational force tends to amplify this effect by producing $\frac{g_{\text{pole}}}{g_{\text{equator}}} > 1$, an effect which is itself proportional to ϵ.

We will next prove (8.161). The attractive force of gravity is the negative gradient of the potential V evaluated at either the pole or at the equator. For each volume element $dx\,dy\,dz$ in the Earth, the contribution to V is

$$dV = G\rho \frac{dx\,dy\,dz}{d}, \tag{8.163}$$

where G is the gravitational constant, ρ is the density of the Earth, and d is the distance between the point x, y, z inside the Earth and the observer on the surface (either at the pole or the equator). If the Earth is considered to be an ellipsoid, the equation of the surface will be

$$\frac{x^2 + y^2}{b^2} + \frac{z^2}{a^2} = 1. \tag{8.164}$$

The Z axis passes through the North Pole. The semiminor axis in the direction of the Z axis is a, in the direction of the equator b. It is useful to adopt scaled dimensionless coordinates: $z = a\zeta$, $y = b\eta$, $x = b\xi$. Then, integrating (8.163), we obtain for an observer at the North Pole:

$$V = Mg_{\text{pole}} = G\rho \frac{b^2}{a} \int_{-1}^{1} d\xi \int_{-\sqrt{1-\xi^2}}^{\sqrt{1-\xi^2}} d\eta\, (I_+ - I_-), \tag{8.165}$$

where

$$I_\pm \equiv \frac{1}{\sqrt{(1 \pm \sqrt{1 - \xi^2 - \eta^2})^2 + \frac{b^2}{a^2}(\xi^2 + \eta^2)}}. \tag{8.166}$$

In principle, we could evaluate the integral exactly for any values of a, b, but the result is complicated and not particularly useful. (The exact result was first obtained by Maclaurin in 1742.) Expanding the double integral above in a power series in ϵ gives two integrals, which can be evaluated using a trigonometric substitution: $\xi^2 + \eta^2 \to r^2$, $r \to \sin 2u$, $0 \leq u \leq \frac{\pi}{4}$, to give

$$M g_{\text{pole}} = \frac{4\pi}{3} G\rho R \left(1 + \frac{3}{10}\epsilon\right). \tag{8.167}$$

If $\epsilon > 0$, we have an oblate ellipsoid, because the attraction of gravity is then a little stronger at the North Pole than would be true for a perfectly spherical Earth of the same density and average radius.

For g_{equator} there is less symmetry, but again we can write the answer as a double integral, and then expand in a power series in ϵ to get

$$M g_{\text{equator}} = G\rho a \int_{-1}^{1} d\zeta \int_{-\sqrt{1-\zeta^2}}^{\sqrt{1-\zeta^2}} d\eta \, (J_+ - J_-) \tag{8.168}$$

with

$$J_\pm \equiv \frac{1}{\sqrt{(1 \pm \sqrt{1 - \zeta^2 - \eta^2})^2 + \eta^2 + \frac{a^2}{b^2}\zeta^2}}. \tag{8.169}$$

The result is

$$M g_{\text{equator}} = \frac{4\pi}{3} G\rho R \left(1 + \frac{\epsilon}{10}\right). \tag{8.170}$$

The attraction of gravity is stronger at the North Pole than at the equator. If we take the ratio $g_{\text{pole}}/g_{\text{equator}}$ and again expand in power series in ϵ, we obtain Newton's result (8.161).

CHAPTER NINE

THE THEORY OF SMALL VIBRATIONS

OVERVIEW OF CHAPTER 9

Consider a mechanical system that has N degrees of freedom. Assume also that the system is close to one of its stable equilibrium points. We will show that this system acts like N independent SHOs, usually with N different frequencies. One or more of these independent oscillations can be present depending on the initial conditions. In a state where only a single oscillation frequency is excited,* the N different degrees of freedom move synchronously at a common *mode frequency*. The ratios between the different displacements for each degree of freedom, known as the *mode displacement ratios*, are an intrinsic characteristic of the normal mode that is oscillating. The amplitude of any particular mode is known as the *normal coordinate*. Each normal coordinate oscillates in time like a single SHO. All possible movements of the system, for sufficiently small displacements from the equilibrium point, can be described as a linear combination of modes.

Why do we concentrate on "small" vibrations for such a system? By definition, if the differential equations of motion are linear, the system is then said to be a *linear system*. Taylor's theorem guarantees that most systems are linear if the displacements are small enough. The motion can then be approximately described by a set of linear differential equations very similar to the equation for a simple harmonic oscillator. We allow an arbitrary but finite number of degrees of freedom. Applications of the theory described in this chapter occur in a wide variety of physical systems, not only in mechanics.

As a concrete example, consider a set of N point masses spaced at equal distances apart, which are constrained to move vertically and are all connected by a horizontal massless string under tension as shown in Figure 9.1. If we observe the vertical motion of this system for small displacements of these masses from their equilibrium points, there is the possibility of having up to N different mode frequencies present. (It is a theorem that the number of possible modes must equal the number of degrees of freedom. If two or more frequencies are identical, it is called *degeneracy*.) For

---

* This is called a *normal mode* or sometimes just a "mode" for short.

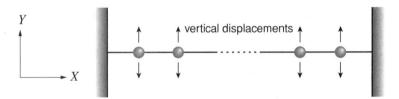

FIGURE 9.1
N point masses connected by a string under tension.

example, if we plucked the string into an arbitrary shape at $t = 0$, the vibration would be very complicated, containing all of the mode frequencies. On the other hand, we could also displace the individual masses by just the right ratios such that only one mode frequency is excited.

Each normal mode is a vibration of all parts of the system at the same frequency. All displacements remain in fixed ratios, with the transverse displacements of each point mass oscillating sinusoidally in synchronism, like a chorus line. (But each leg (degree of freedom) in the chorus line can be in a different position.) For linear systems only, the different vibrational modes are entirely independent of each other. This independence means that the system can have a possible motion that is the sum of the contribution of two or more modes, without any effect of one mode upon another. This is known as *superposition*. The separate modes will vibrate at their individual frequencies indefinitely if there is neither damping nor a nonlinearity present. The importance of superposition was first appreciated by the Bernoullis. The physics of superposition was understood long before it was fully accepted by the mathematicians.

Not only can different frequencies of small oscillations appear at the same time, but *any possible motion* can be described as a linear superposition of the normal modes, at least for sufficiently small amplitudes. The set of modes therefore forms a "complete set."

We will prove all of these results for any system containing a finite number of coupled linear oscillators. These results can also be generalized to linear systems with infinitely many degrees of freedom, such as a vibrating string with a continuously distributed mass, or the theory of the classical electromagnetic field in a resonant cavity.

9.1 TWO COUPLED PENDULUMS

Before developing the general theory of small vibrations, we will discuss a particular case with two degrees of freedom. Two identical pendulums of length l and mass m are suspended side by side as shown in Figure 9.2. They are coupled together with a spring which has the spring constant k. The spring is connected halfway up the pendulums. Assuming a massless spring, the kinetic energy is

$$T = \frac{ml^2}{2}(\dot{\theta}_1^2 + \dot{\theta}_2^2). \tag{9.1}$$

9.1 TWO COUPLED PENDULUMS

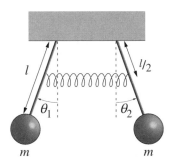

FIGURE 9.2
$\theta_1 < 0$ as shown here.

If there were no coupling, the potential energy for small displacements from the equilibrium at $\theta_1 = \theta_2 = 0$ would be

$$V = \frac{mgl}{2}(\theta_1^2 + \theta_2^2). \tag{9.2}$$

(Where the constant term has been dropped). We would then have two independently oscillating systems, each with a frequency

$$\omega_0 = \sqrt{\frac{g}{l}}. \tag{9.3}$$

Since we have coupling, we can exchange energy between the two systems. If we stretch the spring, there will be an additional potential energy that depends on the relative displacement $\theta_2 - \theta_1$:

$$V_{\text{coupling}} = \frac{k}{2}\left(\frac{l}{2}\right)^2 (\theta_2 - \theta_1)^2. \tag{9.4}$$

After adding this coupling term to the uncoupled potential energy (9.2), standard Lagrangian methods are used to obtain the equations of motion:

$$\begin{aligned}\ddot{\theta}_1 + \frac{g}{l}\theta_1 + \frac{k}{4m}(\theta_1 - \theta_2) &= 0, \\ \ddot{\theta}_2 + \frac{g}{l}\theta_2 + \frac{k}{4m}(\theta_2 - \theta_1) &= 0.\end{aligned} \tag{9.5}$$

Equations (9.5) are a set of coupled second-order linear differential equations. To make these equations simpler, define the dimensionless "coupling parameter" η:

$$\eta \equiv \frac{kl}{4mg}. \tag{9.6}$$

The parameter $\eta = 0$ if there is no coupling. Rewrite the equations of motion (9.5) before solving them:

$$\begin{aligned}\ddot{\theta}_1 + \omega_0^2(1+\eta)\theta_1 - \omega_0^2 \eta \theta_2 &= 0, \\ \ddot{\theta}_2 + \omega_0^2(1+\eta)\theta_2 - \omega_0^2 \eta \theta_1 &= 0.\end{aligned} \tag{9.7}$$

Add and subtract these two equations to obtain two new equations which are not coupled. These can be solved exactly as was done for the SHO. The "symmetric mode," called mode 1, with $\theta_1 = \theta_2$ has the same frequency $\omega_1 = \omega_0$ as the uncoupled oscillators. The spring is not stretched in this mode during the oscillation. In contrast, the "antisymmetric mode," called mode 2, has a higher frequency, which is due to an additional restoring force from the stretched spring. Here $\theta_1 = -\theta_2$ and $\omega_2 > \omega_0$. Intuition could have been used to guess the two modes of vibration and that the antisymmetric mode has a higher frequency than the symmetric mode.

Such a simple case, with this high degree of symmetry, is not always encountered. It is important to learn a general approach. Matrix techniques have to be used in most cases if the number of degrees of freedom is greater than two. We will start by introducing a matrix notation for the problem above. Define a column vector $\binom{\theta_1}{\theta_2}$ and write the equations of motion (9.7) in matrix form

$$\frac{d^2}{dt^2}\binom{\theta_1}{\theta_2} + \omega_0^2 \begin{pmatrix} 1+\eta & -\eta \\ -\eta & 1+\eta \end{pmatrix} \cdot \binom{\theta_1}{\theta_2} = \binom{0}{0}. \tag{9.8}$$

Assume there exists a mode that obeys (9.8). We don't yet know the mode frequency ω. Then, by the definition of a mode, since there is only a single frequency present we can assume a solution of the form

$$\binom{\theta_1(t)}{\theta_2(t)} = \binom{\Theta_1}{\Theta_2} e^{i\omega t}. \tag{9.9}$$

Here Θ_1 and Θ_2 are the mode displacement ratios. Insert this form back into the equations of motion (9.8). The differential equation of the motion has been converted into a set of simultaneous linear equations. To simplify the algebra, define

$$\lambda \equiv \frac{\omega^2}{\omega_0^2}. \tag{9.10}$$

In matrix form, the equations derived by substituting (9.9) into (9.8) become two linear equations for the two constants Θ_1, Θ_2:

$$\begin{pmatrix} -\lambda + 1 + \eta & -\eta \\ -\eta & -\lambda + 1 + \eta \end{pmatrix} \cdot \binom{\Theta_1}{\Theta_2} = \binom{0}{0}. \tag{9.11}$$

An important feature of the two equations implied by (9.11) is that they are homogeneous equations, (i.e., the right side is zero). It is a theorem of linear algebra that for N homogeneous linear equations in N unknowns, the solutions must be identically zero ($\Theta_1 = \Theta_2 = 0$) unless the determinant of the matrix on the left side vanishes. The determinant of the matrix in Equation (9.11) is a quadratic function of λ. It vanishes only for two specific values of λ. These roots of the determinant are proportional to the squares of the mode frequencies. By solving for those values of λ that make the determinant vanish, we will determine the mode frequencies:

$$\text{determinant} \begin{vmatrix} -\lambda + 1 + \eta & -\eta \\ -\eta & -\lambda + 1 + \eta \end{vmatrix} = (1 + \eta - \lambda)^2 - \eta^2 = 0. \tag{9.12}$$

9.1 TWO COUPLED PENDULUMS

This gives a quadratic equation for λ. (If we have N degrees of freedom, we have to find the roots of a polynomial of order N, so there are N frequencies.) In the case we have here, the roots are

$$\lambda_1 = 1, \quad \lambda_2 = 1 + 2\eta. \tag{9.13}$$

The corresponding mode frequencies are (from (9.10))

$$\omega_1 = \omega_0, \quad \omega_2 = \omega_0\sqrt{1 + 2\eta}. \tag{9.14}$$

The mode displacement ratios for the two pendulums can be determined by first noticing that the two equations written in matrix form (9.11) are no longer independent, since the determinant vanishes. This means that you can solve one of them, say the second one, for the ratio of amplitudes:

$$\frac{\Theta_1}{\Theta_2} = \frac{1 + \eta - \lambda}{\eta}. \tag{9.15}$$

Adding subscripts (1) and (2) to label the modes we get

$$\frac{\Theta_{(1),1}}{\Theta_{(1),2}} = 1 \quad \text{(symmetric mode)}, \tag{9.16}$$

$$\frac{\Theta_{(2),1}}{\Theta_{(2),2}} = -1 \quad \text{(antisymmetric mode)}. \tag{9.17}$$

It is important to realize that a given mode is specified only by the *ratio* of amplitudes for the individual oscillators. An overall multiplicative constant is set by the initial conditions. If the number of degrees of freedom is $N > 2$, solving for the displacement ratios corresponding to a definite mode can be done by the method of cofactors to be described in Appendix A. You can also use "row elimination" to solve the N homogeneous equations analogous to (9.11) above.

QUESTION 1: *Normal Modes and Coordinates* First of all, explain the physical meaning of the terms: "normal mode," "mode displacement ratio," and "normal coordinates." Generalize to N degrees of freedom. How many ratios do you need then? Find or invent an example not given in the text and illustrate the use of these terms for that example. Secondly, describe the motion of the coupled pendulums for the symmetric and antisymmetric modes.

Observe that we started with two identical pendulums. If uncoupled, they would have identical frequencies of oscillation. When coupled together, the system as a whole still has two frequencies of oscillation, but now the frequencies are different. The energy of oscillation is distributed between the oscillators for each of the two modes.

We can have motion in which both modes are simultaneously present. Consider the two coupled pendulums of equal mass. Start the pendulums off at $t = 0$ by displacing only one of them, say θ_1 away from zero. This means that the initial condition is a sum of symmetric and antisymmetric modes. These evolve in time independently* and with different oscillation frequencies. The relative phase between the two modes will be π after a time $\frac{\pi}{\Delta\omega}$, where $\Delta\omega = \omega_2 - \omega_1$ is the beat frequency. At that point, for just an instant, the first pendulum stops swinging, and all of the energy has been transferred to the second pendulum, which was initially at rest. As time goes on, the energy will be transferred back and forth at half the beat frequency between the two modes. Similar beats can be observed in more complex systems if more than one mode is excited initially.

QUESTION 2: *Two Coupled Pendulums 1* Why is the relative phase between the symmetric and antisymmetric modes π after a time $\frac{\pi}{\Delta\omega}$? $\Delta\omega \equiv \omega_2 - \omega_1$.

QUESTION 3: *Two Coupled Pendulums 2* Start the two pendulums as follows at $t = 0$: $\theta_1 = A, \dot{\theta}_1 = 0, \theta_2 = 0, \dot{\theta}_2 = 0$. What are the normal coordinates for this case? What are the equations for $\theta_1(t), \theta_2(t)$? Describe the motion of each pendulum as time goes on.

Small Oscillation Theory

The next few sections will guide the reader through the general theory of small oscillations by considering the double pendulum, which consists of one pendulum swinging from the bottom of another. We will do it in a way that can be very easily generalized.

9.2 EXACT LAGRANGIAN FOR THE DOUBLE PENDULUM

The double pendulum system shown in Figure 9.3 will also be used in our later discussion of chaos. For this reason we begin by finding the exact Lagrangian without any small-angle approximation. This exact Lagrangian, which has two degrees of freedom, ϕ_1 and ϕ_2, is

$$T = \frac{1}{2}m(\dot{x}_1^2 + \dot{y}_1^2 + \dot{x}_2^2 + \dot{y}_2^2),$$
$$V = mg(y_1 + y_2), \tag{9.18}$$
$$L = T - V.$$

We can express the Lagrangian in terms of ϕ_1, ϕ_2 and the time derivatives of these angles, using the geometric relations

$$x_1 = l\sin\phi_1, \quad y_1 = -l\cos\phi_1,$$
$$x_2 = x_1 + l\sin(\phi_1 + \phi_2), \quad y_2 = y_1 - l\cos(\phi_1 + \phi_2). \tag{9.19}$$

---

* This is due to the linear property of the equations of motion.

9.2 EXACT LAGRANGIAN FOR THE DOUBLE PENDULUM

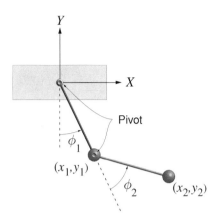

FIGURE 9.3
Origin is at top pivot. Masses are both equal to m; lengths are both equal to l.

See for yourself, taking time derivatives and substituting into the formulas above, that this gives the desired formula in terms of the angles:

$$T = \frac{1}{2}ml^2\left(2\dot{\phi}_1^2 + 2\dot{\phi}_1(\dot{\phi}_1 + \dot{\phi}_2)(\cos\phi_2) + (\dot{\phi}_1 + \dot{\phi}_2)^2\right),$$
$$V = -mgl(2\cos\phi_1 + \cos(\phi_1 + \phi_2)). \tag{9.20}$$

Notice that T is a quadratic form in the time derivatives of the angles. This means that $H = E$. Because there is no explicit time dependence, the total energy E is a constant of the motion.

Small-Angle Approximation

To apply the theory of small vibrations, we need to find the linear approximation to the equations of motion. This is called the "small-angle approximation."

We assume that ϕ_1 and ϕ_2 are sufficiently small displacements from the equilibrium at $\phi_1 = \phi_2 = 0$. We can then approximate both T and V each by a double Taylor series *up to and including quadratic terms in the angles and their time derivatives.* We obtain the approximations below:

$$T \approx ml^2\left(\dot{\phi}_1^2 + \dot{\phi}_1(\dot{\phi}_1 + \dot{\phi}_2) + \frac{1}{2}(\dot{\phi}_1 + \dot{\phi}_2)^2\right),$$
$$V \approx mgl\left(\phi_1^2 + \frac{1}{2}(\phi_1 + \phi_2)^2\right). \tag{9.21}$$

(An unimportant constant in V has been dropped.) Note that there are no linear terms in V; this proves that $\phi_1 = \phi_2 = 0$ is a true equilibrium point. (Why?)

Use Matrix Notation for Complete Generality

At this point it is useful to introduce matrix notation. Use of this notation allows us to write down formulas that are true for any number of degrees of freedom and to prove general theorems about the mode frequencies and mode displacement ratios when we find them. Define the column vector containing the generalized coordinates* by

$$\phi(t) \equiv \begin{pmatrix} \phi_1(t) \\ \phi_2(t) \end{pmatrix}, \quad \tilde{\phi} \equiv (\phi_1, \phi_2) = \text{transpose of } \phi. \tag{9.22}$$

The state of the system is known if all of the displacements ($\phi(t)$s) are known, so ϕ is called the "state vector" of the system, which is a term borrowed from quantum mechanics. In the case of the potential energy we then write

$$V = \tilde{\phi} \cdot \mathbf{v} \cdot \phi = (\phi_1, \phi_2) \cdot \mathbf{v} \cdot \begin{pmatrix} \phi_1 \\ \phi_2 \end{pmatrix} = \sum_{i,j} \phi_i v_{ij} \phi_j. \tag{9.23}$$

The formulas in (9.23) above give three equivalent ways of writing the same expression. The matrix form on the left avoids writing unnecessary subscripts. The symmetric matrix $\mathbf{v}$ can be shown to be

$$\boxed{v_{ij} \equiv \frac{1}{2} \frac{\partial^2 V}{\partial \phi_i \partial \phi_j}\bigg|_{\phi_i, \phi_j = 0}.} \tag{9.24}$$

We have taken this last formula as the definition of the matrix $\mathbf{v}$.

QUESTION 4: Small-Angle Approximation 1 Explain why Equations (9.23, 9.24) represent the Taylor series expansion for the potential energy (9.21). What about the zeroth- and first-order terms? Why do we ignore these terms? What approximation has been made?

For the double pendulum we consider here:

$$\mathbf{v} = \frac{mgl}{2} \begin{pmatrix} 3 & 1 \\ 1 & 1 \end{pmatrix}. \tag{9.25}$$

We intuitively realize that the equilibrium at $\phi_1 = \phi_2 = 0$ is a stable one. To prove that the equilibrium is stable, any deviation from this position must result in an increase in the potential energy. This can be expressed mathematically by the condition

$$\tilde{\phi} \cdot \mathbf{v} \cdot \phi > 0 \tag{9.26}$$

for *any* state vector ϕ in the $N = 2$ degree of freedom configuration space. (Configuration space is not real space but the N-dimensional space of small displacements of generalized

---

* A boldface letter will denote matrix representation just as in Chapters 7 and 8.

9.2 EXACT LAGRANGIAN FOR THE DOUBLE PENDULUM

coordinates.) The mathematicians call any matrix that obeys the condition (9.26) a *positive definite* matrix. To prove that **v** is really a positive definite matrix, the eigenvalues of **v** must all be positive. If they are both positive, **v** is positive definite. In our case the eigenvalues of **v** are positive constants $(2 \pm \sqrt{2})$, so the equilibrium is a stable one, and the potential energy increases for any and all deviations from equilibrium.

Define a matrix of second partial derivatives for the kinetic energy in an analogous way:

$$T = \tilde{\dot{\phi}} \cdot \mathbf{t} \cdot \dot{\phi}, \tag{9.27}$$

$$t_{ij} \equiv \frac{1}{2} \frac{\partial^2 T}{\partial \dot{\phi}_i \partial \dot{\phi}_j} \bigg|_{\dot{\phi}_i, \dot{\phi}_j = 0}. \tag{9.28}$$

QUESTION 5: *Small-Angle Approximation II* Why do positive eigenvalues prove a matrix is positive definite? Why is the left side of Equation (9.26) called a "quadratic form"? (Multiply it out first.) What are the approximations made in deriving Equation (9.27)?

For this example, **t** is

$$\mathbf{t} = \frac{ml^2}{2} \begin{pmatrix} 5 & 2 \\ 2 & 1 \end{pmatrix}. \tag{9.29}$$

The matrix **t** will always be symmetric. Why? The kinetic energy is always positive, so **t** is a positive definite quadratic form in the time derivatives of the angles.

The approximate Lagrangian for small angles can be written in a general matrix notation without explicit indices:

$$L = \tilde{\dot{\phi}} \cdot \mathbf{t} \cdot \dot{\phi} - \tilde{\phi} \cdot \mathbf{v} \cdot \phi. \tag{9.30}$$

Equations of Motion

From the Lagrangian, it is a short step to the equations of motion for the system. The number of equations of motion equals the number of degrees of freedom (two in this case):

$$\frac{\partial L}{\partial \dot{\phi}_i} = 2 \sum_{j=1}^{N} t_{ij} \dot{\phi}_j = 2(\mathbf{t} \cdot \dot{\phi})_i, \quad \frac{\partial L}{\partial \phi_i} = -2 \sum_{j=1}^{N} v_{ij} \phi_j = -2(\mathbf{v} \cdot \phi)_i. \tag{9.31}$$

The equations of motion are

$$t_{11}\ddot{\phi}_1 + t_{12}\ddot{\phi}_2 + v_{11}\phi_1 + v_{12}\phi_2 = 0,$$
$$t_{21}\ddot{\phi}_1 + t_{22}\ddot{\phi}_2 + v_{21}\phi_1 + v_{22}\phi_2 = 0. \tag{9.32}$$

In more compact matrix form these can be written as

$$\mathbf{t} \cdot \ddot{\boldsymbol{\phi}} + \mathbf{v} \cdot \boldsymbol{\phi} = 0. \tag{9.33}$$

Go through the derivation of EOM from the Lagrangian, using (9.21) instead of the matrix notation. Compare to (9.32, 9.33). Try to understand the notation in (9.33) by this comparison.

9.3 SINGLE FREQUENCY SOLUTIONS TO EQUATIONS OF MOTION

Normal Modes

Normal modes are defined as the single frequency vibrations of the entire system. For a certain mode we can write this single frequency condition as

$$\check{\boldsymbol{\phi}}(t) = \boldsymbol{\Phi} e^{i(\omega t - \delta)}. \tag{9.34}$$

$\check{\boldsymbol{\phi}}(t)$ is now a complex N-dimensional column vector,[*] and δ is the common phase of the system determined by the initial conditions. $\boldsymbol{\Phi}$ is the *mode vector* containing the mode displacement ratios. This is a constant vector whose specific values are also determined by the initial conditions. The real part of Equation (9.34) is the physical solution. This is similar to what we did in Chapter 3 (see (3.30)). Equation (9.34) states that each part of the system vibrates with a single common frequency ω.

Substitute the form (9.34) into the EOM (9.33) above to obtain the homogeneous linear equations:

$$[-\omega^2 \mathbf{t} + \mathbf{v}] \cdot \boldsymbol{\Phi} = 0. \tag{9.35}$$

This formula stands for N equations in the N unknown constants $\boldsymbol{\Phi}$. We repeat: *It is a well-known result of linear algebra that N homogeneous equations in N unknowns have no solution other than* $\boldsymbol{\Phi} = 0$ *(a column vector of N zeros)* unless *the determinant of the $N \times N$ coefficient matrix vanishes.* To find a mode we must require that

$$\boxed{\text{determinant}\, |-\omega^2 \mathbf{t} + \mathbf{v}| = 0.} \tag{9.36}$$

The determinant is a polynomial of order N in the variable ω^2. The positive real roots of this polynomial will determine the squares of the mode frequencies. The square root of

---

[*] The notation $\check{\boldsymbol{\phi}}(t)$ is used here to indicate the complex state vector.

9.3 SINGLE FREQUENCY SOLUTIONS TO EQUATIONS OF MOTION

each of these values is the mode frequency, which will be denoted by $\omega_{(i)}$. There should be N modes for N degrees of freedom. This means that the polynomial defined by (9.36) will have *only* positive real roots and not any negative or complex roots. If any two or more mode frequencies are equal, there is a *degeneracy* of modes. We will simplify the theoretical treatment here by assuming that no degeneracy exists.

Notice that solving (9.35) is similar to solving an eigenvalue equation as we did in Chapter 8 when solving for the principal axes. The difference is that here we are finding where the determinant of $|\mathbf{v} - \omega^2 \mathbf{t}| = 0$ instead of where the determinant of $|\mathbf{v} - \lambda \mathbf{1}| = 0$. This is a subtle difference in that $\mathbf{t}$ is definitely not the identity matrix $\mathbf{1}$. However, as we will show in Section 9.6, we can generalize the procedure of finding eigenvalues and eigenvectors to apply it to this situation. Therefore, we can refer to the mode frequencies as generalized eigenvalues and the mode vectors as generalized eigenvectors.

The procedure above will be demonstrated with the double pendulum example. We have

$$-\omega^2 \mathbf{t} + \mathbf{v} = \frac{mgl}{2} \begin{pmatrix} 3 - 5\lambda & 1 - 2\lambda \\ 1 - 2\lambda & 1 - \lambda \end{pmatrix}, \tag{9.37}$$

where $\lambda \equiv \frac{\omega^2 l}{g}$. The squares of the mode frequencies are determined by solving for the values of λ that make the determinant of (9.37) vanish. The determinant of the matrix in (9.37) is proportional to $\lambda^2 - 4\lambda + 2$. The roots are

$$\boxed{\text{mode 1: } \omega_{(1)}^2 = (2 - \sqrt{2})\frac{g}{l},} \tag{9.38}$$

$$\boxed{\text{mode 2: } \omega_{(2)}^2 = (2 + \sqrt{2})\frac{g}{l}.} \tag{9.39}$$

The second mode has a much higher frequency than the first one. The overall frequency scale is set by $\sqrt{\frac{g}{l}}$ as might be expected.

To make the notation clearer, when we want to refer to a particular mode, we will use subscripts enclosed in parentheses, such as (i). This avoids confusion between mode indices and coordinate indices, which will not have parentheses.

How do we know ω^2 will always turn out to be real and positive? A quick intuitive answer (not a proof however) is that if ω^2 is complex, the exponent in $e^{i\omega t}$ will contain a real part. This gives an exponentially growing or decreasing mode amplitude, which would imply that the total energy would not be constant. A rigorous proof is given in Section 9.6.

How to Find the Mode Amplitude Ratios

Now that we have a way to find the mode frequencies, how do we find the "shape" of each mode? In other words, what are the displacements of the different degrees of freedom that are characteristic of this particular mode? The individual displacement ratios for the ith normal mode can be found by solving the set of homogeneous linear equations (9.35) for a particular mode frequency $\omega^2 = \omega_{(i)}^2$:

$$\left[-\omega_{(i)}^2 \mathbf{t} + \mathbf{v}\right] \cdot \mathbf{\Phi}_{(i)} = 0, \quad i = 1, \ldots N. \tag{9.40}$$

Remember that $\mathbf{\Phi}_{(i)}$ is the set of displacement amplitudes for each part of the system corresponding to a single mode. For the double pendulum, $\tilde{\mathbf{\Phi}}_{(i)} = (\Phi_{(i),1}, \Phi_{(i),2})$. For the N degree of freedom system, there will be N values for each $\mathbf{\Phi}_{(i)}$ of the ith mode, $(i = 1, \ldots N) - N^2$ numbers in all. Since the equations are homogeneous, we can always multiply each of the different $\mathbf{\Phi}_{(i)}$s by a different normalization constant if we wish. From (9.40), the solution for the mode vectors of the double pendulum are obtained by solving one of the homogeneous equations for the ratios of the coordinates:

$$\mathbf{\Phi}_{(i)} = C_{(i)}\left(1 - \lambda_{(i)}, 2\lambda_{(i)} - 1\right), \quad \lambda_{(i)} \equiv \frac{\omega_{(i)}^2 l}{g}. \tag{9.41}$$

($C_{(i)}$ is a multiplicative constant, which can be different for each mode.) Next insert the two possibilities for $\omega_{(i)}^2$ (9.38, 9.39) into Equation (9.41) to obtain the two sets of mode vector coordinates. Choosing the constants to be $\frac{1}{2\lambda_{(i)}-1}$ in order to simplify the coordinates of each mode, we obtain, after some algebra,

$$\begin{aligned}\mathbf{\Phi}_{(1)} &= (1 + \sqrt{2}, 1) \quad \text{low frequency mode,} \\ \mathbf{\Phi}_{(2)} &= (1 - \sqrt{2}, 1) \quad \text{high frequency mode.}\end{aligned} \tag{9.42}$$

The two double pendulum modes are sketched in Figure 9.4. The low frequency mode has $\frac{\phi_1}{\phi_2} > 0$. The high frequency mode has $\frac{\phi_1}{\phi_2} < 0$. Thus the angular displacements have opposite signs. You can see that the $\mathbf{\Phi}_{(i)}$ are real numbers. This must be true if the

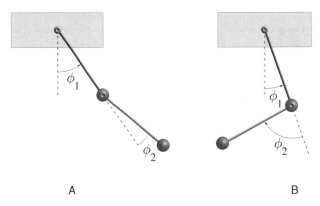

A   B

FIGURE 9.4
A) Low frequency mode: $\frac{\phi_1}{\phi_2} \simeq 2.41$, $\omega \simeq .76\sqrt{\frac{g}{l}}$. B) High frequency mode: $\frac{\phi_1}{\phi_2} \simeq -0.41$, $\omega \simeq 1.85\sqrt{\frac{g}{l}}$.

answer is to make physical sense. (Why?) It follows from the mathematics because only real numbers are involved in solving the homogeneous equations defining the mode ratios.

In the general case we can obtain the mode displacement ratio proportional to the ith mode by taking the cofactors of the top row of the matrix $-\omega_{(i)}^2 \mathbf{t} + \mathbf{v}$. How to do this is explained in more detail in Appendix A. We could instead find the mode displacement ratio by using the technique of row elimination* on the matrix.

> **QUESTION 6: *Double Pendulum*** Describe the motion of the two modes of the double pendulum. Why do you expect two modes? Could there ever be less than two modes? How many modes do you expect for N degrees of freedom? Why?

9.4 SUPERIMPOSING DIFFERENT MODES; COMPLEX MODE AMPLITUDES

To specify an arbitrary motion of the system completely, we must specify N displacements and N velocities at $t = 0$ as initial conditions. There will be N modes, where each mode corresponds to a definite configuration of the different degrees of freedom in the system. The amplitude of each mode can be a complex number, exactly as for the harmonic oscillator in Chapter 3. There is an amplitude and a phase, in all $2N$ numbers, just matching the $2N$ initial conditions to be specified. Since there are as many modes as degrees of freedom, we can express *any* configuration of the system as a superposition of modes with definite complex amplitudes for each mode. Since each possible mode oscillates at a single frequency, a frequency analysis of the motion can reveal how many different modes are present. The N complex amplitudes of the modes comprise a kind of coordinate set. The only time variation in each of these amplitudes is in the exponential factor $e^{i\omega_{(i)}t}$.

We need a systematic way of finding each separate amplitude from the initial conditions for a system, assuming we have already determined the frequencies and the modes themselves. Our method will make use of the fact that not only are the modes generalized eigenvectors of (9.35), but they define a coordinate transformation from "state vectors" to "mode vectors" in which *both* $\mathbf{t}$ and $\mathbf{v}$ are diagonalized. That this could be done is not at all obvious. We will prove later (in Section 9.6) that $\mathbf{t}$ and $\mathbf{v}$ are simultaneously diagonalized after a certain linear transformation.

The different modes are *orthogonal* in the sense of the equation below:

$$\tilde{\boldsymbol{\Phi}}_{(i)} \cdot \mathbf{t} \cdot \boldsymbol{\Phi}_{(j)} = 0, \quad \text{if } i \neq j. \tag{9.43}$$
different modes are orthogonal

We can test the assertion in Equation (9.43) for the double pendulum to see if it is true in

---

* See S. J. Leon, *Linear Algebra*, 2d ed., Chapter 1.

that case:

$$(1+\sqrt{2}, 1) \cdot \begin{pmatrix} 5 & 2 \\ 2 & 1 \end{pmatrix} \cdot \begin{pmatrix} 1-\sqrt{2} \\ 1 \end{pmatrix} = 0. \tag{9.44}$$

(The matrix in the middle is proportional to **t** for the double pendulum.) This orthogonality condition is similar to the orthogonality that we found for the principal axes in Chapter 8. However, it is different because here we must use Equation (9.43) instead of simply multiplying different eigenvectors together. The technical term for this is that **t** is the *metric* for the scalar product of two state vectors in the configuration space.

To prove this orthogonality for N degrees of freedom, use (9.40) for the ith mode:

$$\mathbf{v} \cdot \mathbf{\Phi}_{(i)} = \omega_{(i)}^2 \mathbf{t} \cdot \mathbf{\Phi}_{(i)}, \tag{9.45}$$

and likewise for the jth mode:

$$\mathbf{v} \cdot \mathbf{\Phi}_{(j)} = \omega_{(j)}^2 \mathbf{t} \cdot \mathbf{\Phi}_{(j)}. \tag{9.46}$$

Take the scalar product of the top equation with $\tilde{\mathbf{\Phi}}_{(j)}$ and the bottom equation with $\tilde{\mathbf{\Phi}}_{(i)}$. Since **v** is a real symmetric matrix, we know that $\tilde{\mathbf{\Phi}}_{(j)} \cdot \mathbf{v} \cdot \mathbf{\Phi}_{(i)} = \tilde{\mathbf{\Phi}}_{(i)} \cdot \mathbf{v} \cdot \mathbf{\Phi}_{(j)} = \sum_{\alpha,\beta} (\Phi_{(i)\alpha} \Phi_{(j)\beta}) v_{\alpha\beta}$. Since **t** is also symmetric, after subtracting Equation (9.46) from (9.45), we have

$$0 = (\omega_{(i)}^2 - \omega_{(j)}^2) \tilde{\mathbf{\Phi}}_{(j)} \cdot \mathbf{t} \cdot \mathbf{\Phi}_{(i)}. \tag{9.47}$$

If there is no degeneracy ($\omega_{(i)}^2 \neq \omega_{(j)}^2$) it must be true that $\tilde{\mathbf{\Phi}}_{(j)} \cdot \mathbf{t} \cdot \mathbf{\Phi}_{(i)} = 0$. The proof can be extended to include the case of degenerate modes as well by taking appropriate linear combinations of degenerate mode state vectors.

The normalization of the individual modes should be chosen arbitrarily for each separate mode. Choose

$$\tilde{\mathbf{\Phi}}_{(i)} \cdot \mathbf{t} \cdot \mathbf{\Phi}_{(i)} = 1. \tag{9.48}$$

(Equation (9.48) applies to each of the N modes.)

QUESTION 7: Normalization Use (9.48) to normalize the mode vectors (9.42) for the double pendulum. Prove that the normalized vectors are

$$\frac{1}{\sqrt{ml^2(10+7\sqrt{2})}}(1+\sqrt{2}, 1),$$

$$\frac{1}{\sqrt{ml^2(10-7\sqrt{2})}}(1-\sqrt{2}, 1). \tag{9.49}$$

9.4 SUPERIMPOSING DIFFERENT MODES; COMPLEX MODE AMPLITUDES

As can be seen in Equation (9.49), often the normalization condition (9.48) causes there to be various constants (such as ml^2) in the mode vector. Since the purpose of normalization is to be able to consistently compare mode vectors, often we will set the constants equal to unity to simplify the form of the mode vectors.

With the convention (9.48) for the normalization we can find how to expand an arbitrary complex vibrational state of the system $\check{\phi}(t)$ into normal modes. First, express the arbitrary initial state vector as a linear combination of all N modes at $t = 0$:

$$\text{at } t = 0 \quad \check{\phi}(0) = \sum_{\text{all modes}} A_{(i)}(0) \mathbf{\Phi}_{(i)}, \tag{9.50}$$

where $A_{(i)}(0)$ is the *complex* amplitude of the ith mode at $t = 0$. Then evolve the state vector in time by using what is known about how the modes each individually evolve in time:

$$\check{\phi}(t) = \sum_{\text{all modes}} A_{(i)}(0) e^{i\omega_{(i)} t} \mathbf{\Phi}_{(i)}. \tag{9.51}$$

Call $A_{(i)}(t) \equiv A_{(i)}(0) e^{i\omega_{(i)} t}$ the mode amplitudes. (Notice that here the phase of the system is included in the complex part of $A_{(i)}(0)$. This is similar to the A of (3.25) in Chapter 3.) Finally, take the real part:

$$\phi(t) = \text{Real}[\check{\phi}(t)] = \sum_{\text{all modes}} \text{Real}\big[A_{(i)}(t) \mathbf{\Phi}_{(i)}\big] = \sum_{\text{all modes}} \mathbf{\Phi}_{(i)} \text{Real}\big[A_{(i)}(t)\big]. \tag{9.52}$$

You can use the orthogonality relation (9.43) and the normalization (9.48) to solve for the mode amplitudes in terms of the initial state of the system (9.50):

$$\boxed{A_{(i)}(0) = \tilde{\mathbf{\Phi}}_{(i)} \cdot \mathbf{t} \cdot \check{\phi}(0).} \tag{9.53}$$

(The initial values of the $A_{(i)}(0)$ are in general complex because $\check{\phi}(0)$ is complex. $\mathbf{\Phi}_{(i)}$ and $\mathbf{t}$ are always real.) Equation (9.53) is the inverse relation to expanding in normal modes. It gives a practical formula for expressing an arbitrary initial displacement of the system $\phi(0)$ as a superposition of normal modes.

QUESTION 8: *Mode Orthogonality 1* Prove (9.53). Also, what do we mean when we say that (9.53) is the inverse relation to expanding in normal modes? What do we use Equation (9.53) for?

At an arbitrary time t, the mode amplitudes can be found by the equation

$$\boxed{A_{(i)}(t) = \tilde{\mathbf{\Phi}}_{(i)} \cdot \mathbf{t} \cdot \check{\phi}(t).} \tag{9.54}$$

QUESTION 9: *Mode Orthogonality 2* Exactly what is meant by the term "complex vibrational state" to describe $\check{\phi}(t)$ on the previous page? How would you find the complex state vector $\check{\phi}(0)$, which you know in terms of the initial displacements and velocities, in terms of the normal modes $\Phi_{(i)}$? Prove that this last formula, (9.54), is correct. Hint: Use (9.43) and (9.48).

QUESTION 10: *Mode Orthogonality 3* Define the following terms: superposition of normal modes, orthogonality of modes, and normalization of modes.

⊃ Example

We will use the double pendulum as an example of how to solve for a state vector by expanding in normal modes. The initial conditions will be: $\phi_1(0) = 0$, $\phi_2(0) = \alpha_0$, $\dot{\phi}_1(0) = 0$, $\dot{\phi}_2(0) = 0$. Use (9.53) to find the amplitudes $A_{(i)}(0)$ and use the normalized mode vectors (9.49), but set $ml^2 = 1$ to simplify. The real part of A is proportional to $\phi(0)$ and the imaginary part is proportional to $\dot{\phi}(0)$:

$$\text{Real}[A_{(1)}(0)] = \frac{1}{\sqrt{(10 + 7\sqrt{2})}} (1+\sqrt{2}, 1) \frac{1}{2} \begin{pmatrix} 5 & 2 \\ 2 & 1 \end{pmatrix} \begin{pmatrix} 0 \\ \alpha_0 \end{pmatrix},$$

$$-\omega \, \text{Im}[A_{(1)}(0)] = \frac{1}{\sqrt{(10 + 7\sqrt{2})}} (1+\sqrt{2}, 1) \frac{1}{2} \begin{pmatrix} 5 & 2 \\ 2 & 1 \end{pmatrix} \begin{pmatrix} 0 \\ 0 \end{pmatrix}. \tag{9.55}$$

Doing some algebra to simplify yields

$$A_{(1)}(0) = \alpha_0 \left(\frac{3 + 2\sqrt{2}}{2\sqrt{10 + 7\sqrt{2}}} \right). \tag{9.56}$$

Similarly, for $A_{(2)}$ we obtain

$$A_{(2)}(0) = \alpha_0 \left(\frac{3 - 2\sqrt{2}}{2\sqrt{10 - 7\sqrt{2}}} \right). \tag{9.57}$$

Next use (9.52) to write the state vector in terms of the normal modes:

$$\check{\phi}(t) = A_{(1)} \Phi_{(1)} e^{i\omega_{(1)} t} + A_{(2)} \Phi_{(2)} e^{i\omega_{(2)} t}, \tag{9.58}$$

$$\phi(t) = \text{Real}[\check{\phi}]. \tag{9.59}$$

Equations (9.56)–(9.58) can be simplified to obtain

$$\phi(t) = \frac{\alpha_0 \sqrt{2}}{4} \left[(\sqrt{2 + \sqrt{2}}) \begin{pmatrix} 1 + \sqrt{2} \\ 1 \end{pmatrix} \cos(\omega_{(1)} t) \right. \\ \left. + (\sqrt{2 - \sqrt{2}}) \begin{pmatrix} 1 + \sqrt{2} \\ 1 \end{pmatrix} \cos(\omega_{(2)} t) \right]. \tag{9.60}$$

9.4 SUPERIMPOSING DIFFERENT MODES; COMPLEX MODE AMPLITUDES

The $\omega_{(i)}$s are found from (9.38, 9.39). This is the complete description of the double pendulum in terms of normal modes for small angles near equilibrium and initial conditions as specified above. The angles $\phi_1(t)$ and $\phi_2(t)$ are found by taking the first and second rows of (9.60), respectively.

Normal Coordinates: The Lagrangian and Stability

You can express the original Lagrangian in terms of the *normal coordinates* $\rho_{(i)}$. Define $\rho_{(i)}(t) \equiv \text{Real}[A_{(i)}(t)] = \text{Real}[A_{(i)}(0)e^{i\omega_{(i)}t}]$ so that (9.52) becomes

$$\text{Real}[\check{\phi}(t)] = \sum_{\text{all modes}} \rho_{(i)}(t)\Phi_{(i)}. \tag{9.61}$$

(Equation (9.61) contains only real numbers.) In terms of the ρs, the Lagrangian (9.30) must be

$$L = \frac{1}{2}\sum_{\text{modes}} c_{(i)}\left(\dot{\rho}_{(i)}^2 - \omega_{(i)}^2 \rho_{(i)}^2\right). \tag{9.62}$$

The $c_{(i)}$s are constants that we could calculate from the original form of the Lagrangian if we wished. It makes no difference here if we set them equal to one. (They do determine how much each mode contributes to the total energy.) Then we see that the $\rho_{(i)}(t)$ obey the differential equation

$$\ddot{\rho}_{(i)} + \omega_{(i)}^2 \rho_{(i)} = 0. \tag{9.63}$$

QUESTION 11: *Lagrangian* Prove that the Lagrangian (9.30) can be written in the form of Equation (9.62) and that this form leads to the equation of motion (9.63).

The Lagrangian is now expressed as a sum of independent, uncoupled harmonic oscillators, each having the frequency of a particular mode. This proves that the modes do oscillate *independently* of each other without any coupling between modes. The expansion in terms of normal modes is often very useful. The normal modes represent the natural set of generalized coordinates for any system undergoing small vibrations.

Equation (9.62) is similar to the forms (3.13, 3.14) in Chapter 3 (if we scale out the magnitude of the frequency), which we used to determine the stability of equilibrium points of 1-D systems. We can use (9.62) to determine the stability of many-dimensional linear systems near equilibrium. The sign of the second term will determine the stability. This term comes from diagonalizing **v**. Therefore stability means that all of the eigenvalues of **v**, which is a real symmetric matrix, are positive (9.26). This makes physical sense, because it means that the potential energy of the system increases no matter which direction we make a small displacement. If, however, just one of the eigenvalues of **v** is negative, we have at least one unstable degree of freedom, so the system is unstable. Testing for static equilibrium is easy: Find the eigenvalues of **v** and check that they are all positive.

QUESTION 12: Stability Prove that the requirement that all eigenvalues of **v** be positive is a necessary and sufficient condition for stability near an equilibrium point.

Often, a more complicated system can be approximated by the type of system we have been considering, a set of coupled linear oscillators. The complications are then added in the form of what is hoped are small perturbations, possibly nonlinear or time-dependent terms. The solution to the system without these additional complicating effects is taken as a first approximation to the actual solution, which is then computed by using perturbation theory. At the end, one checks to see that the "small" perturbations are really small. If this is true, then this is a valid procedure. See the section on Lindstedt–Poincaré Perturbation Theory in Chapter 10 for more details.

9.5 LINEAR TRIATOMIC MOLECULE

Sometimes you discover that one or more mode frequencies equal zero. It means that the motion in question is not a vibration at all but a translation or a rotation. When this happens, it is a sign that it is possible to reduce the number of degrees of freedom before you start to solve the problem. This can be done by choosing the right set of coordinates to describe the motion. Below we give an example to illustrate this type of situation.

Consider the CO_2 molecule as drawn in Figure 9.5. Carbon dioxide is a nearly linear molecule. We ask only about vibrations along the X axis. There are three degrees of freedom: the positions of the two oxygen atoms and of the carbon atom along the X axis. The atoms are assumed to be connected by springs which have a spring constant k. Let m stand for the oxygen mass and M for the carbon mass.

At equilibrium, each spring has the length l. The potential energy is

$$V(x_1, x_2, x_3) = \frac{k}{2}(x_1 - x_2 + l)^2 + \frac{k}{2}(x_3 - x_2 - l)^2. \tag{9.64}$$

By taking the partial derivatives of V and setting them equal to zero, we get

$$x_1 - x_2 = -l, \quad x_3 - x_2 = l. \tag{9.65}$$

Notice that we cannot determine all three coordinates, only coordinate differences. We are

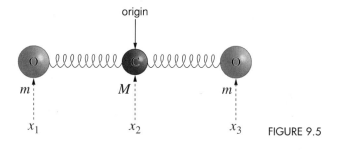

FIGURE 9.5

9.5 LINEAR TRIATOMIC MOLECULE

free to slide the entire molecule along the X axis without doing any work, so the absolute positions have no physical meaning.

We can invent a column vector $\boldsymbol{\delta x}$ that expresses the displacements from the equilibrium positions (like $\boldsymbol{\phi}$ from the previous section):

$$\boldsymbol{\delta x} \equiv \begin{pmatrix} \delta x_1 \\ \delta x_2 \\ \delta x_3 \end{pmatrix}, \tag{9.66}$$

$$\delta x_1 \equiv x_1 + l, \quad \delta x_2 \equiv x_2, \quad \delta x_3 \equiv x_3 - l. \tag{9.67}$$

The potential energy, as a function of small displacements from equilibrium, can be written in matrix notation as

$$V = V(0) + \widetilde{\boldsymbol{\delta x}} \cdot \mathbf{v} \cdot \boldsymbol{\delta x}. \tag{9.68}$$

Now $\mathbf{v}$ from (9.24) is

$$\mathbf{v} \equiv \frac{1}{2} \frac{\partial^2 V}{\partial x_i \partial x_j} = \frac{k}{2} \begin{pmatrix} 1 & -1 & 0 \\ -1 & 2 & -1 \\ 0 & -1 & 1 \end{pmatrix}, \tag{9.69}$$

and $V(0) = 0$ in this case. You should convince yourself that from the form of the kinetic energy,

$$T = \frac{m}{2}\left(\dot{x}_1^2 + \dot{x}_3^2\right) + \frac{M}{2}\dot{x}_2^2, \tag{9.70}$$

one gets immediately

$$T = \widetilde{\boldsymbol{\delta \dot{x}}} \cdot \mathbf{t} \cdot \boldsymbol{\delta \dot{x}}, \tag{9.71}$$

where $\mathbf{t}$ from (9.28) is

$$\mathbf{t} \equiv \frac{1}{2} \frac{\partial^2 T}{\partial \dot{x}_i \partial \dot{x}_j} = \frac{1}{2} \begin{pmatrix} m & 0 & 0 \\ 0 & M & 0 \\ 0 & 0 & m \end{pmatrix}. \tag{9.72}$$

We obtain the matrix $-\omega^2 \mathbf{t} + \mathbf{v}$ by assuming there is only a single mode present: $\boldsymbol{\delta x} = \mathbf{X} e^{i(\omega t - \delta)}$. This matrix should be simplified as much as possible before evaluating the determinant to find the mode frequencies. To simplify it, define the dimensionless ratios

$$r \equiv \frac{M}{m}, \quad \lambda \equiv \frac{\omega^2}{\omega_0^2}, \quad \omega_0^2 \equiv \frac{k}{m}. \tag{9.73}$$

The determinant that has to vanish is

$$\begin{vmatrix} 1-\lambda & -1 & 0 \\ -1 & 2-r\lambda & -1 \\ 0 & -1 & 1-\lambda \end{vmatrix}. \tag{9.74}$$

Setting this determinant equal to zero, we get the polynomial equation for λ:

$$\lambda(r\lambda^2 - 2(r+1)\lambda + (r+2)) = 0. \tag{9.75}$$

Notice that $\lambda = 0$ is a root of this equation. This means one of the mode frequencies is zero. A mode with zero frequency is not an oscillation at all, but corresponds, as we shall see, to translating the whole molecule along the X axis. Had we realized that such a mode would occur, we could have eliminated this degree of freedom from the start by using coordinates relative to the center of mass. There would then be a term $\frac{1}{2}M\dot{x}_{cm}^2$ in the kinetic energy and the center of mass coordinate would not appear in the potential energy at all. Often, prior information can be used to eliminate degrees of freedom, thus reducing the order of the matrix before even starting the mode calculation.

The roots of (9.75) are

$$\lambda_{(1)} = 0, \quad \lambda_{(2)} = 1, \quad \lambda_{(3)} = 1 + \frac{2}{r}. \tag{9.76}$$

This leads to the mode frequencies

$$\omega_{(1)} = 0, \quad \omega_{(2)} = \sqrt{\frac{k}{m}}, \quad \omega_{(3)} = \sqrt{\frac{k}{m}}\sqrt{1 + \frac{2m}{M}}. \tag{9.77}$$

It can be seen experimentally by observing the infrared spectra of triatomic linear molecules that the theoretically predicted vibration frequencies agree quite well.

The next task is to find the mode vectors $\mathbf{X}_{(i)} = (X_{(i),1}, X_{(i),2}, X_{(i),3})$. We use the method of cofactors discussed in Appendix A. The cofactors of the top three elements of our matrix $(-\omega^2\mathbf{t} + \mathbf{v})$ are, in general, proportional to $X_{(i),1}, X_{(i),2}, X_{(i),3}$. These may be multiplied by an arbitrary constant, depending on initial conditions. The cofactors of an element are found by taking the determinant of what is left when the row and the column containing the element are struck out of the matrix. Cofactors for elements from columns 2, 4, 6, ... must be multiplied by -1. We get the cofactors we need for any value of λ:

$$\left(\begin{vmatrix} 2-\lambda r & -1 \\ -1 & 1-\lambda \end{vmatrix}, -\begin{vmatrix} -1 & -1 \\ 0 & 1-\lambda \end{vmatrix}, \begin{vmatrix} -1 & 2-\lambda r \\ 0 & -1 \end{vmatrix} \right). \tag{9.78}$$

Mode 1. Evaluating (9.78) for $\lambda_{(1)} = 0$, the coordinate ratios are

$$X_{(1),1} : X_{(1),2} : X_{(1),3} = 1 : 1 : 1. \tag{9.79}$$

9.6 WHY THE METHOD ALWAYS WORKS

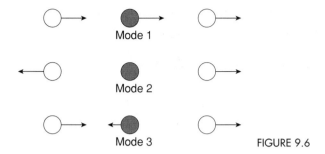

FIGURE 9.6

A picture of how this particular mode vibrates is shown in Figure 9.6. As expected, all atoms move in the same direction in the same amount, so the motion with $\omega = 0$ is a translation and not a true vibration.

Mode 2. Assume $\lambda = 1$. Then

$$X_{(2),1} : X_{(2),2} : X_{(2),3} = -1 : 0 : 1. \tag{9.80}$$

The picture is also shown in Figure 9.6. It is no wonder that the frequency is just the frequency $\sqrt{\frac{k}{m}}$, as the carbon atom doesn't move at all in this mode.

Mode 3. Assume $\lambda = 1 + \frac{2}{r}$. Then

$$X_{(3),1} : X_{(3),2} : X_{(3),3} = 1 : -\frac{2}{r} : 1 \tag{9.81}$$

($\frac{2}{r} = -\frac{2m}{M}$). Figure 9.6 also shows this mode.

QUESTION 13: *Triatomic Molecule* Show that the center of mass remains fixed in mode #3. All three atoms vibrate at the same frequency.

QUESTION 14: *Transverse Modes* We just solved for the longitudinal vibration of the linear triatomic molecule. Thinking qualitatively only, what are the possibilities for the transverse modes of vibration?

9.6 WHY THE METHOD ALWAYS WORKS

We will prove some general theorems about the mode frequencies and mode vectors. Let $\mathbf{\Phi}$ stand for an unknown mode vector for the system. Notice that, if ω^2 is the mode frequency,

$$\mathbf{v} \cdot \mathbf{\Phi} = \omega^2 \mathbf{t} \cdot \mathbf{\Phi}. \tag{9.82}$$

1. Proof that the square of the mode frequency is real: Take the complex conjugate (*) of the equation (9.82) above, remembering that $\mathbf{v}, \mathbf{t}$ are real matrices. Multiply (9.82) by $\tilde{\mathbf{\Phi}}^*$ and the conjugate by $\tilde{\mathbf{\Phi}}$ to obtain the two equations

$$\tilde{\mathbf{\Phi}}^* \cdot \mathbf{v} \cdot \mathbf{\Phi} = \omega^2 \tilde{\mathbf{\Phi}}^* \cdot \mathbf{t} \cdot \mathbf{\Phi}, \quad \tilde{\mathbf{\Phi}} \cdot \mathbf{v} \cdot \mathbf{\Phi}^* = \omega^{*2} \tilde{\mathbf{\Phi}} \cdot \mathbf{t} \cdot \mathbf{\Phi}^*. \tag{9.83}$$

(For any symmetric real matrix $\mathbf{a}$, $\tilde{\boldsymbol{\Phi}}^* \cdot \mathbf{a} \cdot \boldsymbol{\Phi} = \tilde{\boldsymbol{\Phi}} \cdot \mathbf{a} \cdot \boldsymbol{\Phi}^*$.) Subtract these two equations from each other, obtaining

$$0 = (\omega^{*2} - \omega^2)(\tilde{\boldsymbol{\Phi}}^* \cdot \mathbf{t} \cdot \boldsymbol{\Phi}). \tag{9.84}$$

We know for physical reasons that the kinetic energy is always positive, but the generalized eigenvector $\boldsymbol{\Phi}$ might not be a real vector. Suppose $\boldsymbol{\Phi}$ is complex. Break $\boldsymbol{\Phi}$ into its real and imaginary parts:

$$\boldsymbol{\Phi} = \boldsymbol{\alpha} + i\boldsymbol{\beta}, \tag{9.85}$$

$$\tilde{\boldsymbol{\Phi}}^* \cdot \mathbf{t} \cdot \boldsymbol{\Phi} = \tilde{\boldsymbol{\alpha}} \cdot \mathbf{t} \cdot \boldsymbol{\alpha} + \tilde{\boldsymbol{\beta}} \cdot \mathbf{t} \cdot \boldsymbol{\beta} \tag{9.86}$$

(the cross term vanishes). Each of the terms on the right of Equation (9.86) must be positive, so

$$\omega^{*2} = \omega^2 \tag{9.87}$$

if (9.84) is to hold. This proves that ω^2 is real for all the modes. We will show below that if $\boldsymbol{\Phi}$ obeys (9.82), it must be real as well, as is expected for physical reasons.

A second theorem, that $\omega^2 > 0$, is easy to prove. Assume that V is a minimum at the equilibrium point. Then $\tilde{\boldsymbol{\Phi}} \cdot \mathbf{v} \cdot \boldsymbol{\Phi} > 0$ (see Equation (9.26)). Also, we know that kinetic energy is always positive, so $\mathbf{t}$ must be a positive definite matrix. Hence from (9.82)

$$\omega^2 = \frac{\tilde{\boldsymbol{\Phi}} \cdot \mathbf{v} \cdot \boldsymbol{\Phi}}{\tilde{\boldsymbol{\Phi}} \cdot \mathbf{t} \cdot \boldsymbol{\Phi}} > 0. \tag{9.88}$$

Thus $\omega^2 > 0$ is a direct consequence of the fact that V is a minimum. Hence all mode frequencies will be real, since they are square roots of positive quantities.

2. Proof that the mode vectors are real numbers: This proof follows very simply from observing that the mode vectors are solutions to homogeneous equations with real coefficients (9.40); therefore they must be real.
3. Proof that the matrices $\mathbf{t}$ and $\mathbf{v}$ can be simultaneously diagonalized by a single linear transformation: This theorem also leads to a "standard" form for the kinetic and potential energies (9.62) when expressed in terms of the normal coordinates and mode vector basis, a point discussed earlier. To prove this last assertion about how $\mathbf{t}$, $\mathbf{v}$ are simultaneously diagonalized, we need only to point out that the mode vectors can be considered as a new set of basis states in the N-dimensional space of the displacements. To find the actual matrix elements of $\mathbf{t}$ with this new basis we calculate $\tilde{\boldsymbol{\Phi}}_{(i)} \cdot \mathbf{t} \cdot \boldsymbol{\Phi}_{(j)} = t_{ij}$. But the orthogonality property (9.43) means that this matrix element vanishes unless $i = j$, so in the new basis $\mathbf{t}$ is a diagonal matrix. The same can be proved about $\mathbf{v}$ by making use of the relation $\mathbf{v} \cdot \boldsymbol{\Phi}_{(j)} = \omega^2_{(j)} \mathbf{t} \cdot \boldsymbol{\Phi}_{(j)}$ and taking the dot product with $\tilde{\boldsymbol{\Phi}}_{(i)}$ applied on the left to both sides of the equation. Thus both the kinetic and the potential energy matrices are diagonal in the mode vector basis.

9.6 WHY THE METHOD ALWAYS WORKS

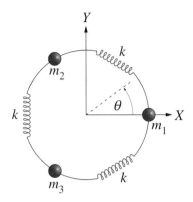

FIGURE 9.7

We have proved that if we have Equation (9.82), the mode frequencies and mode vectors will be real. Changing to the mode vector basis will simultaneously diagonalize both **t** and **v**. Since we will always have an acceptable solution, the method always works.

What to Do with Degenerate Modes

Now we will give an example of degenerate frequencies. Imagine a circular hoop made of some massless wire on which three identical beads of mass m are placed as shown in Figure 9.7. The beads are constrained to move along the wire only. Between each bead is a spring of spring constant k which serves to couple adjacent beads.* This is like the linear triatomic molecule with equal masses constrained to be in a circle. There are three degrees of freedom here, the angular motion of each bead along the wire, denoted by θ_1, θ_2, and θ_3 respectively.

The kinetic energy is

$$T = \frac{m}{2}(\dot{\theta}_1^2 + \dot{\theta}_2^2 + \dot{\theta}_3^2). \tag{9.89}$$

Assume the equilibrium positions of the three beads are at 0, $\frac{2\pi}{3}$, and $\frac{4\pi}{3}$ respectively. The potential energy is then

$$V = \frac{k}{2}\left[\left(\theta_2 - \theta_1 - \frac{2\pi}{3}\right)^2 + \left(\theta_3 - \theta_2 - \frac{2\pi}{3}\right)^2 + \left(\theta_3 - \theta_1 - \frac{4\pi}{3}\right)^2\right]. \tag{9.90}$$

Define the column matrix

$$\delta\theta = \begin{pmatrix} \delta\theta_1 \\ \delta\theta_2 \\ \delta\theta_3 \end{pmatrix}, \tag{9.91}$$

---

* It will be assumed that the springs cause no restriction to the motion of the bead along the wire except for the coupling effect discussed. Also, we will ignore gravitational effects.

where

$$\delta\theta_1 \equiv \theta_1, \quad \delta\theta_2 \equiv \theta_2 - \frac{2\pi}{3}, \quad \delta\theta_3 \equiv \theta_3 - \frac{4\pi}{3}. \tag{9.92}$$

Then we can expand T and V in small displacements $\delta\theta$ about the equilibrium positions. From (9.28), **t** is

$$\mathbf{t} = \frac{m}{2}\begin{pmatrix} 1 & 0 & 0 \\ 0 & 1 & 0 \\ 0 & 0 & 1 \end{pmatrix}. \tag{9.93}$$

Similarly from (9.24) **v** is

$$\mathbf{v} = \frac{k}{2}\begin{pmatrix} 2 & -1 & -1 \\ -1 & 2 & -1 \\ -1 & -1 & 2 \end{pmatrix}. \tag{9.94}$$

Next we want to obtain the matrix $-\omega^2 \mathbf{t} + \mathbf{v}$, assuming there is only a single mode: $\delta\theta = \Theta e^{i(\omega t - \delta)}$. The determinant that has to vanish is

$$\begin{vmatrix} 2-\lambda & -1 & -1 \\ -1 & 2-\lambda & -1 \\ -1 & -1 & 2-\lambda \end{vmatrix}, \tag{9.95}$$

where $\lambda \equiv \frac{\omega^2 m}{k}$. The solutions λ that make the determinant of (9.95) equal to zero are

$$\lambda_{(1)} = 0, \quad \lambda_{(2)} = \lambda_{(3)} = 3. \tag{9.96}$$

These lead to the mode frequencies

$$\omega_{(1)} = 0, \quad \omega_{(2)} = \omega_{(3)} = \sqrt{\frac{3k}{m}}. \tag{9.97}$$

Notice that we have one mode with zero frequency just as with the linear triatomic molecule. Again this is a mode in which all the particles move in the same direction, which corresponds to a rotation of the whole molecule in this case. The other two frequencies are the same, indicating a degeneracy for these two modes.

Next we will solve for the mode vectors to see what types of oscillation we have. Solving for the cofactors of (9.95) we obtain

$$\left(3 - 4\lambda_{(i)} + \lambda_{(i)}^2, 3 - \lambda_{(i)}, 3 - \lambda_{(i)}\right), \tag{9.98}$$

which gives us $(3, 3, 3)$ for the case of $\lambda = 0$, as expected. When normalized (with m set equal to one to simplify the mode vectors), this mode becomes

$$\Theta_{(1)} = \sqrt{\frac{2}{3}}(1, 1, 1). \tag{9.99}$$

However, we obtain (0, 0, 0) for the two degenerate cases of $\lambda = 3$. The degeneracy in the λs leads to an indeterminacy in the mode vectors. If we tried to solve (9.95) for the degenerate cases of $\lambda = 3$ we would have three equations of the form

$$\theta_1 + \theta_2 + \theta_3 = 0. \quad (9.100)$$

We can solve this equation for the two degenerate cases $\Theta_{(2)}$, $\Theta_{(3)}$, by assuming an arbitrary mode vector for $\Theta_{(2)}$ that satisfies (9.100). For example, a simple case would be to chose one of the degrees of freedom to be zero and the other two to be equal and opposite in sign. When normalized (with m set equal to one), this case is

$$\Theta_{(2)} = (0, 1, -1). \quad (9.101)$$

We still need to find the components of $\Theta_{(3)} \equiv (a, b, c)$. We can use the fact that $\Theta_{(3)}$ must also satisfy (9.100), as well as be orthogonal to $\Theta_{(2)}$ and normalized. This gives us three equations respectively:

$$a + b + c = 0, \quad b - c = 0, \quad \sqrt{a^2 + b^2 + c^2} = \sqrt{2}. \quad (9.102)$$

Solving for the three components (a, b, c) we obtain

$$\Theta_{(3)} = \frac{1}{\sqrt{3}}(-2, 1, 1). \quad (9.103)$$

You can use a similar procedure as outlined above to find degenerate mode vectors for any number of degrees of freedom. Start by picking one arbitrarily (but consistent with (9.95)) and then use the normalization and orthogonalization conditions to solve for the others. Since we arbitrarily chose the first of the degenerate cases, it follows that there are an arbitrary number of sets of degenerate mode vectors.

9.7 N POINT MASSES CONNECTED BY A STRING

As a final example, we will work out the original problem of N equally spaced point particles, each of mass m, connected by a massless string under tension τ. Each mass is separated from its adjacent neighbors by a distance d, and the end points of the string are fixed. We will only concern ourselves with vertical displacements here. Call the vertical displacement of the kth particle, y_k, as shown in Figure 9.8.

The kinetic energy is given by

$$T = \sum_{k=1}^{N} \frac{m}{2} \dot{y}_k^2. \quad (9.104)$$

If we neglect gravitational effects, then the potential energy is given by the tension of the string times the sum of the extensions of the string between adjacent masses. The extension

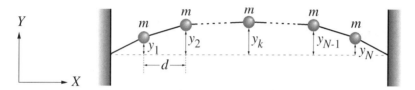

FIGURE 9.8
N point masses connected by a string under tension.

in the string Δl between any two adjacent masses is equal to

$$\Delta l = \sqrt{d^2 + (y_{k+1} - y_k)^2} - d \simeq \frac{1}{2d}(y_{k+1} - y_k)^2, \tag{9.105}$$

where terms higher than second order have been neglected in the approximation. Therefore the potential energy is

$$V = \sum_{k=0}^{N} \frac{\tau}{2d}(y_{k+1} - y_k)^2, \tag{9.106}$$

with the boundary conditions

$$y_0 = 0, \quad y_{N+1} = 0 \tag{9.107}$$

limiting the string at the two ends. Note that we would obtain the same form for T and V if we considered longitudinal vibrations of the spring instead. This will be worked out in a homework problem. Using (9.28) we find

$$\mathbf{t} = \frac{m}{2}\begin{pmatrix} 1 & 0 & 0 & \cdots \\ 0 & 1 & 0 & \cdots \\ 0 & 0 & 1 & \cdots \\ \vdots & \vdots & \vdots & \ddots \end{pmatrix}. \tag{9.108}$$

Similarly, from (9.24) we get

$$\mathbf{v} = \frac{\tau}{2d}\begin{pmatrix} 2 & -1 & 0 & \cdots \\ -1 & 2 & -1 & \cdots \\ 0 & -1 & 2 & \cdots \\ \vdots & \vdots & \vdots & \ddots \end{pmatrix}. \tag{9.109}$$

We obtain the matrix $-\omega^2 \mathbf{t} + \mathbf{v}$ by assuming there is only one mode present:

$$\mathbf{y} = \check{\mathbf{Y}} e^{i\omega t}, \tag{9.110}$$

9.7 N POINT MASSES CONNECTED BY A STRING

where $\mathbf{y}$ and $\mathbf{\check{Y}}$ are the column vectors containing the y_ks and $\check{Y}_k$s respectively. (Notice that this time we did not include a phase term, so $\mathbf{\check{Y}}$ can be complex here. This makes our calculation easier.) Thus we need to find where the determinant of the following matrix is zero:

$$\begin{pmatrix} -\frac{\omega^2 m}{2} + \frac{\tau}{d} & -\frac{\tau}{2d} & 0 & \cdots \\ -\frac{\tau}{2d} & -\frac{\omega^2 m}{2} + \frac{\tau}{d} & -\frac{\tau}{2d} & \cdots \\ 0 & -\frac{\tau}{2d} & -\frac{\omega^2 m}{2} + \frac{\tau}{d} & \cdots \\ \vdots & \vdots & \vdots & \ddots \end{pmatrix}. \quad (9.111)$$

This is no easy task if N is large, although there are many computer algorithms that can help. However, we wish to find a solution for arbitrary N, which can best be found from using the equations of motion (9.33). In our case we wish to solve

$$\mathbf{t} \cdot \ddot{\mathbf{y}} + \mathbf{v} \cdot \mathbf{y} = 0. \quad (9.112)$$

In index notation this is

$$\frac{m}{2}\ddot{y}_k + \frac{\tau}{d}y_k - \frac{\tau}{2d}y_{k-1} - \frac{\tau}{2d}y_{k+1} = 0. \quad (9.113)$$

Substituting (9.110) into the above equation, we obtain, for adjacent particles,

$$\frac{-m\omega^2}{2}\check{Y}_k + \frac{\tau}{d}\check{Y}_k - \frac{\tau}{2d}\check{Y}_{k-1} - \frac{\tau}{2d}\check{Y}_{k+1} = 0. \quad (9.114)$$

We cannot solve Equation (9.114) as is, but if we assume $\check{Y}_k$ is of the form

$$\check{Y}_k = Y e^{i(k\gamma - \delta)} \quad (9.115)$$

we can simplify it. Note that Y is real here, and γ and δ are constants we must find by using our boundary conditions (9.107). Substituting (9.115) into (9.114) and simplifying, we obtain

$$\left(\frac{\tau}{d} - \frac{m\omega^2}{2}\right) - \frac{\tau}{2d}e^{-i\gamma} - \frac{\tau}{2d}e^{i\gamma} = 0, \quad (9.116)$$

which further simplifies to

$$\omega^2 = \frac{4\tau}{md}\left(\sin^2\frac{\gamma}{2}\right). \quad (9.117)$$

Since there are N degrees of freedom, we expect N different values of ω, which depend on the N different values of γ, which we will find from our boundary conditions. Our original y_ks can be rewritten using (9.110, 9.115) as

$$y_k = Y e^{i\omega t} e^{i(k\gamma - \delta)}. \quad (9.118)$$

Using the first of our boundary conditions (9.107) we obtain

$$0 = Ye^{i\omega t}e^{-i\delta}. \tag{9.119}$$

Since we ultimately take the real part of our equations, Equation (9.119) implies that $\delta = \frac{\pi}{2}$. Using the second boundary condition we obtain

$$0 = Ye^{i\omega t}e^{i((N+1)\gamma - \frac{\pi}{2})}. \tag{9.120}$$

Again taking the real part leads to the condition

$$0 = \sin(N+1)\gamma, \tag{9.121}$$

which is satisfied by

$$\gamma = \frac{(i)\pi}{N+1}. \tag{9.122}$$

We have used $(i)\pi$ to indicate the integer i times π. Here i serves as a label for the mode. Substituting this into (9.117) for γ we obtain

$$\omega_{(i)} = 2\sqrt{\frac{\tau}{md}} \sin\left(\frac{(i)\pi}{2(N+1)}\right) \tag{9.123}$$

and (after taking the real part)

$$y_{(i)k} = Y\cos(\omega_{(i)}t) \sin\left(\frac{(i)k\pi}{N+1}\right). \tag{9.124}$$

We will have a continuous string if we take the limit as $N \to \infty$ and $d \to 0$. Then $d(N+1) \to L$, where L is the length of the string, and $m \to 0$ such that $\frac{m}{d} = \rho$, where ρ is the density of the string. Applying these limits to (9.123) we obtain

$$\omega_{(i)} = 2\sqrt{\frac{\tau}{md}} \sin\left(\frac{(i)\pi d}{2L}\right), \tag{9.125}$$

which can be reduced in the limit that d is small to

$$\omega_{(i)} = \frac{(i)\pi}{L}\sqrt{\frac{\tau}{\rho}}. \tag{9.126}$$

This is the usual expression for the frequency of standing waves on a string of length L. $y_{(i)k}$ becomes then

$$y_{(i)} = Y\cos(\omega_{(i)}t)\sin\left(\frac{(i)\pi x}{L}\right), \qquad (9.127)$$

where $kd \equiv x$, the continuous measure of distance along the string.

QUESTION 15: Waves on a String Prove that (9.126, 9.127) are the form for the possible frequencies and amplitudes for standing waves on a fixed string of length L, density ρ, and tension τ.

We can also derive the wave equation by looking at how Equation (9.113) changes as we take the continuous limit. First we rewrite (9.113) as

$$\frac{m}{d}\ddot{y}_k = \frac{\tau}{d}\left(\left(\frac{y_{k+1} - y_k}{d}\right) - \left(\frac{y_k - y_{k-1}}{d}\right)\right). \qquad (9.128)$$

Then as $d \to \infty$, Equation (9.128) becomes

$$\rho\left(\frac{\partial^2 y}{\partial t^2}\right) = \tau\left(\frac{\frac{\partial y}{\partial x}\big|_{x=kd+\frac{d}{2}} - \frac{\partial y}{\partial x}\big|_{x=kd-\frac{d}{2d}}}{d}\right). \qquad (9.129)$$

The term in parentheses on the right simplifies similarly to yield

$$\rho\left(\frac{\partial^2 y}{\partial t^2}\right) = \tau\left(\frac{\partial^2 y}{\partial x^2}\right), \qquad (9.130)$$

which is the familiar wave equation.

SUMMARY OF CHAPTER 9

The steps in solving a small oscillation problem are:
- Eliminate all ignorable coordinates from the Lagrangian to reduce the number of degrees of freedom to the bare minimum necessary. Use any convenient set of generalized coordinates for this. (We will call them q_k.)
- Make a reasonable choice for the coordinates representing the small deviations from the equilibrium point. We will call them ϕ_k. (For example, $\phi_k = \delta x_k$ for the linear triatomic molecule.)

- Find the matrices **t** and **v** from expanding the Lagrangian in the N small deviations and the N velocities in the vicinity of an equilibrium point q_k^0 ($q_k = q_k^0 + \phi_k$). Define the column vector

$$\phi = \begin{pmatrix} \phi_1 \\ \phi_2 \\ \ldots \end{pmatrix}. \tag{9.131}$$

The potential energy is

$$V = \tilde{\phi} \cdot \mathbf{v} \cdot \phi, \tag{9.132}$$

$$\mathbf{v} \equiv \frac{1}{2} \frac{\partial^2 V}{\partial \phi_i \partial \phi_j}\bigg|_{\phi_i,\phi_j=0}. \tag{9.133}$$

Taylor series for V near equilibrium point

If the equilibrium is stable, the potential energy will increase for any displacement, so **v** is a positive definite matrix. The kinetic energy is

$$T = \tilde{\dot\phi} \cdot \mathbf{t} \cdot \dot\phi, \tag{9.134}$$

$$\mathbf{t} \equiv \frac{1}{2} \frac{\partial^2 T}{\partial \dot\phi_i \partial \dot\phi_j}\bigg|_{\dot\phi_i,\dot\phi_j=0}. \tag{9.135}$$

Taylor series for T near equilibrium point

- Close to the equilibrium point, the general form of the Lagrangian is

$$L = \tilde{\dot\phi} \cdot \mathbf{t} \cdot \dot\phi - \tilde{\phi} \cdot \mathbf{v} \cdot \phi. \tag{9.136}$$

The equations of motion are

$$\mathbf{t} \cdot \ddot\phi + \mathbf{v} \cdot \phi = 0. \tag{9.137}$$

- Assume only a single mode $\phi(t) = \mathbf{\Phi} e^{i(\omega t - \delta)}$ is excited. Substitute into the equations of motion to obtain

$$[-\omega^2 \mathbf{t} + \mathbf{v}] \cdot \mathbf{\Phi} = 0. \tag{9.138}$$

- Solve $\det|-\omega^2 \mathbf{t} + \mathbf{v}| = 0$ for the squares of the mode frequencies. Use the method of cofactors or the method of row elimination to find the mode vectors. The N $\mathbf{\Phi}_{(i)}$ vectors are real numbers.

Some properties of small oscillation theory are:

PROBLEMS

- $\omega^2 \geq 0$.
- $\omega^2 = 0$ corresponds to a translation or a rotation.
- The number of modes equals the number of degrees of freedom. Each mode acts like an independent harmonic oscillator.
- The mode displacements for two nondegenerate modes are orthogonal vectors in the sense that $\tilde{\mathbf{\Phi}}_{(i)} \cdot \mathbf{t} \cdot \mathbf{\Phi}_{(j)} = 0, i \neq j$.
- The most general possible motion near to the equilibrium point can be expressed as a superposition of normal modes.
- Each mode amplitude has the time dependence $A_{(i)}(t) = A_{(i)}(0)e^{i\omega_{(i)}t}$.
- The equation $A_{(i)}(0) = \tilde{\mathbf{\Phi}}_{(i)} \cdot \mathbf{t} \cdot \check{\phi}(0)$ determines the complex amplitude from the initial conditions expressed in terms of the complex state vector at $t = 0$.
- An alternative to the complex notation is to consider the normal coordinates $\rho_{(i)}(t)$. These are real numbers that oscillate like separate harmonic oscillators, one for each mode.

PROBLEMS

Various Small Vibration Problems

Problem 1*: *(Pendulum attached to block)* A pendulum with mass m is attached to a block of mass M as shown in Figure 9.9. The block is free to move without friction on a horizontal table. There are two degrees of freedom: X, θ. Find the mode frequencies and mode vectors.

Problem 2*: *(Triple pendulum)* A "triple pendulum" (three pendulums each coupled to the adjacent two) has kinetic energy

$$T = \frac{1}{2}\left(\dot{\theta}_1^2 + \dot{\theta}_2^2 + \dot{\theta}_3^2\right) \tag{9.139}$$

and potential energy

$$V = \frac{1}{2}\left(\theta_1^2 + \theta_2^2 + \theta_3^2 - 2\epsilon[\theta_1\theta_2 + \theta_2\theta_3 + \theta_3\theta_1]\right), \tag{9.140}$$

where m, g, and l have been set equal to one for convenience. What are the $\mathbf{t}$ and $\mathbf{v}$ matrices? Find the mode frequencies. Are there any restrictions on the constant ϵ?

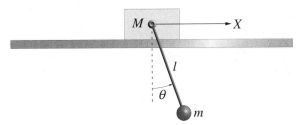

FIGURE 9.9

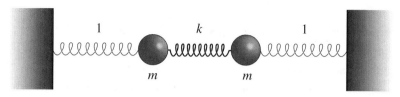

FIGURE 9.10

Then show that a particular set of mode vectors can be chosen to be

$$x_1 = \sqrt{\frac{2}{3}}(\theta_1, \theta_2, \theta_3),$$
$$x_2 = (\theta_1, 0, -\theta_3), \quad (9.141)$$
$$x_3 = \frac{1}{\sqrt{3}}(\theta_1, -2\theta_2, \theta_3).$$

Explain what the "modes" look like. Associate the frequencies with the modes above.

Problem 3: *(Two masses connected by three springs)* Two unit masses ($m = 1$) are connected to three springs so that longitudinal oscillations are possible, as shown in Figure 9.10. The outside springs have a unit spring constant, while the center spring has a spring constant k. Show that the normal mode frequencies for longitudinal oscillations are $\omega_1 = 1$, $\omega_2 = \sqrt{1 + 2k}$. Find the mode vectors as well.

Problem 4: *(Rectangular plate supported by springs)* A rectangular plate of mass m is supported at each corner by a spring of spring constant k (as shown in Figure 9.11. What are the mode frequencies and mode vectors? (Assume only vertical motion of each spring is possible.)

Problem 5*: *(Two coupled oscillators)* For two coupled oscillators with kinetic and potential energies given by $T = ma^2(2\dot{x}^2 - 10\dot{x}\dot{y} + 13\dot{y}^2)$ and $V = mga(5x^2 - 22xy + 25y^2)$, determine the mode frequencies and mode vectors. Prove that the mode vectors are orthogonal. Also find the Euler–Lagrange equations.

Problem 6*: *(Torque-free symmetric top)* Calculate the frequency of small oscillations in θ for a torque-free symmetric top, assuming that it is near to the solution $\theta = $ constant, $\dot{\phi} \equiv \omega_P$, $\dot{\psi} \equiv \Omega$. (Previously we proved, for constant θ, that $\omega_P \cos \theta = \frac{I_3 \Omega}{I_1 - I_3}$ (8.46).) Hint: Use the same Lagrangian we used for the heavy symmetric top (8.93),

FIGURE 9.11

PROBLEMS

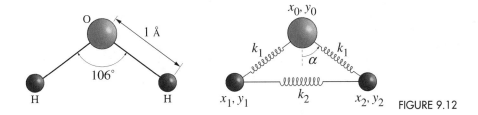

FIGURE 9.12

but without the potential energy due to gravitational torque. How could you excite oscillations in θ about the former equilibrium point? Is the angular momentum still constant?

Problem 7*: *(Water molecule)* (This problem is very long – it might be a class project.) The water molecule can be modeled with two unknown spring constants, as shown in Figure 9.12.

a) If you confine the motion to the plane of the molecule, why are there three vibrational modes and three nonvibrational (i.e., translational or rotational) modes? What motions correspond to the nonvibrational degrees of freedom? What is the general form of the Lagrangian in terms of vibrational and nonvibrational degrees of freedom? Be as specific as you can.

b) The molecule has reflection symmetry about the vertical (Y) axis. If $\mathbf{R}$ is the matrix that sends $x \to -x$, $y \to y$, what is $\mathbf{R}$ explicitly? If ϕ is a column vector expressing a certain mode in terms of a set of vibration coordinates, prove that $\mathbf{R}\phi$ is also a mode, as well as $(\mathbf{1} \pm \mathbf{R})\phi$ (where $\mathbf{1}$ is the identity matrix). Hence prove that the modes must be either odd or even in terms of this reflection, (i.e., either $(\mathbf{1} + \mathbf{R})\phi = 0$ or $(\mathbf{1} - \mathbf{R})\phi = 0$).

c) A useful choice of coordinates is the choice u, v, w in Figure 9.13. How do you find the relative lengths of the arrows in each case? (Hint: Neither the center of mass must move nor the angular momentum change when one or more of these coordinates changes. Why not?) Prove that the motion of the individual molecules are

$$\delta\vec{r}_1 = (\underbrace{-(u+w)\cos\alpha - v\sin\alpha}_{\delta x}, \underbrace{v\cos\alpha - (u+w)\sin\alpha}_{\delta y}),$$

$$\delta\vec{r}_2 = (-(u-w)\cos\alpha + v\sin\alpha, \ v\cos\alpha + (u-w)\sin\alpha), \quad (9.142)$$

$$\delta\vec{r}_0 = 2\frac{M_H}{M_O}(u\cos\alpha, \ -v\cos\alpha + w\sin\alpha).$$

d) Find the $\mathbf{t}$ matrix. Then find the form of the $\mathbf{t}$ matrix in the limit that $M_O \to \infty$ as well. We make this approximation to simplify our calculation.

e) To find the $\mathbf{v}$ matrix, assume

$$\delta V = \delta V_{\text{OH}} + \delta V_{\text{HH}}, \quad (9.143)$$

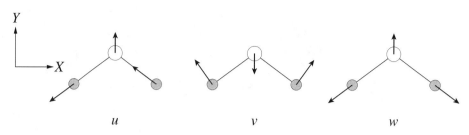

FIGURE 9.13

where

$$\delta V_{\text{OH}} = \frac{k_1}{2}\{\delta l_1^2 + \delta l_2^2\} \tag{9.144}$$

comes from stretching one or both of the O–H bonds of length $\simeq 1$ Å and

$$\delta V_{\text{HH}} = \frac{k_2}{2} l^2 [\delta(2\alpha)]^2 \tag{9.145}$$

comes from changing the bond angle 2α. The complicated part is finding δl_1, δl_2, $\delta(2\alpha)$ in terms of u, v, w, our chosen coordinates. We first express the vectors from the oxygen to the two hydrogen atoms in terms of u, v, w, which are all presumed to be small quantities. Then we project these vectors (by taking the dot product) onto unit vectors, either along the bond directions for δl_1, δl_2 or perpendicular to the bonds (and divided by l) for $\delta(2\alpha)$. Substitute your answers into the above expressions for δV, giving something quadratic in the (small) coordinates. Next, in the limit that $M_O \to \infty$, find $\mathbf{v}$.

f) Find the mode frequencies. Then find the numerical dimensionless mode frequencies ($\frac{\omega_3}{\omega_1}$ and $\frac{\omega_2}{\omega_1}$). Find numerical values for the mode vectors of the three vibrational modes of the water molecule as well. Assume $k_1 = 32k_2$. Interpret the mode vector for the $\omega = 0$ mode. Compare to the experimental values of (in terms of inverse wavelength, which is proportional to frequency) 1,595 cm^{-1}, 3,654.5 cm^{-1}, and 3,755.8 cm^{-1}.

Two Coupled Pendulums

Problem 8: *(Coupled pendulums – different masses)*

a) What happens to the mode vectors and mode frequencies if the pendulum masses in the coupled pendulum problem are not equal? Call them m_1 and m_2 respectively.

b) Suppose pendulum 1 is released from rest with a certain amount of energy E_1. Plot the maximum and minimum fraction of this energy stored in pendulum 2 versus $\frac{m_2}{m_1}$. This result has numerous practical applications.

Problem 9: *(Coupled pendulums – massive spring)* Suppose the spring that couples the two pendulums in the coupled pendulum problem has a small mass m_s. How does the Lagrangian change? What are the frequencies of the modes and the mode displacement ratios in this case?

Problem 10: *(Coupled pendulums – no small-angle approximation)* Write a computer simulation of two coupled pendulums of different masses connected by a massive spring. In the simulation, a small-amplitude approximation should not be made. It should also be possible to vary the individual pendulum masses and the spring mass. Study how energy is exchanged between the two oscillators. What happens if the masses are not equal? Can you derive a formula to predict this for small amplitudes of vibration? Do a spectrum (Fourier) analysis of the motion for various choices of the starting amplitudes. Prove numerically that the modes are what you expect for small amplitudes from the theoretical analysis in Problems 8 and 9. Is the concept of mode still a valid one if the amplitude becomes very large?

Double Pendulum

Problem 11: *(Double pendulum – normal coordinates)* The double pendulum is started at rest, with $\phi_1 = 30°$, $\phi_2 = -30°$. Find the normal coordinates and a formula that gives the two angles as functions of the time.

Linear Triatomic Molecule

Problem 12: *(Linear triatomic molecule – relative coordinates)* For the linear triatomic molecule problem, change the coordinates to the two relative coordinates and the position of the center of mass for the whole molecule. Write the Lagrangian. Since the center of mass is an ignorable coordinate, it can be eliminated from the Lagrangian, either by just ignoring this part or, more rigorously, using the Legendre transformation and constructing a Routhian (new Lagrangian with two degrees of freedom – see (5.95)). Solve the problem with the matrix method and find the mode vectors and their frequencies. Eliminating ignorable coordinates from the start is a very good idea since it reduces the dimensionality of the matrix.

Problem 13: *(Linear triatomic molecule – T and V)* Prove that (9.64) and (9.70) are the potential and kinetic energies. Also, find the Lagrangian and EOM.

Problem 14*: *(Diatomic versus triatomic molecule)* In this question you compare the modes of a diatomic molecule and a triatomic molecule made of the same atoms.

a) Consider the linear diatomic molecule carbon monoxide in Figure 9.14A. The molecule consists of masses M_C and M_O, separated by a spring of constant k. Determine the mode frequencies and mode vectors of the system. The mass ratio $\frac{M_O}{M_C} = \frac{4}{3}$.

b) Now consider a linear triatomic molecule of CO_2 as shown in Figure 9.14B. Assume the spring constants k are the same as in the part a) above. Find T and V,

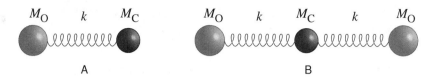

FIGURE 9.14
A) Carbon monoxide atom. B) Carbon dioxide atom.

and then the 3 × 3 symmetric matrices **t** and **v**. Finally, solve for the three possible values of ω^2, the mode frequencies, and find the corresponding modes vectors. Compare to part a).

Matrix Theory

Problem 15: *(Eigenvalues and eigenvectors)* Find the characteristic polynomial from the determinant of the two 3 × 3 matrices **M** below. Also find the eigenvalues and eigenvectors of each. Then find the unitary transformation **U** such that $\mathbf{U M \tilde{U}} = \mathbf{D}$, where **D** is the diagonal matrix consisting of the eigenvalues of **M**. (Hint: You can get **U** from the eigenvectors.) Determine the cofactors of the first row of $\mathbf{M} - \lambda \mathbf{1}$, where λ is one of the three eigenvalues and **1** is the identity matrix.

a)

$$\mathbf{M} = \begin{pmatrix} 1 & 2 & 1 \\ 1 & 0 & -1 \\ 1 & 1 & 1 \end{pmatrix} \tag{9.146}$$

b)

$$\mathbf{M} = \begin{pmatrix} 8 & -8 & -2 \\ 4 & -3 & -2 \\ 3 & -4 & 1 \end{pmatrix} \tag{9.147}$$

Stable Equilibrium

Problem 16: *(Positive definite 1)* Prove that a positive definite **v** matrix gives a *stable* equilibrium. Is the kinetic energy matrix for small deviations from equilibrium always positive definite?

Problem 17: *(Positive definite 2)* Is the matrix $\begin{pmatrix} 1 & -2 & -2 \\ -2 & 1 & -2 \\ -2 & -2 & 1 \end{pmatrix}$ positive definite?

Problem 18*: *(Hénon–Heiles Hamiltonian)* To model the motion of stars in our galaxy, in the 1960s Hénon and Heiles proposed a Hamiltonian with two degrees of freedom, x and y. The Hamiltonian is

$$H = \frac{1}{2}(p_x^2 + p_y^2) + V(x, y). \tag{9.148}$$

PROBLEMS

The potential energy is

$$V(x, y) = \frac{1}{2}\left(x^2 + y^2 + 2x^2 y - \frac{2y^3}{3}\right). \tag{9.149}$$

a) The equilibrium points are the points in 4-D phase space where

$$\dot{x} = \dot{p}_x = \dot{y} = \dot{p}_y = 0. \tag{9.150}$$

Prove that there are four equilibrium points, which depend only on x and y, and plot them on the XY plane. Sketch some approximate equipotential curves in the XY plane.

b) One of these points is a stable equilibrium; the other three are unstable. Find the **v** matrix at all four points and use it to prove this statement and identify which points are unstable. What is $V(x, y)$ close to the three unstable equilibrium points?

c) Prove that the motion is bounded if the total energy $E < \frac{1}{6}$ and unbounded if $E > \frac{1}{6}$. Hint: $E > V$ for positive kinetic energy. $E = V$ is a turning point of the motion. Plot $V(0, y)$ versus y.

The Hénon–Heiles Hamiltonian is a favorite one for demonstrating the onset of Hamiltonian chaos in the event that E is close to, but less than $\frac{1}{6}$.

N Masses Connected by a String

Problem 19: *(N masses – longitudinal vibrations)* Prove that the V (9.106) we found for transverse vibrations of the string is the same as the V for longitudinal vibrations except that $\frac{\tau}{d}$ is replaced by k, the spring constant of the string.

Problem 20: *(3 masses)* Find the different mode frequencies (9.123) and mode vectors (9.124) for the case of $N = 3$ masses. Sketch the shape of the mode vectors.

Problem 21: *(N masses – continuous limit)*

a) Using our results for the continuum limit $N \to \infty$ (9.126, 9.127), assume the continuous string is initially in the shape of a triangle. It has zero displacement at both ends and a displacement s at the center. Express the string shape as a superposition of the mode vectors to find the behavior of the string at any point as a function of time.

b) Show that the equation of motion for the continuous string has solutions that are waves traveling to the right or left with a certain velocity. What is this velocity in terms of the mass density ρ of the string and the tension τ? How do you reconcile this alternative description of the string motion for the case of a triangular deformation at $t = 0$ with the usual description using superposition of modes as in part a)? A violin string, after it slips from the bow, behaves very similarly to what you calculated in part a).

Problem 22: *(N masses – Fourier decomposition)* For the N point masses connected by massless string problem, calculate the restoring force on the kth mass with displacement y_k. Assume the relative displacement is much smaller than the spacing between the different masses. Using this result, do a computer simulation of $N = 5$ masses on a string. Start with a triangular displacement of the string (zero displacement at both ends and a displacement s at the center). Perform a numerical Fourier transform of the displacement of one mass on the string versus time to determine the mode frequencies. How would you isolate a particular mode to determine how the masses are displaced in this mode? How does the result vary if you try a different N?

APPENDIX

WHAT IS A COFACTOR?

To determine what the mode displacement ratios for each part of a system are in a given mode, you must solve a set of N homogeneous linear equations in N unknowns (9.40). Here we discuss one possible way to solve a set of homogeneous linear equations using *cofactors*. Another way is to use row elimination. These methods fail if there is a degeneracy (i.e., if two or more normal modes have the same frequency).

Cramer's rule* can be taken as the definition of the determinant of an $N \times N$ matrix:

$$\det \begin{vmatrix} a & b & c & d \\ \cdot & \cdot & \cdot & \cdot \\ \cdot & \cdot & \cdot & \cdot \\ \cdot & \cdot & \cdot & \cdot \end{vmatrix}. \tag{9.151}$$

The rule says that the determinant of the matrix above is given by

$$\det = a\, \text{Cofactor}(a) - b\, \text{Cofactor}(b) \tag{9.152}$$
$$+ c\, \text{Cofactor}(c) - d\, \text{Cofactor}(d).$$
Cramer's rule

The cofactor of an element of the matrix is defined as the $N-1 \times N-1$ subdeterminant you get by crossing out the row and column of the element whose cofactor you want to find. Cofactors (a), (b), (c), and (d) are shown in Figure 9.15.

---

* This is the rule for evaluating the determinant of a matrix. See Leon, *Linear Algebra*, 2d ed., p. 79 for further discussion of this rule.

APPENDIX WHAT IS A COFACTOR?

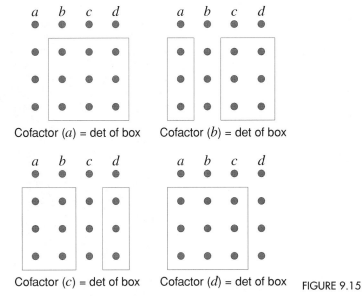

FIGURE 9.15

QUESTION 16: Cofactors Use the method of cofactors to find the determinant of a 3×3 matrix.

Suppose the determinant we want vanishes:

$$\det \begin{vmatrix} a & b & c & d \\ \cdot & \cdot & \cdot & \cdot \\ \cdot & \cdot & \cdot & \cdot \\ \cdot & \cdot & \cdot & \cdot \end{vmatrix} = 0. \tag{9.153}$$

Consider the cofactors as a row vector. Cramer's rule (9.152) implies this vector is orthogonal to (a, b, c, d), the first row of the matrix:

$$(\text{Cofactor}(a), -\text{Cofactor}(b), \text{Cofactor}(c), -\text{Cofactor}(d)) \cdot \begin{pmatrix} a \\ b \\ c \\ d \end{pmatrix} = 0. \tag{9.154}$$

The cofactor vector formed in this way will automatically satisfy the first row of the linear equations we obtain for the mode vector coordinates:

$$\begin{pmatrix} a & b & c & d \\ e & f & g & h \\ \cdot & \cdot & \cdot & \cdot \\ \cdot & \cdot & \cdot & \cdot \end{pmatrix} \cdot \begin{pmatrix} e_1 \\ e_2 \\ e_3 \\ e_4 \end{pmatrix} = \begin{pmatrix} 0 \\ 0 \\ 0 \\ 0 \end{pmatrix}. \tag{9.155}$$

The vector (e_1, e_2, e_3, e_4) stands for any vector that is a solution to Equation (9.155). What about for the second element in (9.155)? If the row vector of cofactors was a solution, then the second element would be proportional to

$$(\text{Cofactor}(a), -\text{Cofactor}(b), \text{Cofactor}(c), -\text{Cofactor}(d)) \cdot \begin{pmatrix} e \\ f \\ g \\ h \end{pmatrix}. \tag{9.156}$$

However, this is the same as evaluating the determinant

$$\begin{pmatrix} e & f & g & h \\ e & f & g & h \\ . & . & . & . \\ . & . & . & . \end{pmatrix}. \tag{9.157}$$

Such determinants (with identical rows or columns) are always equal to zero. We can do this for the rest of the elements $3 \ldots N$ in the same manner as for the second element. Therefore, since the row vector of cofactors satisfies the same equation as the coordinates for the mode vector, it must be proportional to this vector. Up to a multiplicative constant, we have found the mode vectors by this method.

CHAPTER TEN

APPROXIMATE SOLUTIONS TO NONANALYTIC PROBLEMS

OVERVIEW OF CHAPTER 10

Reflecting on the past nine chapters, you may realize that we have only solved a few problems in an analytic form. What about the many other problems that one is sure to encounter in physics? Many of the most interesting problems do not have exact analytic solutions. This chapter will introduce a few methods for dealing with problems of this type. Often we start with a problem we already know how to solve, like the Kepler problem or harmonic oscillator. Then we add on a part, known as a "perturbation," which approximates the more complex problem. To get a more accurate solution, we add on more terms.

If a system in motion is perturbed slightly, does it diverge rapidly from the unperturbed motion or does it oscillate around the unperturbed orbit? In the former case, we say the system is "dynamically unstable"; in the latter case there is "dynamical stability." If we assume the motion, at least initially, is close to the unperturbed motion, we can subtract the perturbed equations of motion from the unperturbed ones, keeping only terms linear in the difference between the perturbed and unperturbed motion. This is known as "linearizing" the equations of motion. There are two ways to introduce a perturbation of the motion. We can either disturb the initial conditions (known as a one-time perturbation) or else add a small change in the Lagrangian, usually in the potential energy.

Perturbations that are periodic functions of the time play an important role in physics, especially for understanding the transition between stable motion and chaotic motion. If the equations of motion for the perturbed system are nonlinear equations, the frequency of the motion doesn't have to equal the natural oscillation frequency in order to create an instability: It can be a multiple of the oscillator frequency – either a *harmonic* (integer multiple) or a *subharmonic* (rational fraction multiple).

In many cases, however, the perturbation theory leads to a linear equation of motion with coefficients that are periodic functions of the time. This special type of linear equation has not been encountered previously. It is called a *Hill equation*. If the coefficients vary sinusoidally, the equation is a *Mathieu equation*. Hill and Mathieu equations have an extremely wide range of applications in physics and engineering. In linear oscillator systems, a periodic variation in the oscillator restoring force can cause *parametric amplification* or *parametric resonance*. A common practical example is that soldiers marching across a bridge who fail to break their stride risk exciting a parametric resonance and thus bringing the bridge down. A simple example of this type of dynamical system with time-dependent parameters in the equation of motion will be a pendulum with a vibrating support point that moves up and down with an adjustable frequency and amplitude. This is a linear parametric oscillator because the equation of motion is linear in θ, but the coefficient multiplying θ is a periodic function of time. A classroom demonstration of such a vibrating pendulum is a very effective way to demonstrate the physical effects of parametric resonance – if possible, such a demonstration should be available to accompany the mathematical treatment.

The only perfectly linear system in physics is the Schrödinger equation of quantum mechanics (even Maxwell's equations get nonlinear corrections from quantum electrodynamics). The Lindstedt–Poincaré perturbation theory, originally developed for astronomical problems, is one method for finding systematic approximations to the motion of nonlinear, yet periodic systems. Our specific example of a nonlinear system will be the Duffing oscillator, which is like a harmonic oscillator but with an additional cubic restoring force as a perturbation. We will explore both the undriven (free) oscillator and also the case of a Duffing oscillator driven by an external force of variable frequency. For this last case, we will employ iteration methods, still another form of perturbation theory.

10.1 STABILITY OF MECHANICAL SYSTEMS

In Chapter 3 we discussed the stability of 1-D scleronomic mechanical systems near static equilibrium points ($\frac{\partial V}{\partial q} = 0$). One could test for stability by looking at the sign of $\frac{\partial^2 V}{\partial q^2}$. We extended this test to many-dimensional linear systems in Chapter 9 by checking that the eigenvalues of the **v** matrix are all positive. Here we will look at the more general question of the stability of a moving system. This is known as *dynamical equilibrium*.

We will focus on perturbations of a modified Keplerian orbit and generalize from this.[*] Recall that for the gravitational force between two bodies (4.43),[†]

$$V_{\text{eff}}(r) \equiv -\frac{k}{r} + \frac{l^2}{2\mu r^2}. \tag{10.1}$$

The equations of motion were derived earlier in Chapter 4. They are (4.44)

$$\mu\ddot{r} = \frac{l^2}{\mu r^3} - \frac{k}{r^2} \equiv -\frac{dV_{\text{eff}}}{dr} \tag{10.2}$$

and (4.38)

$$\mu\dot{\phi} = \frac{l}{r^2}. \tag{10.3}$$

---

[*] For example, the gravitational attraction of the other planets, especially Jupiter, perturbs the orbital motion of the Earth. Will the attraction of Jupiter eventually make the Earth's orbit unstable? See the homework problem.

[†] See Chapter 4 for the definitions of the constant $k > 0$ and the reduced mass μ. l is the angular momentum in the center of mass system.

10.1 STABILITY OF MECHANICAL SYSTEMS

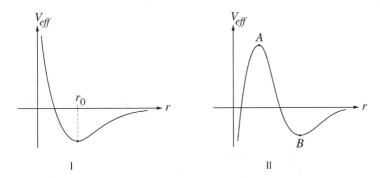

FIGURE 10.1
I) Kepler problem using (10.1) as the potential. II) Kepler problem with $\frac{1}{r^3}$ perturbation to (10.1).

In Figure 10.1(I), we have plotted the effective potential in a gravitational field for the Kepler problem (10.1). Recall that for circular motion the equilibrium point lies at the minimum of this potential, with $r = r_0$, the equilibrium radius. Near the equilibrium point, which is a dynamic, not a static, equilibrium, the potential is like that of a harmonic oscillator, with curvature $\frac{d^2 V_{\text{eff}}}{dr^2}$, evaluated at $r = r_0$. Drawing an analogy to the harmonic oscillator, if we kick the system a little (change $\dot{r}$)*, it should oscillate with a frequency proportional to the square root of the spring constant (i.e., the curvature of V_{eff}). These oscillations will correspond to elliptical orbits.

Note we can generalize (10.2, 10.3) to be

$$\mu \ddot{r} = \frac{l^2}{\mu r^3} - \frac{dV(r)}{dr} \equiv -\frac{dV_{\text{eff}}}{dr} \tag{10.4}$$

for any central force $V(r)$ and the following derivation will still be valid. Also note that the unperturbed orbit $r(t)$ need not be circular.

For sufficiently small perturbations, if the perturbed solution stays close to the unperturbed solution, we can use Taylor's theorem to obtain linear equations of motion for the difference between perturbed and unperturbed motion. This is called "linearization" of the perturbed equations of motion.

To linearize, imagine that we have unperturbed motion $r(t), \phi(t)$ and perturbed motion $\tilde{r}(t), \tilde{\phi}(t)$, which follow the same equations of motion, (10.4, 10.3). The difference lies in slightly different initial conditions. Let's take the special case that the angular momentum is the same for both perturbed and unperturbed orbits. Define $\delta r \equiv \tilde{r} - r$. From (10.4) evaluated at $\tilde{r} = r + \delta r$ subtract (10.4) evaluated at r:

$$\mu \ddot{\delta r} = -\left[\frac{dV_{\text{eff}}}{dr}\bigg|_{r+\delta r} - \frac{dV_{\text{eff}}}{dr}\bigg|_{r}\right]. \tag{10.5}$$

A straightforward application of Taylor's theorem in which terms of order δr^2 and higher are neglected yields the linearized radial equation of motion for small deviations from the

---

* Assume for simplicity that we ignore the case where the perturbation changes the orbital plane.

unperturbed orbit:

$$\mu \ddot{\delta r} + \left.\frac{d^2 V_{\text{eff}}}{dr^2}\right|_r \delta r = 0. \tag{10.6}$$

This is a homogeneous second-order linear differential equation. If the unperturbed orbit is a circular one, $\frac{dV_{\text{eff}}}{dr} = 0$ at $r = r_{\text{eq}}$.* If so, then Equation (10.6) can be written

$$\ddot{\delta r} + \omega_{\text{eq}}^2 \delta r = 0, \tag{10.7}$$

where

$$\omega_{\text{eq}}^2 = \frac{1}{\mu} \left.\frac{d^2 V_{\text{eff}}}{dr^2}\right|_{r=r_{\text{eq}}}. \tag{10.8}$$

Note that the sign of ω_{eq}^2 will indicate whether the perturbed orbit is stable or not. If $\omega_{\text{eq}}^2 > 0$, the perturbed orbit oscillates about the unperturbed orbit. If $\omega_{\text{eq}}^2 < 0$, then the perturbed orbit will diverge exponentially from the unperturbed one. In the latter case, the linear perturbation theory is valid only for a short time interval after the perturbation occurs, before the perturbed motion diverges too far from the unperturbed motion.

For the Kepler case (10.2), we obtain

$$\left.\frac{d^2 V_{\text{eff}}}{dr^2}\right|_r = -\frac{2k}{r^3} + \frac{3l^2}{\mu r^4}. \tag{10.9}$$

For a circular orbit r_0, $l^2 = \mu k r_0$. Substituting this last relation into (10.9) gives $\left.\frac{d^2 V_{\text{eff}}}{dr^2}\right|_{r_0} = \frac{k}{r_0^3}$. From (4.61), if the period of the circular orbit is τ and $\omega_0 \equiv \frac{2\pi}{\tau}$, it can be shown that $\omega_0^2 = \frac{k}{\mu r_0^3}$. By comparing with Equation (10.8), we have proved that small perturbations of the circular Kepler orbits are stable, with the same period as the original orbit. The solutions to (10.7) are sines and cosines, so have bounded variations. In the more general case of Keplerian elliptical orbits, we would have to deal with the time dependence of the coefficient of δr in (10.6). It is best to use the linearity of the "$u \equiv \frac{1}{r}$" equation developed in Chapter 4 (4.49). This equation is already linear with constant coefficients, even for elliptical orbits, so the stability of these is guaranteed too.

Another type of perturbation consists of modifying the potential energy by an additional term or terms. An example of this is the precession of the Kepler ellipses due to the $\frac{1}{r^3}$ correction to the potential from general relativity. Here we will consider an additional potential of the same form: $\frac{b}{r^3}$. The coefficient b is assumed to have arbitrary magnitude and is not necessarily a small perturbation. We will specialize to the circular orbits and study dynamical stability. In Figure 10.1(II), we've plotted V_{eff} for an additional $\frac{1}{r^3}$ attractive

---

* Note that we will use r_{eq} to denote the equilibrium radius of any circular orbit for an arbitrary central force. The parameter r_0 will refer specifically to the Keplerian circular orbit.

10.1 STABILITY OF MECHANICAL SYSTEMS

potential (i.e., $b < 0$). There are now two possible equilibrium points A and B. Note that these points describe circular orbits, but the radii of the circles at points A, B are different from r_0. We guess that at point A the particle will be dynamically unstable, while at point B, stable oscillations will occur if the new equilibrium circular orbit is given a small perturbation, such as an impulse. At point A the additional $\frac{1}{r^3}$ potential cannot be considered small, since it overcomes the repulsive centrifugal angular momentum barrier.

The potential $V(r)$ now takes the form

$$V(r) \equiv -\frac{k}{r} + V_{\text{add}}(r), \qquad (10.10)$$

where $V_{\text{add}}(r)$ is the additional term in the potential that makes the problem deviate from the Kepler problem.

First we need to find the equilibrium condition for a circular orbit, that is, the r_{eq} value that makes $\ddot{r} = 0$. From (10.4) this is the solution to the equation

$$V'(r_{\text{eq}}) = \frac{l^2}{\mu r_{\text{eq}}^3}, \qquad (10.11)$$

where $V'(r_{\text{eq}}) \equiv \frac{dV}{dr}|_{r=r_{\text{eq}}}$.

Both point A and point B in Figure 10.1(II) correspond to solutions to this equation for the potential $V(r) = -\frac{k}{r} + \frac{b}{r^3}$. The quadratic equation that results from inserting this into (10.11) is $r_{A,B}^2 - r_0 r_{A,B} - \frac{3b}{k} = 0$, which has two roots. For real roots, $b > -\frac{kr_0^2}{12}$. If $b > 0$ (repulsive force), only point B can exist because point A would correspond to a nonphysical negative separation of the orbiting bodies. If the added force is attractive, $b < 0$, points A and B will both exist if the condition given above on b is met. Let $b \equiv \epsilon k r_0^2$, thereby defining the new dimensionless constant ϵ. The condition that both points A and B exist now becomes $0 > \epsilon > -\frac{1}{12}$.

We can evaluate the second derivative of the effective potential at $r = r_A$ to confirm our suspicion that the circular orbit at point A is unstable – one slight push and the orbiting masses will experience the catastrophe of falling into an infinitely deep potential well. It can be shown that $r_A < \frac{r_0}{2}$ and that the second derivative of the effective potential at point A is $\frac{k}{r_A^3}(2 - \frac{r_0}{r_A})$. This is always negative at point A; hence the instability of this orbit is confirmed by actual calculation of the second derivative. An unstable orbit means that the solutions to the linearized equation of motion for δr (10.7) are exponentials – a small perturbation of the equilibrium orbit leads to an exponential divergence away from the original circular orbit.

Point B is the new stable circular orbit after the addition to the potential is turned on. Equation (10.7) can be used to check that the modified Kepler potential does have a stable circular orbit at point B.

QUESTION 1: Modified Kepler Potential Consider the Kepler potential V_{eff} modified with an additional $\frac{1}{r^2}$ potential. What effect on the equilibrium positions for circular orbits would you expect for attractive and repulsive forces of this kind? Verify the restrictions on the strength of the $\frac{1}{r^2}$ term needed for the existence of a stable circular orbit.

QUESTION 2: *Stability of Inverse Cubed Perturbation* Prove that the two roots of (10.11) are $r_{A,B}^2 - r_0 r_{A,B} - \frac{3b}{k} = 0$. Then prove that $\frac{d^2 V_{\text{eff}}}{dr^2}|_{r=r_{\text{eq}}}$ is $\frac{k}{r_{A,B}^3}(2 - \frac{r_0}{r_{A,B}})$. Finally, check that point A is unstable and point B is stable.

Linearized equations of motion must be used with care so that neglected terms do not become large and render conclusions invalid. Another comment concerns the use of the "δ" notation. This notation meant something entirely different in Chapter 2 where it was used in connection with the variational calculus to represent an arbitrary, but infinitesimally small, function of time. Here it stands for the solution to linearized perturbative equations of motion. The meaning is quite different, and these two uses of "δ" notation should not be confused with each other.

10.2 PARAMETRIC RESONANCE

In this section we will examine what happens when the linearized equation contains periodic time-varying coefficients.

A very simple physics toy consists of a rigid plastic pendulum attached to a stick which can be held in the hand as shown in Figure 10.2. By shaking the stick up and down, one can excite an oscillation in the pendulum, which quickly builds up enough amplitude to spin it rapidly around the stick. To produce the desired effect, one has to adjust both the frequency and the amplitude of shaking – this determines the rate at which the pendulum gains energy and the ultimate angular rotation speed. What frequencies and amplitudes of shaking give unstable motion (continuous rotations rather than oscillations)? What determines which direction the pendulum will spin? We will answer these questions below.

If the stick is held steady, the equation of motion (in the limit of small deflections) is

$$\ddot{\theta} + \frac{g}{l}\theta = 0. \tag{10.12}$$

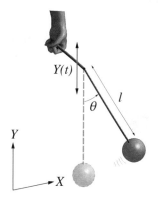

FIGURE 10.2
Pendulum with driven support.

10.2 PARAMETRIC RESONANCE

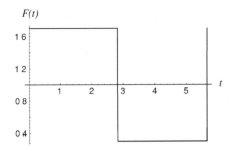

FIGURE 10.3
$F(t)$ = restoring force for $x = .9, r = .7$.

By shaking the stick, one accelerates the pendulum up or down, effectively adding to or offsetting the effect of gravity. Taking $\sqrt{\frac{l}{g}}$ as the unit of time, let's assume that we can write Equation (10.12) with a variable $g = g(t)$ as

$$\ddot{\theta} + (1+r)\theta = 0, \quad 0 \le t < \pi x,$$
$$\ddot{\theta} + (1-r)\theta = 0, \quad \pi x \le t < 2\pi x. \tag{10.13}$$

Subsequent periods of $2\pi x$ just repeat the same sequence of a restoring force proportional to $(1+r)$ for one half of the period and $(1-r)$ for the second half. There are two variable parameters: x, which determines the period of shaking, and r, which determines the variation in the amplitude of the restoring force for the pendulum. We plot the restoring force for one set of x, r in Figure 10.3. Unstable solutions for the linearized Equations (10.13) will give exponentially diverging values for $\theta(t)$. Physically these solutions correspond to the pendulum spinning continuously around the stick.

The Hill Equation

This simple toy illustrates a type of equation of motion that occurs widely in physics. For small deflections of the pendulum, Equation (10.13) is a second-order linear differential equation of the type very similar to the simple, damped harmonic oscillator, except that the coefficients depend periodically on the time:

$$\ddot{q} + a(t)\dot{q} + b(t)q = 0. \tag{10.14}$$

Hill equation: $a(t)$, $b(t)$ are periodic

Normally the coefficient of $\dot{q}$ is zero, so we deal with an equation of the form (10.14) with $a(t) = 0$. The periodicity means $b(t+T) = b(t)$ for the period T of shaking the stick. In our case (10.13), $a(t) = 0$, and $b(t) = 1 + r$ for $0 \le t < \pi x$ and $b(t) = 1 - r$ for $\pi x \le t < 2\pi x$.

A linear equation of this type occurs often when a system exhibiting periodic motion is perturbed in some way. This type of equation was first derived by G. W. Hill to describe the effect of perturbations on the orbit of the Moon, and it occurs in many other places in physics, including the quantum motion of electrons in a periodic potential of a crystal. The band theory of solids is based on a similar equation, as is the theory of propagating electromagnetic waves in a periodic structure. There are applications to technology as well – parametric amplifiers are an example. The theory of beam stability in strong focusing circular accelerators is still another application of the Hill equation. For unstable situations such as the spinning pendulum, the basic idea is to set up oscillations in a resonant system by a periodic variation of the restoring force in the system.

For a simple harmonic oscillator, which has constant coefficients in the equation of motion, the origin of time can be shifted by any amount without changing the equation. For a Hill equation, (10.14), the equation only remains invariant if we shift the time by a single period T. Nevertheless, important consequences can be deduced from this fact, as we will see.

Floquet Theorem

The *Floquet Theorem* says that complex solutions of (10.14) can be found that can be put in the form

$$q(t) = P(t)e^{\pm i\mu t}, \quad P(t+T) = P(t). \tag{10.15}$$

In other words, the general solution is a periodic function of period T times an exponential containing a constant μ. In practice μ must be determined by solving the equation, often by numerical integration on a computer. If μ is real, there is bounded motion. In that case, μT is the *phase advance* of the oscillator in one period of the restoring force coefficient.

The key fact is this: We can determine the solution of any second-order equation (including Equation (10.14)) for all time by specifying the value and the time derivative (position and velocity) of the solution at the start ($t = 0$). Furthermore, if the equation of motion is linear, the solution and its time derivative at any later time can be expressed as linear combinations of just two independent solutions. Imagine two possible sets of initial conditions: $\theta_c(0) = 1, \dot{\theta}_c(0) = 0$ for one solution, which we call "cosinelike" and $\theta_s(0) = 0, \dot{\theta}_s(0) = 1$, which we call "sinelike." The most general solution is a superposition of these two special solutions with arbitrary constant coefficients, $\theta(0), \dot{\theta}(0)$:

$$\theta(t) = \theta(0)\theta_c(t) + \dot{\theta}(0)\theta_s(t). \tag{10.16}$$

(Prove for yourself that this is the most general form of the solution to Equation (10.14).) Define the constants $A, B, C,$ and D at the time $t = T = 2\pi x$:

$$\begin{aligned} \theta(T) &= \theta(0)\theta_c(T) + \dot{\theta}(0)\theta_s(T) \equiv A\theta(0) + B\dot{\theta}(0), \\ \dot{\theta}(T) &= \theta(0)\dot{\theta}_c(T) + \dot{\theta}(0)\dot{\theta}_s(T) \equiv C\theta(0) + D\dot{\theta}(0), \\ A &\equiv \theta_c(T), \quad B \equiv \theta_s(T), \quad C \equiv \dot{\theta}_c(T), \quad D \equiv \dot{\theta}_s(T). \end{aligned} \tag{10.17}$$

Write Equations (10.17) in terms of a matrix

$$\mathbf{M} \equiv \begin{pmatrix} A & B \\ C & D \end{pmatrix}, \tag{10.18}$$

10.2 PARAMETRIC RESONANCE

which takes us from an arbitrary starting point at $t = 0$ to the solution and its derivative at $t = T = 2\pi x$, which is one period of the restoring force:

$$\begin{pmatrix} \theta \\ \dot{\theta} \end{pmatrix}_T = \mathbf{M} \begin{pmatrix} \theta \\ \dot{\theta} \end{pmatrix}_{t=0}. \tag{10.19}$$

Since Equation (10.14) is periodic, it is invariant to translations of the time axis by the period T. Therefore, we must have exactly the same matrix to carry us from $t = (n-1)T$ to $t = nT$:

$$\begin{pmatrix} \theta \\ \dot{\theta} \end{pmatrix}_{t=nT} = \mathbf{M}^n \begin{pmatrix} \theta \\ \dot{\theta} \end{pmatrix}_{t=0}. \tag{10.20}$$

QUESTION 3: Matrix M1 Why is the matrix that takes the solution from $t = 0$ to $t = T$ the same as the matrix that takes us between any two of the subsequent periods of the support motion?

The question of stability then depends on whether the matrix elements of $\mathbf{M}^n$ remain finite as n increases without limit, or instead blow up.

In most cases, the actual values of A, B, C, and D must be determined by a numerical integration. Liouville's theorem requires that the determinant of $\mathbf{M} = AD - BC = 1$. Usually this fact is used to check the accuracy of the numerical integration. If there were a damping term in the original equation, the determinant would be less than 1.

QUESTION 4: Matrix M2 Explain why Liouville's theorem implies that the determinant of $\mathbf{M}$ is one.

In the simple example of the pendulum on a stick, the solutions and the matrix $\mathbf{M}$ can be determined analytically, since they are combinations of sines and cosines during the period when the force is constant and proportional to $1 \pm r$. The form of the matrix is given as the product of two matrices, the first of which transports the solution for the first half period, the second for the second half of the period. The matrix in this case is

$$\mathbf{M}(x, r) = \begin{pmatrix} \cos \Gamma_- & \frac{\sin \Gamma_-}{\sqrt{1-r}} \\ -\sqrt{1-r} \sin \Gamma_- & \cos \Gamma_- \end{pmatrix} \cdot \begin{pmatrix} \cos \Gamma_+ & \frac{\sin \Gamma_+}{\sqrt{1+r}} \\ -\sqrt{1+r} \sin \Gamma_+ & \cos \Gamma_+ \end{pmatrix}. \tag{10.21}$$

We define

$$\Gamma_\pm \equiv x\pi\sqrt{1 \pm r}. \tag{10.22}$$

QUESTION 5: Matrix M3 Prove that the terms of the matrix $\mathbf{M}$ are given by (10.21, 10.22).

To find out if $\mathbf{M}^n$ remains finite as n increases, all we have to do is diagonalize $\mathbf{M}$ by finding the eigenvalues. Suppose $\mathbf{M}_d$ is the diagonal matrix. Then there must be a matrix $\mathbf{U}$ such that

$$\tilde{\mathbf{U}} \cdot \mathbf{M}_d \cdot \mathbf{U} = \mathbf{M} \quad \text{with } \tilde{\mathbf{U}} \cdot \mathbf{U} = \mathbf{I}. \tag{10.23}$$

Raising this to the nth power we get

$$\mathbf{M}^n = \tilde{\mathbf{U}} \cdot \mathbf{M}_d^n \cdot \mathbf{U}. \tag{10.24}$$

The nth power of $\mathbf{M}_d$ is just the diagonal matrix with all of the eigenvalues raised to the nth power. Even though the eigenvalues may be complex (they are the solutions of a quadratic equation), if the absolute magnitude of all the eigenvalues is equal to or less than 1, the solution will be stable. If one of the eigenvalues is greater in magnitude than 1, the solution will be unstable. If we chose an initial direction in phase space along this unstable eigenvector, we will just be moved along a ray with increasing distance from the origin, without changing direction in phase space.

To find the eigenvalues of $\mathbf{M}$, we solve the quadratic equation below for λ:

$$\lambda^2 - (A + D)\lambda + 1 = 0. \tag{10.25}$$

The "1" is really the determinant $AD - BC$. The *Trace* (Tr) of a matrix was defined earlier as the sum of its diagonal elements:

$$Tr\,\mathbf{M} \equiv A + D = \theta_c(T) + \dot{\theta}_s(T). \tag{10.26}$$

It is also the sum of the eigenvalues, which must have a product of 1 as well, by Liouville's Theorem. The solution is

$$\lambda_\pm = \frac{Tr\,\mathbf{M}}{2} \pm \sqrt{\left(\frac{Tr\,\mathbf{M}}{2}\right)^2 - 1}. \tag{10.27}$$

If $|Tr\,\mathbf{M}| < 2$, the solutions are complex. In fact, the two eigenvalues that are the solutions of (10.25) are complex conjugates of each other and lie on the unit circle, so their magnitudes are both unity, and the solutions are stable. If, however, $|Tr\,\mathbf{M}| \geq 2$, the solutions are real. Since the product of the eigenvalues is 1 (no damping case), one of them must be greater than 1 if $|Tr\,\mathbf{M}| > 2$. The solutions are, say λ and $\frac{1}{\lambda}$, with $|\lambda| > 1$. Since λ^n increases in magnitude without limit if n is increased, this must correspond to unstable motion. If $|Tr\,\mathbf{M}| = 2$, then both $\lambda_\pm = 1$, which gives periodic motion with a period of oscillatory motion equal to the period of the equation of motion, T. This is the only case in which the period of stable motion is the same as the period of the restoring force.

The test for stability is now only a simple check of whether the magnitude of the trace of $\mathbf{M}$ is greater than 2 (unstable) or less than or equal to 2 (stable). In the case of stability,

10.2 PARAMETRIC RESONANCE

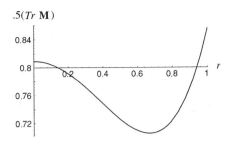

FIGURE 10.4
$x = 0.9$, $.5(Tr\,\mathbf{M})$ is given by (10.28).

if we set $\lambda \equiv e^{i\mu T}$, thereby defining the constant μ, we have the same constant appearing in the Floquet Theorem (10.15). The trace of the matrix is $\lambda + \frac{1}{\lambda} = 2\cos(\mu T)$. Here μ is called the *characteristic exponent*. The period of oscillatory motion is defined as $T_{\text{oscillator}} = \frac{2\pi}{\mu}$. In general, $T_{\text{oscillator}} \neq T$.

In the shaken stick example, $T = 2\pi x$. The term $\mu T = 2\pi \mu x$ is the phase advance of the stick pendulum oscillator per period of the force oscillation. The ratio of oscillation frequency to restoring force frequency is μx. The parameter x is known and μ must be determined from sine- and cosinelike solutions to the Hill equation. We chose the example of the stick because it is easy to solve the equations analytically for the sine- and cosinelike solutions. One half of the trace of the matrix $\mathbf{M}(x, r)$ given by (10.21, 10.22) is

$$\cos 2\pi\mu x = \cos \Gamma_- \cos \Gamma_+ - \frac{1}{\sqrt{1-r^2}} \sin \Gamma_+ \sin \Gamma_-. \tag{10.28}$$

In Figure 10.4 we fix the shaking frequency relative to the free pendulum frequency by choosing $x = 0.9$, and plot $.5(Tr\,\mathbf{M})$ versus the shaking amplitude r. The range of variation for r has been chosen to be $0 \leq r \leq 1$ to keep the solutions in terms of sines and cosines. As we can see, the motion for $x = 0.9$ remains stable for all values of r in this range.

In Figure 10.5, choosing $r = 0.7$ as a representative value, we start from a unit vector along the θ axis in phase space, with $\dot{\theta} = 0$, and apply the matrix (10.21) eighty times to show the phase space locations of the next eighty periods of the force. We are looking at the continuous motion in the phase space with a "stroboscope" which samples the position every period. Sampling in this way at fixed time intervals is called a *Poincaré section*. We see that the motion tends to concentrate in the region of only eight points in phase space. If we ran enough iterations so that we would be looking at the long-term behavior, the points would gradually broaden into a continuous closed curve over the plane. If the ratio of the oscillation frequency to the force frequency (μx) were a rational number (i.e., were a ratio of integers – here it is close to $\frac{7}{8}$), the points would not turn into a curve but would remain stable as points. We will have much more to say about this phenomenon in the next chapter. Here we only observe that the motion remains bounded in phase space, as it must if the trace of the matrix lies between -2 and $+2$.

We now shake the stick a little more slowly – closer to the natural period of the pendulum, which is 2π in these time units. The period is lengthened by increasing x from 0.90 to

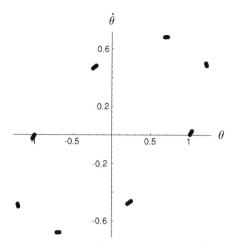

FIGURE 10.5
Poincaré section for $x = 0.9, r = 0.7$.

0.95. In Figure 10.6 we again plot half of the trace of the matrix versus r, the amplitude of shaking. Now we see how the motion becomes increasingly unstable as r increases above about 0.7, since $.5(Tr\,\mathbf{M})$ exceeds 1 at that point. Instead of an oscillatory solution, we get growing and decaying exponentials because μ becomes imaginary. The eigenvalue of the matrix with the positive exponent determines the rate of growth.

Ten periods of the unstable solution are shown in Figure 10.7 for contrast with the previous stable solution phase plot in Figure 10.5. The magnitude of the unstable solution grows without bound. In real life, as with our pendulum, when the amplitude becomes large, we must take into account the action of the nonlinear terms that were ignored in deriving (10.13). The actual spinning of the pendulum must be described by a more exact equation. Notice that we have an amplifier – the direction of spinning is not determined by the force, since there is no "intrinsic" direction of rotation. Instead, any slight asymmetry in the initial conditions is amplified when the solution becomes unstable. Notice also that the unstable solution is only achieved if x (period) and r (amplitude) lie within certain interdependent limits.

Figure 10.8 shows the contours of constant $Tr\,\mathbf{M}$. The large white area in the middle is the region where $.5(Tr\,\mathbf{M}) < -1$, while the white area at the top corresponds to

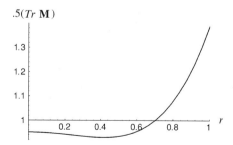

FIGURE 10.6
$x = 0.95$, $.5(Tr\,\mathbf{M})$ is given by (10.28).

10.2 PARAMETRIC RESONANCE

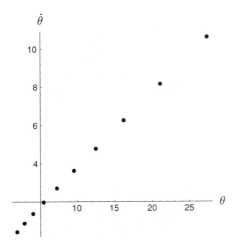

FIGURE 10.7
Poincaré section for $x = 0.95, r = 0.7$.

$.5(Tr\,\mathbf{M}) > 1$. Both of these are regions of instability. The gray areas are stable. One can see that even a very small amplitude of shaking the pendulum near $x = 0.5$ leads to unstable motion. The value $x = 0.5$ corresponds to shaking the pendulum up and down at *twice* its natural frequency. Close to this shaking frequency, friction will limit the smallest unstable r value. Without friction, even the smallest shaking would lead to instability. This is the phenomenon of parametric resonance.

The Mathieu Equation

Another good example of parametric resonance is a pendulum that has its top support vibrating sinusoidally up and down with an amplitude $Y(t) = Y_0 \cos \Omega t$. This leads to a special case of the Hill equation, called the Mathieu equation. This equation first arose in a study of waves in an elliptical cavity, but it has many applications. Leave

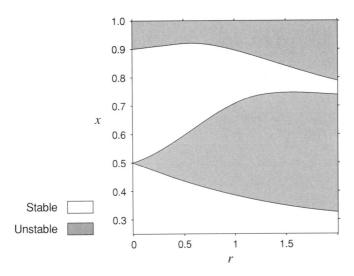

FIGURE 10.8

the driving amplitude Y_0 and the driving frequency Ω arbitrary to begin with, although you might guess that something special happens when $\Omega = 2\omega_0$, where ω_0 is the natural frequency of the pendulum for small oscillations, $\sqrt{\frac{g}{l}}$. To derive the Lagrangian if the support is shaken, first find the kinetic energy. Let x be the horizontal displacement of the pendulum, y the vertical displacement from equilibrium, and $Y(t)$ the function that describes the shaking as shown in Figure 10.2:

$$x = l \sin\theta \approx l\theta,$$
$$y = Y(t) + l(1 - \cos\theta) \approx Y(t) + \frac{l\theta^2}{2}. \tag{10.29}$$

Differentiate with respect to time and again assume θ is small to get

$$\dot{x} = l\dot{\theta},$$
$$\dot{y} = \dot{Y}(t) + l\theta\dot{\theta}. \tag{10.30}$$

The kinetic energy is

$$\frac{m}{2}(\dot{x}^2 + \dot{y}^2) = \frac{m}{2}(l^2\dot{\theta}^2 + 2l\dot{Y}\theta\dot{\theta}) \tag{10.31}$$

(neglecting higher than quadratic terms in θ, $\dot{\theta}$, and the $\dot{Y}^2$ term). The potential energy is

$$mg\left(Y(t) + l\frac{\theta^2}{2}\right). \tag{10.32}$$

Forming the Lagrangian, and applying the Euler–Lagrange equations, we obtain

$$\ddot{\theta} + \left(\frac{g + \ddot{Y}}{l}\right)\theta = 0. \tag{10.33}$$

Next change the time variable to $\tau = \frac{\Omega t}{2}$. (Let $\ddot{\theta}$ mean now $\frac{d^2\theta}{d\tau^2}$.) We get

$$\frac{\Omega^2}{4}\ddot{\theta} + \frac{g - Y_0\Omega^2 \cos 2\tau}{l}\theta = 0. \tag{10.34}$$

Defining some new parameters

$$a \equiv \left(\frac{2\omega_0}{\Omega}\right)^2, \quad q \equiv \frac{2Y_0}{l}, \tag{10.35}$$

we obtain

$$\ddot{\theta} + (a - 2q\cos 2\tau)\theta = 0. \tag{10.36}$$

This equation is the special form of the Hill equation known as a *Mathieu equation*. The parameter a is positive and equals 1 if we drive the system at twice its natural frequency:

10.2 PARAMETRIC RESONANCE

$\Omega = 2\omega_0$. The sign of q is unimportant, since it can be changed merely by shifting the phase of the drive. For convenience we can assume that q is positive.

We can also show that we get the same basic equation if the pendulum starts from the unstable equilibrium where it is inverted, but that the sign of a now becomes negative. (θ measures the deviation from π if the pendulum is upside down.)

QUESTION 6: *Upside-Down Pendulum* Find the equation of motion for the upside-down pendulum with a driven support near the point $\theta = \pi$. Find the form of the Mathieu equation in terms of the variables defined above.

The solutions to the Mathieu equation are called "Mathieu functions." These cannot be expressed in closed form in terms of simpler functions, but only as infinite series. We can find out if the solutions are stable or unstable by using the Floquet theory and the **M** matrix discussed above.

Using numerical integration of Equation (10.36), $Tr\,\mathbf{M}$ was computed for a wide range of (a, q) values, and the boundary lines between stable and unstable regions were plotted as seen in Figure 10.9. For $\Omega = 2\omega_0$ ($a = 1$) there is no stable region at all. An arbitrarily small driving amplitude Y_0 (i.e., value of the parameter q defined in Equation (10.35)) will drive an unstable parametric resonance.

There are smaller bands of instability for Ω equal to any even multiple of ω_0. The first three unstable regions are shown in Figure 10.10. The nth region starts at $a = n^2, q = 0$. The upper curve defining the region corresponds to $Tr\,\mathbf{M} = 2$, and the lower curve to $Tr\,\mathbf{M} = -2$. These curves correspond to even and odd periodic solutions. The shaded areas in between the solutions are the unstable regions, which grow larger with increasing amplitude q.

Airplane wings have an infinite number of such unstable resonances which could be driven by small vibrations. The designer of an airplane wing must be certain that there is no parametric instability, not only for the linear terms such as for the Hill equation, but also for the nonlinear terms as well.

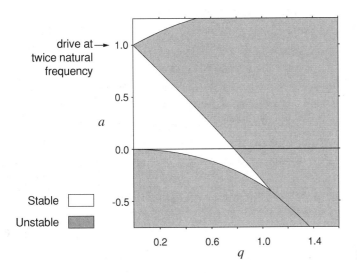

FIGURE 10.9

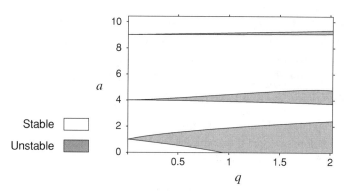

FIGURE 10.10
Mathieu equation: first three unstable regions. Note that the region from 0 to 1.5 is the same as Figure 10.9.

Another interesting feature of Figure 10.9 is for $a < 0$, the case of the upside-down pendulum. We see from the plot that, if q is sufficiently large, the motion is *stable*. If we suspend a pendulum upside down, and drive it with a very high frequency and a finite amplitude in the correct range, the pendulum will oscillate about its vertical position in a stable manner. This is a marvelous experiment to see demonstrated.

If damping is present, as it always is in a real system, the chief effect is to require a finite threshold drive amplitude q (i.e., Y_0) for instability. The lines converging at $a = n^2$ are rounded off, and there is a small stable region between the vertical axis and the points of instability onset.

10.3 LINDSTEDT–POINCARÉ PERTURBATION THEORY

Consider the case of a "generic" anharmonic (i.e., nonlinear) oscillator, whose period we solved for in Chapter 4. We can use Lindstedt–Poincaré perturbation theory to find the amplitude function $q(t)$ as well as the period. We will see that the nonlinearity both shifts the frequency as a function of the amplitude and distorts the trajectory $q(t)$, introducing harmonics of the shifted linear oscillator frequency.

The Lindstedt–Poincaré perturbation theory is used in quantum mechanics to find the energy levels and wave function of a perturbed system. (But it is usually called only "perturbation theory" in quantum mechanics texts, so the unsuspecting student may not realize the wider applications of this method.) In classical mechanics, we can use Lindstedt–Poincaré perturbation theory to solve periodic systems. For example, one might use perturbation theory to predict the effect of the other planets on the Earth's motion. The method assumes that we start with a problem we can solve exactly and then modifies it a little bit. We will demonstrate the theory for the Duffing oscillator, but the method is completely general.

A nonlinear oscillator with a cubic restoring force (i.e., a quartic potential) is called a Duffing oscillator. The exact EOM is

$$\ddot{q} + q + \epsilon q^3 = 0, \qquad (10.37)$$

10.3 LINDSTEDT–POINCARÉ PERTURBATION THEORY

where ϵ is a measure of the strength of the nonlinear term. For simplicity we will assume the initial conditions

$$q(0) = A, \quad \dot{q}(0) = 0. \tag{10.38}$$

Try at first an expansion of $q(t)$ in powers of the strength of ϵ. Note that this is known as *the method of successive approximation*. Write

$$q(t) = q_0(t) + \epsilon q_1(t) + \epsilon^2 q_2(t) + \cdots = \sum_{i=0}^{\infty} \epsilon^i q_i(t). \tag{10.39}$$

Insert this expression into the EOM and look at the coefficient of each separate power of ϵ (i.e., ϵ^i). If you think of ϵ as a variable, then each coefficient must vanish separately, since the different powers are linearly independent functions. We thus obtain an infinite set of coupled differential equations, one for each order i. These have the property that if we know $q_0(t)$, we can solve for $q_1(t)$; then $q_0(t), q_1(t)$ determine $q_2(t)$, etc. So it seems that we can proceed order by order to find a power series expansion of the solution. Substituting (10.39) into (10.37), we get the first three equations for the coefficients of $\epsilon^0, \epsilon^1, \epsilon^2$:

$$\ddot{q}_0 + q_0 = 0, \tag{10.40}$$
$$\ddot{q}_1 + q_1 = -q_0^3, \tag{10.41}$$
$$\ddot{q}_2 + q_2 = -3q_0^2 q_1. \tag{10.42}$$

The solution to (10.40) is $q_0 = A \cos t$. We will then use this solution to solve for $q_1(t)$ in Equation (10.41), etc.

Treat the nonlinear terms in the EOMs as if they were inhomogeneous driving terms and put them on the right side of the equality sign (as we have done in (10.40, 10.41, 10.42)). Then, for any order (q_1, q_2, etc.), we have the form

$$\ddot{q}_k + q_k = F_k(t). \tag{10.43}$$

The periodic solutions for equations of this type can be found by a general application of the Green's function method discussed in Chapter 3 (3.71). Note that we want solutions with a vanishing derivative at $t = 0$. The Green's function method gives the unique solution

$$q_k(t) = \int_0^t \sin(t - t') F_k(t') \, dt'. \tag{10.44}$$

Using (10.44), the solution to $q_1(t)$ is (as you can check)

$$q_1(t) = A^3 \left(-\frac{3}{8} t \sin t - \frac{1}{32} \cos t + \frac{1}{32} \cos 3t \right). \tag{10.45}$$

Note that $q_1(0) = 0$ so that $q_0 + q_1$ also equals A at $t = 0$.

The solution (10.45) is not correct – it is wrong! We know the motion is periodic, yet the attempt to find a power series solution blows up proportional to the time. (The next order, q_2, blows up as t^2.) The trouble can be traced to the fact that q_0^3 contains a term*

---

* Using the identity $\cos^3 t = \frac{3 \cos t + \cos 3t}{4}$.

$\frac{3}{4}\cos t$, which acts as a "resonant" driving term leading to nonperiodic solutions in q_1. We call the nonperiodic terms *secular terms*. We can't get a periodic solution with conserved energy from this approach.

A way out of this predicament was found by Lindstedt and Poincaré. They realized that one is trying to expand a function with a period varying with the amplitude A in terms of functions that have periods of 2π. Instead of the previous approach, they suggested making a change of independent variable at the same time as the power series expansion of the solution. This allows the period to change with amplitude, but it makes for a more complicated expansion. However, it has the virtue that it works. Usually we need only the lowest orders, since we are restricted to small amplitudes. Try

$$s \equiv \omega t, \tag{10.46}$$
$$\omega \equiv 1 + \epsilon \omega_1 + \epsilon^2 \omega_2 + \cdots. \tag{10.47}$$

Once again

$$q(s) = \sum_{i=0}^{\infty} \epsilon^i q_i(s). \tag{10.48}$$

This defines the series for the solution we want. With the extra freedom we get from the additional constants ω_1, ω_2, etc., we can remove all secular terms from the equations for the q_i. With $\ddot{q} \equiv \frac{dq}{ds}$ now, the EOM (10.37) becomes

$$\omega^2 \ddot{q} + q + \epsilon q^3 = 0. \tag{10.49}$$

Insert both power series (10.47, 10.48) and again isolate different powers of ϵ:

$$\ddot{q}_0 + q_0 = 0, \tag{10.50}$$
$$\ddot{q}_1 + q_1 = -q_0^3 + 2q_0\omega_1, \tag{10.51}$$
$$\ddot{q}_2 + q_2 = -3q_0^2 q_1 + 2(q_1 + q_0^3)\omega_1 + q_0(2\omega_2 - 3\omega_1^2). \tag{10.52}$$

The solution to (10.50) is still $q_0 = A\cos s$. Inserting this into (10.51) and using the trigonometric identity mentioned earlier, we obtain

$$\ddot{q}_1 + q_1 = \left(2A\omega_1 - \frac{3}{4}A^3\right)\cos s - \frac{A^3}{4}\cos 3s. \tag{10.53}$$

If we set the coefficient of the resonant driving term $\cos s$ to be zero, we will eliminate the secular term. We get

$$\omega_1 = \frac{3}{8}A^2, \tag{10.54}$$

which then leads to the solution

$$q_1 = -\frac{A^3}{32}(\cos s - \cos 3s). \tag{10.55}$$

There are no secular terms here! The period, to first order in the nonlinear perturbation using (10.47), is

$$T = 2\pi \left(1 - \frac{3}{8}\epsilon A^2\right). \tag{10.56}$$

Just for the record

$$\omega_2 = -\frac{21}{256} A^4 \tag{10.57}$$

and

$$q_2 = \frac{A^5}{1024}(23\cos s - 24\cos 3s + \cos 5s). \tag{10.58}$$

To second order in ϵ:

$$T = 2\pi \left(1 - \frac{3}{8}\epsilon A^2 + \frac{57}{256}(\epsilon A^2)^2 + \cdots\right). \tag{10.59}$$

You can compare this with the formula in Chapter 4 for the period of a physical pendulum (4.13).

The algebra can be done almost trivially to even higher orders if one uses a computer algebra program. By hand it is rather daunting, but not impossible if one has the stamina. If the nonlinear effects become too large, we can't use this method, but have to resort to numerical methods. Also, as we will explore in a homework problem, Lindstedt–Poincaré theory does not work well with systems containing transient responses like those in a damped oscillator. This is because we assume the zeroth-order solution (10.50) has a constant amplitude, which is only true for the steady-state part of such solutions.

10.4 DRIVEN ANHARMONIC OSCILLATOR

As discussed in Chapter 3, a linear oscillator always responds at the driving frequency, once transients are damped out. The oscillator response grows very large when the driving frequency approaches the natural frequency of the oscillator – this is resonance. In contrast, if a small nonlinearity is present, periodic solutions still occur and resonant instabilities are now possible for rational frequency ratios between the driving frequency and the natural oscillator response. This phenomenon plays a key role in the route to chaos for Hamiltonian systems. In the next chapter we will study the chaotic motion of such a system.

Here we will study a one-dimensional anharmonic oscillator driven by an external periodic force of the form $f \cos \omega t$. Here f and ω are considered to be parameters, fixed at any given time, but variable by us in order to study the oscillator's response to this time-dependent driving force.

We will choose one special anharmonic oscillator, the Duffing oscillator, which could have a little linear damping as well:

$$\ddot{q} + \frac{\dot{q}}{Q} + q + \epsilon q^3 = f \cos(\omega t). \tag{10.60}$$

If $\omega = 1$ and $\epsilon \to 0$, the oscillator is being driven at the natural resonant frequency of a linear oscillator. This problem was treated in detail in Chapter 3. One obtains a symmetrical Lorentzian curve of amplitude versus driving frequency (see Figure 3.9). The maximum amplitude and the width of the resonance curve are determined by the damping Q.

There are at least three new physical effects from the presence of the nonlinear term:

a) If the driving frequency is tuned from above the resonance toward lower frequencies, the amplitude will suddenly jump to a larger value at a certain frequency. However, if the driving frequency is increased from low to high frequency, nothing will happen until a much higher frequency is passed, in which case the amplitude suddenly decreases. The oscillator "remembers" whether the driving frequency ω is being tuned higher or lower. This phenomenon is called *hysteresis*.

b) Under certain conditions subharmonic frequencies (such as $\frac{\omega}{3}$) can be generated. The possibility of subharmonics means that there is a significant response even if we drive the oscillator at three times the natural resonant frequency. Only certain driving amplitudes will then lead to stable solutions. In the section on subharmonic resonance, we will demonstrate a general method for using the linear Hill equation to test for the stability of the nonlinear oscillations.

c) A nonlinear system can be used as a "mixer." If excited by two different frequencies, the output contains oscillations at the sum and the difference of the original exciting frequencies. Although the practical uses of this effect are many – most radio receivers employ nonlinear mixer devices in their circuitry – we only mention the effect here and will not consider it further.

Harmonic Analysis

First we will look at a specific case, obtained by numerical integration of the equations of motion (10.60) on a computer. To the precision this can be done,* we obtain exact solutions. Choose $\epsilon = \frac{1}{10}$ and $f = 1$. Also choose $Q \to \infty$ (i.e., the damping is turned off). Drive the oscillator right at the resonant frequency of the linear part ($\omega = 1$) and make a phase plane plot $(q, \dot{q})$ for the solution obtained by numerical integration, as shown in Figure 10.11.

Without the nonlinear term, we would have a perfect circle, with $q(t) = \frac{f}{1-\omega^2} \cos \omega t$. The amplitude would be infinite for the case of zero damping if $\omega = 1$, the resonant frequency. From the appearance of the plot in Figure 10.11, in the case of a nonlinear equation, we see a distorted circle of finite amplitude, even without damping. We will show that the nonlinear term means that we have to solve a cubic equation to find this finite starting value.

Assume that the solution is periodic – this means that it can be expanded in a Fourier series.[†] We can first use the symmetries of (10.60) with no damping to simplify our

---

* In our case, about six significant figures.
† Kaplan, *Advanced Calculus*, 3rd ed., p. 469ff.

10.4 DRIVEN ANHARMONIC OSCILLATOR

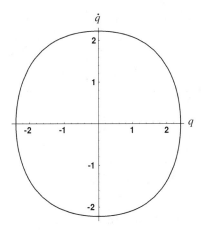

FIGURE 10.11
Duffing oscillator (10.60) with $\epsilon = \frac{1}{10}$, $f = 1$, and $Q \to \infty$.

expansion. Time reversal symmetry implies that the Fourier series is a series of cosines. By using the fact that (10.60) is also invariant with respect to the transformation $t \to t + \pi, q \to -q$, we can also deduce that only the odd frequencies will appear:

$$q(t) = \sum_{n=1,3,5,\ldots} A_n(\omega) \cos n\omega t. \tag{10.61}$$

QUESTION 7: *Using Symmetry* First prove that (10.60) is invariant under the following transformations: a) $t \to -t$, b) $t \to t + \pi, q \to -q$. Then use these symmetries to prove that we only need to consider odd cosine terms in our Fourier series expansion.

Determine $A_1(\omega = 1) = 2.356$ by a numerical integration of the product $q(t) \cos \omega t$ over time from 0 to $\frac{2\pi}{\omega}$ multiplied by $\frac{\omega}{\pi}$. This *Fourier analysis* isolates the A_1 term. Then look at the remainder $q(t) - A_1 \cos \omega t$, which we plot in Figure 10.12. The part of $q(t)$ not varying with the same frequency as the driving term closely resembles a third harmonic of the driving frequency, that is, it is approximately equal to $A_3 \cos 3\omega t$. This is the second term in (10.61) as we expected by our symmetry arguments. In this case $A_3(\omega = 1) = .046$.

Purely numerical work does not easily give us the more complete picture supplied by an analytic solution, but it suggests that we can assume a solution of the form of an infinite

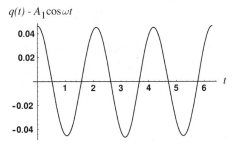

FIGURE 10.12

series in $\cos n\omega t$, $n =$ odd integer. We expect the dominant terms to be

$$q(t) = A_1 \cos \omega t + A_3 \cos 3\omega t + \cdots . \tag{10.62}$$

Inserting this expression into the original equation (10.60) (with the damping still absent), we obtain

$$(1 - \omega^2)A_1 \cos \omega t + (1 - 9\omega^2)A_3 \cos 3\omega t + \cdots$$
$$+ \epsilon \left(\frac{3}{4} A_1^3 \cos \omega t + \frac{1}{4} A_1^3 \cos 3\omega t + \cdots \right) = f \cos \omega t. \tag{10.63}$$

(Here "..." stands for other harmonics, which can be ignored, due to their being very small. The largest neglected term is $\epsilon A_1^2 A_3$.) We have used the trigonometric identity $\cos^3 x = \frac{3 \cos x + \cos 3x}{4}$ in (10.63).

The coefficients of each nth harmonic, $\cos(n\omega t)$, must vanish, since we can isolate each one of them by Fourier analysis. Thus, Equation (10.63) implies that, to this level of approximation,

$$(1 - \omega^2)A_1 + \frac{3}{4}\epsilon A_1^3 = f \tag{10.64}$$

and

$$(1 - 9\omega^2)A_3 + \frac{1}{4}\epsilon A_1^3 = 0. \tag{10.65}$$

We can check to see how well each of the above equations is obeyed for the case we solved numerically. Using $A_1 = 2.356$, $\omega = 1$, $f = 1$, and $\epsilon = .1$ in (10.64), we obtain $0.98 \approx 1$. If we then solved (10.65) for A_3, we would obtain .041 instead of the correct value, .046. Greater accuracy could be obtained by keeping more terms in the Equation (10.62). The method we have used here is called *harmonic analysis*.

The amplitude A_1 from the solution to Equation (10.64) for $f = 1$ and no damping is plotted in Figure 10.13, over the frequency interval $0.5 \leq \omega \leq 2$. Notice that nothing special happens when we cross the linear resonance $\omega = 1$. An explanation of the peculiar shape of this curve is given below, when we have also added a variable damping term. One might consider how to interpret the case of three solutions at the same frequency.

Perturbation Theory

Instead of harmonic analysis, we can use the more general perturbation theory.* To avoid secular terms in the solution for the free oscillator, we had to assume that the frequency $\omega = \omega(\epsilon)$ was a power series in ϵ (10.47), in combination with assuming that

---

* C. Hayashi, *Nonlinear Oscillations in Physical Systems*, Princeton University Press, 1985, p. 20.

10.4 DRIVEN ANHARMONIC OSCILLATOR

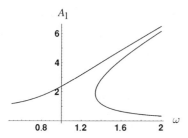

FIGURE 10.13
Solutions to the damped anharmonic oscillator (10.64) without damping.

the solution itself was also a power series in ϵ (10.48). The perturbation theory for the driven oscillator *cannot* assume a variable frequency, since the driving frequency is a fixed constant ω. Instead we introduce an auxiliary constant μ (which we later set equal to one), and assume that both the solution and a variable phase shift can be expanded as a power series in μ:

$$q = q_0 + \mu q_1 + \mu^2 q_2 + \cdots,$$
$$\delta(\mu) = \delta_0 + \mu \delta_1 + \mu^2 \delta_2 + \cdots. \tag{10.66}$$

We introduce μ into the differential equation by first rewriting the original equation (10.60) (without damping) as

$$\ddot{q} + \omega^2 q = (\omega^2 - 1)q - \epsilon q^3 + f \cos \omega t. \tag{10.67}$$

Then we treat the entire right-hand side as a perturbation of strength μ:

$$\ddot{q} + \omega^2 q = \mu(\omega^2 - 1)q - \mu \epsilon q^3 + \mu f \cos \omega t. \tag{10.68}$$

Make a change of time variable $\tau \equiv \omega t - \delta(\mu)$. The equation can be rewritten in the form

$$\ddot{q} + q = \mu\left(1 - \frac{1}{\omega^2}\right)q - \mu \frac{\epsilon}{\omega^2} q^3 + \mu \frac{f}{\omega^2} \cos(\tau + \delta(\mu)). \tag{10.68}$$

(Now, due to the change of variable, $\ddot{q} \equiv \frac{d^2 q}{d\tau^2}$.) The form of the equation is chosen to give a "left-hand side" that has the driving term as a solution. This will allow us to identify secular terms arising from the periodic driving force, and cancel them by choosing an infinite series of arbitrary constants. The method* can be used for any nonlinear equation with an inhomogeneous driving term. In the spirit of previous perturbation theory calculations use (10.66) to expand (10.68) in a power series in the auxiliary variable μ. The different

---

* This is one of several possible iteration methods. See Mickens, *An Introduction to Nonlinear Oscillations*, Cambridge University Press, 1981, p. 130ff.

powers of μ give a hierarchy of equations, which we can solve progressively. The power series expansion up to terms in μ^2 gives three equations:

$$\ddot{q}_0 + q_0 = 0, \tag{10.69}$$

$$\ddot{q}_1 + q_1 = \left(1 - \frac{1}{\omega^2}\right)q_0 - \frac{1}{\omega^2}\left(\epsilon q_0^3 - f\cos(\tau + \delta_0)\right), \tag{10.70}$$

$$\ddot{q}_2 + q_2 = \left(1 - \frac{1}{\omega^2}\right)q_1 - \frac{1}{\omega^2}\left(3\epsilon q_0^2 q_1 + \delta_1 f \sin(\tau + \delta_0)\right). \tag{10.71}$$

The term in (10.71) proportional to δ_1 comes from the power series for the phase shift $\delta(\mu)$ (10.66). We choose the solutions of (10.69)–(10.71) that have $\dot{q}_i(0) = 0$, $i = 0, 1, 2$, so that $q(\tau)$ will be an even function of τ. After solving the equations above, we obtain the physical solution to the original undamped, forced, anharmonic oscillator by setting $\mu = 1$.

The first equation (10.69) has the solution $A_1 \cos \tau$, since $\dot{q}_0(0) \equiv 0$. Expand the right side of (10.70) using trigonometric identities:

$$\left(1 - \frac{1}{\omega^2}\right)A_1\cos\tau + \frac{1}{\omega^2}\left(-\frac{3}{4}\epsilon A_1^3 \cos\tau + f\cos\tau\cos\delta_0\right.$$
$$\left. - f\sin\tau\sin\delta_0 - \epsilon A_1^3 \frac{1}{4}\cos 3\tau\right). \tag{10.72}$$

The coefficients of $\cos\tau$ and $\sin\tau$ must vanish on the right side of every equation including this one, in order to avoid secular terms. This requirement applied to (10.70) gives two equations:

$$\sin\delta_0 = 0, \quad \cos\delta_0 = 1, \tag{10.73}$$

$$(\omega^2 - 1)A_1 - \frac{3}{4}\epsilon A_1^3 + f = 0. \tag{10.74}$$

Equation (10.74) is the same equation found previously by the method of harmonic balance, Equation (10.64). Without the now vanquished secular terms, the linear inhomogeneous equation for q_1 is

$$\ddot{q}_1 + q_1 = -\frac{\epsilon}{4\omega^2}A_1^3 \cos 3\tau. \tag{10.75}$$

We can use (10.44) to solve (10.75). We can also add an arbitrary $\cos\tau$ term, which is a solution to the homogeneous equation. For the function q_1 we find

$$q_1(\tau) = b_1 \cos\tau + \frac{\epsilon}{32\omega^2}A_1^3 \cos 3\tau. \tag{10.76}$$

This gives an expression for the lowest-order approximation to the third harmonic in the solution. The unknown constant b_1 is used to eliminate secular terms in the next order,

10.4 DRIVEN ANHARMONIC OSCILLATOR

TABLE 10.1 PERTURBATION THEORY FORMULA FOR THE FIRST THREE HARMONIC COEFFICIENTS. A_1 IS FOUND FROM SOLVING (10.74) AND b_1 FROM SOLVING (10.78). c_1 IS NOT REPRODUCED HERE

Harmonic	q_0	q_1	q_2
1	A_1	b_1	c_1
3		$\frac{\epsilon A_1^3}{32\omega^2}$	$\frac{\epsilon A_1^3}{256\omega^2}\left(-1 + \frac{1}{\omega^2}\right) + \frac{3\epsilon^2 A_1^5}{512\omega^4} + \frac{3\epsilon A_1^2 b_1}{32\omega^2}$
5			$\frac{\epsilon^2 A_1^5}{1024\omega^4}$

where we find

$$\delta_1 = 0, \tag{10.77}$$

$$b_1\left(-1 + \frac{1}{\omega^2} + \frac{9\epsilon}{4\omega^2}A_1^2\right) + \frac{3\epsilon^2}{128\omega^4}A_1^5 = 0. \tag{10.78}$$

The full expression for q_2 is too messy to reproduce here (see Table 10.1). It contains corrections to the third harmonic. An additional small fifth harmonic term also appears:

$$q_2^{\text{fifth harmonic part}}(\tau) = \frac{\epsilon^2 A_1^5}{1024\omega^4} \cos 5\tau. \tag{10.79}$$

Further corrections to the harmonics arise in higher orders of perturbation theory, so there is no simple correspondence between a harmonic coefficient and the order, although the lowest order in which a given harmonic occurs is a good approximation for this case since we have rapid convergence.

The comparison between the coefficients of the harmonics in the solution for $q(\tau)$ obtained by numerical integration on a computer and the perturbation sum to second order for the first three harmonics can be seen in Table 10.2. Even though we stopped at only second order, the remarkable precision and rapid convergence of the perturbation theory in this example is clear from Table 10.2. The constant c_1 would be determined by requiring

TABLE 10.2 COMPARISON OF THE NUMERICAL INTEGRATION OF THE UNDAMPED, DRIVEN, ANHARMONIC OSCILLATOR WITH PERTURBATION THEORY HARMONIC COEFFICIENTS $q_0 + q_1 + q_2$ AS LISTED IN TABLE 10.1 (WITH $f = 1$, $\epsilon = \frac{1}{10}$, AND $\omega = 1$)

Harmonic	Numerical Integration	q_0	q_1	q_2	Sum
1	2.35555	2.37126	−0.01389	c_1	$2.35737 + c_1$
3	0.04563		0.04167	.00366	.04533
5	0.00083			.00073	.00073
>5(sum)	1.17×10^{-5}				

the secular terms in q_3 to vanish. However, this correction to the first harmonic is extremely small, $\approx .002$.

In summary, the method of harmonic balance gives a quick solution for the lowest approximation. The perturbation method outlined above is quite general for any type of nonlinear oscillator driven by a fixed frequency. It gives a systematic method for finding the higher-order approximations to any order. The algebraic complexity of higher orders can be handled by using computer algebra. Had there been any damping, the phase shift would have played a more important role, just as for the driven linear oscillator.

Hysteresis

The amplitude of the oscillator response varies as the driving frequency is changed, according to the solution to a cubic equation (10.64), in the absence of damping. Most oscillators have some damping in real life, and we will now look at how this modifies the response. If we have damping, the differential equation becomes (10.60) with Q finite.

Fortunately, we do not need the powerful apparatus we developed for the perturbation theory to make a semiquantitative theory of how the oscillator response varies with driving frequency. We do need to take into account the general fact that, when damping is present, a phase shift develops between the driving signal and the oscillator response. We assume an approximate solution of the form

$$q(t) = a \cos \omega t + b \sin \omega t. \tag{10.80}$$

Insert this form into (10.60) and drop all terms for harmonics higher than the first. The result is

$$a\left(1 - \omega^2 + \frac{3\epsilon r^2}{4}\right) + \frac{b\omega}{Q} = f,$$
$$b\left(1 - \omega^2 + \frac{3\epsilon r^2}{4}\right) - \frac{a\omega}{Q} = 0, \quad r^2 \equiv a^2 + b^2. \tag{10.81}$$

Treat the above equation (10.81) as two simultaneous equations for a, b. From the solution calculate r^2. It is not necessary to find a and b separately, since we are not interested in the phase shift. We get

$$r^2 = \frac{f^2}{\left(1 - \omega^2 + \frac{3\epsilon r^2}{4}\right)^2 + \frac{\omega^2}{Q^2}}. \tag{10.82}$$

Equation (10.82) is a cubic equation for r^2. We can compare $|r(\omega)|$ for two different curves. The upper curve in Figure 10.14 was computed for $Q = 10$ and the lower curve for $Q = 4$ ($\epsilon = \frac{1}{10}$ and $f = 1$ as before). The dotted portion of each curve is unstable; the solid part is stable. This reveals the reason for the difference in behavior when sweeping the driving frequency up or down. The amplitude always follows the stable curve. Choose $Q = 10$. If we sweep ω up over the resonance, at about $\omega = 1.8$, the amplitude suddenly drops to the lower branch of the curve. This is different from the case of sweeping the frequency down, since the oscillator follows the lower solid curve as long as possible, and

10.4 DRIVEN ANHARMONIC OSCILLATOR

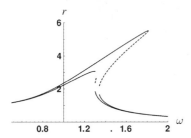

FIGURE 10.14
Solutions to the damped anharmonic oscillator.

then, at $\omega \approx 1.3$, jumps discontinuously to the upper branch. These transitions occur at different frequencies, which results in the observed hysteresis effect. If, instead, $Q \to \infty$, the curve would produce Figure 10.13, with multiple separated branches. This situation is unphysical, however, since for real oscillators there is always damping at some level. Hence the curves are joined somewhere at a frequency above the resonance.

The direction of "tilt" of the resonance curve depends on the sign of ϵ. For negative ϵ, the resonance curve "tilts" towards lower frequencies.

Subharmonic Oscillations

We want to demonstrate the existence of *subharmonic* oscillations. These are oscillations of the nonlinear system with periods that are an integral multiple of the period of the driving force. In other words, the oscillator changes with a frequency of $\frac{\omega}{n}$, with $n =$ an integer. We will use the Hill equation to study whether or not these oscillations are stable. For simplicity, let's assume that the nonlinear oscillator is undamped. The equation of motion is assumed to be

$$\omega^2 \ddot{q} + q + \epsilon q^3 = f \cos 3\tau \tag{10.83}$$

(where we have made the substitution $\tau = \omega t$). Note that for convenience we have redefined the driving frequency as 3ω. This is only done to avoid fractional trigonometric identities. Thus we are going to look for subharmonics that are at $\frac{1}{3}$ of the driving frequency. We could also look for subharmonics at $\frac{\omega}{5}, \frac{\omega}{7}$, etc. Note that because of the symmetry of Equation (10.83) we need only look for odd subhamonics.

Next we will assume that there *are* subharmonics at $\frac{1}{3}$ the driving frequency. Therefore we guess that there is a solution of the form

$$q = a \cos \tau + b \sin \tau + \frac{f}{1 - 9\omega^2} \cos 3\tau, \tag{10.84}$$

where a and b are amplitudes proportional to the subharmonic amplitude at one third of the driving frequency. The coefficient of the response at the driving frequency is chosen as $\frac{f}{1-9\omega^2}$ to satisfy the linear part of the equation.

We will check to see if our solution is real by inserting (10.84) into (10.83). We obtain a long messy expression after using standard trigonometric identities, so we will not reproduce the whole expression here. To eliminate secular terms, the coefficients of sin τ and cos τ must vanish. This will give us two equations for the two unknown subharmonic amplitudes a, b. Setting the coefficient of cos τ equal to zero gives

$$a\left[\frac{1-\omega^2}{\epsilon} + \frac{3f^2}{2(1-9\omega^2)^2} + \frac{3}{4}(a^2+b^2)\right] = \frac{3f}{4(1-9\omega^2)}(b^2-a^2). \tag{10.85}$$

Setting the coefficient of sin τ to zero gives

$$b\left[\frac{1-\omega^2}{\epsilon} + \frac{3f^2}{2(1-9\omega^2)^2} + \frac{3}{4}(a^2+b^2)\right] = \frac{3abf}{2(1-9\omega^2)}. \tag{10.86}$$

Numerous terms of frequency $3, 5, 7, 9$ are neglected if we call (10.84) the solution to (10.83). We choose to ignore these further corrections, just as we did in the last section. Multiplying Equation (10.86) by a and Equation (10.85) by b, and subtracting we get $b(b^2 - 3a^2) = 0$. Either $b = 0$ or $a = \pm\frac{b}{\sqrt{3}}$. It is sufficient to consider only the case $b = 0$, because the other two solutions correspond to rotating this solution by $\frac{2\pi}{3}$ or by $-\frac{2\pi}{3}$ in the (a, b) plane. This symmetry can be seen from the original Equation (10.83), which is unchanged under translations of the time origin by $\pm\frac{2\pi}{3}$.

If $b = 0$, a real solution for a can be found:

$$a = \frac{-f}{2(1-9\omega^2)} \pm \frac{1}{2}\sqrt{\frac{D(f)}{3}} \tag{10.87}$$

if the condition

$$D(f) \equiv 16\frac{(\omega^2-1)}{\epsilon} - 21\frac{f^2}{(1-9\omega^2)^2} > 0 \tag{10.88}$$

is obeyed. This implies that if $\epsilon > 0$, the oscillator must be driven above its natural frequency ($\omega = 1$) and below a certain limiting amplitude f. Equation (10.87) is that of a tilted ellipse centered on the origin in the (f, a) plane. For Figure 10.15, we assume $\epsilon = \frac{1}{10}$ and $\omega = 1.1$, and we plot the relation between the subharmonic amplitude a and the driving amplitude f. The condition that $D(f) > 0$ means that $0 < |f| \leq 12.51$ for subharmonic oscillations to occur.

The remaining question concerns the stability of the solution. Return to the original Equation (10.83), and consider a perturbed solution: $q = q_0 + \xi$. Here q_0 is the solution we

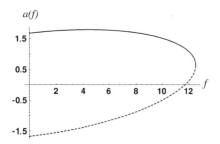

FIGURE 10.15
Real solutions to (10.83), where $\epsilon = .1$, $\omega = 1.1$. Only $f > 0$ is plotted here.

SUMMARY OF CHAPTER 10

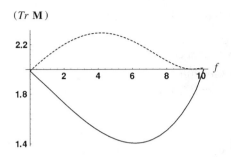

FIGURE 10.16

have just found, $q_0(\tau) = a(f)\cos\tau + \frac{f}{(1-9\omega^2)}\cos 3\tau$. (There is still no damping present.) Subtract the equation for q_0 from the equation for q. Neglecting all terms higher than the first order in ξ allows us to write a linear equation with periodic coefficients (a Hill equation):

$$\omega^2\ddot{\xi} + (1 + 3\epsilon q_0^2)\xi = 0. \tag{10.89}$$

In the case of interest, to learn whether we have stability, we integrate the sinelike and cosinelike solutions of (10.89) over a single period of q_0: $0 \leq \tau \leq 2\pi$. We take $a(f)$ to be either of the two solutions in Equation (10.87). We then plot the trace of the **M** matrix (10.18) and declare that stability exists if $|Tr\,\mathbf{M}| \leq 2$.

The dotted curve in Figure 10.16 corresponds to

$$a = \frac{-f}{2(1-9\omega^2)} - \frac{1}{2}\sqrt{\frac{D(f)}{3}}.$$

This part of the solution is *unstable*. The solid line corresponds to the solution

$$a = \frac{-f}{2(1-9\omega^2)} + \frac{1}{2}\sqrt{\frac{D(f)}{3}},$$

which is stable. The solid line in Figure 10.15 corresponds to the solid line solution in Figure 10.16 and thus shows the physically possible relationship between driving amplitude and subharmonic amplitude. The symmetry transformation $f \to -f, a \to -a$ shows that the stable and unstable solutions are just reflections of the two elliptical arcs shown for positive f.

This example gives an idea of how the Hill equation enters the picture when one tries to determine the stability of periodic motion in a case where the equations of motion are nonlinear.

SUMMARY OF CHAPTER 10

Dynamic Stability

- The stability of a system disturbed by a perturbation of initial conditions can be investigated by "linearizing," that is, subtracting the unperturbed equations from the perturbed

equations and keeping only linear terms in the difference between the perturbed and the unperturbed motion.
- For scleronomic systems in dynamic equilibrium with a central force, if the system gets an impulsive kick (change $\dot{r}$), but the angular momentum l remains constant, stability depends on the sign of the curvature of V_{eff} at the equilibrium point.

Parametric Resonance

- Nonlinear oscillator systems with periodic motion can be linearized into the Hill equation:

$$\ddot{q} + a(t)\dot{q} + b(t)q = 0, \tag{10.90}$$

where $a(t)$ and $b(t)$ are both periodic functions. Stability of the motion is reduced to the question of the stability of (10.90).
- To determine if the Hill equation is stable, we find the trace of the matrix $\mathbf{M}$ that takes us from an arbitrary starting point at $t = 0$ to the solution and its derivative at $t = T$, the period of $b(t)$:

$$\begin{pmatrix} \theta \\ \dot{\theta} \end{pmatrix}_{t=T} = \mathbf{M} \begin{pmatrix} \theta \\ \dot{\theta} \end{pmatrix}_{t=0}. \tag{10.91}$$

If $|Tr\,\mathbf{M}| \leq 2$, the system is stable against perturbations, whereas if $|Tr\,\mathbf{M}| > 2$ then the system is unstable.
- The equation of motion for a pendulum whose top support point varies as $Y(t) = Y_0 \cos \omega t$ leads to the Mathieu equation

$$\ddot{\theta} + (a - 2q\cos 2\tau)\theta = 0, \tag{10.92}$$

where $\tau = \Omega t/2$, $a \equiv (2\omega_0/\Omega)^2$, and $q \equiv 2Y_0/l$.

Lindstedt–Poincaré Perturbation Theory

- Many slightly nonlinear systems can be solved by treating the nonlinearity as a perturbation and making power series expansions in the perturbation strength ϵ (where $s \equiv \omega t$):

$$\omega \equiv 1 + \epsilon \omega_1 + \epsilon^2 \omega_2 + \cdots, \tag{10.93}$$

$$q(s) = \sum_{i=0}^{\infty} \epsilon^i q_i(s). \tag{10.94}$$

- By equating like terms of ϵ, and choosing the ω_is so that the secular terms vanish, the individual q_is can be obtained, in a series that grows increasingly complicated for successively higher orders in the strength of the nonlinearity.

Driven Anharmonic Oscillator

- The approximate harmonic amplitudes of a driven anharmonic oscillator can be found most easily by the method of harmonic balance. This entails expanding the solution

in a Fourier series in harmonics of the driving frequency and equating like coefficients.
- We can find a more precise solution to a driven anharmonic oscillator with iterative methods of perturbation theory. First rewrite the equation so that the left side has a solution with the same frequency as the driving term. Then treat the right side as a perturbation using an auxiliary constant μ. (See the text.) Finally make the change of variables $\tau \equiv \omega t - \delta(\mu)$, and expand both $q(t)$ and $\delta(\mu)$ as power series in μ and equate powers of μ. This is similar to what is done in the Lindstedt–Poincaré theory, but the frequency remains constant.
- Hysteresis phenomena can only occur if a nonlinear term in the EOM is present, such as in the Duffing oscillator.
- With certain driving amplitudes, subharmonic frequencies (rational fractions of the driving frequency) can be generated. To test for the existence of such solutions, start by first assuming a subharmonic solution exists and then solving for a real, stable solution. Stability is investigated by using a Hill equation.

PROBLEMS

Stability

Problem 1*: *(Stability in a central force)* A particle moves in a potential of the form $V(r) = \frac{\beta}{r^n}$.

a) What condition on β and n is necessary for circular orbits to exist? Hint: Use the radial equation of motion, (4.41).

b) "Linearize" the equation of motion for small radial deviations from the original circular orbit, $\delta r(t)$. Discuss what restriction is placed on the power n if the circular orbit is dynamically stable. This restriction should not depend on β or r.

Problem 2: *(Stability against angular momentum changes)* Investigate a one-time perturbation of the initial conditions E_0, l_0, where the initial state is a circular Kepler orbit. First perturb the angular momentum $l \to l_0 + \delta l$. By again assuming $r \to r_0 + \delta r$ and linearizing the equations of motion, show what effect this has on the circular orbit. Next perturb the energy $E \to E_0 + \delta E$. Again show what effect this has on the circular orbit, this time using the energy equation.

Problem 3*: *(Stability of the Kepler problem for elliptical orbits)* Find a condition on the stability of elliptical orbits for the Kepler problem. Again assume $r \to r_0 + \delta r$ and linearize the orbit equation, (4.41), for an elliptical orbit $r_0(t)$ with eccentricity $\epsilon \neq 0$. What kind of linear differential equation do you obtain? Compare the linearized "r" equation with the linearized "u" equation (4.49). Decide which equation is easiest to use. Are the elliptical orbits of planets (including ours) stable? We certainly hope so!

Problem 4: *(Stability for a screened potential)* Investigate the stability of circular orbits for a screened potential ($\beta > 0$):

$$V(r) = -\frac{\beta}{r}e^{-\frac{r}{a}}. \tag{10.95}$$

First simplify the equation of motion by choosing the unit of length to be a, and change the unit of energy so that the effective potential becomes

$$V_{\text{eff}}(r) = \frac{1}{r^2} - \frac{\beta' e^{-r}}{r}, \tag{10.96}$$

where β' is the potential strength multiplied by a positive scale factor. Show that there exists a critical minimum value, β'_{critical}, such that if $\beta' > \beta'_{\text{critical}}$, two circular orbits exist. Discuss the stability of these orbits and find the numerical value of β'_{critical}. What happens to the circular orbits as β' increases?

Problem 5: *(Computer investigation of stability)* Imagine that you have a particle orbiting a force center with an attractive $\frac{k}{r}$ potential. The solution to the orbit equation (4.49) using the variables ($u = \frac{1}{r}$, ϕ) in the orbital plane is an ellipse. If an additional potential of the form $V_{\text{add}} = \frac{C}{r^n}$ is combined with the original $\frac{k}{r}$ potential, one obtains a modified equation

$$\frac{d^2 u}{d\phi^2} + u = \frac{1}{p} + nC' u^{n+1}. \tag{10.97}$$

Here p is the constant $\frac{l^2}{\mu k}$, where l is the angular momentum and μ the reduced mass with the dimensions of length. Choose the unit of length so that $p = 1$. Write a computer program to numerically integrate (10.97) for values of n and $C' = \frac{\mu C}{l^2}$, which you can input as test values, and plot the orbits that result. Take a range of ϕ up to several times 2π.

a) Check your program by plotting the ellipse you get if $C = 0$.
b) Try different n values to generate "rosettes" – precessing "ellipses." How does the sign of the force – attractive or repulsive – affect the direction of the precession?
c) If $n = 2$ and C is large and negative, the angular momentum barrier can be overcome. The orbit will become unstable and spiral into the origin. Find the threshold value of C' for this.
d) Choose $n \neq 2$ and numerically find the point of instability as the magnitude of C' is increased. Check that this agrees with analytic calculations.

Parametric Resonance

Problem 6: *(Driving a pendulum)* Let the support point of a pendulum be driven up and down vertically at twice the natural frequency of the pendulum (ω_0). The motion

PROBLEMS

of the support point is

$$Y_{\text{support}}(t) = Y_0 \cos(2\omega_0 t). \tag{10.98}$$

The period of the support motion is $T = \frac{\pi}{\omega_0}$.

a) Write a computer program to find $\theta(T)$, $\frac{d\theta(T)}{dt}$ in terms of $\theta(0)$, $\frac{d\theta(0)}{dt}$, where $T = 1$ period of the vertical motion. Assume that the pendulum is undamped and its length is 2 meters. What period does this give for $Y_0 = 0$?

b) Now express your numerical calculation in terms of the matrix $\mathbf{M}$ that connects the phase space coordinates at $t = 0$ to those at $t = T$. Find the least value of Y_0 that makes the pendulum unstable. Hint: Take the trace of your matrix and plot this versus Y_0 for values near zero. Require that $|Tr\,\mathbf{M}| > 2$.

c) Add a little damping, so that the Q of the pendulum for $Y_0 = 0$ is 20. This Q is realistic for a typical child's swing, for example. Find the minimum driving amplitude for instability in this case.

d) Give a physical argument based on energy lost versus energy gained per cycle (using approximate solutions) to explain why the presence of damping creates a threshold for the parametric instability.

Problem 7: *(Upside-down driven pendulum)* You have a little pendulum 3 inches long. The support or pivot is being driven up and down at a variable frequency Ω with an amplitude of ± 0.25 inches. If the pendulum is placed upside down, $a < 0$. When driven at a sufficiently high frequency, it becomes stable in this position, and just oscillates back and forth.

a) Find q as defined in Equation (10.35).

b) For $-.1 < a < -.02$ (variable driving frequency), find the lowest frequency at which the pendulum just becomes unstable. What is this frequency in hertz?

c) Suppose the driving frequency corresponds to $a = -.1$. What are the eigenvalues of the $\mathbf{M}$ matrix? Give a physical interpretation of the largest eigenvalue and convert this to a growth rate in s^{-1}.

Problem 8*: This problem deals with two different ways to move the support pivot of a pendulum.

a) *(Driving a pendulum horizontally)* A pendulum of length l has its support point driven *horizontally* at frequency Ω as shown in Figure 10.17A. Find the Lagrangian and the equations of motion if the pendulum is executing small oscillations about $\theta = 0$, a vertical downward position. Show that the equation of motion in this approximation is identical with that of a driven harmonic oscillator and hence that the position is stable.

b) *(Driving a pendulum with a wheel)* Now a pendulum of length l has the support point located on the rim of a wheel of radius R which is rotating clockwise at angular frequency Ω as shown in Figure 10.17B. Find the Lagrangian and the equations of motion if the pendulum is executing small oscillations about

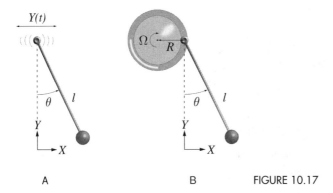

FIGURE 10.17

$\theta = 0$. Show that one obtains a Mathieu equation, like (10.36). Find expressions for a and q in terms of the parameters l, R, g, and Ω. Would the pendulum always be stable? Explain.

Problem 9: *(How does a child pump a swing?)* Suppose a child is standing on a swing. If a kind bystander starts the swing with a small amplitude, it is then possible for the child to amplify this motion by alternatively shortening and lengthening the ropes. (We will not consider any forward or backward motion here.) Assume the child raises the ropes at the bottom of the swing and lowers them at the highest parts of the swing. If this is done at *twice* the natural frequency ω, it will be a very efficient way to amplify the original motion. Imagine that the swing is like a pendulum, but with variable length $l(t)$. Define $l(t) \equiv \bar{l}(1 + \delta(t))$, where $\bar{l}$ is the average pendulum length and $\delta(t)$ is a function describing how the child changes the length of the string. We assume $\delta(t) \ll 1$.

a) Prove that the equation of motion for small oscillations is

$$\frac{d}{dt}(l^2 \dot\theta) + gl\theta = 0. \tag{10.99}$$

We will define the following initial conditions $\theta(0) = A$, $\dot\theta(0) = 0$ for simplicity.

b) Next make the substitution $\phi(t) = \theta(t)l(t)$, as well as the scaling to dimensionless time: $\tau \equiv (\sqrt{\frac{g}{\bar{l}}})t$. Prove that we obtain the following differential equation, which has eliminated the first derivative term:

$$\ddot\phi + \left(\frac{1 - \ddot\delta(\tau)}{1 + \delta(\tau)}\right)\phi = 0. \tag{10.100}$$

c) Next make the approximation that δ is small enough to only keep first-order terms in δ. Also, assume that the ϕ value is very close to the harmonic solution $\phi_0 = Al(0)\cos\tau$. (This means that we are looking only for relatively short times after the start of pumping.) Finally, assume the child changes the length as the function $\delta(t) = \delta_0 \cos 2\tau$, where δ_0 is a constant that measures the maximum variation of the relative length of the swing when the child is pumping. This is probably not exactly what the child does, but it does approximate the situation described above.

Using the method of successive approximations out to only the first correction term, prove that the equation we obtain is

$$\ddot{\phi} + \phi = -\frac{3}{2}\delta_0 Al(0)(\cos\tau + \cos 3\tau). \quad (10.101)$$

d) Solve Equation (10.101) by using a Green's function integral or by some other method, such as guessing the correct form of the solution. Show that you obtain a term that increases with time.

e) Find the increase in magnitude of the swing angle after half a period $\Delta\theta = |\theta(\pi)| - \theta(0)$. Prove that it is a positive number. Thus the child is able to increase his amplitude of swinging by the method described above. The child pumping the swing corresponds to an instability, but the actual amplitude is limited by the nonlinear terms we have neglected. In pumping a swing, if the pump is too feeble, then because of damping friction, nothing will happen. You have to exceed a certain amplitude to get parametric amplification. In a more precise calculation, you could calculate the rate of growth for the unstable solution to (10.100) numerically by applying the Floquet theory of the Hill equation of which (10.100) is a special case.

Note that current research suggests that for small amplitudes, the case of a child swinging (either sitting or standing) is dominated by a driven oscillator effect due to shifting the center of mass back and forth at the swing frequency, rather than a parametric oscillator effect. As the amplitude gets larger, the parametric effect begins to dominate. This is why you rarely see children swinging as described above. (See *American J. of Phys.* **64** (3), pp. 215–220, ibid **58** (5), pp. 463–467, and ibid **38** (7), pp. 920–922.)

Problem 10: *(Pumping a swing II)* Make a physical analysis of a child pumping a swing as discussed above. Assume the swing is shortened Δy at the bottom of the motion, and released at the top of the motion. What happens to the energy? To the angular momentum? To the angular velocity? Analyze one period to show that the existing motion is amplified, and calculate by how much.

Problem 11: *(An asteroid perturbed by Jupiter – is it stable?)* Parts c) and d) of this problem require some numerical work with a computer. Any planet or asteroid in our solar system is perturbed by being attracted to Jupiter, as well as to the Sun. (Jupiter is by far the most important of the extra solar gravitational effects.) We will focus on asteroids here. The additional contribution to the gravitational potential is (neglecting constant terms)

$$V_{\text{Jupiter}} = -\frac{Gm_a M_J}{|\vec{r}_a - \vec{r}_J|} \approx -\frac{Gm_a M_J r_a}{r_J^2}\cos(\phi_a - \phi_J), \quad (10.102)$$

where G is the gravitational constant, m_a is the mass of the asteroid, M_J is the mass of Jupiter, and $|\vec{r}_a - \vec{r}_J|$ is the distance from the asteroid to Jupiter. With the coordinate system origin located at the center of mass of the sun, ϕ_a, ϕ_J are the angles in plane polar coordinates of the asteroid and Jupiter, respectively, in what we assume is their common plane. (We neglect a shift in the center of mass caused by the finite mass of Jupiter which is about $\frac{1}{1000}$ of the Sun's mass.) In (10.102), we have neglected terms in the potential of $O[\frac{r_a^2}{r_J^2}]$ and higher.

a) Assume that Jupiter moves in a circular orbit with a frequency ω_J. (The period is 11.86 years.) The angle $\phi_a - \phi_J$ between Jupiter and the asteroid is thus $\phi_a - \frac{\omega_J}{\omega_a}\phi_a$, if we also assume the asteroid moves in a nearly circular orbit with frequency ω_a. Show that the ($u = \frac{1}{r_a}, \phi = \phi_a$) form of the radial equation is

$$\frac{d^2u}{d\phi^2} + u - \frac{1}{p} + \frac{1}{p}\frac{M_J}{M_\odot}\left(\frac{1}{ur_J}\right)^2 \cos(x\phi) = 0, \qquad (10.103)$$

where p, the radius for (circular) asteroid orbits, is given by $p = a_a(1 - \epsilon^2) = a_a'$. $x \equiv 1 - \frac{\omega_J}{\omega_a}$. $M_\odot$ is the solar mass. We are neglecting a small effect of the torque exerted by Jupiter in the azimuthal equation of motion which averages to zero, thereby allowing us to treat this as a one-dimensional problem.

b) Linearize Equation (10.103) to study the stability of small deviations in the circular asteroid orbit. Prove that the linearized equation has the form of a Mathieu equation:

$$\frac{d^2\delta u}{d\tau^2} + [a - 2q\cos 2\tau]\delta u = 0, \qquad (10.104)$$

with $a \equiv (\frac{2}{x})^2$, $q \equiv (\frac{2p}{xr_J})^2(\frac{M_J}{M_\odot})$, and $\tau \equiv \frac{x\phi}{2}$.

c) There are nonlinear resonances leading to instability if $a = n^2, n = 1, 2, 3, \ldots$. We will first investigate whether $n = 3$ could be important for driving asteroids out of the asteroid belt. If $a = 9$, then $x = \frac{2}{3}$, which means that the period of Jupiter is three times the period of the asteroid. This is called a "3:1" nonlinear resonance. Recall from Chapter 4 that Kepler's Third Law means that the square of the period is proportional to the cube of the radius, so at the 3:1 resonance, the ratio $\frac{p}{r_J} = \frac{1}{3^{\frac{2}{3}}} = 0.4807$, while $q = .002$. The perturbation is very weak, but the resonance makes the motion unstable. Prove that if Jupiter's semimajor axis is 5.203 in AU (astronomical units equal to the Earth's semimajor axis), then the unstable asteroid radius for the 3:1 resonance is 2.5 AU.

d) Let σ stand for the ratio of the asteroid's radius to Jupiter's radius. It can be shown that

$$x(\sigma) = 1 - \sigma^{\frac{3}{2}}, \quad q(\sigma) = .0038\frac{\sigma^2}{(1-\sigma^{\frac{3}{2}})^2},$$

$$a(\sigma) = \frac{4}{(1-\sigma^{\frac{3}{2}})^2}. \qquad (10.105)$$

It is difficult to compute $Tr\,\mathbf{M}$ because q is so small. An easier approach to the solution is to take advantage of the small q to make an approximation to the shape of the boundaries ($|Tr\mathbf{M}| = 2$) of the unstable regions in the graph of a versus q. For example, in Figure 10.10, we can approximate the lines emanating from $a = 1$, the first unstable region of the Mathieu equation, by*

$$a(q) = 1 \pm q + O[q^2]. \tag{10.106}$$

The shape of the boundaries of the third unstable region of the Mathieu equation* are

$$a(q) = 9 + \frac{q^2}{16} \pm \frac{q^3}{64}. \tag{10.107}$$

Scale up the strength of $q(\sigma)$ by a factor of 1,000 to verify, using the formula (10.107) for numerical calculations, that the range of unstable radii for the scaled-up q is $.485255 \leq \sigma \leq .497367$. Since the width around the unstable point decreases like q^3, the width for the actual value of q from Jupiter's gravity is approximately $\Delta\sigma \approx 8.7 \times 10^{-12}$. This small value, which comes from the weakness of the effect of Jupiter, makes it hard to understand the gap in the distribution of asteroids around 2.5 AU, even though several billions of years have elapsed. Of course the solutions we have found do not include the phenomenon of chaos, since chaotic motion does not exist in a purely linear system. See the next chapter for a detailed discussion of chaos.

Driven Anharmonic Oscillators

Problem 12: *(Duffing oscillator)* Solve the driven Duffing oscillator numerically, and demonstrate the hysteresis effect. Make a resonance plot of amplitude versus frequency using your results.

Lindstedt–Poincaré Perturbation Theory

Problem 13: *(Solving differential equations I)* Prove that you get Equation (10.45) by solving Equation (10.41). You can use (10.44) or any other method for solving differential equations, for instance, variation of parameters.

Problem 14: *(Solving differential equations II)* Prove that you obtain Equations (10.55, 10.58) by solving Equations (10.51, 10.52) respectively. You can use (10.44) or any other method for solving differential equations.

Problem 15: *(Linear damping as a perturbation)* Treat the case of linear damping as a perturbation to the free linear oscillator equation:

$$\ddot{q} + \frac{1}{Q}\dot{q} + q = 0 \rightarrow \ddot{q} + \frac{\epsilon}{Q}\dot{q} + q = 0. \tag{10.108}$$

---

* C. Hayashi, *Nonlinear Oscillations in Physical Systems*, Princeton Univ. Press, 1985, p. 88. Note that Hayashi's q differs from our q.

a) First make the substitutions (10.46, 10.47, 10.48) and equate like powers of ϵ, up to and including ϵ^2.
b) Next solve the three differential equations you obtain, making sure to set the coefficients of the secular terms to zero.
c) Can you think of what went wrong in the theory? Hint: Lindstedt–Poincaré theory is not used to solve for transient responses. Why?

Problem 16: *(Squared perturbation)* Solve the case of a "squared" perturbation to the free linear oscillator using the Lindstedt–Poincaré perturbation theory. The equation is

$$\ddot{q} + q + \epsilon q^2 = 0. \tag{10.109}$$

Find the corrections out to second order in ϵ. Note: You can also solve this problem by using the method of successive approximations. Comment on why it is possible to use either method here.

Problem 17: *(Van der Pol equation)* Solve the Van der Pol equation

$$\ddot{q} + q - \epsilon(1 - q^2)\dot{q} = 0. \tag{10.110}$$

using the Lindstedt–Poincaré perturbation theory. Find the corrections out to second order in ϵ.

Precession of Kepler Ellipses from General Relativity

(Note: This discussion applies to the following two problems.)
General relativity gives a new type of equation of motion for a point mass orbiting a stationary massive object of mass M. The equation of motion comes from requiring the planet to follow a geodesic path in space-time curved by the gravitational potential of the Sun. The theory gives a correction of magnitude $\frac{v^2}{c^2}$ to the gravitational potential. This small correction causes the Kepler ellipse to gradually rotate or precess its spatial orientation in the orbital plane. For the planet Mercury, $\frac{v}{c} = 1.6 \times 10^{-4}$, so the correction to the potential is $\frac{v^2}{c^2} = 2.5 \times 10^{-8}$. The effect[*] of this small correction on the precession of the perihelion (point closest to the Sun) of the ellipse is not negligible, and it has been measured for Mercury's orbit with great precision! Begin by using the transformation $u = \frac{1}{r}$ for the dependent variable and ϕ rather than the time as the independent variable. It is useful to review the derivation of (4.48, 4.49) now.

The equation derived from general relativity[†] is very similar to (4.49) when recast in terms of u, ϕ.

$$\frac{d^2 u}{d\phi^2} + u - \frac{1}{p} = 3mu^2 \quad \text{(general relativity)}. \tag{10.111}$$

---

[*] A more complete discussion of this effect can be found in the book by B. F. Schutz, *A First Course in General Relativity*, Cambridge Univ. Press, 1985, p. 280ff.
[†] W. Pauli, *Theory of Relativity*, English translation, Pergamon Press, 1959, p. 166ff.

(The length $p \equiv \frac{l^2}{\mu k}$.) However, it contains an extra term proportional to the "geometrized" mass of the Sun, $m \equiv \frac{GM_\odot}{c^2} = 1.477$ km. Here $M_\odot$ is the solar mass in kilograms. By combining the constants in m, we obtain a convenient measure of the mass that is the source of the gravitational potential.

Problem 18: *(Using successive approximations)* For convenience, rewrite the equation of motion (10.111) as

$$\frac{d^2u}{d\phi^2} + u - \frac{1}{p} = 3\lambda m u^2. \tag{10.112}$$

Assume that the solution can be expanded in powers of λ: $u = u_0 + \lambda u_1 + \lambda^2 u_2 + \cdots$. Use the method of successive approximation to find the equations for u_0, u_1, u_2, by separating the different powers of λ to give $O[m^0]$, $O[m]$, $O[m^2]$ terms. (Actually $\lambda = 1$. It is only introduced to have a dimensionless expansion parameter for the perturbation theory, which is really an expansion in powers of the perturbation strength proportional to the small quantity m.) Prove that you get secular terms (i.e., the solution is not periodic in ϕ). Why can this result not represent a physically correct solution?

Problem 19: *(Using Lindstedt–Poincaré theory)* Instead of successive approximation, use the perturbation method of Lindstedt–Poincaré. The equation of motion is written by introducing a new independent variable $s = \omega\phi$: Modify the frequency by another power series in λ: $\omega(\lambda) = 1 + \lambda\omega_1 + \lambda^2\omega_2 + \cdots$. The equation of motion (10.112) becomes

$$\frac{d^2u}{ds^2} + \frac{1}{\omega^2}\left(u - \frac{1}{p}\right) = 3\frac{\lambda m}{\omega^2}u^2. \tag{10.113}$$

a) Use the Lindstedt–Poincaré theory with Equation (10.113) to find another set of differential equations for $u_i(s)$, $i = 0, 1, 2$ up through second order.
b) Solve these equations to prove that the first-order correction to ω is

$$\omega_1 = -\frac{3m}{p}. \tag{10.114}$$

The semimajor axis of a Kepler ellipse was shown in Chapter 4 to be related to p by $p = a(1 - \epsilon^2)$. Find ω_1 for the first three planets. Use Table 4.9 for pertinent data.

c) The term $\delta\phi$ is the precession of the Kepler ellipse due to the perturbation, expressed in units of arc seconds/per century. The orbit passes through one complete period when the variable s increases by 2π. The angle of the perihelion in polar coordinates must advance from the starting point, defined as $\phi = s = 0$ to $\phi = \frac{2\pi}{\omega} \approx \frac{2\pi}{1+\omega_1} \approx 2\pi(1-\omega_1)$ if we neglect second-order terms. If the orbit is plotted on a polar plot, the perihelion will be reached again after one complete trip around the ellipse when $\phi = 2\pi + \delta\phi$, where $\delta\phi = -2\pi\omega_1$ is the advance of the perihelion in one orbital period. A positive $\delta\phi$ means the perihelion moves in the direction of the planet's motion. Prove the general formula for general relativity:

$$\delta\phi = \frac{6\pi m}{a(1-\epsilon^2)}. \tag{10.115}$$

This $\delta\phi$ is in units of radians/revolution. Find $\delta\phi$ in units of arcsec/century for the first three planets. Again use Table 4.9.

d) Choose the solutions for $u_1(s)$ and $u_2(s)$ so that $u_1(0) = u_2(0) = 0$. Use computer algebra or else exercise dogged perseverance to obtain the second-order frequency correction:

$$\omega_2 = -\frac{3m^2}{4p^2}(18 + 5\epsilon^2). \tag{10.116}$$

What is the ratio $\frac{\omega_2}{\omega_1}$? If a hypothetical planet or a star is assumed to be orbiting a black hole in a nearly circular orbit at 100 times the distance the Earth is from the Sun, how many solar masses would the black hole have to be before the second-order correction became 10% of the first-order correction?

CHAPTER ELEVEN

CHAOTIC DYNAMICS

INTRODUCTION AND OVERVIEW OF CHAPTER 11

Three hundred years ago, Isaac Newton discovered that the motion of dynamical systems with N degrees of freedom could be described by N second-order differential equations. These differential equations provide us with a mathematical road map, giving directions about the motion of a system for each successive time interval. Since the system's motion in each interval of time is connected smoothly to the motion in the preceding time interval, Newton was convinced that the equations of this motion would have solutions that change smoothly as the initial conditions are varied, that is, would be analytic functions of the time and the initial conditions. Generations of physicists shared Newton's belief that all mechanical problems would have analytic solutions. By the 1830s, Lagrange and Hamilton had improved the analytical techniques for finding the equations most appropriate to a particular physical system. If analytic solutions to a particular problem could not be found, it was thought that only a cleverer, more sophisticated approach was needed. The concept of the "clockwork universe" was accepted after Newton. Such a universe is completely determined by the initial conditions to move along smooth paths for the rest of time, just as the planets seemed to move in perpetual ellipses around the Sun. Laplace was a particular champion of this universal determinism, a view his contemporaries did not hesitate to extend to everything, not only to the problems of mechanics.

Who would have expected, metaphorically speaking, that following a smooth continuous road on a road map would end one up in China instead of California if the starting point is changed by 0.1 millimeter? We now know that the eighteenth and early nineteenth century theorists hoped in vain for analytic solutions to all mechanics problems. In fact, only very special problems can be solved analytically. It does not mean that that Newton's laws for nonrelativistic classical mechanics are wrong. They predicted chaotic motion all along. Poincaré was the first person to see this.

Chaos was discovered because a prize of 2,500 crowns* was proposed for the answer to one of four outstanding questions in mathematical physics. One question

---

* The prize was to be awarded on the sixtieth birthday of King Oscar II of Sweden and Norway, on January 21, 1889.

was: "Prove that the solar system is stable." Poincaré submitted a paper over 200 pages long in which he showed that an analytic solution was not possible, even to the simpler problem of only three gravitating bodies, one of which had a mass too small to affect the motion of the other two. Poincaré won the prize, but there was a scandal because, after he was declared the winner, it was discovered that there was an important error in his proof. The error was corrected before the award was made, however.*

In 1892, Poincaré made the following observation after studying the 3-body problem:

> ... it may happen that small differences in the initial conditions produce very great differences in the final phenomena. A small error in the former will produce an enormous error in the latter. Prediction becomes impossible, and we have the fortuitous phenomenon.

In reaching the conclusion quoted above, Poincaré discovered deterministic chaos. However, his discovery was overshadowed by the discoveries of quantum mechanics and relativity. Poincaré's, discovery was ignored by mainstream physics until the 1960s. By doing computer experiments with systems of differential equations, it was discovered that even very simple systems can become chaotic. The implications of this discovery for many fields, including meteorology, chemistry, biology, and even perhaps the stock market, are widespread and of fundamental importance. Even dripping faucets can exhibit chaos! In this chapter we will study two of the simplest mechanical systems exhibiting chaotic motion.

"Deterministic chaos" has a special meaning in physics. It implies completely deterministic motion – each time step depends only on the motion at previous times in a well-defined way. The distinctive new feature is that long-term prediction is impossible without perfect knowledge of the initial conditions. Even a small change in the initial conditions produces a completely new solution at later times. Practical prediction for long times is impossible. Chaos has nothing to do with random noise, although there is a superficial resemblance when the motion is observed. For chaotic motion, two trajectories that are initially arbitrarily close in phase space will diverge exponentially in time from each other. All memory of the fact that these two trajectories had nearly identical starting points is ultimately lost. The key property is "exponential divergence." For nonchaotic motion, nearby trajectories diverge at most linearly with the time.

In the last thirty years considerable progress has been made in classifying and understanding the motion of nonlinear systems. We have also changed our point of view on what constitutes a "solution." Physicists now concede that analytic solutions will never be found for many interesting dynamical systems. Instead, the emphasis has shifted to the geometric picture of phase space flow over long periods of time. Perhaps Poincaré's greatest contribution to physics was to lead physicists away from a search for analytic solutions towards a study of the geometry of phase trajectories. This geometric point of view has already been encountered in Chapter 6, with the discussion of invariant tori.

There are certain "standard" routes to chaos as the energy or other parameter is increased in strength. There exist surprisingly universal characteristics of this

---

* For more details, see *Newton's Clock* by Ivars Peterson, W. H. Freeman, 1993, p. 143ff.

phenomenon and these have been the subject of intensive research. Simplified mathematical models are very useful for investigating these universal features, since the computation time can be significantly reduced if there is no need to provide a realistic simulation of a particular physical system. Numerical "experiments" using computers now play a key role in increasing our understanding.

There is an important distinction between chaos in conservative systems and chaos in dissipative systems. In the former, phase volume must be conserved. A small bundle of phase trajectories must be squeezed exponentially in one or more directions if it is expanded exponentially in other directions as a result of chaos. If the motion is bounded, due to energy conservation, the phase trajectories must be constantly folded back onto the phase space.

For conservative motion, an important phenomenon is *phase-locking* onto a *nonlinear resonance*. If the ratio of two frequencies associated with the motion approaches an integer or a rational fraction, nonlinearities can cause phase-locking of nearby trajectories onto this resonance. (This ratio is called the *winding number*,[*] as was discussed in a preliminary way in Chapter 6.) As a parameter in the equations of motion is changed (it could be the energy or another parameter in the Hamiltonian), the strengths of competing nonlinear resonances can increase, such that two of the resonances overlap. The system does not "know" which of the two resonances to lock onto when two such resonances overlap, and thus it becomes chaotic. It no longer follows a smooth regular trajectory in phase space, which is the hallmark of an integrable system, but moves chaotically in phase space. The smooth curve on the invariant torus is broken by the competition between the nonlinear resonances. This is the basic source of chaos. Nonchaotic motion is often called *regular motion*, whereas chaotic motion is called *irregular motion*. An integrable system has only regular motion, whereas a nonintegrable system can have both regular and irregular motion, depending on the initial conditions.

For dissipative systems there must be an external source of energy to maintain the motion. Phase volumes will always contract, even as the trajectory bundle expands exponentially in one or more directions. In many cases, trajectories relax to a single one-dimensional curve in phase space, known as a "periodic attractor." In the example of the linearized, damped driven pendulum considered in Chapter 3, this curve was called the steady-state solution. If the nonlinear case is considered, it is possible for the system to have a chaotic or *strange attractor*. This means that the motion in phase space occupies a set with noninteger dimension, known as a *fractal*. These terms will be explained in more detail later.

As examples, we will study the motion of two different systems:

A) The double pendulum is a conservative system with a time-independent Hamiltonian. There are two degrees of freedom. Chaos in this system can be produced either by varying the strength of gravity or by increasing the energy. We will study what happens as the energy is increased.

B) As an example of a dissipative system, we study a single pendulum driven by an external periodic force. Friction in the pendulum makes it dissipative. Although there is only one degree of freedom, chaos can be observed because

---

[*] A synonym for winding number is "rotation number."

the restoring force is nonlinear (proportional to $\sin\theta$). The Hamiltonian is time dependent because of the driving force. We can produce chaos by increasing the amplitude of the external driving force. This is the simplest mechanical system exhibiting chaotic behavior.

After studying the richly complex behavior of these two different systems, we will look into a 1-D mathematical model, the logistic map. This map captures some of the generic features of chaos and the approach to chaos. This discussion appears in the appendix to the chapter. Other systems and maps are proposed as possible student projects.

The defining characteristic of deterministic chaos – extreme sensitivity to initial conditions – is known as the "butterfly effect," because it was first (re)discovered* in attempts to simulate turbulence associated with the weather by integrating a set of nonlinear equations on the computer. The rather fanciful notion is that the weather can involve chaotic motion – if a single butterfly flaps its wings, then by waiting long enough, such a small perturbation can have a major effect on the world's weather, rendering long-term weather prediction by computers hopeless.

11.1 CONSERVATIVE CHAOS – THE DOUBLE PENDULUM: A HAMILTONIAN SYSTEM WITH TWO DEGREES OF FREEDOM

Poincaré spent a great deal of his time investigating the motion of systems with two degrees of freedom, and so shall we. We will specialize to the double pendulum as shown in Figure 11.1, which illustrates many of the generic features common to all two degree of freedom systems.

Let α stand for the angle between the top pendulum and the vertical direction, while β stands for the angle between the top pendulum and the bottom pendulum, as shown in the figure (a change from our previous notation for the double pendulum in Chapter 9). Suspend one of the two pendulums so that the bottom part is hanging straight down and the top part makes an angle α with the vertical ($\beta = -\alpha$). Then the energy is all potential energy, with zero kinetic energy. We will use this initial angle to reference the total energy, so we will talk about an energy corresponding to $\alpha(0) = 120°$, rather than $E = -1.5$. It evokes a more physical picture to use the initial angle as a label for the energy.

The Hamiltonian and the exact equations of motion can be found from the Lagrangian for this system (9.20), which was derived in Chapter 9. If we divide (9.20) by mgl and change the time scale to $t \to t\sqrt{\frac{g}{l}}$, we obtain a dimensionless form of the equation. The corresponding canonical angular momenta will be called l_α and l_β respectively. From the scaled form of (9.20) we find them to be

$$l_\alpha = [(3 + 2\cos\beta)\dot\alpha + (1 + \cos\beta)\dot\beta],$$
$$l_\beta = [(1 + \cos\beta)\dot\alpha + \dot\beta]. \tag{11.1}$$

---

* Lorenz's personal recollections of his discovery appear in "On the Prevalence of Aperiodicity in Simple Systems," *Global Analysis*, ed. M. Grmela and J. Marsden, Springer-Verlag, 1979.

11.1 CONSERVATIVE CHAOS – THE DOUBLE PENDULUM

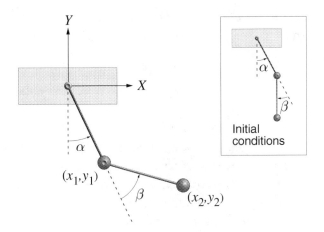

FIGURE 11.1
Origin is at top pivot. Masses are both equal to m; lengths are both equal to l. The total energy is defined by the initial conditions shown ($\alpha(0) = -\beta(0)$).

The exact Hamiltonian is

$$E = H(\alpha, l_\alpha, \beta, l_\beta) = -2\cos\alpha - \cos(\alpha + \beta)$$
$$+ \frac{l_\alpha^2 - 2(1 + \cos\beta)\, l_\alpha l_\beta + (3 + 2\cos\beta)l_\beta^2}{3 - \cos 2\beta}. \tag{11.2}$$

Hamilton's equations of motion are

$$\dot{\alpha} = 2\frac{l_\alpha - (1 + \cos\beta)l_\beta}{3 - \cos 2\beta}, \tag{11.3}$$

$$\dot{l}_\alpha = -2\sin\alpha - \sin(\alpha + \beta), \tag{11.4}$$

$$\dot{\beta} = 2\frac{-(1 + \cos\beta)l_\alpha + (3 + 2\cos\beta)l_\beta}{3 - \cos 2\beta}, \tag{11.5}$$

$$\dot{l}_\beta = -\sin(\alpha + \beta) - 2\sin\beta\frac{(l_\alpha - l_\beta)l_\beta}{3 - \cos 2\beta}$$
$$+ 2\sin 2\beta\frac{l_\alpha^2 - 2(1 + \cos\beta)l_\alpha l_\beta + (3 + 2\cos\beta)l_\beta^2}{(3 - \cos 2\beta)^2}. \tag{11.6}$$

QUESTION 1: *Hamiltonian* Derive (11.2) from (9.20). Don't forget to use the scaling.

If α and β are sufficiently small at $t = 0$, then $E \approx -3$. In that case, the equations of motion will be well approximated by the linear equations (9.21). In this limit, we have a set of coupled *linear* equations of motion with two degrees of freedom, so the motion will be a superposition of two modes, each with its characteristic frequency. These frequencies were derived in Chapter 9 (9.38, 9.39) to be $\omega_1 = \sqrt{2 - \sqrt{2}}$, $\omega_2 = \sqrt{2 + \sqrt{2}}$. The winding number, which is defined to be the ratio $\Omega \equiv \frac{\omega_2}{\omega_1}$ of the two mode frequencies, is $1 + \sqrt{2}$. If we do not restrict the amplitudes to be small, no analytic solution is possible. The equations have to be solved numerically.

Is It Really Chaotic?

There is a dramatic way to show the physical reality of chaos in a classroom demonstration. Two identical uncoupled double pendulums are raised to a starting position, with the lower pendulums hanging down vertically as described in Figure 11.1 ($\beta(0) = -\alpha(0)$). As the energy is increased, the motion changes dramatically. Below the threshold for chaos, the motion of the two pendulums is quite close if sufficient care is taken to start the two systems as nearly identical as possible. Upon increasing the starting value $\alpha(0)$ beyond about 70°, the two pendulums no longer swing in approximately the same way. After a while the two motions seem to become totally independent and uncorrelated. However much care was taken in making the starting conditions identical, there will always be a very small difference. This difference is magnified exponentially if chaos is present – soon there is no relation between one pendulum and the other. As the starting energy is increased even further, the exponential divergence of the two motions occurs in progressively shorter times after the pendulums are released.

Let's try a computer simulation of the experiment just described. The computer integrates the equations of motion (11.3)–(11.6). To test for chaos, we compared two numerical solutions for the angle of the lower pendulum, β, which differ by .001° in the starting angle of the upper pendulum. (Using the lower pendulum gives a more sensitive test of chaos.) Starting values of 50° and 90° were tried. For each case, the difference in the two values of β, $\delta\beta$, was plotted as a function of time. (Actually we took the $\log_{10}|\delta\beta|$, so that exponential growth in this quantity would appear as a straight line on the graph.)

The case of a 50° starting angle (Figure 11.2A) is not chaotic. The difference between starting at 50° and 50.001° remains approximately constant, with $|\delta\beta| \sim 10^{-5}$. The case of $\alpha(0) = 90°$ (Figure 11.2) is definitely chaotic. The very small initial difference of .001° is magnified exponentially as time passes. It becomes so large that $|\delta\beta| > 10$ by the end of the integration time. This large value for β is due to rotation of the bottom pendulum. We can say that the threshold angle for chaotic motion lies between 50° and 90°.

We will investigate the double pendulum's "quasiperiodic route to chaos," which is typical of Hamiltonian (nondissipative) systems. First we must develop some useful mathematical tools. The double pendulum will also be used as an illustration of the application of these tools.

11.2 THE POINCARÉ SECTION

A useful way of visualizing the motion was invented by Poincaré. For a conservative system of two degrees of freedom, such as the double pendulum, the motion in 4-D phase space $(\alpha(t), l_\alpha(t), \beta, l_\beta(t))$ must be confined to a three-dimensional "energy

11.2 THE POINCARÉ SECTION

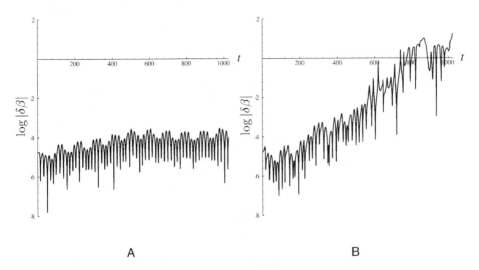

FIGURE 11.2
Testing for chaos. A) $\alpha(0) = 50°$, showing little deviation from initial difference. B) $\alpha(0) = 90°$, showing significant deviation from initial difference.

shell" as a consequence of energy conservation. The topology of this motion is that of a 2-torus, as discussed in Chapter 6. This is difficult to visualize. Poincaré suggested taking a stroboscopic picture of this motion by recording the β and l_β values every time $\alpha = 0$ and $\dot{\alpha} > 0$. This means we are looking at the motion of the lower pendulum for successive complete oscillations of the upper pendulum. (Of course we could also record the upper pendulum for successive oscillations of the lower one, but we will not choose to do this here.) $\alpha = 0$ is a three-dimensional hyperplane* in the 4-D phase space. We can imagine the phase trajectory winding in and out of this hyperplane as shown in Figure 11.3.

Each time the $\alpha = 0$ hyperplane is crossed with positive $\dot{\alpha}$, record β, l_β as a point on a two-dimensional plot. Repeat this process to obtain a collection of points known as a Poincaré section.† In the limit of an infinite number of periods, some of these sets merge into continuous curves, while others consist of a finite number of points. What determines this? If the ratio of frequencies is irrational, the points never exactly repeat, and the curve is filled in. If, however, the winding number $\Omega = \frac{r}{s}$, the ratio of two integers, then after s periods of the upper pendulum, the lower one will have had exactly r periods and will be back at the exact same starting point. If there are s discrete points on the Poincaré section, we know the denominator in the rational fraction $\frac{r}{s}$. Determining r is done by seeing how many times the points "wind" around a center after s periods. A rational winding number can be determined simply by counting.

For the double pendulum example, in the linear approximation, the winding number is irrational and equal to $1 + \sqrt{2}$. The integer part cannot be determined from a Poincaré

---

* That is, it comprises the parts of 4-D phase space satisfying a linear relationship between the four coordinates.
† An alternative name is *surface of section*, which is preferred by some authors.

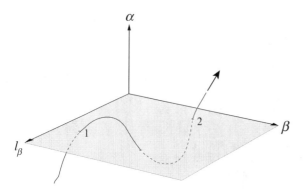

FIGURE 11.3
A phase trajectory crossing $\alpha = 0$ with $\dot\alpha > 0$ at the points labeled 1 and 2. (This is a schematic rendition of 4-D space.)

section. Values of $\Omega = 2.4142\ldots$ and $\Omega = 0.4142\ldots$ give the same curve on the Poincaré section.

QUESTION 2: *Winding Number 1* Draw a trajectory of a 2-D conservative system having a winding number of $\frac{1}{2}$. Compare to what you will see on the Poincaré section. (Remember that the motion is confined to a 2-torus.) Now try $\frac{2}{5}$. Then generalize to the case $\frac{r}{s}$.

We start by looking at the Poincaré section for the double pendulum for a starting angle $\alpha(0) = -\beta(0) = 1°$, which may still be considered a small angle where the linear approximation works well. To produce Figure 11.4 and all subsequent Poincaré sections, the initial conditions are varied. The total energy has the value $E = H(\alpha(0), 0, -\alpha(0), 0)$.

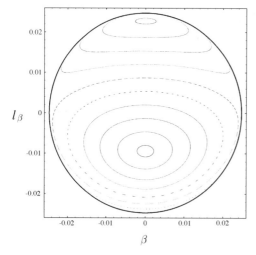

FIGURE 11.4
Poincaré section for $\alpha(0) = -\beta(0) = 1°$.

11.2 THE POINCARÉ SECTION

The boundary curve also appears on these plots. It corresponds to $\dot{\alpha} = 0$. It can be shown from the equations of motion that being on the boundary implies that $l_\alpha = (1 + \cos\beta)l_\beta$. All points on the Poincaré section must lie within the boundary curve $E = H(0, (1 + \cos\beta)l_\beta, \beta, l_\beta) = -2 + \frac{l_\beta^2}{2} - \cos\beta$. This curve is closed for $E < -1$, which is a starting angle of 90°. Above this energy, it is possible for the lower pendulum to rotate completely through an angle of 2π, so the boundary curve becomes two separate curves bounded by the lines $\beta = \pm\pi$.

QUESTION 3: Boundary First show from (11.3)–(11.6) that being on the boundary implies that $l_\alpha = (1 + \cos\beta)l_\beta$. Then prove that all points on the Poincaré section must lie within the bounday curve $E = H(0, (1 + \cos\beta)l_\beta, \beta, l_\beta) = -2 + \frac{l_\beta^2}{2} - \cos\beta$.

Two "centers" are visible in Figure 11.4. These centers correspond to *fixed points* of the motion, meaning that the lower pendulum is synchronized exactly with the upper one. This happens when the motion takes place in one of the two modes. For different initial conditions, not corresponding to the fixed points, the motion of the lower pendulum describes curves similar to ellipses (at least close to the fixed points). We will refer to the curves on the Poincaré section as "invariant tori," but of course these are only subsets of the complete tori. "Invariant" in this context means that the motion never leaves its torus.

Each of the curves will have a winding number close to .4142.... We know this means that the high frequency mode executes 2.4142... oscillations for one oscillation of the low frequency mode. Two frequencies are present, except at the fixed points, where only a single frequency governs the motion of both pendulums, since they are synchronous. Because the winding number is the ratio of number of oscillations of the lower pendulum per oscillation of the upper pendulum, it is 1 at the fixed point. The winding number can be calculated as we will describe next.

QUESTION 4: Winding Number 2 Why can we not determine the integer part of a winding number from a Poincaré section?

Poincaré Sections Are Area-Preserving Maps

We can think of successive points $\beta(n), l_\beta(n)$ for $n = 1, 2, 3, \ldots$ as a two-dimensional map. A point at $\beta(n), l_\beta(n)$ is mapped after one oscillation of the top pendulum into another point at $\beta(n + 1), l_\beta(n + 1)$. *A crucial property of conservative systems is that the Poincaré section is an area-preserving map.* This follows directly from the fact that Hamiltonian motion has the symplectic property. (See the discussion in the appendix to Chapter 6.) This means that the sum of the projected areas in phase space, $\oint l_\alpha \, d\alpha + \oint l_\beta \, d\beta$, is preserved as a function of time. Recall that time development can be regarded as an infinite series of infinitesimal canonical transformations. If we consider a parallelogram formed by two infinitesimal vectors emanating from $\beta(n), l_\beta(n)$, all in the plane $\alpha = 0$, then $\oint l_\alpha \, d\alpha = 0$ around this infinitesimal area (because $\alpha = 0$ and hence $d\alpha = 0$). For the images of these two vectors on the next iteration of the map, the symplectic property means in this case that $\oint_n l_\beta \, d\beta = \oint_{n+1} l_\beta \, d\beta$. Therefore, any infinitesimal area near the nth point in the Poincaré section must be mapped into an equal area near to $(n+1)$th point. Thus we have for the determinant of the Jacobian of the 2-D map transformation:

$$\left| \frac{\partial [\beta(n+1), l_\beta(n+1)]}{\partial [\beta(n), l_\beta(n)]} \right| = 1. \tag{11.7}$$

Finding this Jacobian is a bit involved. It will be explained below when we discuss the stability of small excursions from fixed points.

In the linear case, for small angles, the energy $E = \omega_1 I_1 + \omega_2 I_2$ (see (6.107)). In this same linear approximation, the area within a closed curve on the Poincaré section in the variables (β, l_β) is either $2\pi I_2$ or $2\pi I_1$ depending on whether the curve encircles the mode 1 fixed point or the mode 2 fixed point. We can find I_1 and I_2 for a given set of initial conditions at constant E from the Poincaré sections. First find the area $/2\pi$ on the (β, l_β) Poincaré section that would correspond to either I_1 or I_2. Then the area $/2\pi$ on the (α, l_α) section would correspond to the other value of the action, I_2 or I_1 respectively. If these areas $/2\pi$ are plotted on the I_1, I_2 plane, they should give a straight line at constant energy E, with a slope equal to the winding number.

This much can be seen analytically in the linear approximation. The analysis also can be extended using numerical means to higher energies where the motion is no longer purely a combination of linear oscillations with two frequencies. If we plotted I_2 versus I_1 for many different initial conditions at constant energy, we would obtain a curve whose slope would give us the winding number. However, there is an easier way to obtain the fractional part of the winding number which will be explained later.

Poincaré Sections for the Double Pendulum

Using a computer simulation, we can show what happens as the energy of the double pendulum is increased. In Figure 11.5A, the starting angle of the top pendulum, and thus the potential energy (which equals the total energy at $t = 0$) has been increased to $30°$. Once again, two fixed points of period 1 (synchronous motion of both pendulums) are visible. Although the motion is decidedly nonlinear, this possible type of motion is not destroyed. If one observes the encircling tori, one sees places where the "curves" consist of a discrete number of dots. The effect of the nonlinearity is to "detune" the winding number away from the irrational linear value. The change in the winding number is different in different parts of the (β, l_β) plane now. In Figure 11.5A we also see two small islands, which are due to the formation of a period 2 nonlinear resonance. Although it is not visible in the figure, in the center of each of these two small islands is one of the fixed points of this resonance, which has $r = 1, s = 2$.

If the energy is increased still further, more complex behavior becomes visible. A Poincaré section for an energy of $60°$ is shown in Figure 11.5B. In addition to encircling

11.3 KAM TORI: THE IMPORTANCE OF WINDING NUMBER

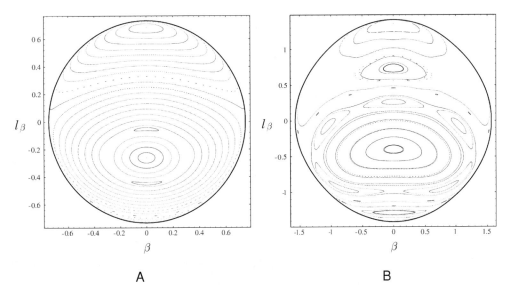

FIGURE 11.5
Poincaré section for A) $\alpha(0) = -\beta(0) = 30°$ and B) $\alpha(0) = -\beta(0) = 60°$.

tori, there are areas with chains of islands. In the next few sections we will investigate what causes these areas to develop.

11.3 KAM TORI: THE IMPORTANCE OF WINDING NUMBER

For most initial conditions and most Hamiltonians, we observe an abundance of smooth, invariant curves on the Poincaré section. One cannot necessarily find explicit analytic solutions, but the motion can be confidently predicted to remain for all time on the torus. This smooth nonchaotic behavior is regular motion. If the motion is always regular for *all* possible initial conditions, the system is an integrable system. Recall that for a system with N degrees of freedom to be integrable, it is a necessary and sufficient condition that N compatible constants of the motion exist. Usually the existence of these constants is guaranteed by symmetries of the Hamiltonian.

Since the time of Poincaré, there has been intense theoretical interest in what happens to these smooth curves on the Poincaré section if a small perturbation H_1 is added, which breaks the symmetry of an integrable Hamiltonian H_0. We could imagine examining the tori while increasing the strength of the perturbation, which takes the form

$$H = H_0 + \epsilon H_1. \tag{11.8}$$

(Poincaré called understanding the nature of solutions to (11.8) the "fundamental problem of dynamics.") The perturbation strength parameter ϵ is gradually increased from $\epsilon = 0$ to some finite value. What is observed is that most of the original tori are deformed but remain unbroken. A few tori are "broken," with only a finite number of points from each broken torus surviving as fixed points of nonlinear resonances. New tori are formed that

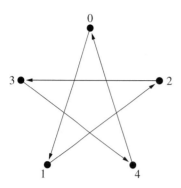

FIGURE 11.6
Fixed points for a 2/5 resonance.

encircle the new fixed points, forming "islands." The sequence of fixed points repeats every s iterations of the map, having gone r times around the set. These fixed points and the island tori around them are the nonlinear resonances of order s.

Let us define T as an operator that stands for one iteration of the Poincaré section map. A succinct description of this transformation can be written: $T(\beta(n), l_\beta(n)) \mapsto (\beta(n+1), l_\beta(n+1))$. The sth iteration is $T^s(\beta(n), l_\beta(n)) \mapsto (\beta(n+s), l_\beta(n+s))$.

The numbers on Figure 11.6 refer to successive iterations of a particular map: T, T^2, T^3, T^4, T^5, starting and ending on point 0. The resonance in this case has period 5. In the strictest sense of "fixed point," each point above is a fixed point of T^5, but we will say that each member of the set of five points is a fixed point of T, meaning that iterations of the map will return to each of these points in some definite order. On surrounding island tori, the map would jump from a point in the vicinity of one fixed point to another, in the same order as the fixed points themselves. Choose a center (say the average) and calculate the ratio of the number of times we "wind" around this center in the limit of a large number ($5N$) of map iterations. Any curve trapped on the island tori will average $2N$ complete rotations; hence the winding number is $\frac{2}{5}$ for *all* the island tori, not only for the fixed points.

Winding numbers can be calculated in this manner for any curve on the Poincaré section, wherever the motion is regular. We again need to look for how many map iterations are needed to "return" to a point on the Poincaré section. Quotes are used around "return" since Poincaré sections with irrational winding numbers will never exactly return to the same point. Even if the winding number is a rational fraction, we may never exactly return to the same point on a torus that encloses a fixed point. But in the limit of a large number N of map iterations, the number of rotations per iteration will approach a definite limit. In practice, to determine the winding number we can count how many map iterations are needed to minimize the distance from the starting point. If we count the number of rotations in the Poincaré section for that number of map iterations and then divide by the number of map iterations, we obtain the winding number, at least in the limit of an infinite number of iterations. A few thousand iterations will usually give a very good approximation to the winding number for regular motion.

11.3 KAM TORI: THE IMPORTANCE OF WINDING NUMBER

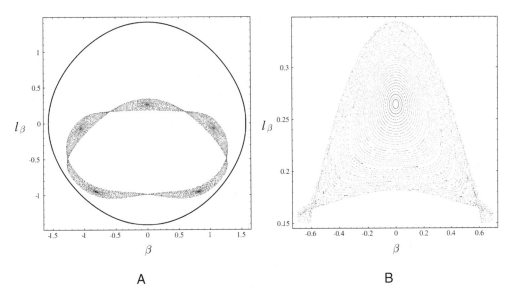

FIGURE 11.7
A) Poincaré section for $\alpha(0) = -\beta(0) = 60°$ with conditions restricted to be near the $\Omega = \frac{2}{5}$ nonlinear resonance. B) A closeup view of one island in the 2/5 resonance.

Figure 11.7A shows an $\Omega = \frac{2}{5}$ resonance in the double pendulum system at an energy corresponding to $60°$ ($E = -2$). We see what appears to be a boundary "curve" weaving its way around the set of period 5 fixed points. There is a close resemblance to the separatrix of a single pendulum. Like a single pendulum, the crossing points represent equilibrium points, also with period 5, but unstable. The manifold, like the pendulum separatrix, is really a limit of realizable phase trajectories. The phase trajectories themselves cannot ever cross, but manifolds that are limits of phase trajectories can and do cross.

The fixed points at the center of the small tori are *elliptic* points or *centers*. Motion near to them is stable, moving as regular motion on invariant tori. The periodic fixed points at the (almost) crossing of the separatrices are unstable, and are called *hyperbolic* (saddle) points. We magnify part of Figure 11.7A in Figure 11.7B to give some idea of the complexity of the motion for one of the islands.

A very important theorem, known as the KAM theorem, was developed from 1954 to 1967 by Kolmogorov, Arnol'd, and Moser. They proved that most of the H_0 tori remain unbroken under the perturbation in Equation (11.8) above. The proof involved studying the convergence of Lindstedt series, which are power series in powers of the strength ϵ of the perturbation. The KAM theorem states that the set of invariant tori that are not destroyed by the perturbation has a finite measure (i.e., occupies a finite area in the Poincaré section). For sufficiently small values of ϵ, most invariant tori are not destroyed but are only deformed by the perturbation. The motion on these tori remains quasiperiodic, with the number of independent frequencies equal to the number of degrees of freedom. Furthermore, the tori that do break have rational winding numbers. A torus with winding number of $\frac{2}{3}$ dissolves into three fixed points with accompanying sets of smaller tori or islands with finite area. One with a winding number $\frac{4687}{33102}$ may be broken, but the area of the islands surrounding the

33102 fixed points is so small it is invisible on a reasonable scale for plotting the Poincaré section.

What causes this breakup of the tori? As the strength of the perturbation is increased, the winding numbers change due to the "tuning" of the oscillations by the perturbation. This causes formerly irrational winding numbers to shift to rational ones and break into a finite number of periodic fixed points. The "most" irrational numbers are the last to have their tori broken in this way. The basic reason why rational winding numbers have broken tori is that the perturbation can have a cumulative effect. If the winding number is $\frac{r}{s}$, with r, s relatively prime integers (i.e., they contain no common factors), then the motion is repeated every s oscillations. Perturbative effects no longer average out, but can accumulate, with major effects on the dynamics. In nature there are many practical examples of these nonlinear resonances. As will be described later, an asteroid that goes around the Sun three times while Jupiter orbits once is said to be in a $1:3$ resonance with Jupiter (i.e., $r = 1, s = 3$). The gravitational effect of Jupiter can be cumulative, causing large effects on the orbit from a relatively weak perturbation.

11.4 IRRATIONAL WINDING NUMBERS

The last KAM tori to break are those with the "most" irrational winding numbers.

Some numbers are more irrational than others! For example, $\frac{\sqrt{5}-1}{2}$ is more irrational than either e or π. To see why this is true, we need to understand the concept of a *continued fraction expansion*, which is used to obtain a sequence of rational approximates to an irrational number.* If we call σ an irrational number, it can be shown that, for truncated decimal expansions,

$$\left| \sigma - \frac{r}{s} \right| < \frac{1}{s}, \tag{11.9}$$

which implies that the approximation improves as the integer s increases. A better representation is to use continued fractions, writing σ as

$$\sigma = a_0 + \cfrac{1}{a_1 + \cfrac{1}{a_2 + \cfrac{1}{a_3 + \cfrac{1}{a_4 + \cdots}}}}. \tag{11.10}$$

Here a_0 is a positive or negative integer, possibly 0, while the a_i are positive integers. This expansion is unique. It is derived by subtracting the integer part of σ (a_0), reciprocating the result, subtracting the integer part (which equals a_1), reciprocating, etc. For π, we have

---

* Our discussion here follows that of Berry. See the article "Regular and Irregular Motion" by M. V. Berry; reference in bibliography.

11.4 IRRATIONAL WINDING NUMBERS

the continued fraction expansion

$$\pi = 3 + \cfrac{1}{7 + \cfrac{1}{15 + \cfrac{1}{292 + \cfrac{1}{1 + \cdots}}}}. \qquad (11.11)$$

QUESTION 5: *Continued Fractions* Work out for yourself the continued fraction for π.

Successively closer rational approximations $\frac{r_n}{s_n}$ to σ are found by truncating the series at a_n. This yields a rational fraction, with $s_n > s_{n-1}$. The approximations alternate in being greater and less than σ. It can be shown that*

$$\left| \sigma - \frac{r_n}{s_n} \right| < \frac{1}{s_n s_{n-1}}. \qquad (11.12)$$

The slowest possible convergence would occur if all the a_n beyond a certain point were to equal 1. If $\sigma = \frac{\sqrt{5}-1}{2} = .618033988\ldots$, the continued fraction expansion is

$$\frac{\sqrt{5}-1}{2} = 0 + \cfrac{1}{1 + \cfrac{1}{1 + \cfrac{1}{1 + \cfrac{1}{1 + \cdots}}}}. \qquad (11.13)$$

In this case, σ is called the *golden mean*. It is supposed to be the most pleasing proportion for a rectangle, a fact related to the success of 3×5 index cards. Numbers of this type, with all 1s in the continued fraction past a certain point, are the "most irrational" numbers in the sense defined above.

KAM proved that the tori that are destroyed by the perturbation in (11.8) above are those satisfying

$$\left| \Omega - \frac{r}{s} \right| < \frac{K(\epsilon)}{s^{2.5}}, \qquad (11.14)$$

where Ω is the winding number and K is a constant that depends on the perturbation strength ϵ. Unless the perturbation is strong, $K \ll 1$. Equation (11.14) means that we must delete an interval of length $\frac{K}{s^{2.5}}$ about each winding number that is a rational fraction.

---

* According to Berry, for $\sigma = \pi$, the approximate $\frac{r_3}{s_3} = \frac{355}{113}$, which differs from π by $< 3 \times 10^{-7}$, was already known to Lao-Tze (604–531 B.C.).

TABLE 11.1 RATIONAL APPROXIMATIONS TO SELECTED IRRATIONAL NUMBERS

σ	d (75 terms)
$\sigma_1 = \frac{\sqrt{5}-1}{2}$	3.8×10^{-32}
$\sigma_2 = \sqrt{3}$	1.2×10^{-43}
$\sigma_3 = \pi$	6.2×10^{-76}

In the interval 0 to 1, the total length of all intervals deleted can be (over)estimated to be K (see the question below). This was proved by KAM for the exponent 2.5. However, since for 2.5 some of the intervals overlap, it was later proved that the exponent 2.5 could be replaced by $2 + \mu$, where μ is any positive value.

Since K, the measure of all possible winding numbers, is much less than one, the remaining unbroken tori will have finite measure. Motion on these tori will be confined there for all time. The excluded intervals of winding numbers become the regions of stable islands with rational winding numbers. In between these islands chaotic motion could exist. You can see this in Figure 11.7B. We will explore this aspect later in more detail. Since rational fractions are everywhere dense among the irrational numbers, the picture grows increasingly complicated as we magnify the scale.

As a numerical experiment, we investigated the generalization of condition (11.14) for three different irrational numbers. We found the continued fraction for 75 terms and reduced this to an ordinary fraction of the form $\frac{r}{s}$. We then formed the difference $d \equiv |\sigma - \frac{r}{s}|$. The result is shown in Table 11.1. We see that π is more accurately approximated by a rational fraction of a given length than either of the other two numbers. For the same number of terms in the continued fraction, $\frac{\sqrt{5}-1}{2}$ is less well approximated than $\sqrt{3}$, the difference from the best rational approximation being more than eleven orders of magnitude larger.

QUESTION 6: *Deleted Intervals* Prove that the total length of all the deleted intervals $\frac{K}{s^{2.5}}$ in the interval from 0 to 1 is estimated to be $\leq K$. Sum over r from 1 to $s - 1$, and then over s from 1 to ∞.

KAM revealed the crucial role of nonlinear resonances in mechanics. We go beyond KAM in applying these ideas to the double pendulum, where there is no "perturbation," but the constant K now depends on the energy E. The excluded intervals, corresponding to broken tori with winding numbers near to rational winding numbers, grow as the energy and hence the effect of the nonlinearity increases. The low-order resonances have a wider band of destroyed tori than high-order (large s) resonances. Motion near the edge of the resonance is very hard to study, since there is a thin layer of chaos nearby.

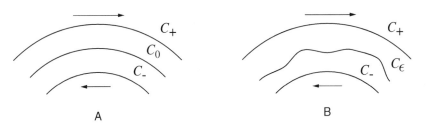

FIGURE 11.8
Close-up of one area of the Poincaré section A) before perturbation is turned on and B) after perturbation is turned on.

For Hamiltonian motion, KAM theory appears to be valid well beyond the original context. Of course, the smallest amount of random noise or dissipative interactions with the environment will destroy the delicate features of the more complicated tori, while leaving the main features, such as low-order resonances, intact.

The great significance of the KAM theorem is that it answered the question: Under what conditions is the motion in a conservative system stable over arbitrarily long times? Tests for integrability still remain an open question. We know nothing general that can reveal whether or not a given Hamiltonian is integrable. Only special methods in special cases can prove integrability and thus the total absence of chaos.

11.5 POINCARÉ–BIRKHOFF THEOREM

How does a smooth torus break up into a necklace of islands surrounding s elliptic points joined together by s hyperbolic points? A clear picture of how this can happen as a perturbation strength is increased was discussed by Poincaré and later by Birkhoff. Let us first imagine that we have $\epsilon = 0$, and a set of tori with increasing winding numbers. Part of the Poincaré section looks like Figure 11.8A. The curves represent different KAM tori. If we followed how the points on one of the curves were plotted, there would be a rotation to either the left or the right. In other words, the next time a point was plotted near a previous point, it would be consistently either to the left or to the right of that previous point, corresponding to a counterclockwise or clockwise rotation, respectively.

Now select a reference frame such that the outside curve C_+ rotates clockwise relative to the middle curve C_0, while the rotation of the inner curve C_- is counterclockwise relative to C_0. Let us assume that the winding number of the middle curve is $\Omega_0 = \frac{r}{s}$, which is a rational fraction. In order to obtain the correct relative rotation rates it must be true that $\Omega_- < \Omega_0 < \Omega_+$. If we look at the T^s, T^{2s}, ... iterations of the map, *all* of the points on C_0 are fixed points of T^s. Thus the C_0 curve is frozen in place under T^s, while C_+ rotates clockwise and C_- anticlockwise.

After the perturbation is turned on (Figure 11.8B), all the curves are distorted, but for clarity we will continue to draw $C_\pm$ as circles. If the perturbation is sufficiently small, the relative rotations of $C_\pm$ remain the same: Although the winding number changes, the same inequality is kept for the three curves. Proceeding along a radius from the center past C_- towards C_+, there must therefore be a unique radial point between the two curves $C_\pm$ at

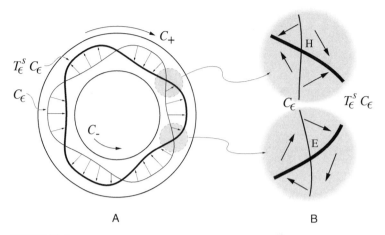

FIGURE 11.9
Graphical illustration of Poincaré–Birkhoff Theorem. E stands for an elliptic fixed point and H for a hyperbolic fixed point.

which no rotation takes place. The locus of all such points is drawn as C_ϵ. The *angular* coordinates of C_ϵ are unchanged under a mapping $T_\epsilon^s C_\epsilon$.

Next consider Figure 11.9A. The heavy curve is $T_\epsilon^s C_\epsilon$. We can draw little arrows to show the radial movement from C_ϵ to $T_\epsilon^s C_\epsilon$. We find that C_ϵ and $T_\epsilon^s C_\epsilon$ must intersect each other in an even number of points. This is because the two curves must have the same enclosed area, since a Poincaré section is an area-preserving map. The intersection points are on both C_ϵ and $T_\epsilon^s C_\epsilon$, so they must be fixed points of the mapping T_ϵ^s. If x is an intersection point, then $x, T_\epsilon x, T_\epsilon^2 x, \ldots, T_\epsilon^{s-1} x$ must all be fixed points of T_ϵ^s. Since each intersection leads to a fixed point, the number of fixed points must also be even, with half of them being elliptic (11.9B, bottom) and half of them hyperbolic (11.9B, top). Thus the minimum number of fixed points is $2s$, but it may also be a multiple of this.

> **QUESTION 7: *Fixed Points*** We just stated that we must have an even number of fixed points because C_ϵ and $T_\epsilon^s C_\epsilon$ must have the same enclosed area. Prove this for yourself by using some examples.

Hyperbolic points and elliptic points must alternate, as can be seen in Figure 11.9B, using arrows to again indicate flow direction under repeated iterations of the map.

Figure 11.10B shows the *stable* (incoming) manifolds called H_+ and *unstable* (outgoing) manifolds called H_- of the hyperbolic fixed point. The arrows show the direction of circulation around fixed points.

Nonlinear Resonances of the Double Pendulum

We show another part of the Poincaré section for $60°$, $E = -2$ in Figure 11.11A. Not shown are the two period 1 points that remain with us at this energy or the other resonances that are also present (see Figure 11.5B).

11.5 POINCARÉ–BIRKHOFF THEOREM   441

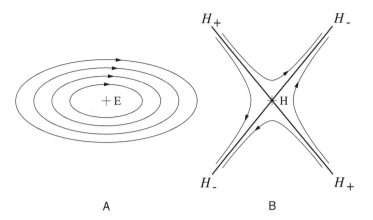

FIGURE 11.10
Magnified views of A) elliptic and B) hyperbolic fixed points.

Enlargement of the region of the upper one of the two elliptic fixed points of T^2 shows two new interesting features (see Figure 11.11B). One can notice that the ellipses surrounding the central point have, in some cases, broken again into secondary resonances. If one counts, say, four secondary resonances, there would, inside of each small "island," be a fixed point of $(T^2)^4 = T^8$. This would be an example of a case where s is not a prime number. The mechanism of torus destruction is exactly the same as already described above by the Poincaré–Birkhoff theorem.

The second feature visible is the chaotic region of "random" points near the two hyperbolic fixed points. In Figure 11.11B, the hyperbolic fixed point is located at $\beta =$

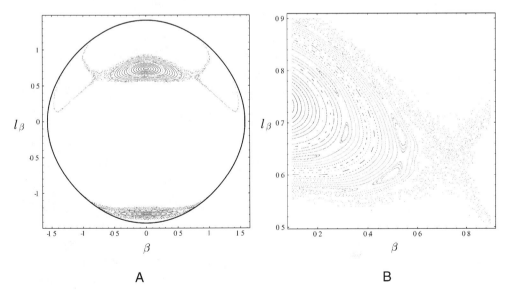

FIGURE 11.11
A) Period 2 resonance for $\alpha(0) = -\beta(0) = 60°$. B) Enlarged 1 : 2 resonance region.

.7380, $l_\beta = .6189$. For an integrable system, we would have smooth curves approaching a single separatrix in this region.

It is this feature that distinguishes nonintegrable systems such as the double pendulum from integrable systems like a single pendulum. What we see here is the first visible appearance of chaos for a nonintegrable system. Successive iterations of the map move from point to point in an apparently random fashion, within a limited region. We call this "local chaos" since it is confined to a definite and relatively small part of the phase space, as can be seen by looking at the entire Poincaré section for 60°, as shown in Figure 11.5B. What Gutzwiller* calls "an intimate mixture of elliptic and hyperbolic behavior" is also called "soft" chaos. Later, at still higher pendulum energies, we will see chaos spread throughout phase space. This is "global" or "hard" chaos. (This distinction will become important when we discuss possible chaotic motion within the solar system.) The chaotic region seen here is quite thin, but it can be expected to expand at higher energy, destroying more and more of the regular motion on the smooth KAM tori as the energy increases.

> **QUESTION 8:** *Local Chaos* Does it make sense that a limited region of a Poincaré section is chaotic but other regions are not? What does this imply for the system? How do we get to the chaotic regions? How do we get to the nonchaotic regions?

In a nonintegrable system, there is always an exponentially thin layer of chaos near the unstable/stable manifolds ($H_\pm$) connecting the hyperbolic fixed points. By exponentially thin, we mean the thickness disappears to negligible thickness, as the energy is lowered. It is the imperfection of this connection from one hyperbolic point to another that "causes" chaos. We will discuss this more below.

The hyperbolic fixed points of Figure 11.11A have period 1, rather than the period 2 we expected from the Poincaré–Birkhoff theorem. Near the boundary of the hyperbolic region, the motion is affected by conflicting resonances. It is this clash of colliding resonances (between period 1 and period 2) with incompatible winding numbers, that causes the local chaos to lead to more global chaos.

11.6 LINEARIZING NEAR A FIXED POINT: THE TANGENT MAP AND THE STABILITY MATRIX

We learned in Chapter 10 that small perturbations on the motion of a periodic dynamical system can be described by a linear equation such as the Mathieu or Hill equation. Due to the periodic nature of the functions in the Hill equation, stability could be determined by integrating the equation over a single period. The fixed points in our Poincaré sections

---

* See M. C. Gutzwiller, *Chaos in Classical and Quantum Mechanics*, pp. 85–86.

11.6 LINEARIZING NEAR A FIXED POINT: THE TANGENT MAP AND THE STABILITY MATRIX

TABLE 11.2 SUMMARY OF MAJOR 60° RESONANCES. THESE WINDING NUMBERS ARE STILL NOT FAR FROM THE SMALL OSCILLATION VALUE OF 0.4142...

Period s	r	Ω
1	1	–
1	1	–
2	1	0.5
5	2	0.4
9	4	0.444...

are periodic solutions of the pendulum equations of motion. Therefore we can employ this linearization technique here for the same purpose: If we deviate infinitesimally from a fixed point of period s, we wish to know whether successive iterations of T^s remain close to the fixed point.

Consider either the 4-D phase trajectory or, since it is easier to visualize, the 2-D map for the Poincaré section, with $(\alpha, l_\alpha, \beta, l_\beta) \equiv (0, l_\alpha^*, \beta^*, l_\beta^*)$ at the fixed point. Imagine an infinitesimal "four-vector" representing a short excursion from the initial fixed point: $\delta z(0) \equiv (0, \delta l_\alpha, \delta \beta, \delta l_\beta)$. (If $\delta \beta, \delta l_\beta \neq 0$ but $\delta \alpha = 0$, then in general $\delta l_\alpha \neq 0$ in order to keep the energy constant.) By Taylor's theorem and the assumed continuity of the map, the map T must produce an image of the displaced point which is infinitesimally close to the image of the fixed point, and which depends linearly on the components of the original infinitesimal displacement vector. Thus there must exist a matrix $\mathbf{M}$ obeying the equation

$$\delta \mathbf{z}(t) = \begin{pmatrix} \delta\alpha(t) \\ \delta l_\alpha(t) \\ \delta\beta(t) \\ \delta l_\beta(t) \end{pmatrix} = \mathbf{M}_4(t) \begin{pmatrix} \delta\alpha(0) \\ \delta l_\alpha(0) \\ \delta\beta(0) \\ \delta l_\beta(0) \end{pmatrix} = \mathbf{M}_4(t)\, \delta \mathbf{z}(0). \quad (11.15)$$

The matrix $\mathbf{M}$ is called the *tangent map* or sometimes the *monodromy matrix*. In the event we stay close to a periodic fixed point, we will give it the more prosaic name of *stability matrix*. A subscript on the $\mathbf{M}$ matrix will indicate the dimension of the map. $\mathbf{M}_4(0) = \mathbf{I}$, with $\mathbf{I}$ being the 4-D identity matrix.

The differential equation for $\mathbf{M}_4(t)$ can be obtained by differentiating the equations of motion ((11.3)–(11.6)). We collect them together by writing

$$\frac{d}{dt}(\delta \mathbf{z}) = \mathbf{J}\, \delta \mathbf{z}. \quad (11.16)$$

$\mathbf{J} \equiv \frac{\partial \dot{z}}{\partial z}$, which is the 4×4 matrix Jacobian $\mathbf{J}$ of the four functions $\dot{z}$ with respect to $z \equiv (\alpha, l_\alpha, \beta, l_\beta)$:

$$\mathbf{J} = \begin{pmatrix} \frac{\partial \dot{\alpha}}{\partial \alpha} & \frac{\partial \dot{\alpha}}{\partial l_\alpha} & \frac{\partial \dot{\alpha}}{\partial \beta} & \frac{\partial \dot{\alpha}}{\partial l_\beta} \\ \frac{\partial \dot{l_\alpha}}{\partial \alpha} & \frac{\partial \dot{l_\alpha}}{\partial l_\alpha} & \cdots & \cdots \\ \frac{\partial \dot{\beta}}{\partial \alpha} & \cdots & \cdots & \cdots \\ \frac{\partial \dot{l_\beta}}{\partial \alpha} & \cdots & \cdots & \cdots \end{pmatrix}. \quad (11.17)$$

Each of the sixteen elements of $\mathbf{J}$ consists of one of the possible second partial derivatives of the Hamiltonian with respect to the four phase space variables. In can be shown that the trace of $\mathbf{J}$ must vanish for the determinant of $\mathbf{M}_4$ to remain constant. This happens automatically if the motion is conservative. (Note that $\det |\mathbf{M}_4| = 1$ by Liouville's theorem.)

QUESTION 9: *Jacobian* By using Taylor's theorem in four variables, prove that $\delta \dot{\mathbf{z}} \equiv \dot{\mathbf{z}}(\mathbf{z} + \delta \mathbf{z}) - \dot{\mathbf{z}}(\mathbf{z}) = \mathbf{J} \, \delta \mathbf{z}$.

A basis for the space of all possible infinitesimal four-vectors can be established by setting up four linearly independent four-vectors at $t = 0$. This basis can be propagated to an arbitrary time t using Equation (11.16) for each basis vector. The equation describing the time development of the complete set of infinitesimal basis vectors can be written as a single matrix equation:

$$\frac{d\mathbf{M}_4}{dt} = \mathbf{J}(\mathbf{z})\mathbf{M}_4. \quad (11.18)$$

Equation (11.18) is a collection of four equations, one for each basis vector, of the type (11.16.) We can calculate $\mathbf{M}_4(t)$ by integrating this set of four coupled linear equations with variable coefficients (due to $\mathbf{J}$). To carry out this integration, the solution itself, $\mathbf{z}(t)$, must be known.

In the most general case, this is almost all that can be said. If we go from one fixed point through all other fixed points in the same sequence of length s and back to the starting point, in general $\mathbf{M}_4$ will be a definite nonsingular matrix with unit determinant. We do know, however, that it has at least one eigenvalue (i.e., *characteristic multiplier*) equal to one. The eigenvector with unit eigenvalue is the infinitesimal vector along $\dot{\mathbf{z}}$ (i.e., the tangent vector) because its time derivatives are only functions of the coordinates through the appropriate first derivatives of the Hamiltonian. Thus if the coordinates are periodic with period s, then we will also return to the same tangent vector after a period s. Therefore the matrix $\mathbf{M}_4$ must have an eigenvector with unit eigenvalue. We also know, although it is not proved here, that characteristic multipliers come in reciprocal pairs. They must also be either real or complex conjugates of each other. This means that there must be two multipliers equal to 1, not only one.

11.6 LINEARIZING NEAR A FIXED POINT: THE TANGENT MAP AND THE STABILITY MATRIX

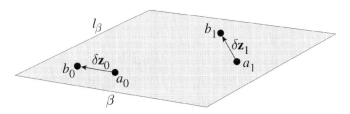

FIGURE 11.12
P_2 from M_4. $\delta z_1 = P_2 \, \delta z_0$.

Let us assume we have obtained M_4 by numerical integration of the linear first-order differential equations implied by (11.18). How do we obtain the 2×2 stability matrix for the Poincaré section, known as P_2? Take two adjacent points a_0, b_0 which are on the β, l_β Poincaré section and are separated by the infinitesimal vector δz_0 as shown in Figure 11.12. The coordinates of a_0 are (β, l_β) and the coordinates of b_0 are $(\beta + \delta\beta, l_\beta + \delta l_\beta)$. Then integrate the equations of motion until the next Poincaré section points a_1, b_1 are obtained. Determine the displacement δz_1. The relationship between δz_0 and δz_1 is given by P_2. However, note that the two points a_1 and b_1 do not necessarily occur at the same time, even if a_0 and b_0 did. Since the mapping preserves area, the determinant of P_2 must be 1. Therefore the two multipliers must be reciprocals of each other, since the determinant is the product of the characteristic multipliers.

QUESTION 10: P_2 from M_4 How do you actually calculate P_2 from M_4, since the two trajectories on the Poincaré section in Figure 11.12 do not hit the plane at the same time?

The eigenvalues (multipliers) of a 2×2 real matrix with unit determinant are completely determined by the trace (Tr) of the matrix. We solved this in Chapter 10 when we discussed the Floquet Theory (see (10.27)). There are three possibilities:

1. Elliptic fixed point (see Figure 11.13A): $-2 < Tr \, P_2 < 2$. The multipliers are complex and lie on the unit circle: $\lambda \equiv e^{i\mu}$, $\frac{1}{\lambda} \equiv e^{-i\mu}$ ($0 \leq \mu < 2\pi$). In this case a small deviation from the fixed point follows an elliptical and therefore bounded path around the fixed point. Define the "tune" Q,* which can be calculated directly from the trace: $\cos(2\pi Q) \equiv \cos\mu \equiv \frac{1}{2} Tr$. The tune range is $0 < Q < 1$. $\frac{N}{Q}$ iterations of the map will move N times around the ellipse. It is possible to calculate the exact shape and tilt of the ellipse from the matrix elements of P_2. The stable KAM tori very near the elliptic fixed point can easily be proved to be the invariant ellipses of the linear transformation P_2.

---

* This is not the winding number we discussed previously, since the fixed point is taken as the center, not the average, but it is a kind of "local" winding number. Q is also not to be confused with the quality factor for damped oscillators.

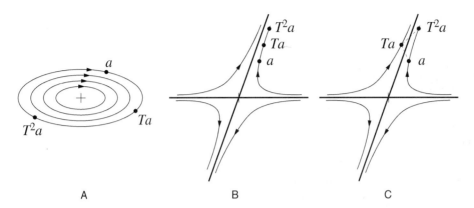

FIGURE 11.13
Phase trajectories near the three kinds of fixed points.

2. "Saddle" hyperbolic fixed point (see Figure 11.13B): $2 < Tr\,\mathbf{P}_2$. The multipliers $\lambda, \frac{1}{\lambda}$, are both positive real numbers. One is greater than 1 in magnitude; the other less. The fixed point is unstable, with hyperbolic phase trajectories. The limiting case corresponds to two intersecting straight lines, just as with the ordinary pendulum. One of these lines is the direction of the eigenvector corresponding to the multiplier that is greater than one. Points on or near to this line are mapped away from the fixed point at an exponential rate (powers of $\lambda > 1$). This line is called the unstable manifold H_-. Along the other eigenvector line, the multiplier being less than one means that points are mapped towards the fixed point. This line is called the stable manifold H_+.

3. "Flip" or "reflection" hyperbolic fixed point (see Figure 11.13C): $Tr\,\mathbf{P}_2 < -2$. This is also an unstable case, but the points after T^s appear alternately on opposite branches of the hyperbolas because the eigenvalues are negative. There is a stable and an unstable manifold as in case 2. This type of hyperbolic fixed point is associated with period-doubling transitions. Iterating the map twice converts this type into the saddle type for T^{2s}.

11.7 FOLLOWING UNSTABLE MANIFOLDS: HOMOCLINIC TANGLES

Why does chaos develop near hyperbolic points? We can start off with a point displaced from a hyperbolic fixed point by a very small distance along one of the eigenvectors that define the unstable manifold. Successive iterations of the map will produce points that move outwards from the fixed point. If the starting point is sufficiently close to the fixed point, the first few iterations move along a straight line in the direction of the eigenvector of the tangent map. (For flip hyperbolic points, we have to iterate the map twice to get back to the same curve.)

In the case of an integrable Hamiltonian, the stable and unstable manifolds join smoothly onto each other as shown in Figures 11.14A and B. In Figure 11.14A, the unstable and stable manifolds are really different parts of the same continuous curve. Successive iterations of a small starting eigenvector of the tangent map will march along the unstable

11.7 FOLLOWING UNSTABLE MANIFOLDS: HOMOCLINIC TANGLES 447

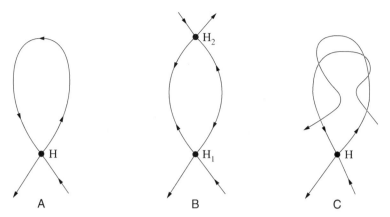

FIGURE 11.14
Integrable hyberbolic manifolds versus nonintegrable hyperbolic manifolds (after Berry, pp. 57–58).

manifold H_-, away from the fixed point, with lengths initially, at least, increasing as $1, \lambda, \lambda^2, \lambda^3, \ldots$ with each iteration of the map ($\lambda > 1$). The points then return along the stable manifold H_+ of the fixed point, approaching it at distances $\ldots, \lambda^{-3}, \lambda^{-4}, \lambda^{-5}, \ldots$; therefore, in the limit of an infinite number of map iterations we dive back into the fixed point again. In Figure 11.14B, the unstable manifold of one hyperbolic fixed point becomes the stable manifold of another, different hyperbolic fixed point. In both of these integrable cases, the connection between the manifolds is smooth, which means the motion is regular or nonchaotic.

Consider instead a scenario in which this smooth joining of manifolds does not occur. In making successive iterations of the initial point on the unstable manifold, the stable manifold of the next (or same) fixed point is crossed but not joined, as illustrated in Figure 11.14C. This failure to join smoothly the stable with the unstable manifolds is the crucial difference between regular (integrable, nonchaotic) and irregular (nonintegrable, chaotic) trajectories. In Figure 11.14C several crossings of the unstable and the stable manifolds are shown. Poincaré was able to show that, if one crossing exists, there must be an infinite number of such crossings. Furthermore, neither curve can cross itself, which makes the motion extremely complicated. Also, the area enclosed between successive crossings of the two curves must always be the same. Figure 11.14C thus represents a greatly oversimplified view of what actually must happen.

We will prove the impossibility of self-crossing first. In Figure 11.15 we draw a hypothetical self-crossing of H_-, the unstable manifold of a hyperbolic fixed point. Call the point at the crossing a. Also choose the points b, c, d as one follows the arrow around the loop. The mapped points Ta, Tb, Tc, Td must follow each other as shown, because the map is continuous. But although d is next to a, Td is farthest from Ta. This contradicts the necessary continuity that must map nearest neighbors into nearest neighbors. Self-crossing of the manifold must not be possible. The same is true for H_+.

However, H_- and H_+ *can* cross each other. If the manifolds belong to different fixed points, the crossing point is a *heteroclinic* point. If $H_\pm$ belong to the same fixed point, a crossing point is called a *homoclinic* point.

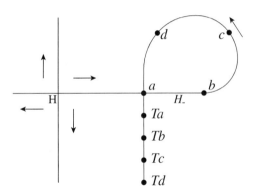

FIGURE 11.15
Self-crossing is impossible.

The next step is to prove that if there is one crossing of H_+ with H_-, there must be an infinite number of such points. If we have a point a on H_- and also on H_+, consider $T^n a$, the nth image of a under the map. Since the manifolds are, by definition, mapped into themselves, $T^n a$ must also belong to both manifolds. Hence one crossing means an infinite number of crossings. Since the two curves must cross each other an infinite number of times, but cannot cross themselves, there must be increasingly numerous and violent oscillations. In fact between any two crossing points there must be a dense set of such points, (i.e., there is an intersection arbitrarily close to any point in the interval between two points).

It is the existence of this "*homoclinic tangle*" (or heteroclinic tangle) that creates the chaotic regions appearing first in the vicinity of hyperbolic fixed points as the energy or perturbation is increased. The extreme complexity of the situation of irregular motion in the region of a hyperbolic fixed point is what led Poincaré to his discovery of chaotic motion in classical dynamics. Poincaré made the famous and often-quoted comment:*

> The intersections form a kind of lattice, web, or network with infinitely tight loops; neither of the two curves (H_+ and H_-) must ever intersect itself, but it must bend in such a complex fashion that it intersects all the loops of the network infinitely many times.
>
> One is struck by the complexity of this figure which I am not even attempting to draw. Nothing can give us a better idea of the complexity of the three-body problem and of all problems in dynamics...

How do we explain the appearance of this chaotic motion in phase space as the energy is increased in the example of the pendulum? Initially, the chaotic tangle of trajectories is limited to an exponentially small area of phase space, too "thin" to be observed in the calculations, which at very low energies are dominated entirely by KAM tori and two stable fixed points corresponding to the modes of the linear theory. As the energy increases, the chaotic tangle's regime expands.

---

* Jackson, A. E., *Perspectives of Non-linear Dynamics*, vol. II, p. 55.

11.8 LYAPUNOV EXPONENTS

Nothing really restricts us to the vicinity of fixed points. We can investigate how a small initial displacement anywhere in phase space or on the Poincaré section is squeezed, stretched, or rotated as the mapping proceeds. Liouville's theorem still tells us that phase space is preserved by Hamiltonian motion. But we do lose the advantage of periodicity: $\mathbf{M}_4$, $\mathbf{P}_2$ can be determined by integrating over a single period only if we stay near a fixed point.

If $\lambda_1, \lambda_2, \ldots$ are the Lyapunov exponents, we have to obtain them by taking a limit as $t \to \infty$. This amounts to assuming an average rate of rotating, squeezing, or stretching of the phase space per unit time. Cast into a more precise mathematical formula, Lyapunov exponents are defined to be proportional to the logarithms of the eigenvalues of the stability matrix (now more appropriately called the monodromy matrix, since we don't need to stay close to a fixed point). We define the set of Lyapunov exponents for continuous time in the four-dimensional phase space as

$$(\lambda_1, \lambda_2, \lambda_3, \lambda_4) \equiv \lim_{t \to \infty} \frac{1}{t} \ln[\text{Eigenvalues } \mathbf{M}_4(t)]. \tag{11.19}$$

For the Poincaré section (where N is the number of map iterations):

$$(\lambda_1, \lambda_2) \equiv \lim_{N \to \infty} \frac{1}{N} \ln[\text{Eigenvalues } \mathbf{P}_2(N)]. \tag{11.20}$$

These definitions can be applied to chaotic orbits as well as to regular motion. According to Oseledic's multiplicative ergodic theorem,* the limit defined by Equations (11.19, 11.20) will always exist and be independent of the initial values and the time step. For Hamiltonian motion, the sum of the Lyapunov exponents must be zero. For dissipative motion, this restriction is relaxed, but the sum must be negative. Chaotic motion implies that one of the Lyapunov exponents is real and positive. The monodromy matrix maps infinitesimally close points in phase space, so having real, positive Lyapunov exponents is equivalent to saying that small deviations in initial conditions diverge exponentially from each other, which was our original definition of chaotic motion.

QUESTION 11: *Lyapunov Exponents* Why must the sum of the Lyapunov exponents be zero for Hamiltonian motion? Hint: Use the fact that $\det \mathbf{M} = 1$, which follows from Liouville's theorem.

A practical difficulty in calculating Lyapunov exponents for some conservative systems is illustrated in Figure 11.16. In Figure 11.16A, we compare two trajectories with zero initial kinetic energy and $\alpha(0) = 90°$, which differ initially by $\delta\beta = 10^{-8}$ at $t = 0$. The

---

* A reference to Oseledic's original paper is given in *Chaos in Dynamical Systems* by Edward Ott, p. 132.

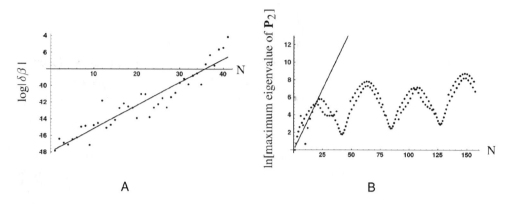

FIGURE 11.16
Comparison of two methods for finding Lyapunov exponents. N = number of Poincaré section points, $E = -1.0(90°)$. A) Numerical integration of the double pendulum equations. The largest Lyapunov exponent is measured to be .277 (on the Poincaré section). B) Numerical calculation of the eigenvalues of $\mathbf{P}_2$ at each Poincaré section point. Note that the line here is the same as that in A.

motion for this initial condition is fully chaotic (see the next section). A continuous stretching in one or more directions in phase space eventually singles out the largest Lyapunov exponent as a dominant term. The straight line represents the best linear fit to the logarithm of the ratio $|\frac{\delta\beta(t)}{\delta\beta(0)}|$. The motion is chaotic, with the largest Lyapunov exponent estimated to be 0.277.

There are some practical pitfalls. First of all, if the two points are chosen too close together, the finite precision of the computer will lead to wrong answers. Choosing the points too far apart also leads to wrong answers if the system is chaotic, because any two points cannot diverge arbitrarily since we have a finite volume of phase space. Not only "stretching," but also "folding" takes place on chaotic orbits for them to remain in the finite volume of available phase space.

In Figure 11.16B, the logarithm of the largest eigenvalue of $\mathbf{P}_2$ for each successive point on the Poincaré section is calculated. Actually carrying out this procedure numerically requires special techniques which we will describe later. Although there is a steady upward trend in the eigenvalues of $\mathbf{P}_2$, the time dependence of the logarithm of the eigenvalues is more complicated than a simple linear relationship. The same linear fit as in Figure 11.16A is shown. By comparison, we see that the method used to find the Lyapunov exponent in Figure 11.16A is not accurate. Since the Lypanunov exponent is only defined in the limit $N \to \infty$, there is no contradiction between Figure 11.16B and the existence of a definite value for the exponent. Separate fits to the local maxima and to the local minina of the points in the figure give approximately .018 for the Lyapunov exponent – still a positive number, but more than one order of magnitude smaller than obtained with the naive linear fit in Figure 11.16B. Longer integration times are required to obtain an accurate value. After a sufficiently long integration time, eventually, a good straight line fit to the data in Figure 11.16B must become possible. The slope of this line will give the correct Lyapunov exponent. This example illustrates some possible difficulties in calculating these exponents for conservative systems. A damped system converges more rapidly when one tries to apply (11.20), as we will see.

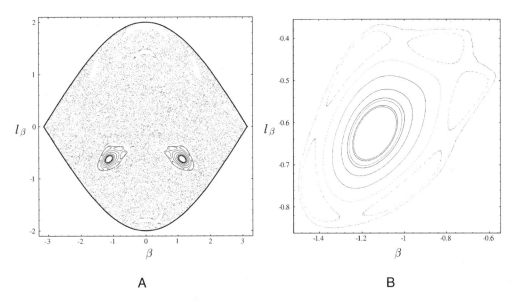

FIGURE 11.17
A) Poincaré section for $\alpha(0) = -\beta(0) = 90°$ ($E = -1$). B) A closer view of the region around the period 1 fixed point on the lower left of A.

11.9 GLOBAL CHAOS FOR THE DOUBLE PENDULUM

Now proceed to even higher energy, corresponding to a starting angle of 90°. The Poincaré section has become now mostly chaotic, but several strong resonances persist, as seen in Figure 11.17A. A small but finite area is still occupied by this regular (nonchaotic) motion.

At lower energies, there are "KAM barriers" – KAM tori that divide the whole phase space and isolate one region from another. Since no phase trajectory can cross another, the smooth tori prevent chaotic mixing over all of phase space and keep the chaotic regions "local."* As the energy is raised these last barriers are broken, there is "global" chaos, and almost no part of phase space is then immune to being visited by a chaotic orbit. Still, as can be seen in Figure 11.17A, there remain small regions of the phase space that are not chaotic.

Summary of Hamiltonian Systems: Integrable versus Nonintegrable

To return to a question we asked earlier, given a particular Hamiltonian, how can we tell whether it is integrable or not?[†] The difference between integrable (and therefore not subject to chaos) and nonintegrable can be due to a very small change in the Hamiltonian. There is a conjecture that the system is integrable if the differential equation possesses certain types of singularities for complex time. This idea was first introduced in 1888 by S. V. Kovalevskaya, who used it to discover a new integrable case of the Euler equations

---

* This is not true for more than two degrees of freedom, where diffusion of chaos occurs throughout phase space, albeit very slowly. This is called Arnold diffusion. It is of enormous practical importance for accelerator design.
† A discussion of proofs of integrability appears in the book by Michael Tabor listed in the bibliography.

for the motion of a rigid body about a fixed point.* But the question about integrability remains open in many important cases, and we must resort to computer simulations. Simulations are not the same as mathematical proofs, nor do they help us find the constants of the motion explicitly in the event the system is an integrable one. The question is connected to several practical physics problems. In fact, nonintegrable systems appear to be far more common than integrable ones. Is the solar system stable? (More about this below.) Will the large proton accelerator known as the LHC, which is being constructed at CERN in Geneva, Switzerland, actually work? It is currently not possible to answer these questions in the affirmative using the capability of even the most sophisticated computer simulations.

11.10 EFFECT OF DISSIPATION

In Chapter 3, we studied damped simple harmonic oscillators driven by an external energy source. When the external driving force was turned on, initially there could be present a combination of "transient" solutions to the homogeneous free oscillator equation together with a "steady-state" solution. The transient signal disappeared after a few damping periods. If the driving signal were sinusoidal, the steady-state solution would oscillate sinusoidally at the driving frequency. The equilibrium amplitude at long times was determined by the balance between energy gains from the external driving signal and the losses from dissipation.

When nonlinear forces are present, the general situation is qualitatively very similar to the linear case. The resonant or chaotic motion we have been describing earlier for conservative systems become transient effects, which disappear after sufficiently long times. The solution is damped towards a particular subset of the phase space called an "attractor." For a driven simple harmonic (linear) oscillator, the attractor is an elliptical path in phase space, which repeats itself with the frequency of the external driving force. A stroboscopic picture of the motion will give a single fixed point in phase space if the strobe frequency matches the frequency of the driving force.

The phase volume occupied by a set of these oscillators with different initial conditions is not preserved, since Liouville's theorem is not valid in a dissipative system. The time dependence of a volume element dV in phase space† is given by

$$dV(t) = dV(0) \exp\left[\int Tr\,\mathbf{J}\,dt\right]. \qquad (11.21)$$

$\mathbf{J}$ is the Jacobian of the time derivatives as defined in (11.17) (for $2N$ dimensions we obtain a $2N \times 2N$ matrix). For conservative motion the trace of $\mathbf{J}$ is automatically zero, so the phase volume remains constant. For dissipative systems energy is constantly lost

---

* Jackson, A. E., *Perspectives of Nonlinear Dynamics*, p. 296ff.
† Note that the dimension of the "volume" in phase space is equivalent to the dimension of the phase space (i.e., twice the number of the degrees of freedom).

and thus the trace of **J** will be negative. For example, friction losses are due to exciting a large number of molecules in close proximity to the mechanical system. We model these losses by adding a phenomenological* term to the differential equation, a term that destroys the symmetry under time reversal. ($Tr\,\mathbf{J} < 0$ is an equation that reverses sign if $t \rightarrow -t$.)

Our interest in dissipative dynamical systems will focus on the nature of the steady-state solution, that is, the attractor. Initial conditions in whole areas of phase space will collapse onto a single attractor. These areas are called *basins of attraction*. For a periodically driven system with one degree of freedom, the attractor will be a set of less than two dimensions in phase space. The Poincaré section of the attractor will consist of a set of points in the two-dimensional phase plane of the section. The attractor can in some chaotic cases be a "fractal," a set with fractional dimension. If the number of degrees of freedom is 1, the dimension of the fractal attractor will be higher than that of a curve (1-D) but less than the whole phase space (2-D). An attractor that lies on such a fractal "curve" is called a *strange attractor*. Chaos in such circumstances involves exponential separation of neighboring trajectories as before, but these trajectories must lie on the (bounded) attractor, the long time solution for the motion. These conflicting requirements produce a very complicated set of points in the Poincaré section. Once again, computer simulation is to be our tool for disentangling and classifying what happens.

11.11 DAMPED DRIVEN PENDULUM

A damped pendulum under the influence of a periodic external drive force has all the necessary features to illustrate dissipative chaos. The damping force will be linear – proportional to $\dot{\theta}$ – but the pendulum motion contains nonlinear effects from a $\sin\theta$ restoring force, which is *not* approximated by θ, as done previously in Chapter 3 (see (3.66)). The equation of motion for such a pendulum can be derived by standard Lagrangian methods, by incorporating an additional nonconservative force of friction. In terms of θ and $p \equiv \dot{\theta}$, the equation of the motion is

$$\dot{\theta} = p,$$
$$\dot{p} = -\omega_0^2 \sin\theta - \frac{p}{Q} + F(t), \qquad (11.22)$$
$$F(t) = \text{external driving force} = f\sin(\omega t).$$

We will choose the time scale so that the natural resonant frequency for small oscillations $\omega_0 \equiv 1$. Notice that the frequency of the external driving force, ω, can be considered to be a variable parameter. As in Chapter 3, Q is the quality factor. We will fix $Q \geq 2$ in our numerical studies, so that transient effects are rapidly damped out. The nonlinearity of

---

* The definition of "phenomenological" in a physics context is a purely formal term or equation which reproduces the observed phenomenon (here friction) and which is used in an ad hoc fashion without being derived from more fundamental theory.

the sin θ term "folds" the phase trajectories back onto the finite region of available phase space and causes chaos to develop.

QUESTION 12: *Phase Space Flow Equations* Find the equation of motion for a pendulum driven by an external force $F(t) = f \sin(\omega t)$ and with a damping term $\frac{\dot\theta}{Q}$ added in the equation of motion. Make no linear approximations in this case. Then prove that (11.22) is equivalent to the equation of motion you found.

The Jacobian **J** for this case is

$$\mathbf{J} = \begin{pmatrix} 0 & 1 \\ -\cos\theta & -\frac{1}{Q} \end{pmatrix}. \tag{11.23}$$

The trace of **J** is seen to be negative, and Equation (11.21) implies that any finite area in the θ, p plane shrinks exponentially as $e^{-\frac{t}{Q}} \to 0$, as $t \to \infty$.

We will fix the driving frequency ω and vary the drive amplitude f. Because the system has only one degree of freedom, we can plot the attractors (after several damping times) and look at them directly. When the motion becomes too complicated, we can resort again to the method of Poincaré sections which we used so extensively in the section on Hamiltonian chaos. A Poincaré section in this case is made by recording the position in phase space of the pendulum at intervals spaced by one drive period. You can think of the Poincaré section as what we would see if we flashed a strobe light once for each oscillation of the driving force, in phase with this force. At sufficiently small driving amplitudes, the pendulum, being a linear system, must oscillate at the frequency of the driving signal, so we obtain a single point on the Poincaré section.

We begin by fixing $\omega = \frac{2}{3}$, $Q = 2$, and starting with $f = 1$. Figure 11.18A shows the entire phase trajectory on a p, θ plot. This is a graph of the attractor, after the initial conditions have been damped out. No matter where the initial starting point lies in the phase space, the motion ends up on the attractor. This driving amplitude is large enough to cause significant deviations from the circular phase trajectory familiar from the linear

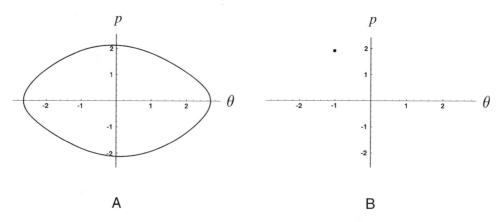

A   **B**

FIGURE 11.18
Damped driven pendulum with $f = 1$, $\omega = \frac{2}{3}$, and $Q = 2$. A) Phase trajectory. B) Poincaré section.

11.11 DAMPED DRIVEN PENDULUM

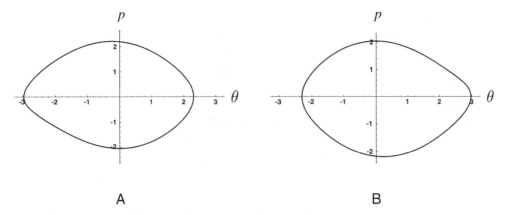

FIGURE 11.19
Damped driven pendulum with $f = 1.03$, $\omega = \frac{2}{3}$, and $Q = 2$. A) Phase trajectory for certain initial conditions. B) Phase trajectory for other initial conditions.

approximation. Figure 11.18B shows the Poincaré section, which is a snapshot of the phase trajectory taken at times $t = \frac{2\pi n}{\omega}$, $n = 1, 2, \ldots$. It is still true that the motion responds with the period of the driving force.

Breaking the Symmetry

The equations of motion (11.22) are invariant under the transformation $\theta \to -\theta, p \to -p, t \to t + \frac{\pi}{\omega}$. At first sight this seems to imply that the solution itself must be unchanged under this transformation, since Figure 11.18A has the reflection symmetry of the transformation.

QUESTION 13: *Symmetry Breaking* Prove that (11.22) is invariant under the transformation $\theta \to -\theta, p \to -p, t \to t + \frac{\pi}{\omega}$.

However, since the equations are invariant under the transformation, we can use this symmetry transformation to produce a second physically valid solution from the first one. *The solution and the transformed solution need not be the same, but if one exists, the symmetry of the equations of motion implies that the other must also exist.* To understand what this means, make a slight increase in the drive amplitude f to $f = 1.03$. Now, depending on the initial conditions, two attractors can exist, as shown in Figures 11.19A, B. (Remember that the attractor is the curve to which the motion goes after all memory of initial conditions is damped out.) The two curves differ by just the reflection (and shift of the time axis) described above. The threshold for breaking the left–right symmetry is $f \approx 1.0045(1)$. (The "1" in parenthesis means an uncertainty of 1 unit in the last place.)

These mirror images of each other are both equally valid attractors for the symmetric equations of motion. Despite the complete damping of starting transients, some memory of the initial conditions must cause the motion to select one curve or the other. Once the final attractor is approached, the motion stays on that curve forever. This is a beautiful example of an important phenomenon in physics: *broken symmetry*.

◯ Example: Symmetry Breaking

The laws of physics are completely symmetric between matter and antimatter, with one minor exception: Time reversal violation in the neutral K meson decays. How can it be then that there are no galaxies or even stars made of antimatter? Current experiments continue to search for the presence of antimatter in the universe in cosmic rays, but no primordial antimatter has yet been detected. This great asymmetry between matter and antimatter in our world is thought by some theorists to be due to broken symmetry at the moment of the Big Bang, when the existing particles condensed out of the gluon–quark–lepton "soup." Metaphorically speaking, it is as if we found ourselves traveling on Figure 11.19B because of an accident of the initial conditions from which all other information has been lost due to damping.

There are many other examples in nature of broken symmetry. There are two kinds of quartz crystals, left- and right-handed (mirror images). How does nature, starting from the basic silicon dioxide molecule, decide to build a right- or left-handed crystal structure? An even more intriguing mystery is why amino acids and sugars associated with living systems have only one "handedness," while it is not difficult to produce the opposite handedness in the laboratory. Dextrose is a sugar easily assimilated and metabolized by the body, but the mirror twin levulose cannot be digested. No reason is known for this striking biological asymmetry, but this preference of living systems for one over the other has caused speculation that all life comes from a single common ancestor.

Period Doubling Sequences and the Onset of Chaos

Let us continue to increase f, this time to $f = 1.07$ as shown in Figures 11.20A, B. The pendulum no longer repeats its motion every driving force period, but instead repeats every *other* driving period. This phenomenon can only occur in a system with nonlinear forces. It is called *period doubling*. The Poincaré section now has two distinct points, which alternate as the motion is sampled every period of the driving force. The threshold force amplitude for period doubling to period 2 is $f = 1.0665(1)$.

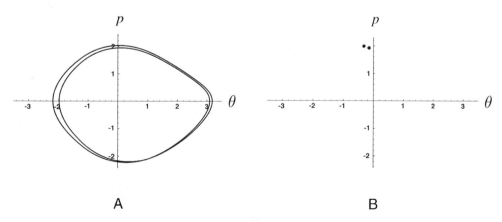

FIGURE 11.20
Damped driven pendulum with $f = 1.07$, $\omega = \frac{2}{3}$, and $Q = 2$. A) Phase trajectory. B) Poincaré section.

11.11 DAMPED DRIVEN PENDULUM

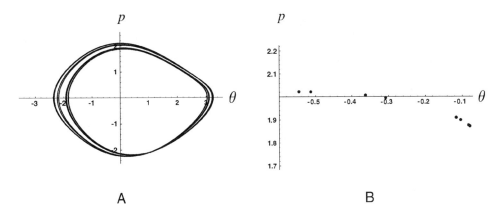

FIGURE 11.21
Damped driven pendulum with $f = 1.0825$, $\omega = \frac{2}{3}$, and $Q = 2$. A) Phase trajectory. B) Magnified Poincaré section.

We will continue by increasing f still further. We pass a period 4 cycle at $f = 1.0793(1)$ and find a period 8 cycle at $f = 1.0821(1)$. The phase trajectory for $f = 1.0825$ is shown in Figure 11.21A. Although it is difficult to distinguish all the curves in the phase trajectory, the trajectory repeats itself after eight drive periods. This can be seen clearly in the magnified view of a portion of the Poincaré section, shown in Figure 11.21B.

Table 11.3 gives the threshold f_n for each step in the period doubling sequence found by computer simulation of the damped driven pendulum. We notice that the f_n become closer as n increases. Chaos develops above $f \geq f_\infty$. (The two mirror symmetric solutions have identical thresholds.)

This sequence of period-doubling force strengths has a universal character. It is an example of a "period-doubling cascade," one of the most common routes to chaos.* This occurs in many physical systems such as dripping faucets, balls bouncing on vertically oscillating surfaces, nonlinear electronic circuits, and sometimes in beating hearts. The universal nature of this "route to chaos" was discovered in the late 1970s by M. J. Feigenbaum, who predicted that the sequence would have the form $f_n = f_\infty - \frac{C}{\delta^n}$, where $\delta = 4.6692016\ldots$ is a universal constant, which is applicable to a very general and broad range of chaotic systems. The values of C and f_∞, the threshold for chaotic motion, depend on the system being studied.

To estimate the threshold for chaotic motion in our damped, driven pendulum system, we can use the information in Table 11.3. By plotting f_n versus δ^{-n}, using the universal value of δ discovered by Feigenbaum, we should obtain a straight line. The result is shown in Figure 11.22, and the extrapolated value of f_∞, the threshold for chaos, is in Table 11.3. A numerical test with $f = 1.0829$ reveals that the motion has become totally chaotic. We will discuss period doubling again when we investigate the logistic map in the appendix.

---

* The other routes to chaos are not discussed here. See, for example, the discussion of intermittency in the book *Order within Chaos*, by Bergé et al.

TABLE 11.3 THRESHOLD DRIVING AMPLITUDES FOR PERIOD DOUBLING

n	Period 2^n	Approximate threshold f f_n
1	2	1.0665
2	4	1.0793
3	8	1.0821
4	16	1.0827
∞	∞	1.08289 (fit)

Lyapunov Exponents

The existence of chaos is determined by the Lyapunov exponents. If one of the exponents is positive, the motion is chaotic. Although the phase space volume decreases like $e^{-\frac{t}{Q}}$, one direction is stretched (if $\lambda_1 > 0$) like $e^{\lambda_1 t}$, while the orthogonal direction is contracted (if $\lambda_2 < 0$) like $e^{\lambda_2 t}$. One can think of an ellipse, of steadily shrinking area, with one axis growing or shrinking like the first eigenvalue, depending on whether the motion is chaotic or not, while the second, orthogonal axis shrinks proportional to the second eigenvalue.

In principle, we can use the 2-D equivalent of (11.19) to find the Lyapunov exponents. However, at long integration times, $e^{-\frac{t}{Q}}$, the determinant of $\mathbf{M}_2$, becomes an extremely small number. Any attempt to use the straightforward methods, or even to calculate the $\mathbf{M}_2$ matrix, is doomed to fail, even with the use of double precision calculations.

This practical problem was solved in a very clever way by Wolf et al.* Starting with the identity matrix as an initial condition, after integrating $\mathbf{M}_2$ for each time step using (11.18), take the two columns of the resulting matrix as two vectors: $\{\mathbf{M}_2(1, 1), \mathbf{M}_2(2, 1)\}$, $\{\mathbf{M}_2(1, 2), \mathbf{M}_2(2, 2)\}$. Find an orthonormal basis for these vectors where N_1 and N_2 are the normalizations of the first and second vector, respectively. Use a new $\mathbf{M}_2$ matrix, whose two columns are these normalized and orthogonal vectors as an initial value for $\mathbf{M}_2$ in the next time step. After integrating over a large number of periods to average over the motion, λ_1 and λ_2 can be found from

$$\lambda_1 = \frac{\sum_i \log N_1(i)}{T}, \quad \lambda_2 = \frac{\sum_i \log N_2(i)}{T}, \quad (11.24)$$

where the index i refers to the ith time step, and T is the elapsed time since the start of the integration.

As the strength of the driving force is increased, two distinct chaotic regions can be seen in Figure 11.23 where $\lambda_1 > 0$. Also, $\lambda_1 = 0$ at the points where period doubling or symmetry breaking bifurcations take place. Lyapunov exponents provide us with an accurate

---

* A. Wolf, J. B. Swift, H. L. Swinney, and J. A. Vastano, "*Determining Lyapunov Exponents from a Time Series,*" Physica **16D** (1985), pp. 285–317.

11.11 DAMPED DRIVEN PENDULUM

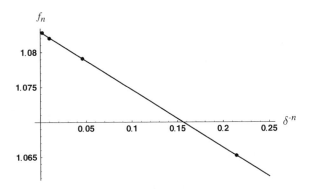

FIGURE 11.22
Test of universal period doubling.

means for determining the important transitions in this problem. For example, it has been estimated from the Lyapunov exponent calculation that the threshold for chaos is $f_\infty = 1.08240(1)$. We could use this computational capability to make a more precise test of the Feigenbaum constant for the period-doubling cascade of the damped driven pendulum.

Bifurcation Diagrams

Another way to visualize the route through period doubling to chaos is to make a "bifurcation diagram." Plot the driving force strength on the horizontal axis and the values of $p = \dot{\theta}$ from the Poincaré sections on the vertical axis. Until the period doubles, we obtain a smooth curve of the fixed points on the $\dot{\theta}$ versus f diagram. After the point for period doubling, this curve undergoes a "pitchfork bifurcation" into two curves for period 2, because there are now two points on the Poincaré section. We obtain four curves for period 4, etc. In the chaotic region, there is no discernible structure, because the attractor fails to settle down to a finite number of fixed points on the Poincaré section.

A bifurcation diagram is useful because it gives a panoramic view of the qualitative features of the motion. Chaotic and nonchaotic regions of the motion are clearly seen in Figure 11.24. One can correlate the appearance of a positive Lyapunov exponent with the

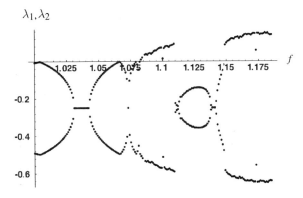

FIGURE 11.23
Lyapunov exponents λ_1, λ_2 versus f for the region $1.00 \leq f \leq 1.175$.

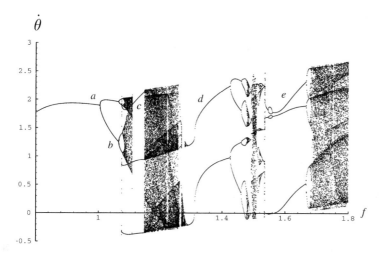

FIGURE 11.24
Birfurcation diagram for damped driven pendulum with $\omega = \frac{2}{3}$, $Q = 2$, and $.8 \leq f \leq 1.8$.

chaotic parts of the bifurcation diagram. It is evident that chaos can disappear again for higher driving force strengths.

We will first explore the nonchaotic regions, labeled as a, b, c, d, and e in Figure 11.24. The point a is the symmetry breaking point, at $f \approx 1.0045$. The system undergoes a period-doubling cascade, starting at the point b, and becoming chaotic soon afterwards. A closer look at this part of the bifurcation diagram appears in Figure 11.25.

Region c has period 3. This is a phenomenon unrelated to period doubling. We take a closer look at the phase trajectory of the attractor in Figure 11.26A at a typical value in region c: $f = 1.117$. In Figure 11.26B we plot $\theta(t)$ versus time as well as the driving force

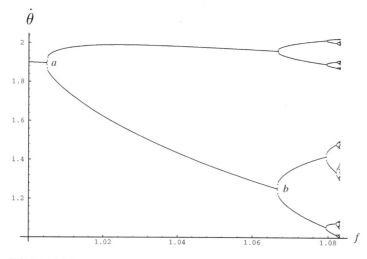

FIGURE 11.25
Close-up of birfurcation diagram of Figure 11.24 in the region $1 \leq f < 1.083$.

11.11 DAMPED DRIVEN PENDULUM

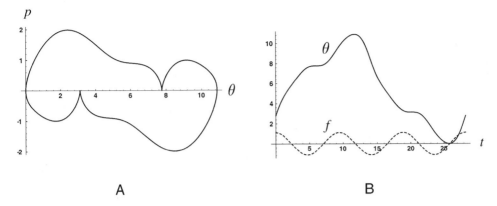

FIGURE 11.26
Region c of damped driven pendulum with $f = 1.117$, $\omega = \frac{2}{3}$, and $Q = 2$. A) Phase trajectory. B) θ versus t (solid line) and f versus t (dotted line).

versus time. It can be seen that the solution is perfectly periodic, with a period three times the driving period. The pendulum actually rotates more than one complete revolution and then rotates back again during each cycle such that there is no net rotation. The attractor is the same for any starting point (i.e., there is no symmetry breaking here). Region e exhibits similar behavior, as expected since it too has three curves.

For region d, two curves are visible. Unlike the regions c and e, the curves are each period 1. In the top curve of region d in Figure 11.24, the pendulum rotates once per driving cycle in the counterclockwise direction, while in the lower curve, the rotation is clockwise, again with period 1. This is known as "phase-locking."* The initial conditions divide into basins of attraction, one of which is attracted to the upper curve, and the other is attracted to the lower curve.

The *winding number* Ω plays the same role as it did for Hamiltonian systems. It is defined by

$$\Omega \equiv \lim_{t \to \infty} \frac{\theta(t) - \theta(0)}{\omega t}. \tag{11.25}$$

For the upper curve, $\Omega = -1$, and for the lower curve in the bifurcation diagram, $\Omega = 1$. In Figures 11.27A, B, we show the $\Omega = -1$ case for two periods of the driving frequency. For the case in the figures, $f = 1.4$. The pendulum rotates by -2π for each driving period.

Phase-locking onto a resonance can also be found for narrow ranges of f in the vicinity of 1.22 and from 1.495 to 1.497. In these cases $\Omega = \pm\frac{2}{3}$, which means that the pendulum rotates twice for every three driving periods. Solutions exist for either sense of rotation.

Strange Attractors

At least four chaotic regions can be seen in Figure 11.24. For such chaotic motion, the period can be considered to be infinite. This does not mean that the motion will

---

* Also called "mode-locking."

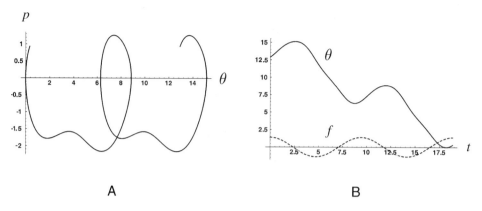

FIGURE 11.27
Region d of damped driven pendulum with $f = 1.4$, $\omega = \frac{2}{3}$, and $Q = 2$. A) Phase trajectory. B) θ versus t (solid line) and f versus t (dotted line).

"fill up" all of phase space however, even if we run the system for an infinitely long time. The Poincaré section for a large number of points ($f = 1.5$, $Q = 4$) is shown in Figure 11.28. This strange shape does not change appreciably if we increase the integration time. However, the detailed shape does depend on the choice of f and Q.

For the attractor in Figure 11.28, the Lyapunov exponents are $\lambda_1 = 0.164$, $\lambda_2 = -0.414$. This means that in some unspecified (and not even fixed) direction, points spread apart at the rate $e^{.164t}$ from their original separation, while in a direction orthogonal to the first one, they approach each other by a factor $e^{-.414t}$. Any circle gradually gets squashed into a longer and longer ellipse if the motion is chaotic, and the area of the ellipse diminishes by $e^{-\frac{t}{Q}} = e^{-.25t}$. Since the total energy is finite, the attractor cannot grow without bound but must be continuously folded back onto itself. This object is called a *strange*

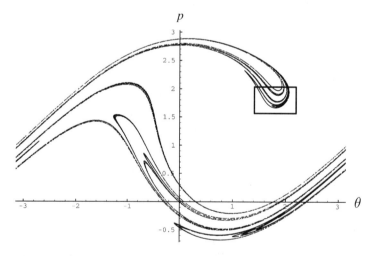

FIGURE 11.28
Poincaré section for damped driven pendulum with $f = 1.5$, $\omega = \frac{2}{3}$, and $Q = 4$.

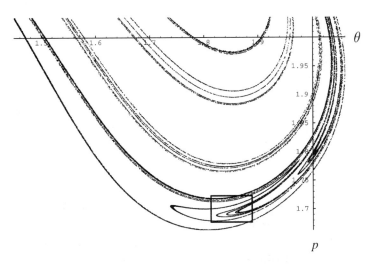

FIGURE 11.29
Close-up of Figure 11.28.

attractor. To discuss it quantitatively, we need to develop some more mathematical ideas about fractals.

The box in the Figure 11.28 outlines an area we magnify in Figure 11.29 just to give a hint of the endless self-similarity of this attractor as a result of the folding. The box in this figure could be magnified further, and so on indefinitely. One would always encounter an infinite number of points, which do not fill the plane.

11.12 FRACTALS

Fractals are mathematical point sets with fractional dimension.

The dimensionality D of a space is usually defined in elementary mathematics as the number of coordinates one must specify in order to determine a unique point in that space. Thus defined, D can take on only integer values: $D = 0, 1, 2, \ldots$. It is useful to redefine dimension in a way that retains the original meaning but allows D to take on values corresponding to any positive real number (and zero). It turns out that there are several ways in which this extended definition of dimension can be done. We will adopt what is called the "box-counting" or "capacity" dimension. This is a simplified version of what mathematicians call the "Hausdorff dimension."*

To find the dimension of an object in the box-counting scheme, we ask how many "boxes" are needed to cover the object, where the "boxes" are of the appropriate type for coverage (lines, squares, cubes, etc). We then repeatedly decrease the size of the boxes. The dimension of the object is based on how the number of covering boxes scales with the

---

* See Edward Ott, *Chaos in Dynamical Systems*. In most cases the Hausdorff dimension gives the same result as the box-counting dimension.

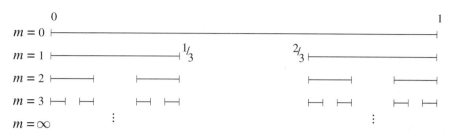

FIGURE 11.30
Cantor set.

length of the side of the box. For example, if we wish to cover a line of length l, we would need one box (line) of length l, two boxes of length $\frac{l}{2}$, ..., 2^m boxes of length $\frac{l}{2^m}$, ... We define ϵ_m to the the length of the side of the mth box: $\epsilon \equiv \frac{l}{2^m}$. Therefore the number of boxes $N(\epsilon_m)$ scales as $N(\epsilon_m) = \frac{l}{\epsilon_m}$. In two dimensions, we need one box (square) of area l^2, four boxes of area $(\frac{l}{2})^2$, ... 2^{2m} boxes of area $(\frac{l}{2^m})^2$, ... Here $N(\epsilon_m) = (\frac{l}{\epsilon_m})^2$. Generalizing to D *integer* dimensions, $N(\epsilon_m) = (\frac{l}{\epsilon_m})^D$. From this we can define the dimension D to be

$$D = \lim_{m \to \infty} \frac{\log(N(\epsilon_m))}{\log(l) + \log(\frac{1}{\epsilon_m})}. \tag{11.26}$$

As $m \to \infty$, the length of the system l becomes negligible, and so we write

$$D = \lim_{m \to \infty} \frac{\log(N(\epsilon_m))}{\log(\frac{1}{\epsilon_m})}. \tag{11.27}$$

This definition of dimension can be used for noninteger dimensions as well. As an example, consider the set of real numbers in the interval from 0 to 1. The length of any *finite* collection of points on this interval is zero, because a point has no dimension. A point can be contained in any finite interval, no matter how small. Now remove the open* middle one third of this interval. One obtains two closed intervals, each of length $\frac{1}{3}$, for a total length of $\frac{2}{3}$. Now repeat this process a total of m times. The results are indicated schematically in Figure 11.30.

The total length $L(m)$ of the remaining set after m iterations is $L(m) = (\frac{2}{3})^m$. This goes to zero as $m \to \infty$. In this limit, the points that remain are called the *Cantor set*. Despite having zero length, the Cantor set contains an uncountable infinity of points. This means it can be put into a one-to-one correspondence with the set of real numbers on the interval 0–1. To see this, in the $m = 1$ case, assign 0 if a point is in the first interval 0–$\frac{1}{3}$ or 1 if it lies in the second interval $\frac{2}{3}$–1. For the second ($m = 2$) division again assign 0 or 1 depending on whether the point lies to the left or to the right of the excluded open interval. In this way, each Cantor set member corresponds to an infinite series of zeros and ones. But this is the binary representation of *any* point x for $0 \leq x \leq 1$. $x = .011010001\ldots$ specifies

---

* "Open" means don't include the end points, which here are $\frac{1}{3}$ and $\frac{2}{3}$.

11.12 FRACTALS

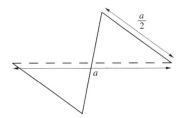

FIGURE 11.31

both a unique Cantor set point and a unique number x if interpreted differently. Thus the number of Cantor set points is uncountable (i.e., they cannot be put in correspondence with the integers) and is of the same order of infinity as the real numbers.*

To find the dimension, at the mth step, choose a one-dimensional box of length $\epsilon_m \equiv (\frac{1}{3})^m$. It takes $N(\epsilon_m) = 2^m$ boxes to cover the set. Using (11.27) for the Cantor set, $D = \frac{\log 2^m}{\log 3^m} = 0.6309\ldots$. This number remains constant as we pass to the limit $m \to \infty$. Notice that this is less than 1 (line) and greater than 0 (finite collection of points).

⊃ **Example: How Long Is the Coastline of Britain?**[†]

The answer is that it depends on the scale on which we look. From the Moon, Britain might appear to be a smooth, albeit rather irregularly shaped object. One could take a picture and estimate the length with a ruler. With improved resolution, we might become aware of more complexity, with river inlets and promontories jutting out into the sea. Still higher resolution reveals the rocky nature of a strip of the beach, giving an even greater estimate of the length. How can we quantify this process and express the dependence of length on the scale? First we introduce a process for approximating a coastline in which a continuous curve is generated, a curve that has an infinite length, and yet does not occupy a finite area.

First draw Figure 11.31 in which we replace a line of length a with three connected straight lines of length $\frac{a}{2}$.

Next replace the three lines of length $\frac{a}{2}$ by nine lines of length $\frac{a}{4}$ as shown in Figure 11.32A. Continue this process again to get $3^3 = 27$ line segments, as shown in Figure 11.32B.

After six iterations of the folding and shrinking process, you get $N = 3^6 = 729$ line segments, as shown in Figure 11.33. This process can be continued indefinitely. The result resembles a rugged coastline, with increasingly complex structure visible on a smaller and smaller scale.

As with the Cantor set, to find the dimension we will cover the set with lines and count the number needed.[‡] To simplify the dimension calculation, set the overall length scale a equal to one. If ϵ_m is the length of a single line segment, after m iterations, we need $N = 3^m$ segments of length $\epsilon_m \equiv (\frac{1}{2})^m$ to cover the curve. The fractional dimension is thus (from (11.27))

$$D = \lim_{m \to \infty} \frac{\log 3^m}{\log 2^m} = \frac{\log 3}{\log 2} = 1.584963. \qquad (11.28)$$

---

* For more information about orders of infinity, see Rucker, R., *Infinity and the Mind*, Birkhauser, Boston, 1982.
[†] This type of fractal construction was developed by Prof. D. Henderson at Cornell.
[‡] Note that we could also use squares, but we obtain a more exact dimension here by using lines.

A   B   FIGURE 11.32

The constant folding and shrinking have produced a fractal "curve" of infinite length. The dimension is larger than a line ($D = 1$) and smaller than an area ($D = 2$).

QUESTION 14: *Magnification* An equivalent way of looking at dimension is to ask: "If we magnify a self-similar object by a scale factor M, how many copies of the set do we get?" If we magnify a line segment by 2 we get 2 copies, so the dimension is 1. For a square, magnification by 2 gives $2^2 = 4$ copies. The dimension of a square is 2. Apply the same reasoning to a cube: Doubling the size gives 8 copies – hence a dimension of 3. The rule is $M^D = N$ copies. Find the dimensions of the Cantor set and British coastline using this reasoning.

Folding and shrinking occur in the phase space trajectories of dynamical systems, so we might expect the strange attractor of our pendulum (Figure 11.28) to be a fractal. To find its dimension we can use little squares of side ϵ_m and area ϵ_m^2. Let $N(\epsilon_m)$ be the number of squares or "boxes" containing one or more points of the strange attractor. For $m = 0$, we could take a single square filling the entire graph in Figure 11.28. (To simplify, after finding the maximum and minimum $\theta, \dot{\theta}$ values, the points on the strange attractor should be scaled to fill a square of side $\epsilon = 1$.) Then, in successive steps, use boxes in which the side is cut in half from the previous step. Make a plot of $\log N(\epsilon_m)$ versus $\log \frac{1}{\epsilon_m}$. According to Equation (11.27), for large m, the slope of this line should be a constant equal to D.

The results are shown in Figure 11.34. A straight line fits all but the last two points. This indicates a power law $N \sim \epsilon^{-D}$ with $D \approx 1.39$. The strange attractor is an object between a curve $D = 1$ and an area in the plane $D = 2$. The last two points deviate from the straight line only because we calculated a *finite* number of points on the strange attractor.

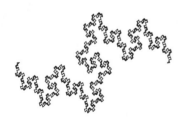

FIGURE 11.33

11.12 FRACTALS

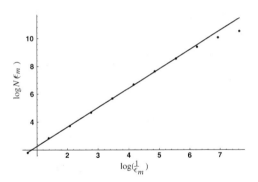

FIGURE 11.34
Finding the dimension of a strange attractor for the damped driven pendulum with $f = 1.5$, $\omega = \frac{2}{3}$, and $Q = 4$.

There are a number of other ways to define fractional dimension.* It seems reasonable that the Lyapunov exponents would be closely involved with the scaling properties of the attractor. One other proposed definition of fractional dimension is called the *Lyapunov dimension* D_L, which can be defined for a system of any number of degrees of freedom. Defined in the case of one degree of freedom (like our pendulum) with Lyapunov exponents $\lambda_1 > 0 > \lambda_2$,

$$D_L = 1 + \frac{\lambda_1}{|\lambda_2|}. \quad (11.29)$$

For the case ($f = 1.5$, $Q = 4$) we have been discussing, $D_L = 1.396$ according to (11.29), in rather satisfactory agreement with what we obtained by the box-counting method. There is no proof why $D_L \approx D_{\text{box}}$, but it is conjectured that they will agree in most cases where the points on the fractal are uniformly distributed.

To see if this agreement is maintained for other chaotic regions of the double pendulum, we repeated the calculation of the Lyapunov exponents and the fractal dimension of the strange attractor for other force strengths f and damping constants Q. The results are shown in Table 11.4. We can see that $D_L \approx D_{\text{box}}$, within a few percent, except for the first case, which is just barely chaotic. An accurate straight line fit on a log–log plot could not be obtained in this case. The box-counting dimension would be accurately determined if the number of points were substantially increased. Weaker damping (larger Q) and stronger driving forces lead to higher values for the fractional dimension, as you can see from the table.

As we have discovered, chaotic dissipative systems yield strange attractors of fractional dimension. Magnification of a portion of the strange attractor reveals a self-similar structure, which itself can be further magnified, until a practical limit is reached due to either the finite number of points being considered or the precision of the computer. In a

---

* These are discussed at length in more advanced textbooks, such as *Chaos in Dynamical Systems*, by Edward Ott.

TABLE 11.4 FRACTAL DIMENSION FOR THE DRIVEN, DAMPED PENDULUM. D_{box} IS OBTAINED BY THE SAME FIT TO POWER-LAW SCALING AS WAS DONE IN FIGURE 11.34

Strength f	Damping Q	D_{box}	λ_1	λ_2	D_L
1.10	2	fails to fit	0.011	−0.511	1.022
	4	1.34	0.143	−0.393	1.364
	8	1.52	0.164	−0.289	1.567
1.20	2	1.20	0.132	−0.632	1.209
1.50	2	1.21	0.113	−0.613	1.179
	4	1.39	0.164	−0.414	1.396

more theoretical and strictly mathematical sense, the self-similar structure is always there at any level of magnification. The strange attractor is a naturally occurring set of points in phase space that can be produced experimentally or computationally as the infinite time limit of a driven damped pendulum or other chaotic dissipative dynamical system.

11.13 CHAOS IN THE SOLAR SYSTEM

To a good approximation, the solar system is a Hamiltonian system with many degrees of freedom, including those from the nine planets, their satellites, and the asteroid belt between Mars and Jupiter. The system is controlled by the universal law of gravitational attraction. The stability of the solar system is a question that has plagued scientists since Newton's time. Newton was concerned that the perturbations of the other planets would destroy the regularity found when only considering the interactions between the individual planets and the Sun. Newton suggested divine intervention as a stabilizing mechanism. Laplace was the first to perform a serious study of solar system stability at the end of the eighteenth century. He used perturbation theory to show that, compared to the Sun, the effects of the other planets on a planet would vanish to first order. Strongly believing in the deterministic predictability of Newton's laws, he noted that[*]

> Given ... an intelligence which could comprehend all the forces ... for it, nothing would be uncertain and the future, as the past, would be present before its eyes.

As discussed earlier in this chapter, Poincaré answered the unresolved challenge to prove the stability of the solar system by looking at the higher-order corrections to Laplace's perturbation theory. Although Poincaré looked at a much more simplified problem, he discovered that an analytic solution to the solar system was impossible, due to a sensitivity on initial conditions. However, recent advances in chaos theory have guided physicists and astronomers back to this question of solar-system stability. It turns out that much of the solar system is weakly chaotic. We will focus specifically on the asteroid belt and the nine planets.

---

[*] Laplace, P. S., *A Philosophical Essay on Probabilities*, Transl. by F. W. Truscott and F. L. Emory, Dover, 1951.

11.13 CHAOS IN THE SOLAR SYSTEM

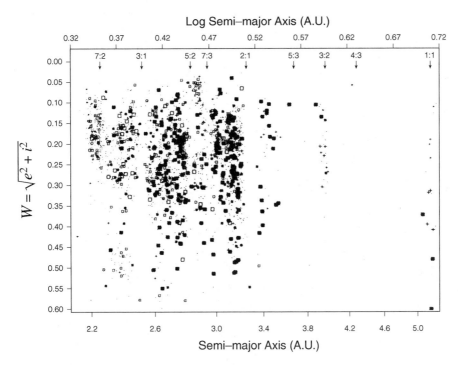

FIGURE 11.35

Distribution of asteroids in the asteroid belt. w is a quantity related to eccentricity. (From *Asteroids*, ed. T. Gehrels, Univ. of Arizona Press, 1979.)

There is room for another planet in the space between Mars and Jupiter. Instead we find many asteroids, all in orbit around the Sun. However, the distribution of asteroids is nonuniform: We see several gaps in Figure 11.35, known as Kirkwood gaps. Kirkwood was the first to notice that these gaps occur near orbital period resonances with Jupiter. For example, any asteroid at a distance of 2.5 AU from the Sun will have an orbital period $\frac{1}{3}$ times that of Jupiter's orbital period of 11.86 years. (Jupiter orbits at a radius of 5.203 AU.) The asteroid belt is depopulated here, and it is known as the $3:1$ Kirkwood gap to denote its commensurability. Other gaps exist at the resonances $2:1$, $5:2$, $7:3$, etc. as seen the Figure 11.35. However, as Wisdom notes*

> The mere association of a gap with a resonance does not itself explain the formation of the gap. Nature herself provides the counter-example: there are gaps at some resonances and enhancements at others.

We see clusters of asteroids at the $3:2$, $4:3$, and $1:1$ resonances. Wisdom was the first to apply chaos theory and use numerical calculations to sort this out.

As we learned in the Hamiltonian chaos section, KAM tori with rational winding numbers are the first to break. Similarly, Wisdom's calculations show that a chaotic zone

---

* Wisdom, J., "Chaotic Behavior in the Solar System," in *International Conference on the Physics of Chaos*, ed. M. Duong-Van, 1987.

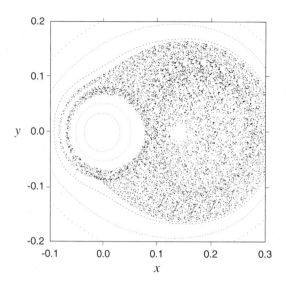

FIGURE 11.36
Surface of section for the orbit shown in Figure 11.37. Here $x = \epsilon \cos(\omega - \omega_J)$, and $y = \epsilon \sin(\omega - \omega_J)$, where ϵ is the eccentricity, and ω and ω_J are the longitudes of perhelia for the asteriod and Jupiter, respectively. (From Wisdom, J., "Chaotic behavior in the solar system," in *International Conference on the Physics of Chaos*, ed. M. Duong-Van, 1987.)

exists in the vicinity of the 3 : 1 resonance. This can be seen in the surface of section in Figure 11.36. An asteroid at a 3 : 1 resonance may have a small eccentricity for a long time, traveling along in one of islands close to the origin. However, eventually it could diffuse* to the large chaotic area to the right of the origin, and its eccentricity will increase. The eccentricity versus time has been calculated in Figure 11.37 for a typical 3 : 1 asteroid. Any orbit at this radius with eccentricity $\epsilon > .3$ will cross the orbit of Mars. Thus there is a finite possibility that such an asteroid will strike the surface of Mars. It is theorized that most asteroids in the 3 : 1 region periodically experienced such an eccentricity increase and that close encounters or even collisions with Mars eventually created the 3 : 1 Kirkwood gap.

Also, from Wisdom's numerical studies we learn that chaos may play a role in bringing some of these asteroids in the 3 : 1 region to the surface of the Earth. After enhancing his numerical calculations to allow for three-dimensional motion, and including the perturbations to Jupiter's motion by the other planets, Wisdom found that some asteroid eccentricities could become larger than .6. Any $\epsilon > .6$ is an Earth-crossing orbit.

How about the other resonances? It turns out that the 5 : 2 gap is similar to the 3 : 1 gap.[†] However, it is harder to prove that the 7 : 3 and 2 : 1 gaps will lead to Mars-crossing orbits,

---

* Since the dimensions of phase space are greater than two, Arnold diffusion allows for local chaos to spread slowly throughout the phase space.
† Ferraz-Mello, S., "Kirkwood Gaps and Resonant Groups," in *Asteroids, Comets, Meteors 1993*, ed. A. Milani.

11.13 CHAOS IN THE SOLAR SYSTEM

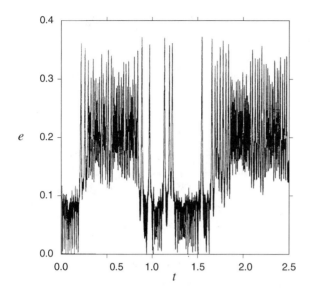

FIGURE 11.37
Eccentricity versus time for a typical asteroid in the 3 : 1 resonance. Time is measured in millions of years. (From Wisdom, J., "Chaotic behavior in the solar system," in *International Conference on the Physics of Chaos*, ed. M. Duong-Van, 1987.)

since they need higher eccentricities to reach Mars. Including the perturbations due to other planets can make a few of the asteroids reach Mars, but the evidence is not yet conclusive.

The 3 : 2 and 4 : 3 resonances are strangely populated. Integrations at these resonances show a relatively stable zone for small eccentricities $\epsilon < .25$, that is, a zone in which the positive Lyapunov exponent is small compared to that of the other resonances. This could account for their staying power. The 1 : 1 resonance is another problem entirely, since it is unlikely for those asteroids to develop an eccentricity high enough for a Mars-crossing orbit.

Over the past few years, numerical studies of the long-term evolution of the solar system have been made. Time step size is a major limiting factor in evolving the planets over the time scales comparable to the lifetime of the solar system, that is, 12 billion years (Gyr) (past and future included). The longest nine-planet integration to date was done by Sussman and Wisdom,[*] who used a time step of 7.2 days to integrate out to 100 million years (Myr). The longest secular theory (to be described below) was done by Laskar,[†] who reached 25 Gyr by using a time step of 250 years. Since the work on this subject is relatively recent, as noted by Tremaine[‡] "The conclusions described below must be treated cautiously."

Sussman and Wisdom found that Pluto's orbit has a positive Lyapunov exponent of $\lambda = \frac{1}{20} \text{Myr}^{-1}$. This implies a *Lyapunov time* (i.e., time scale) of 20 Myr before neighboring trajectories begin to diverge significantly. The orbital motion of Pluto is quite complicated,

---

[*] Sussman, G. J., and Wisdom, J., *Science* **257**, p. 56, 1992.
[†] Laskar, J., *Astronomical Astrophysics*, **287**, p. L9, 1994.
[‡] Tremaine, S., "Is the Solar System Stable?", in *Frontiers of Astrophysics, Proceedings of the Rosseland Centenary Symposium*, ed. P. B. Lilje, et. al., 1995.

with many oscillation frequencies. However it seems to have a fairly regular semimajor axis, eccentricity, and inclination with the plane of the solar system* over the range of integration (845 Myr in another calculation by Sussman and Wisdom using only the outer five planets). The other four outer planets were also found by Sussman and Wisdom to have Lyapunov times in the range of 5–20 Myr.

Laskar was able to integrate for much longer times by using secular theory, in which he averaged the solar system EOMs over the short-term motion of the planets along their orbits and then integrated over the long-term trends such as the precession frequencies of the orbits. The idea is that these long-term perturbations will be more significant, since they act for a longer duration. However, Laskar still had a 15-degree-of-freedom problem[†] (he left out Pluto) which led to 150,000 algebraic terms to be integrated. Surprisingly, this procedure is still more efficient than the full-blown 9-planet integrations, as can be seen by Laskar's much longer evolution time.

Laskar found that the four inner planets are also chaotic with a Lyapunov time of 5 Myr. The chaos in the inner planets could be traced to two resonances, one between Mars and the Earth and the other related to Mercury, Venus, and Jupiter. Since Laskar did not find chaos in the five outer planets, it is suspected that the outer planets' chaos (excluding Pluto) is due to fast orbital motion, rather than slow precession.

It seems that the entire solar system is chaotic! Luckily, the effect of these chaotic exponents is less dire than we might imagine at first. The implication that the positive Lyapunov exponents do have is that they make long-term predictability of the solar system impossible. The dependence on initial conditions is so severe that an error of 15 meters in the initial position of the Earth will be an error of 150 million kilometers after 100 Myr. Even the impulse imparted by the launch of a rocket will change the position of the earth by $\simeq .5$ degrees after 100 Myr. Predictability is completely gone after about 100 Myr, and we see that our high-tech numerical calculations have led us back to Poincaré's conclusion.

The conservative nature of the solar system keeps this chaos in check. We must remember that we have a Hamiltonian system in which chaos can be local, that is, confined to a certain region. Two trajectories may diverge from each other, but they are confined to a specific part of phase space. Arnold diffusion would cause more catastrophic chaos, such as the ejection of planets, but this has been shown to occur on a time scale larger than the age of the solar system.[‡]

The numerical evolution confirms this. The eccentricities, inclinations, and semimajor axes of the planets vary regularly over the 100 Myr integration (as do those of the five outer planets over the 845 Myr integration). Laskar did find large and irregular variations over his 25 Gyr time span in the inner planets, especially Mars and Mercury, the least massive of the planets.

---

* Rather surprisingly, the motion of the planets is coplanar, except for Pluto which has an orbital plane tilted 17° from the plane of the Earth's orbit.
[†] See Laskar, J., *Celestial Mechanics and Dynamical Astronomy* **64**, pp. 115–62, 1996 for details.
[‡] Tremaine, S., "Is the Solar System Stable?", in *Frontiers of Astrophysics*, *Proceedings of the Rosseland Centenary Symposium*, ed. P. B. Lilje et al., 1995.

11.13 CHAOS IN THE SOLAR SYSTEM

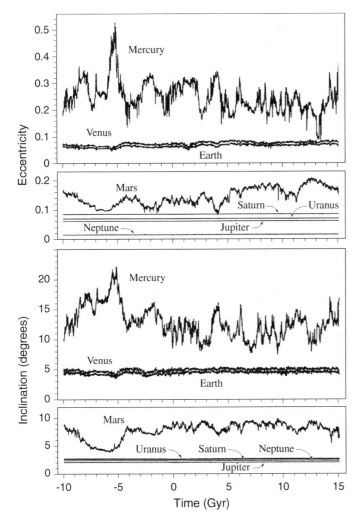

FIGURE 11.38
Numerical integration of Laskar's secular theory for the motion of the solar system (excluding Pluto). For each planet, the maximum value obtained over intervals of 10 Myr is plotted for eccentricity and inclination. (From Laskar, J., *Celestial Mechanics and Dynamical Astronomy* **64**, pp. 115–62, 1996.)

Although Laskar's 25 Gyr integrations are inaccurate predictors of our own solar system, they do show one possible evolution scenario. Thus long-term solar system predictions become probabilistic. What Laskar has found is that the maximum eccentricities and inclinations of the four inner planets do vary over 25 Gyr, whereas the outer four planets (excluding Pluto) have essentially constant eccentricity and inclination, as shown in Figure 11.38. The diffusion of maximum eccentricities of Earth and Venus vary by only 0.02, but for Mars it is 0.12 and for Mercury it is as high as 0.5. However, angular momentum conservation seems to keep the eccentricities in check. As the eccentricity of Mercury increases, it is seen to decrease for the other three planets.

Mercury's eccentricity changes significantly, but not enough for a close encounter with Venus. However, when Laskar altered the initial conditions of the planets, he found that there was a finite probability that Mercury would cross the orbit of Venus in less than 5 Myr, perhaps being either ejected or colliding with Venus. Earth and Mars could get very close to each other, but not enough for disaster.

One conclusion is that perhaps other planets were ejected from the solar system in the past, particularly small ones. This means that the solar system could have looked quite different in the past. Currently the solar system is "full," that is, there is no room for other planets. This is said because any additional planets would cause the solar system to be more unstable. Based on the above results, the term "marginal stability" is coined for our solar system. It is even surmised by Laskar that marginal stability is better than a very stable system with only one or two planets. A stable system would be crowded with debris from the original formation which can't be cleaned out by the few orbiting planets. Our system's planets have effectively removed such debris due to the fact that it is weakly chaotic.

Laskar also investigated the chaotic obliquity (tilt axis) of the planets and found that without the Moon, the Earth's obliquity would change by 60° in less than 2 Myr. The implications for the environment could be catastrophic. Luckily, the Moon keeps us stable at an obliquity of $23.3 \pm 1.3°$. Mars, not having a large stabilizing satellite, will experience an obliquity change of 25° over 25 Myr. Perhaps one necessary condition for an hospitable planet is the presence of a very large satellite as a stabilizing mechanism.

As we can see, the solar system is weakly chaotic and amazingly complex. This is both a curse and a blessing, since the chaos helps clear out unwanted debris, yet allows for a finite probability of catastrophic events like ejection. These results were unexpected, especially by those believers in the deterministic probability of Newton's laws. As remarked by Sir James Lighthill* on the 300th anniversary of Newton's *Principia*:

> We are all deeply conscious today that the enthusiasm of our forebears for the marvellous achievements of Newtonian mechanics led them to make generalizations in this area of predictability which, indeed, we may have generally tended to believe before 1960, but which we now recognize were false. We collectively wish to apologize for having misled the general public by spreading ideas about the determinism of systems satisfying Newton's laws of motion that, after 1960, were proven to be incorrect . . .

STUDENT PROJECTS

Several projects involving computer work with chaotic systems are presented here. Typically, they were done in one or two weeks by a group of three or four students, at least one of whom had sufficient computer expertise to write the necessary programs. The results were then presented orally and/or as a written report to the rest of the class. In general, the

---

* Lighthill, J., *Proc. Roy. Soc. (London)*, **A407**, p. 35, 1986.

STUDENT PROJECTS

projects involving integration of differential equations are somewhat more difficult to carry out than the ones involving maps. All of these projects involve fairly extensive numerical calculations.

Maps

Problem 1: *(The logistic map)* As described in the appendix, the logistic map is a 1-D map defined by

$$x_{j+1} = rx_j(1 - x_j), \quad 0 < x_j < 1, \quad 1 < r < 4, \tag{11.30}$$

a) Construct the bifurcation diagram $\{x_j\}$ versus r, for $3 < r < 4$. Start with $x_1 = \frac{2}{3}$ and run off about 1,000 points to damp the transients; then plot all subsequent values for, say, another 1,000 points. You will need to take very small steps in r as the limit r_∞ of chaos is approached.

b) Find the bifurcation points $r_1, r_2, \ldots, r_n, \ldots$, where r_1 is the value of r for the first bifurcation and r_n is the point at which the period 2^n map appears. Find the value of

$$\frac{r_n - r_{n-1}}{r_{n+1} - r_n} \tag{11.31}$$

in at least two cases. Feigenbaum showed that this limit is the same as that for the supercycles R_n. What is the largest value of r_n you can find numerically?

c) Chaos starts at $r_\infty = 3.5699\ldots$. Calculate the Lyapunov exponent as discussed in the appendix:

$$\lambda(r) \equiv \lim_{m \to \infty} \frac{1}{m} \sum_{i=1}^{m} \ln |f'(x_i^*)| \tag{11.32}$$

for a small range around r_∞, showing that λ goes positive at the threshold for chaos.

d) Plot $f^{(3)}(x) \equiv f(f(f(x)))$ versus x for values of r just below and above $r = 1+\sqrt{8}$ to show what happens when chaos suddenly disappears and the stable period 3 orbit appears. Numerically find the fixed points by solving $f^{(3)}(x) = x$ for $r = 1+\sqrt{8}$. Show that there are three double roots and one single root, besides the trivial one at $x = 0$. Which are the stable fixed points? Do the same for r near, but not equal to, $1 + \sqrt{8}$. What happens to the roots as r increases through the threshold value for the period 3 orbit?

Problem 2: *(Bouncing ball)* A simplified model of a ball bouncing on a vertically vibrating table can be approximated by the following 2-D map:

$$\begin{aligned} x_{n+1} &= x_n + y_n, \\ y_{n+1} &= ay_n - b\cos(x_n + y_n), \end{aligned} \tag{11.33}$$

where x_n is the impact time (taken modulo 2π), y_n is proportional to the velocity of the ball immediately after impact, a represents the damping in the system, and b is related to the frequency of the oscillating table. (For a more detailed discussion see Moon, F. C., *Chaotic Vibrations*.)

a) Find the Jacobian of the map and show that a conservative map must have $a = 1$, whereas a dissipative map has $a < 1$. For the work below, set $a = \frac{1}{2}$ and vary b as the control parameter.

b) Let $b = 6$. Start with an initial point, say $(3, 3)$, then compute and plot at least 10,000 points on the attractor y_n versus x_n. (More points will give a better result.) Save these points for part c.

c) Compute the dimension of this fractal attractor by the box-counting method. Suggestion: Normalize your data so that they fall within $[0, 1] \times [0, 1]$. Let $\epsilon_m \equiv \frac{1}{2^m}$, $m = 1, 2, 3, \ldots$. Set up an all-zero $2^m \times 2^m$ matrix. If any of your points falls in a box corresponding to a particular element of the matrix, make that element 1. By taking finer and finer meshes, you ought to be able to get the value of D from (11.27). There should be a region on a graph of $\ln N(\epsilon_m)$ versus $\ln \frac{1}{\epsilon_m}$ that is a straight line, the slope of which gives you D. However, if you take ϵ_m too small, the fact that your data are really just a finite set of points will cause $D \to 0$.

d) From your results for $b = 6$, verify that the map is chaotic and has a strange attractor. For $b > 2$ there exists a period-doubling sequence. Make a bifurcation plot of x_n versus b (throwing away the first 100 points to get onto the attractor). Obtain the first few approximate values for the Feigenbaum constant $\delta \approx 4.669\ldots$, as a test of period doubling universality.

Problem 3: *(The standard map)* The standard map is the prototype of the map for the Poincaré section of a conservative system. The map $(\theta_n, I_n) \to (\theta_{n+1}, I_{n+1})$ is given by the equations

$$I_{n+1} = I_n + k \sin \theta_n,$$
$$\theta_{n+1} = \theta_n + I_{n+1}.$$
(11.34)

A physical example with this map is a pendulum that gets kicked with an impulse proportional to $\sin \theta$ at equal time intervals. The pendulum undergoes uniform rotation between kicks (no gravity). It is assumed that the unperturbed system ($k = 0$) is expressed in angle–action variables, (θ, I). The motion of the system keeps the action constant, while the angle changes during each interval between kicks by an amount that depends on the action. The lines of constant action may be plotted on an I versus θ graph. (Both I and θ are taken modulo 2π.) These lines are the KAM tori, with winding numbers that depend on the action. When a nonlinear perturbation is added by making $k \neq 0$, the tori are distorted, and, as predicted by the KAM theorem, the ones with rational or nearly rational winding numbers are broken. The winding number is

STUDENT PROJECTS

the average increase in θ:

$$\Omega \equiv \langle\theta\rangle \equiv \lim_{n\to\infty} \frac{(\theta_n - \theta_1)}{n}. \tag{11.35}$$

a) Prove that the Jacobian of the map is 1 and thus that the map preserves area. When $k = 0$, for what values of I is Ω a rational number, giving periodic orbits?

b) Increase k so that $k = 0.6$. Start the map off at a number of I_1 values (you can choose $\theta_1 = 0$) and make at least several hundred points/initial value. Observe what happens to tori with both rational and irrational winding numbers. In particular, discuss what happens in the region $I = \pi$ and near other places where the winding number is rational.

c) Keep on increasing k. The last KAM torus disappears for $k > 0.9716\ldots$. The winding number of this torus is the golden mean, $\Omega = \frac{\sqrt{5}-1}{2}$. After the last torus disappears, there is no further barrier to global chaos. Demonstrate these effects with suitable 2-D plots of I versus θ.

Continuous Time

Problem 4: *(The Lorenz equations)* The modern discovery of chaos began in the early 1960s with work by E. Lorenz of the Meteorology Department at the Massachusetts Institute of Technology. Hoping to do a numerical simulation of weather patterns, Lorenz considered a horizontal fluid layer heated from below. As the temperature of the bottom is increased, convection currents in the fluid become turbulent and irregular. Lorenz made a simplified model of this problem, in which he expressed three physical quantities as a set of nonlinear differential equations in the variables x, y, z. He was quite astonished that he could not reproduce his results if he varied the starting conditions slightly. At first Lorenz believed that this was due to the finite precision of his computer, but he found that adding more significant figures did nothing to fix the problem. Eventually he realized that it was a fundamental property of the differential equations themselves. The Lorenz equations are

$$\frac{dx}{dt} = 10(y - x),$$
$$\frac{dy}{dt} = -xz + rx - y, \tag{11.36}$$
$$\frac{dz}{dt} = xy - \frac{8}{3}z.$$

The control parameter is r, with the interesting region being $0 < r < 30$. A good starting point is very near to, but not exactly on, the origin. Notice that the "phase space" is three dimensional, so trajectories can be plotted as if they were in real space.

a) Near the origin, what is the determinant of the Jacobian for the infinitesimal time transformation $t \to t + dt$? Is phase volume conserved or does it shrink?

b) For very small values of (x, y, z), make a linear analysis of stability by plotting the real part of the eigenvalues of $\mathbf{J}$ verus r. Recall that to find $\mathbf{J}$ and $\mathbf{M}$, you have to linearize the equations of "motion" (11.36) near to the origin. The time development then comprises a set of matrix equations with constant coefficients from (11.18). The eigenvalues of $\mathbf{M} = e^{t\mathbf{J}}$ are $e^{\lambda_i t}$, where the λ_i are the eigenvalues of the matrix $\mathbf{J}$.

c) Find $r_{\text{threshold}}$, the minimum r value for chaos in this system.

d) The route to chaos followed here is called the *intermittency route*. Plot the motion in the (x, y) plane or the $(x, \dot{x})$ plane. Also try some plots of $x(t)$. Describe what happens as r is increased, keeping the starting point fixed. Is there period doubling? Is there an attractor?

e) Make a 3-D plot of the trajectory in x, y, z. Start from *very* near the origin. Set $r = 28$. The result is a strange attractor called the *Lorenz attractor*. Try varying the starting point slightly. What happens? What is the fractal dimension of this strange attractor?

Problem 5: *(Hénon–Heiles potential)*

a) Solve Problem 9.18, to introduce yourself to the Hénon–Heiles Potential.

b) Integrate the equations of motion. Find the Poincaré section for $x = 0, \dot{x} > 0$ and also plot the boundary, which occurs at $\dot{x} = 0$. Plot the points on the Poincaré section in the (y, p_y) plane. ($E = \frac{1}{12}$ gives a nice-looking Poincaré section.) Find the threshold for chaos and observe whether the chaotic trajectories are local or global as the energy is raised.

c) The appearance of chaos proves that the Hamiltonian (9.148) is not integrable. Now change the sign of the last term (proportional to y^3) in the potential, (9.149). Again look at the Poincaré section. Verify that this seemingly small change leads to complete integrability at all energies where the motion is bounded.

Problem 6: *(The restricted three-body problem: Earth, Jupiter, and Sun)* The Earth has such a small effect on the Jupiter–Sun system, it is natural to ask what the motion of such a dust mote as the Earth would be in the presence of two massive bodies that obey Kepler's laws. This is the so-called *restricted three-body problem*. Poincaré discovered chaos while analyzing this problem, winning the King of Sweden's prize for proving the nonexistence of analytic solutions and hence the impossibility of obtaining an analytic proof of the stability of the solar system as a whole. The mass of Jupiter is about 0.1% that of the Sun. The mass of the Earth is assumed to be negligible, without any effect on the Sun, which is fixed at the origin or on Jupiter. Both Jupiter and Earth are assumed to move in the same plane. The Earth's motion can be described by

$\vec{r}(t) = (x(t), y(t))$ in that plane. (Assuming that the center of mass is at the center of the Sun, so the Sun remains in a fixed position, is not really a good assumption when Jupiter's mass is increased to be comparable to that of the Sun, but we will ignore this complication to keep the equations of motion simple.)

a) Verify the following 2-D equation of motion for the Earth under the influence of both Jupiter and the Sun:

$$\ddot{\vec{r}} = -4\pi^2 \left(\frac{\vec{r}}{r^3} + \xi \frac{\vec{r} - \vec{r}_J(t)}{|\vec{r} - \vec{r}_J|^3} \right) \tag{11.37}$$

Here ξ is the ratio of Jupiter's mass to the solar mass: $\xi \equiv \frac{M_J}{M_\odot}$, $x_J = r_J \cos(\omega_J t)$, $y_J = \sin(\omega_J t)$, $\omega_J = \frac{2\pi}{11.86}$, $r_J = 5.203$. The units in (11.37) are AU (1 AU = mean distance to Earth from the Sun) and Earth years. Assume in (11.37) that Jupiter follows a circular orbit around the Sun with a period of 11.86 Earth years. The periodic term from the gravitational potential between Earth and Jupiter appears as an explicitly time-dependent forcing term in the differential equation.

b) Write a computer program to integrate Equation (11.37) for at least 50 Jupiter years. As a test of your integration program, set $\xi = 0$ and check that Earth actually does follow a circular orbit in a stable fashion for that period of time. (Try at least two different planetary starting radii for this test.)

c) Numerically solve for the motion of the Earth through one Jovian year. Start the Earth one AU from the Sun. Does the Earth return to its initial position? Plot the Earth's motion and show that it is not much perturbed by Jupiter.

d) Increase Jupiter's mass by a factor of 40. Notice that the Earth's new orbit looks like motion on a torus projected onto the plane of the motion. Increase the mass of Jupiter further and determine whether the system remains stable and appears to be quasiperiodic or becomes chaotic. Produce some graphical examples of the effect on Earth's orbit for different Jupiter masses up to a mass of Jupiter equal to the solar mass.

e) Try the same thing for asteroids that have resonant orbits with Jupiter as given in Table 11.5. Also try some nonresonant orbits and compare.

TABLE 11.5 RESONANT PERIODS WITH JUPITER

Resonance	Period (yr.)	Radius (AU)
2:1	5.93	3.278
5:2	4.744	2.825
3:1	3.953	2.501

Problem 7: *(Entrainment)* Instead of maintaining the beat frequency observed with purely linear systems, nonlinear oscillators with nearly equal frequencies can become locked to a constant phase relation at a single frequency. This phenomenon is known as *entrainment*. Hysteresis is sometimes involved if one changes the coupling or the uncoupled oscillator frequencies. In practical situations this is sometimes a blessing and sometimes a drawback. (See the discussion by A. B. Pippard in *The Physics of Vibration*.) Entrainment was first observed by Christian Huygens in 1665, who observed that two clocks mounted on the same support bracket would pull each other into synchronism. It might be the first report of a mechanical effect due to nonlinear dynamics. From our treatment in this chapter, we recognize entrainment as a nonlinear resonance of a two-degree-of-freedom dynamical system.

Consider here a single vertical pendulum hanging from a support which can be moved horizontally with an amplitude $Y(t) = f \cos \omega t$ as shown in Figure 10.17A. For simplicity, we assume the mass m and length l of the pendulum, and its natural frequency ω_0, are all 1. The problem resembles the damped driven pendulum treated in the text.

a) Show that the equations of motion for the angle θ of this pendulum from the vertical are

$$\frac{d\theta}{dt} = p - \dot{Y} \cos \theta,$$

$$\frac{dp}{dt} = -\frac{1}{Q}\dot{\theta} - (1 + \dot{Y}\dot{\theta}) \sin \theta. \tag{11.38}$$

If $\dot{\theta}$ is substituted into the right side of the second Equation in (11.38), you will see that there is a term proportional to $\dot{Y}^2$, so if $\dot{Y} = f \sin(\omega t)$, with constant f and ω, second harmonics of the shaking frequency ω can appear.

b) Obtain the Hill equation for small oscillations, and use it to guess which driving frequencies might lead to instabilities.

c) Write a computer program to integrate the equations (11.38) numerically. (Choose $Q \approx \infty$ so the transients are not damped.) Plot the phase trajectory for a few values of the the driving strength for a frequency $\omega \approx 1$, to achieve phase locking for sufficiently strong coupling (f value). You might want to first sample at the driving frequency and look at a Poincaré section of p versus θ.

d) For a coupling that will phase lock, try sweeping the driving frequency to see if there is hysteresis, that is, if the behavior of the system depends on whether you are sweeping the frequency up or down.

e) Can the system become chaotic? Set Q equal to 2 or 4 in order to strongly damp transients, and look at the Poincaré section. Is there a period-doubling route to chaos for a fixed ω and variable f? If chaos does occur, what is the fractal dimension of the strange attractor? (Assume some definite f value above the chaos threshold.)

APPENDIX

THE LOGISTIC MAP: PERIOD-DOUBLING ROUTE TO CHAOS; RENORMALIZATION

We introduce a system with a single variable x, where time is measured in discrete intervals, $j = 1, 2, \ldots$. At each time step, the new value x_{j+1} of this variable is determined as a function of the previous value x_j: $x_{j+1} = f(x_j)$. Despite the simplicity of the function $f(x)$ in the example here, when the function $f(f(f(x)))\ldots$ is iterated indefinitely, a richly complex structure can be generated, one containing the essence of physical and mathematical chaos. Our example of such a one-dimensional map $x_{j+1} = f(x_j)$ is the *logistic map*, where

$$f(x) \equiv rx(1-x), \quad 0 < x < 1, \quad 1 < r < 4, \tag{11.39}$$

Here r is called the *control parameter*.

Numerical calculations can be carried out with a computer for many generations to show what happens for different choices of r. We are interested in long-term behavior of the map; the initial conditions settle down onto the attractor just like for the damped driven pendulum. It can be seen that increasing r from $r = 2$ to $r = 4$ makes a qualitative change in the successive map values. In Figure 11.39A, we see that for small values of r, the map tends to a limit (i.e., a *fixed point*) and we call this map a 1-cycle. In Figure 11.39B, we see the map alternating between two points, called a 2-cycle. An m-cycle is the periodic sequence of the attractor, m numbers long. Although m can be any integer, we will focus on cycles that are multiples of two. In Figure 11.39C we see that chaos has appeared for $r = 4$, because the relation between successive map values appears to be random. A small difference in the initial conditions is found to grow exponentially. The importance of this map is that it is the simplest example of a universal type of chaotic behavior.

QUESTION 15: *Logistic Map* This is an easy map to calculate on a pocket calculator for different values of r. A good starting point is $x_1 = 0.5$, but you can use any starting

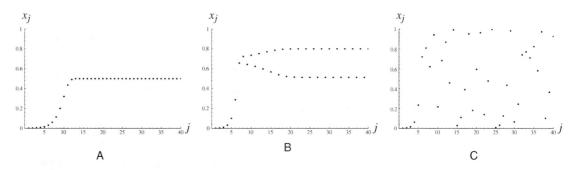

FIGURE 11.39
A) Logistic maps for $r = 2.0$; B) $r = 3.2$; C) $r = 4.0$.

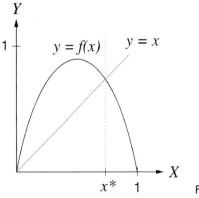

FIGURE 11.40

point. Try the values $r = 2, 2.5, 2.9, 3.2, 3.5$. What are the m-cycles of the map? For what values of r do they change? You will have to ignore the first few iterations, which do depend on your starting point. Now find a friend with another brand or model of calculator and try the value $r = 3.6$. Each of you should use exactly the same starting point. Compare your answers after, say, 10 iterations of the map.

Focusing on the 1-cycle first, we find a fixed point for that value of $x = x^*$ satisfying

$$x = f(x) = rx(1-x). \tag{11.40}$$

Excluding the trivial fixed point at $x^* = 0$, the solution to (11.40) is $x^* = 1 - \frac{1}{r}$, which will be stable or unstable depending on $|f'(x^*)|$ (see below). If a plot is made of the two equations $y = rx(1-x)$ and $y = x$, the fixed point x^* must lie on the intersection of the curve and the line as shown in Figure 11.40.

We will now prove that the condition for "orbits" (sequences of $x_j, x_{j+1}, x_{j+2}, \ldots$) to converge on a fixed point is that $|f'(x^*)| < 1$. Consider a point near the fixed point x^*: $x_1 = x^* + \delta x$ ($\delta x \ll 1$). Taylor series expand the first iteration $x_2 = f(x_1)$ around $x = x^*$ to obtain $x_2 = x^* + f'(x^*)\delta x + O[\delta x^2]$. For the second iteration, use the chain rule to calculate the Taylor series approximation for $x_3 = f(f(x_1))$. The derivative of $f(f(x))$ with respect to x is $f'(f(x))f'(x)$. For the fixed point this is $f'(x^*)^2$. The Taylor series for x_3 is therefore: $x_3 = x^* + f'(x^*)^2 \delta x + O[\delta x^2]$. For the case of x_j, we obtain

$$x_j = x^* + [f'(x^*)]^{j-1}\delta x + O[\delta x^2]. \tag{11.41}$$

This means that convergence of x_j to x^* as $j \to \infty$ can occur only if $|f'(x^*)| < 1$, and we have a *stable* fixed point. If $|f'(x^*)| > 1$, x^* is an *unstable* fixed point, with exponential divergence of small deviations away from x^*. If $|f'(x^*)| = 1$, we have *neutral* stability, the significance of which will be determined below. The case $f'(x^*) = 0$ is a special condition with ultrarapid convergence. In that case, the fixed point is said to be *superstable*.

If the map $f(x)$ alternates between two points, as in a 2-cycle, then the second iterate of the map $f(f(x))$ has two fixed points. If you start from one of the two points of the 2-cycle for $f(x)$, you arrive back at the same point by applying $f(f(x))$. Of course the

APPENDIX THE LOGISTIC MAP

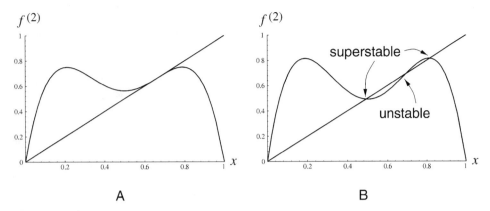

FIGURE 11.41
$f^{(2)}$ versus x for A) $r = 3.0$; B) $r = 3.2360679$. The line $y = x$ is also drawn in each case.

fixed points of $f(x)$ are also fixed points of $f(f(x))$. We need to define a new notation for repeated iterations of $f(x)$: $f(f(x)) \equiv f^{(2)}(x) = r^2 x(1-x)(1 - rx(1-x))$. To look for the fixed points of $f^{(2)}(x)$, plot $y = f^{(2)}(x)$ and $y = x$. For a 2-cycle, fixed points $x = x^*$ satisfy the equation $x = f^{(2)}(x)$, and now the stability depends on the value of $|f^{(2)\prime}(x^*)|$.

For $r = 2$, there is only one stable fixed point of $f^{(2)}$ at $x^* = \frac{1}{2}$, and the graph looks similar to Figure 11.40. As r increases, the shape of $f^{(2)}(x)$ changes, causing the location and number of fixed points to change. For example, when $r \equiv r_1 = 3$, there are three identical fixed points with neutral stability ($|f^{(2)\prime}(x^*)| = 1$), as shown in Figure 11.41A. This is the value of r for which a *bifurcation* or *period doubling* takes place. We can find this point analytically by requiring that $f^{(2)}(x) = x$ and $f^{(2)\prime} = 1$. These two equations are solved to give $r = 3$, $x^* = \frac{2}{3}$. One might generalize that bifurcations to a $2m$-cycle of f will involve two stable fixed points and one unstable fixed point of $f^{(2m)}$ emerging from each stable fixed point of $f^{(m)}$ (an m-cycle of f) as r increases.

The slope of the graph of $f^{(2m)}$ must have the absolute value 1 at the point where each stable fixed point bifurcates and the period doubles. This implies that the Lyapunov exponent is zero at the bifurcation point, as we will see. This kind of bifurcation is called a *pitchfork bifurcation*. If one plots the fixed point(s) against r as r increases through $r = 3$, the emergent unstable fixed point can be plotted as a dotted line and the stable fixed points as solid lines as shown in Figure 11.42.

Make a further increase in the control parameter to $r \equiv R_1 = 3.2360679\ldots$, another special r value shown in Figure 11.41B. Here $f^{(2)\prime}(x_i^*) = 0$ for the two stable fixed points. These are superstable fixed points of period 2. In between these stable fixed points is an unstable fixed point. All three points can be found by solving $f^{(2)}(x) = x$. By the chain rule

$$f^{(2)\prime}(x_1^*) = f^{(2)\prime}(x_2^*) = f'(x_1^*) f'(x_2^*). \tag{11.42}$$

The derivative of $f^{(2)}$ is the same at either stable fixed point. If we have a superstable fixed point, one of the derivatives must be zero; hence one of the points must be 0.5, since $f'(0.5) = 0$. The two stable points are 0.5 and 0.80902.

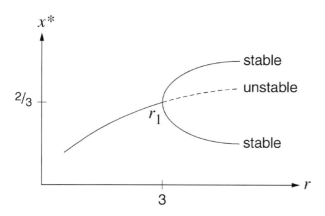

FIGURE 11.42
Pitchfork bifurcation.

What happens when r increases even further? As shown in Figure 11.43, each prong of the "pitchfork" bifurcates again, and the map has a 4-cycle (i.e., the period doubles again to four times its original value). As r continues to increase, the m-cycle periods are 8, 16, 32, 64, ... At $r = r_\infty = 3.5699456...$, the orbit has an infinite period: It is a chaotic orbit. Thus there is a sequence – a period-doubling cascade – of bifurcation points $\{r_1, r_2, r_3, ..., r_n, ..., r_\infty\}$. These are points where period-doubling bifurcations occur, leading to cycles of length 2^n (i.e., 2^n-cycles).

The superstable cycle (supercycle) of the same period occurs for r values just above the period-doubling bifurcations; there is a similar sequence of supercycle points $\{R_1, R_2, R_3, ..., R_n, ...\}$ with periods $2, 4, ..., 2^n, ...$. It will be proved that the ratios of successive

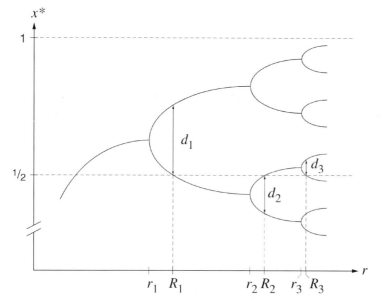

FIGURE 11.43
Schematic of a bifurcation diagram for the logistic map.

superstable intervals and the ratios of successive sizes of the "forks" at the supercycles (d_i in Figure 11.43) have a universal behavior seen in many physical systems, including the driven damped pendulum. This was discovered by Mitchell Feigenbaum in the 1970s. Figure 11.43 is sometimes called a "fig tree" diagram after him.

We can generalize what has been learned so far about the logistic map:

1. As the control parameter r is increased in the interval $3 < r < 3.5699456\ldots$, there is "period doubling" with "pitchfork bifurcations" increasing the number of m-cycles by a factor of 2 as each bifurcation occurs.
2. Fixed points x^* of the map for an m-cycle obey the equation

$$x = f^{(m)}(x), \tag{11.43}$$

with the mth iterate of the map defined by

$$f^{(m)}(x) \equiv \underbrace{f(f(\ldots f(x)\ldots))}_{m \text{ times}}. \tag{11.44}$$

They are the intersection of the line $y = x$ with the curve $y = f^{(m)}(x)$.

3. By the chain rule, the derivative of the mth iterate of the map at the fixed point is the product over the cycle of the derivatives at each fixed point in the cycle. This means that all the fixed points for a given value of r have the same derivative:

$$f^{(m)\prime}(x_i^*) = \prod_{\text{all f.p.}} f'(x_1^*) f'(x_2^*) \ldots f'(x_m^*). \tag{11.45}$$

4. *Stable* fixed points have $|f^{(m)\prime}(x^*)| < 1$. *Unstable* fixed points have $|f^{(m)\prime}(x^*)| > 1$. *Neutral* fixed points have $|f^{(m)\prime}(x^*)| = 1$. *Superstable* fixed points have $|f^{(m)\prime}(x^*)| = 0$ and include the point $x^* = \frac{1}{2}$, for which $f'(\frac{1}{2}) = 0$. Ultrarapid convergence of the map for points near the fixed points of supercycles is the result.

Lyapunov Exponent for the Logistic Map

In keeping with our general definition of Lyapunov exponent, we assume that the logistic map, after many iterations, leads to some kind of exponential behavior for the tangent map (i.e., the small deviations from the fixed points). For an m-cycle, we define the Lyapunov exponent λ from

$$x_{j+1} - x^* \to e^{\lambda m j}(x_1 - x^*) \quad \text{as } j \to \infty, \tag{11.46}$$

where x^* is any of the set $\{x_i\}$ of fixed points of $f^{(m)}$ and j counts the number of times we return to that fixed point. Thus defined, λ is a measurement of the amount of stretching per iteration, mj, since we need to iterate m times to cycle back to the fixed point, which we return to j times. Generalizing (11.41) to an m-cycle, we can iterate the $f^{(m)}$ map near x^*, now one of the m fixed points of $f^{(m)}$:

$$x_{j+1} - x^* = [f^{(m)\prime}(x^*)]^j (x_1 - x^*). \tag{11.47}$$

Equating (11.46) with (11.47), and using (11.45), we find the Lyapunov exponent to be

$$\lambda \equiv \frac{1}{m} \sum_{i=1}^{m} \log |f'(x_i^*)|. \tag{11.48}$$

We see immediately that, at a bifurcation, where $|f^{(2^n)\prime}(x_i^*)| = 1$, the Lyapunov exponent is zero. For a supercycle, $\lambda = -\infty$, since the first derivative of the map vanishes.

This definition (11.48) can be extended to the chaotic region. In this region, since the period is infinite, Equation (11.48) becomes an infinite sum over all points on the attractor:

$$\lambda \equiv \lim_{m \to \infty} \frac{1}{m} \sum_{i=1}^{m} \log |f'(x_i^*)|. \tag{11.49}$$

We must assume that this sum approaches a limit; in the computer we can take only a finite number of points. Although the existence of a limit is plausible, proving it is beyond the scope of this book.

What Feigenbaum Did: Universal Period Doubling and Renormalization

Fix your attention on the superstable points in the fig tree diagram of Figure 11.43. Feigenbaum showed that, in the limit $n \to \infty$, the ratios of the d_ns, $\frac{d_n}{d_{n+1}}$, and the ratios of superstable intervals, $\frac{R_{n+1}-R_n}{R_n-R_{n-1}}$, obey universal laws giving us two universal constants. Period-doubling cascades of this type are observed in many physical systems existing in nature, including not only mechanical systems capable of chaos, but chemical and biological systems as well.

We know that the slope of $f^{2^n}(x)$ is zero for $r = R_n$ and $x = \frac{1}{2}$. This suggests a certain similarity near the point $x = \frac{1}{2}$ for maps of different R_n and correspondingly different order. Let us test this out with $f^{(2)}$, $f^{(4)}$, $f^{(8)} \ldots f^{(64)}$. We will set the control parameter equal to R_n for the superstable 2^n-cycle in the test cases and tabulate the results for d_n, the distance from $\frac{1}{2}$ to the nearest "prong" of the pitchfork for the period 2^n.

The numerical results in Table 11.6 suggest that the R_n intervals shrink by a ratio that approaches a constant $\frac{1}{\delta} \approx .214$ as $n \to \infty$. There is also numerical evidence to believe

TABLE 11.6 SUPERSTABLE 2^n-CYCLES OF THE LOGISTIC MAP. NOTE THAT $d_n < 0$ IF THE OTHER BRANCH OF THE "PITCHFORK" IS LESS THAN $\frac{1}{2}$

n	2^n-cycle	R_n	$\frac{R_{n+1}-R_n}{R_n-R_{n-1}}$	d_n	$\frac{d_n}{d_{n+1}}$
1	2	3.2360679		.30901698	-2.65474
2	4	3.4985617	.213642	$-.11640185$	-2.53188
3	8	3.5546415	.214446	.04597440	-2.50867
4	16	3.5666676	.214192	$-.01832618$	-2.50411
5	32	3.5692435	.214216	.00731843	-2.50364
6	64	3.5697953		$-.00292311$	

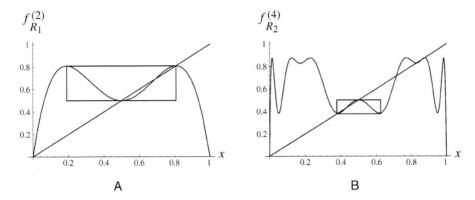

FIGURE 11.44
A) Supercycle of period 2. B) Supercycle of period 4. The line $y = x$ is also drawn in each case.

that a constant $\alpha \approx 2.5$ exists such that

$$-\alpha d_{n+1} \approx d_n \quad \text{and} \quad \lim_{n \to \infty} (-\alpha)^n d_{n+1} \to \text{a definite limit} \equiv d. \quad (11.50)$$

The constant α is the Feigenbaum constant. What is unexpected is that these scaling constants depend only on very general features of the function used in the mapping.

Next we outline Feigenbaum's method for calculating the value of the constant α using renormalization ideas. The self-similarity at successive supercycles is evident by comparing a portion of the curves $f^{(2)}(x)$ for the supercycle of period 2 (denoted by $f_{R_1}^{(2)}$) and $f^{(4)}(x) = f^{(2)}(f^{(2)}(x))$ for the supercycle of period 4 (denoted by $f_{R_2}^{(4)}$). The boxed parts of the two curves of $f_{R_1}^{(2)}$ and $f_{R_2}^{(4)}$ in Figure 11.44 appear to be related by a similarity transformation, in which the box for the period 2 curve is transformed by a scale demagnification and an inversion of both x and y into the box for the period 4 curve. The similarity includes both the iterated function curves and their intersection points with the line $y = x$. The box in Figure 11.44A is larger by a factor $-\alpha = -2.502807876$ than the box in Figure 11.44B. Only the regions inside the boxes, near $x = \frac{1}{2}$, $y = \frac{1}{2}$, are related in this way.

The reader should convince her- or himself that, given any function $g(x)$, the same function magnified with respect to the x and y axes by a factor β is given by the function $g_{\text{magnified}}(x) \equiv \beta g(\frac{x}{\beta})$. If β is negative, the axes are both inverted.

QUESTION 16: *Scaling Transformation* Prove that you get a function magnified in the x and y direction by using the transformation $g_{\text{magnified}}(x) \equiv \beta g(\frac{x}{\beta})$ on $g(x) = x$ and $g(x) = x^2$.

In order to concentrate on the region near the fixed point $x = \frac{1}{2}$, shift the origin to $(\frac{1}{2}, \frac{1}{2})$ by defining the function $y_n(u, R_n) \equiv f_{R_n}^{(2^n)}(\frac{1}{2} + u) - \frac{1}{2}$, where $u = x - \frac{1}{2}$. Hence $y_1(u, R_1)$ refers to the period 2 supercycle and $y_2(u, R_2)$ to the period 4 supercycle. Parameter u is restricted to the region inside the box in Figure 11.44. The property of self-similarity says that if we demagnify and invert the graph for $y_n(u, R_n)$ we obtain a function almost identical to $y_{n+1}(u, R_{n+1})$. If exact self-similarity existed, this would mean that

$$y_1(u, R_1) = -\alpha y_2\left(\frac{u}{-\alpha}, R_2\right) = -\alpha y_1\left[y_1\left(\frac{u}{-\alpha}, R_2\right), R_2\right]. \quad (11.51)$$

The right-hand side of (11.51) comes from the definition of y_n and the property that $f^{(2^{n+1})}$ is the functional composition of $f^{(2^n)}$ with itself. We might conjecture that the postulated scaling relationship (11.51) between period 2^n and 2^{n+1} holds for all supercycles:

$$y_n(u, R_n) = -\alpha y_{n+1}\left(\frac{u}{-\alpha}, R_{n+1}\right) \quad \text{(conjecture)}. \quad (11.52)$$

The self-similarity is, however, only true in the limit $n \to \infty$. Intuitively we might guess that the difference between R_n and R_{n+1} can be neglected in this limit. This would give us a self-similar functional equation of the form

$$y_\infty(u) = -\alpha y_\infty\left(y_\infty\left(\frac{u}{-\alpha}\right)\right). \quad (11.53)$$

As we will see below, this guess will turn out to be correct when the limit is taken more carefully.

Consider the iterated map $f_{R_{n+1}}^{(2^{n+1})} = f_{R_{n+1}}^{(2^n)}[f_{R_{n+1}}^{(2^n)}(x)]$. Two iterations of the $f_{R_{n+1}}^{(2^n)}$ map give us a whole period and hence return us to the starting point. We therefore call the $f_{R_{n+1}}^{(2^n)}$ map, the period 2 map at the supercycle $r = R_{n+1}$. It must be true therefore that one iteration of the period 2 map puts us on the other branch of the pitchfork containing $\frac{1}{2}$ (i.e., $f_{R_{n+1}}^{(2^n)}(\frac{1}{2}) = \frac{1}{2} + d_{n+1}$) (see Figure 11.43). Adopting the more compact notation introduced above, we say $d_{n+1} = y_{n+1}(0, R_{n+1})$, while $0 = y_n[y_n(0, R_{n+1}), R_{n+1}]$. These last equations say that iterating the period 2 map once takes us to $u = d_{n+1}$. Iterating the period 2 map twice brings us back to $u = 0$. Consider the set of all such period 2 maps for supercycles with periods 2^n. Take the limit of these with the appropriate magnification as $n \to \infty$. This limit defines the new function $g_1(u)$:

$$g_1(u) \equiv \lim_{n \to \infty} (-\alpha)^n y_n\left(\frac{u}{(-\alpha)^n}, R_{n+1}\right). \quad (11.54)$$

The limiting process is not restricted to period 2. We can consider the period 2^i sequences for each $n > i$ supercycle, and take the limit of these as $n \to \infty$. Define in this way a whole class of functions, one for each integer value of the index $i = 1, 2, \ldots$:

$$g_i(u) \equiv \lim_{n \to \infty} (-\alpha)^n y_n\left(\frac{u}{(-\alpha)^n}, R_{n+i}\right). \quad (11.55)$$

That $r = R_{n+i}$ means that there is a superstable $2^{n+i} = 2^n 2^i$ cycle. Starting from $x = \frac{1}{2}$, $f_{R_{n+i}}^{(2^n)}$ must be iterated 2^i times to return to $x = \frac{1}{2}$. We are assuming that, as $n \to \infty$, the maps of fixed period 2^i approach a self-similar limit in a constantly shrinking region near $x = \frac{1}{2}$ (i.e., $u = 0$).

For a concrete example, consider the maps of period 4 ($i = 2$). We want to compare the sequence of functions

$$-\alpha y_1\left(\frac{u}{-\alpha}, R_3\right), \quad (-\alpha)^2 y_2\left(\frac{u}{(-\alpha)^2}, R_4\right), \quad \ldots$$

Next, using the functional properties of the map, g_{i-1} can be expressed in terms of g_i:

$$g_{i-1}(u) = -\alpha g_i\left[g_i\left(\frac{u}{-\alpha}\right)\right]. \tag{11.56}$$

To prove (11.56), return to the definition of g_i as the limit in Equation (11.55) above. We must also recall that, for the same $r = R_{n+i}$ value, $y_n(u, R_{n+i}) = y_{n-1}[y_{n-1}(u, R_{n+i}), R_{n+i}]$. Then

$$\begin{aligned}
g_{i-1}(u) &= \lim_{n\to\infty} (-\alpha)^n y_n\left(\frac{u}{(-\alpha)^n}, R_{n+i-1}\right) \\
&= \lim_{n\to\infty} (-\alpha)(-\alpha)^{n-1} y_n\left(\frac{u}{(-\alpha)(-\alpha)^{n-1}}, R_{n+i-1}\right) \\
&= \lim_{n\to\infty} (-\alpha)(-\alpha)^{n-1} y_{n-1} \\
&\quad \times \left[\frac{1}{(-\alpha)^{n-1}} \underbrace{(-\alpha)^{n-1} y_{n-1}\left(\frac{1}{(-\alpha)^{n-1}} \frac{u}{(-\alpha)}, R_{n+i-1}\right)}, R_{n+i-1}\right] \\
&= -\alpha g_i\left[g_i\left(\frac{u}{-\alpha}\right)\right].
\end{aligned} \tag{11.57}$$

The "underbrace" identifies $g_i(\frac{u}{-\alpha})$, in the limit $n \to \infty$. What Equation (11.56) says is that we must iterate a period 2^i subsequence twice to generate the period 2^{i-1} subsequence for any value of n. For example, if $i = 2$, any g_2 map must be iterated four times to give the full period of the 2^n period map, for a given n value. We must iterate g_2 twice to generate the g_1 map, which must be iterated itself twice to get the 2^n period map.

Now take the limit as $i \to \infty$ to obtain the fixed point equation:

$$g(u) = (-\alpha)g\left[g\left(\frac{u}{-\alpha}\right)\right]. \tag{11.58}$$

If the self-similar function defined by (11.58) is known on a small, but finite interval, it can be scaled using Equation (11.58) to find the function for a larger interval.

For numerical answers, assume that $g(u)$ has the power series expansion

$$g(u) = 1 + b_1 u^2 + b_2 u^4 + \cdots. \tag{11.59}$$

In actual calculations, we keep terms up to u^{16}. The Taylor series expansion of the left side minus the Taylor series expansion of the right side of (11.58) is evaluated at $u = 0$,

TABLE 11.7 TAYLOR SERIES FOR UNIVERSAL SCALING FUNCTION $g(u)$

Order of Term (i)	Coefficient of u^{2i}	Our Result	Feigenbaum's Result*
0	defined as 1	1	1
2	b_1	-1.52763	-1.527632997
4	b_2	1.04816×10^{-1}	$1.048151943 \times 10^{-1}$
6	b_3	2.6705×10^{-2}	$2.670567349 \times 10^{-2}$
8	b_4	-3.52686×10^{-3}	$-3.527413864 \times 10^{-3}$
10	b_5	8.12044×10^{-5}	$8.158191343 \times 10^{-5}$
12	b_6	2.55029×10^{-5}	$2.536842339 \times 10^{-5}$
14	b_7	-2.65167×10^{-6}	$-2.687772769 \times 10^{-6}$
16	b_8	-6.27444×10^{-8}	—

*Feigenbaum's data were obtained from Feigenbaum, M. J., *J. Stat. Phys.*, **21**, p. 699ff, 1979.

with as many even-order derivatives kept as the unknowns $\alpha, b_1, \ldots, b_8$. A reasonable guess is made, and the nine nonlinear polynomial equations in nine unknowns are solved numerically. The result for α is $\alpha = 2.50291$. We can also check that $\alpha = -\frac{1}{g(1)}$ directly from the power series. The power series coefficients $b_1, \ldots, b_8$ are tabulated in Table 11.7. With more extensive calculations, Feigenbaum obtained $\alpha = 2.502907876$. His values are also listed in Table 11.7. An accurate knowledge of $g(u)$ can be obtained in the interval $0 \le u \le 1$ by using the information in Table 11.7.

Feigenbaum showed that the constant α and the function $g(u)$ are *universal* for all maps $f(x)$ having a quadratic "bump" (i.e., maxima) in the middle. For example, a measurement of the intervals between drips of a leaking faucet, when each succeeding time interval is plotted as a function of the preceding interval, yields a "map" of this type with a period-doubling sequence whose scaling constant (now named after Feigenbaum) is $\alpha = 2.502907876\ldots$. This prediction has been verified by actual experiments on leaky faucets. It also applies to the example we have used of the forced damped pendulum, where the map connects succeeding points on a Poincaré section. There are many other examples of "period doubling" universality to be found in nature.

Feigenbaum also showed* that the R_n obey an asymptotic scaling equation for large n:

$$R_n = R_\infty - \frac{C}{\delta^n}. \tag{11.60}$$

Here C is a constant specific to the map, which can be calculated, and $\delta = 4.6692016\ldots$ is another universal Feigenbaum scaling constant, which is actually related to α, being approximately $\alpha^2 - \alpha$. Note that $\frac{1}{\delta} = 0.214169\ldots$ agrees quite well with the apparent limit of the ratios of successive $R_n - R_{n-1}$ intervals listed in Table 11.6.

This is not where the similarity between the logistic map and the forced damped pendulum ends. We close this section by showing the bifurcation plot for the logistic map

---

* Feigenbaum, M. J., *J. Stat. Phys*, **19**, p. 25ff, 1987.

APPENDIX THE LOGISTIC MAP

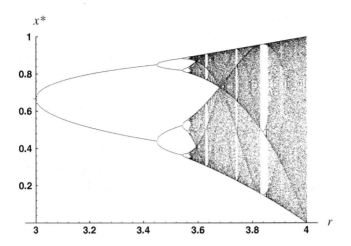

FIGURE 11.45
Bifurcation diagram for logistic map.

in Figure 11.45. The reader is invited to compare this plot with Figure 11.24 for the pendulum, which was obtained by integrating the equations of motion for a particular choice of Q. In the bifurcation plot for the logistic map, we vary the control parameter r between $r = 3$ and $r = 4$, whereas in the pendulum case we varied the driving force strength over an appropriate range.

As with the pendulum, the chaos abruptly ends and a period 3 "window" appears for $r = 1 + \sqrt{8}$, and then again the motion becomes chaotic. In Figure 11.46 we magnify the center of the period 3 window ($3.84 < r < 3.855$ and $0.425 < x^* < .575$). It is a tiny fraction of the bifurcation diagram. One sees in Figure 11.46 a miniature copy, almost an exact replica, of the whole diagram for $3 \leq r < 4$ (Figure 11.45). The structure of Figure 11.45 is very complicated. There are an infinite number of windows with periodic orbits of arbitrary period mixed in among the chaotic bands. All this complexity arises from iterating a simple quadratic map function!

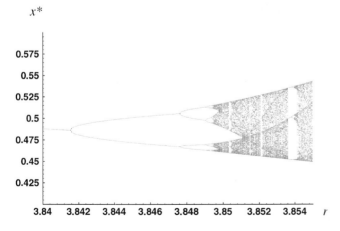

FIGURE 11.46
Magnification of Figure 11.45.

There are several references where a more extensive discussion of the logistic map can be found. A general reference is *Fractals, Chaos, Power Laws*, by Manfred Schroeder. More technical is the textbook *Chaos in Dynamical Systems*, by Edward Ott. Another good reference is the book *Deterministic Chaos – An Introduction*, by H. G. Shuster (second revised edition).

Maps are an example of "dynamics" with a discrete time variable. The logistic map is the simplest possible nontrivial map. And yet the full complexity seen in the bifurcation diagram for the damped driven pendulum is reproduced in the bifurcation diagram for the logistic map. This reveals that chaos is a general property of many dynamical systems, even the simplest ones, and is not due to a specific choice of physical example.

CHAPTER TWELVE

SPECIAL RELATIVITY

OVERVIEW OF CHAPTER 12

By the end of the nineteenth century it had gradually become clear that a contradiction existed between the classical mechanics of Newton and the electromagnetic theory of Maxwell. Both theories were brilliantly successful when applied in their respective areas of natural phenomena, but it was impossible to reconcile them. The problem became apparent when Maxwell predicted that light was an electromagnetic wave which moved through space at the fixed velocity of $c \approx 3 \times 10^8$ m/s. His theory connected the experimental value of c with a number that could be determined from constants obtained from table-top experiments involving static electric and magnetic fields in capacitors and inductors. We quote from Maxwell's original paper on the subject:

> In air or vacuum $\mu = 1$, and therefore ... V = 193,088 mi/s. The velocity of light in air, as determined by M. Fizeau, is ... V = 195,647 mi/s. The velocity of transverse undulations in our hypothetical medium, calculated from the electro-magnetic experiments of MM. Kohlrausch and Weber, agrees so exactly with the velocity of light calculated from the optical experiments of M. Fizeau, that we can scarcely avoid the inference that *light consists in the transverse undulations of the same medium which is the cause of electric and magnetic phenomena.*

(Note: Maxwell calculated the rate of propagation of transverse vibrations through an elastic medium which he called the "ether.")

In the late nineteenth century physicists imagined that light, like sound, must travel through a medium. It was difficult for them to conceive how energy could be transmitted through a vacuum. If a medium for transmitting light did exist, then it should be possible to detect the Earth's motion through the ether, unless the ether was "dragged" along with the moving Earth. However, neither the motion of the Earth nor the "ether-dragging" existed, as was shown by experiments. A puzzle emerged, causing a crisis among theoretical physicists. In the homework problems we will investigate some of the experiments that revealed this crisis.

Albert Einstein had wondered since the age of 16 what it would be like to travel at a velocity of c beside a light wave. Would one observe "frozen" static electric and magnetic fields? According to Maxwell's theory, such fields could not exist in a vacuum without charges present. In 1905, Einstein solved the problem, which turned out to be a misunderstanding of the nature of time itself. Newton's concept of absolute time was replaced by a concept of time that can be different for different observers. Simultaneity turned out to be relative: Events occurring at different places, which appear to be simultaneous to one observer, are not simultaneous to a second observer moving with respect to the first one.

Einstein constructed his theory, known as the Special Theory of Relativity* from only two postulates:

> 1. *No observer can detect his absolute velocity through space. Physics is the same in all inertial frames.* This is the *Principle of Relativity* as first clearly stated by Galileo. It remains a basic principle in Einstein's theory as well.
> 2. *The speed of light is the same for observers in different inertial frames.*

According to the second postulate, the velocity of light, in contrast to the velocity of sound in air, *does not depend in any way on the relative motion between the observer and the source of the light wave*. To see how counterintuitive this result is, imagine that a very fast airplane is landing at an airport with its landing lights turned on. Both the pilot and the air traffic controller on the ground would see the light traveling at the same speed c. A second airplane traveling toward the first one would also see the light moving at c in its own reference frame!

We have so much faith in Einstein's theory that c is no longer an experimental quantity. The velocity of light was *defined* to be 299,792,458 m/s in 1983 during the 17th General Conference on weights and measures. Since the second is defined to be 9,192,631,770 periods of radiation from a particular radiative transition in a Cesium-133 atom, the meter can be experimentally determined from these two constants.

On this foundation Einstein built a theory that has withstood many experimental tests. The speed of light c is the "speed limit" which can never be exceeded. In some modern high energy accelerators, the velocity of electrons differs from c by only an amount $<5 \times 10^{-11} c$, and yet the predictions of special relativity are seen to be perfectly valid.

This chapter will reveal the far-reaching consequences of Einstein's two postulates for physics. Newton's mechanics will be seen to be an approximation valid only for slowly moving objects ($v \ll c$). However, the Lagrangian technique and the variational principles we have developed from Newtonian mechanics can still be used effectively for Einstein's *relativistic mechanics*.

You may have had some previous exposure to special relativity. We think, nevertheless, that considerable benefit is obtained from following the logic closely, rather than accepting relativity as dogma and proceeding directly to the mathematical formalism. It is easy to lose sight of the physics in this way. In fact, Poincaré and Lorentz had discovered the correct mathematical formulas before Einstein! Poincaré even understood the importance of the two postulates. He introduced the concept of

---

* The General Theory of Relativity was developed later to incorporate gravity.

12.1 SPACE-TIME DIAGRAMS

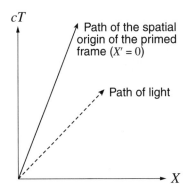

FIGURE 12.1
Space–time plot in the unprimed frame.

"local time" in order to explain the impossibility of detecting motion through the ether. The reason why Einstein is credited with discovering relativity is only revealed if the mathematical formalism plays a secondary role to the physical interpretation in the story which follows. The most elegant way to formulate relativity theory mathematically was developed by Minkowski. We will introduce the modern version of this formalism only after the logical development of the physics is outlined.

12.1 SPACE–TIME DIAGRAMS

It has been our previous custom to plot displacement on the Y axis versus time on the X axis. To conform to the accepted practice when discussing relativity, we will instead plot ct, which has the dimensions of distance, along the Y axis and displacement x along the X axis. The displacement versus time plot of a light wave that starts out at $t = 0$ and moves down the $+X$ axis is represented by the dashed line in Figure 12.1. The slope of this line is unity owing to the units we have chosen. The successive crests of the light wave (maxima of the electric field) could be shown as lines parallel to the dashed line and displaced up the time (cT) axis. For simplicity, only one crest of the wave is shown as it moves down the $+X$ axis at the velocity of light.

To distinguish this particular "reference frame" or coordinate system, we will call it the "unprimed" frame. A second reference frame moves at a constant velocity v with respect to the unprimed frame. This "primed" frame is also an inertial frame, with space and time coordinates x', t'. *For any observer in the primed frame, the same light wave will also have a unit slope* (second postulate). If the spatial origins of the two frames coincide at $t = t' = 0$, then $x' = 0$ implies that $x = 0$ at $t = 0$. (Newton assumed absolute time everywhere in space. Except at the "event" (point in the space–time diagram) when the two origins coincide, we will allow for the possibility that $t' \neq t$.) A point fixed at the spatial origin in the primed coordinate system will obey the equation $x = vt = (\frac{v}{c})ct$, as measured in the unprimed system. If we define the dimensionless ratio $\beta \equiv \frac{v}{c}$ and if $v < c$, the slope of the space–time line of the primed system's spatial origin in the unprimed coordinates is $\frac{1}{\beta} > 1$, as shown by the slanted solid line in Figure 12.1.

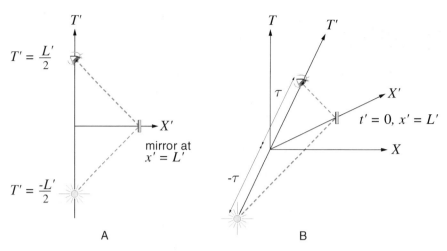

FIGURE 12.2
A) Reflection in a mirror (primed system). B) Space–time plot in the unprimed system.

We want to be able to convert one observer's description into the other observer's description of any event or sequence of events (trajectory in the space–time diagram).* If we remain sitting on the spatial origin of the primed reference system, $x' = 0$, then the only variable is t'. Thus the space–time trajectory ("world line") of $x' = 0$ is the time axis in the primed reference system. Different points along this trajectory correspond to different values of t' – perhaps ticks on a clock located in the origin of the primed system. These events appear to be at different places, moving with a uniform velocity v in the unprimed system.

What about the space–time trajectory of the X' axis (i.e., all the points along the X' axis at the moment $t' = 0$ in the primed frame)? Imagine that a mirror is set up at some distance L' from the origin. At the time $t' = -\frac{L'}{2c}$ a light wave emerges from the origin $x' = 0$, travels to the mirror at $x' = L'$, is reflected, and returns to the origin at $t' = +\frac{L'}{2c}$. A space–time plot of this sequence of events in the *primed* observer's reference frame is shown in Figure 12.2A.

From now on, when we refer to "time" t or t' in either frame, we will mean ct or ct', respectively. This is equivalent to adopting units in which $c = 1$, and we thereby avoid having to write factors of c everywhere. In these units, the velocity of the primed frame from the perspective of the unprimed frame is β. We will revert to putting in the c explicitly later, when it appears to make things clearer.

In the unprimed frame, the path of the light ray must have a slope of unity, by Einstein's second postulate. We do not yet know the scale factor from one coordinate system to the other, but we know the space–time trajectory of the T' axis in the unprimed system has a

---

* The other two space dimensions y, z are omitted in this discussion for the sake of simplicity. Only one-dimensional motion is considered at first. The more general case will be considered later.

12.1 SPACE–TIME DIAGRAMS

slope, by Einstein's second postulate, of $\frac{1}{\beta}$. Let us assume the light starts at the unknown[*] time $-\tau$ down the time (T') axis, travels with unit slope to the point $x' = L'$, $t' = 0$, and returns, again with unit magnitude slope to the unknown, but symmetrically located, time τ on the T' axis. Using this information, the same sequence of events graphed above in Figure 12.2A in the primed system is also graphed in Figure 12.2B in the *unprimed* system.

Extending the two light propagation lines (heavy dashes) from the points $\pm \tau$ along the T' axis gives a unique point in the space–time diagram where the light changes direction. This is the point $x' = L'$, $t' = 0$ as plotted by the unprimed observer. If now the location of the mirror in the primed frame is varied (vary L'), we trace out a line that must represent the locus of all points with $t' = 0$ and hence is the X' axis as labeled in the diagram. We can find the equation of this line and hence its slope.

Using the notation (x, t) for the coordinates of any event on the diagram, the starting point of a light ray is a vector of length τ along a line with slope $\frac{1}{\beta}$; hence $\tau = -\frac{\tau}{\sqrt{1+\beta^2}}(\beta, 1)$. We add to this a vector whose length we will determine and whose slope is unity: $r(1, 1)$. This takes us to the point where the light reverses direction. It then follows a path $s(-1, 1)$ to the point $\frac{\tau}{\sqrt{1+\beta^2}}(\beta, 1)$. Here r, s are considered to be the unknowns. They can be found by solving two simultaneous equations obtained from the vector equation

$$-\frac{\tau}{\sqrt{1+\beta^2}}(\beta, 1) + r(1, 1) + s(-1, 1) = \frac{\tau}{\sqrt{1+\beta^2}}(\beta, 1). \tag{12.1}$$

This equation expresses in a formula the path of the heavy broken line shown in Figure 12.2B. Solving the vector Equation (12.1) gives $r(\tau)$ and $s(\tau)$ explicitly:

$$r = \frac{1+\beta}{\sqrt{1+\beta^2}}\tau, \quad s = \frac{1-\beta}{\sqrt{1+\beta^2}}\tau. \tag{12.2}$$

The unknown value of τ is a function of the mirror distance from the origin in the primed frame, $\tau = \tau(L')$. The point at which the light is reflected is given in unprimed coordinates by

$$(x_{\text{reflected}}, t_{\text{reflected}}) = \frac{\tau}{\sqrt{1+\beta^2}}(1, \beta). \tag{12.3}$$

Equation (12.3) implies that the X' axis is the straight line with slope β passing through the origin. This is the solid line labeled X' in Figure 12.2B. We still do not know the actual connection between the two sets of coordinates, because we do not know the single unknown function $\tau(L')$, except that $\tau(0) = 0$. We will see in the homework problems that τ is proportional to $\frac{L'}{2}$, but the proportionality factor is not one.

The two primed coordinate axes appear as oblique straight lines in the unprimed system. They are not perpendicular. Their slopes are reciprocals. If α is the angle between these

---

[*] In the primed system this time is $-\frac{L'}{2}$, but we do not yet know the space–time distance τ that this corresponds to in the space–time diagram for the unprimed observer.

two primed axes viewed in the unprimed system, it is a straightforward geometric exercise to show that

$$\tan\frac{\alpha}{2} = \frac{1-\beta}{1+\beta}. \tag{12.4}$$

As the velocity v increases towards c, $\beta \to 1$. The angle between X' and T' axes, as viewed in the unprimed frame, decreases, and both axes approach a line with unit slope passing through the origin.

QUESTION 1: *Mirror Reflection 1* Which of the dashed lines in Figure 12.2B corresponds to the vector $r(1, 1)$? Which to $s(-1, 1)$? Also, derive Equations (12.3, 12.4).

QUESTION 2: *Mirror Reflection 2* Explain the physical meaning of the space–time distance $\tau \neq \frac{L'}{2}$. How could you prove that it is proportional to L'?

12.2 THE LORENTZ TRANSFORMATION

We can now find the explicit transformation rules for coordinates between the two frames. We already know that the T' axis as it appears in the unprimed frame is proportional to the vector $(x, t) \sim (\beta, 1)$, while the X' axis lies along the vector $(1, \beta)$ in the unprimed frame. The general transformation must be a linear one, since the coordinate axes transform into straight lines. Let us express an arbitrary point (x, t) in the unprimed coordinates as a linear combination of vectors $(1, \beta)$ and $(\beta, 1)$ along the X' and T' axes respectively:

$$(x, t) = \gamma(\beta)(1, \beta)x' + \tilde{\gamma}(\beta)(\beta, 1)t', \tag{12.5}$$

which can be written as two equations:

$$x = \gamma(\beta)x' + \tilde{\gamma}(\beta)\beta t', \quad t = \gamma(\beta)\beta x' + \tilde{\gamma}(\beta)t'. \tag{12.6}$$

The two scale factors $\gamma(\beta)$, $\tilde{\gamma}(\beta)$ are still unknown, but the transformation must have this general form. As indicated, the scale factors can depend on velocity β.

The situation between the two frames is symmetrical. If the unprimed observer sees the primed frame moving down the $+X$ axis at velocity β, the primed observer must see the unprimed observer moving down the X' axis at a velocity $-\beta$. This means that the transformation from $(x, t) \to (x', t')$ should have the same form as (12.6), except for replacing β by $-\beta$:

$$x' = \gamma(-\beta)x - \tilde{\gamma}(-\beta)\beta t, \quad t' = -\gamma(-\beta)\beta x + \tilde{\gamma}(-\beta)t. \tag{12.7}$$

We will also use a symmetry argument to prove that γ and $\tilde{\gamma}$ are even functions of β. Looking at Figure 12.2B, we see that reversing β just flips the diagram around the T

12.2 THE LORENTZ TRANSFORMATION

(vertical) axis, so τ remains the same. If we try to calculate τ from the transformation (12.6), we set $x' = 0$ and get the vector $x = \tilde{\gamma}(\beta)\beta t', t = \tilde{\gamma}(\beta)t'$. Now reverse the sign of β. This new vector, which represents the new T' axis, will not become a mirror image of the old T' axis under the transformation $\beta \to -\beta$ unless $\tilde{\gamma}$ is an even function of β. We can find a similar relation for γ by using (12.7). Thus $\gamma(\beta) = \gamma(-\beta)$ and $\tilde{\gamma}(\beta) = \tilde{\gamma}(-\beta)$.

Using this property, along with substituting (12.6) into (12.7), we obtain two equations each with x' and t' on the left side and zero on the right side. Since x' and t' may take on arbitrary values, all four coefficients of x', t' must vanish in the two equations. From one of these equations, we obtain the two conditions

$$(\gamma^2 \beta - \tilde{\gamma}\gamma\beta) = 0, \quad (1 - \tilde{\gamma}^2 + \gamma\tilde{\gamma}\beta^2) = 0. \tag{12.8}$$

We can solve (12.8) to obtain a unique form for the scale factors $\gamma, \tilde{\gamma}$ and a unique form for the transformation law:

$$\gamma = \tilde{\gamma} = \frac{1}{\sqrt{1-\beta^2}}. \tag{12.9}$$

The other two coefficients from the other equation will also vanish if (12.9) is true, so all the conditions are met by (12.9). The transformation laws we have just found are:

$$\boxed{\begin{aligned} x' &= \gamma(x - \beta t), & t' &= \gamma(t - \beta x), & (12.10) \\ x &= \gamma(x' + \beta t'), & t &= \gamma(t' + \beta x'). & (12.11) \\ & \text{Lorentz transformation} & & \end{aligned}}$$

If the velocity between the two frames $\beta = 0$, it is true that $\gamma(0) = \tilde{\gamma}(0) = 1$ as expected. One can also see from the formula (12.9) that if the velocity were greater than c (i.e., $\beta > 1$) complex coordinates would be obtained – a nonsensical result. Later we will see why it is impossible to accelerate a massive particle to $\beta \geq 1$.

QUESTION 3: *Lorentz Transformations* Compare (12.10) to (12.11). What is the difference? How do you know which form to use? For a particular situation of two different observers moving at relative velocity β with respect to each other, how do you know what the sign of β is in the transformations? Hint: Look at the origin of the other observer's coordinate system. It that motion consistent with the transformation you picked?

Now we will use the transformation law to discuss some relativistic effects.

1. *Time dilation – moving clocks slow down:* Take a clock in the primed system which is permanently attached to the origin, $x' = 0$. By the transformation law of Equation (12.10), $x = \beta t$ for the unprimed observer – the clock location moves down

the X axis with velocity β as expected. Looking at the second equation, we obtain a curious effect: $t' = \frac{1}{\sqrt{1-\beta^2}}(t - \beta^2 t) = \sqrt{1-\beta^2}\, t$. If we mark off equal time intervals on the T' axis, the elapsed time intervals in T are a factor of $\gamma > 1$ longer. The moving clock slows down, and this effect is called *time dilation*. Any process will be lengthened in time by the same factor of γ. This effect is due to the nature of time itself, not changes in forces, etc. There are many practical consequences and the effect has been observed in a multitude of different experimental circumstances. For example, a radioactive particle moving rapidly will have a longer lifetime than it would at rest. Any process that did not slow down in exactly this way could be used to determine one's absolute velocity, in violation of the first of Einstein's postulates.

> **QUESTION 4: *Time Dilation*** In what sense do we mean the "moving" clock slows down? In terms of two observers moving with velocity β with respect to each other, whose clock does the first observer see slow down? How about the second observer?

2. *Lorentz contraction – moving rods become shorter*: A related effect is the shortening of moving rods. Suppose that we measure the position of the two ends of a rod at the same time in the primed frame, say $\Delta t' = 0$, and that $\Delta x' = D'$, the length of the rod. Using (12.11), the difference in the x coordinates *measured at the same time in the unprimed frame*, say $\Delta t = 0$, is $\Delta x = D = \sqrt{1-\beta^2}\, D'$. The rod is shortened by a factor of $\sqrt{1-\beta^2} = \frac{1}{\gamma} < 1$. One must understand that the two measurements of length in the two frames are different measurements. To determine D', we must find the position of the ends of the rod at a certain time t'. To find D, we have to measure the positions in the unprimed frame at a certain fixed value of t. Due to the relativity of simultaneity, these times are not the same in the two frames.

3. *The relativity of simultaneity – events simultaneous in one frame are not in another*: To see this effect from the transformation law, we will further consider the measurement of the positions of the ends of a rod of length D' at the time $\Delta t' = 0$. In the unprimed frame, using the transformation law (12.11), the time interval between the measurements made in the primed frame is $\Delta t = \gamma \beta D' = \frac{\beta}{1-\beta^2} D$. The departure from simultaneity of the two measurements in the primed frame, as seen by the observer in the unprimed frame, is proportional to the separation of the ends of the rod. In the last formula, we have expressed this in terms of the (contracted) length of the moving rod, as seen by the unprimed observer.

These effects are reciprocal, since either frame may be considered to be at rest. Each observer sees the other's moving clocks slow down and measuring rods contract. But whose clock is *really* slow? This kind of question will be shown to arise because of a failure to specify exactly how the clock rate is determined. The correct answer is that each clock slows down to the other observer in a predictable way that depends only on the relative velocity of the clock and the observer. We will deal with another similar paradox, the so-called "twin paradox," in the Appendix.

4. *The transformation of space–time areas*: The Jacobian of the transformation from the unprimed to the primed coordinates is the matrix of partial derivatives:

$$\mathbf{J}(\beta) = \begin{pmatrix} \gamma & -\beta\gamma \\ -\beta\gamma & \gamma \end{pmatrix}. \qquad (12.12)$$

The determinant of this Jacobian is 1. For any transformation, the determinant gives the relation between an "area" in (x, t) space and the corresponding area in (x', t') space:

$$dx'\, dt' = \text{determinant}[\mathbf{J}]\, dx\, dt = dx\, dt. \qquad (12.13)$$

A square with sides of length 1 in the unprimed frame transforms into a parallelepiped with elongated sides of length greater than 1 in the primed frame. Equation (12.13) means that we can confidently assert that the parallelepiped's area is 1 in the primed frame. *Requiring* the Jacobian to have unit determinant could have been an alternative way to find the velocity dependence of the gamma factor, since the reciprocal relationship between the two frames must preserve the space–time area in making the transformation. If this were not the case, there would not be complete symmetry, and one inertial frame would be different from the other. Looking backwards would not have been equivalent to looking forwards.

The symmetry between space and time in the Lorentz transformation led Minkowski to propose that nature is four rather than three dimensional, with the four dimensions being (ct, x, y, z). A Lorentz transformation bears more than a passing resemblance to a rotation in this 4-D space. To quote from Minkowski's very famous 1908 Cologne lecture:*

The views of space and time which I wish to lay before you have sprung from the soil of experimental physics and therein lies their strength. They are radical. Henceforth space by itself and time by itself, are doomed to fade away into mere shadows, and only a kind of union of the two will preserve an independent reality.

12.3 SIMULTANEITY IS RELATIVE

The relativistic effects discussed above (time dilation, length contraction, simultaneity of relativity) can also be proved using space–time diagrams.

For example, we can perform an experiment illustrating how different observers will come to different conclusions about whether or not two spatially separated events are simultaneous. Arrange two light sources along the X axis, one at $x = D$ (called source B) and the other at $x = -D$ (called source A). Let each source emit a pulse of light which is detected at the origin ($x = 0$) at a time $t = 0$. The two flashes of light are seen

---

* Quoted by A. Pais, '*Subtle Is The Lord...*' *The Science And Life Of Albert Einstein*, p. 152.

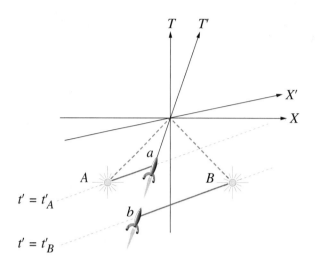

FIGURE 12.3
Finding the supernova times and distances.

simultaneously by an observer located at the origin in the unprimed frame. This observer then calculates from the distance and the velocity of light when the flashes actually occurred. If you want to add a little drama, assume that we are the unprimed observers, and that we observe two supernovas located 180° apart, each at a distance of 10 light years from us. The times at which the flashes occurred are calculated to be $t_{A,B} = -D$ earlier (remember $c \equiv 1$). We conclude that the time interval $\Delta t \equiv t_B - t_A = 0$. Both stars exploded at the same time, according to us. The two explosions are represented in Figure 12.3 on a space–time diagram in the unprimed frame.

A rocket ship passes by the Earth, moving towards light source B and away from light source A at a velocity β. At the particular moment when both flashes are detected, the rocket ship is at $x = x' = t = t' = 0$. The crew of the rocket ship also sees both flashes simultaneously and calculates when the supernovas exploded. The conclusion from Figure 12.3 is that the rocket ship crew will determine that supernova A exploded at time $t'_A > t'_B$, the time supernova B exploded, even though $t_A = t_B$. Were the explosions simultaneous or not? Both the Earth-bound observers and the rocket ship crew are right. There is no preferred inertial frame.

Did A's explosion trigger an explosion in B? A rocket ship traveling in the opposite direction would conclude that explosion B occurred before explosion A, which makes it impossible that explosion A caused explosion B. This paradox is resolved by the fact that *no physical influences can be propagated faster than the velocity of light*. (See the homework problem on tachyons, however.) Neither explosion could have caused the other. Any signal propagated from A to B would be a line of unit slope, which, if drawn on Figure 12.3, must have originated long before the explosion of A in order to reach B.

How far were the stars from the rocket ship observers at the different times they exploded? (The stars are moving at $-\beta$ to the observers in the rocket ship if they are stationary for us.) The distances to the rocket *in the frame of the rocket ship observers*

12.4 WHAT HAPPENS TO y AND z IF WE MOVE PARALLEL TO THE X AXIS?

at the two different moments of explosion are shown as heavy solid lines in Figure 12.3. They differ just by the same ratio as the transit times of the light. Since both stars move at the same velocity in the primed frame, they remain the same distance apart at all times. The rocket crew explains that they see the flashes at the same time, because although the supernovas occurred at different times, the light traveled different distances to them from each star. (In Figure 12.3, line Aa is shorter than line Bb.)

QUESTION 5: *Simultaneity* Explain how Figure 12.3 shows that events that are simultaneous, but spatially separated, in the unprimed frame are not simultaneous in the primed frame.

QUESTION 6: *Space–Time Diagram* Draw and describe a space–time diagram for the supernova explosions from the viewpoint of the primed frame. Indicate on your diagram the unprimed T and X axes and the events A and B.

12.4 WHAT HAPPENS TO y AND z IF WE MOVE PARALLEL TO THE X AXIS?

We haven't yet discussed what happens to y and z, the two coordinates transverse to the motion down the X axis. Start in the primed frame and there construct a clock consisting of two parallel mirrors, separated along the Y' axis by a distance $\frac{\Delta y'}{2}$. A pulse of light is emitted from a hole in one of the mirrors, travels to the second mirror, and is reflected. The light travels back to the first mirror and the time interval of the round trip is recorded, while a new light pulse is emitted. The clock "ticks" every $\Delta t' = \frac{\Delta y'}{c}$ seconds.

Now observe this same clock from the unprimed frame. The clock moves down the X axis with a velocity β. Since the light returns to the same place in the primed frame (call it $x' = 0$) every time the clock ticks, the Lorentz transformation predicts $\Delta t = \gamma \Delta t'$. This result must be independent of whatever happens to the distance between the mirrors under the transformation. How does the unprimed observer calculate the interval between ticks? He sees that the light travels along a vertical distance Δy and a horizontal distance $v\Delta t$. From the Pythagorean theorem, the total distance traveled by the light in the unprimed frame is $\sqrt{\Delta y^2 + v^2 \Delta t^2}$. Since the unprimed observer also thinks that the velocity of light is c, the time Δt is just this total distance divided by c. Solving for Δt (and setting $c = 1$ again), we obtain

$$\Delta t = \frac{1}{\sqrt{1 - \beta^2}} \Delta y. \qquad (12.14)$$

From the time dilation predicted by the Lorentz transformation we also know

$$\Delta t = \gamma \Delta t' = \gamma \Delta y'. \qquad (12.15)$$

Upon comparing (12.14) with (12.15), we conclude $\Delta y' = \Delta y$. Lengths perpendicular to the direction of motion are not affected by the transformation:

$$y' = y, \quad z' = z. \tag{12.16}$$

transformation of perpendicular lengths

Qualitatively this makes sense because if something moves down the X axis, only x is a function of t, and y and z are constant. There is no comparable problem with the measurement of simultaneous events separated in the y and z directions as exists from separation in the x direction.

12.5 VELOCITY TRANSFORMATION RULES

As a direct consequence of the Lorentz transformation, Equations (12.10, 12.16), we can deduce the relationship between the velocity of a moving object in one frame and the velocity of the same object in a second frame. Say a particle moves down the X' axis in the primed frame with a velocity $u'_x \equiv \frac{dx'}{dt'}$. A formula for $u_x \equiv \frac{dx}{dt}$ can be derived from the Lorentz transformation by replacing x by dx, etc. and dividing the equation for the transformation to dx by the transformation to dt. We obtain

$$u_x = \frac{u'_x + v}{1 + \frac{v u'_x}{c^2}}. \tag{12.17}$$

This formula agrees with various intuitive limits. If $u'_x = 0$, then $u_x = v$ as it must. If $u'_x = c$, then $u_x = c$ also. Anything that moves at the velocity of light in one frame must do so in the other frame as well. This result applies to a massless neutrino as well as to a light wave, so perhaps we should not refer to c as the "velocity of light" but instead as the "ultimate speed." We can jokingly refer to the formula (12.17) when $u'_x = c$ as $c + c = c$.

What about velocity in other directions? Employing the Lorentz transformation in differential form again we have

$$u_y = \sqrt{1 - \beta^2} \, \frac{u'_y}{1 + \frac{v u'_x}{c^2}}. \tag{12.18}$$

A similar equation applies to u_z. These formulas are much more complicated than the prerelativity Galilean transformation laws $u_x = u'_x + v$, $u_y = u'_y$, etc. In particular, transverse velocities are no longer invariant from one frame to another. A particle heading straight up the Y' axis is seen to have a y velocity slower by the factor $\frac{1}{\gamma}$ in the unprimed frame. This is due to time dilation. In the limit where $c \to \infty$ (i.e., all velocities are small compared to c), we recover the nonrelativistic Galilean formulas from the correct relativistic ones.

QUESTION 7: *Velocity Transformation Rules* Derive (12.17, 12.18) explicitly using $u'_x \equiv \frac{dx'}{dt'}$ and $u_x \equiv \frac{dx}{dt}$, etc. Be careful though, because there is a factor of $c = 1$ hidden in these derivatives. Where is it?

12.6 OBSERVING LIGHT WAVES

Our next goal is to find out how observers in relative motion will differ on their description of the energy and frequency of a light wave.

Transformation Law for Frequency of a Light Wave

Although the speed of a light wave is the same for all observers, the color is not. There is a Doppler shift towards the blue (higher frequency) if the observer travels towards the oncoming wave and towards the red (lower frequency) if the observer travels in the direction of the light wave. We may picture the crests of the transverse electric field in the wave on the space–time diagram of Figure 12.4. These wave crests are the two-dimensional surfaces of constant phase, which become lines when projected into the (x, t) plane.

At $x = 0$, the crests are separated by 2π in phase. They occur at the times t_n given by $\omega t_n = -2\pi n$ ($n = -2, -1, 0, 1, 2, \ldots$) or $t_n = -\frac{n}{f}$, as shown in Figure 12.4. Here f is the frequency of the wave, as observed by the unprimed observer. The phase is constant along a line of slope $+1$, so that the equation of the nth crest with phase $2\pi n$ is $(x - t)_n = \frac{n}{f}$.

In Figure 12.5, this situation is plotted on an X, Y projection of ordinary 3-D XYZ space, from the perspective of two different observers, each viewing the *same* physical light

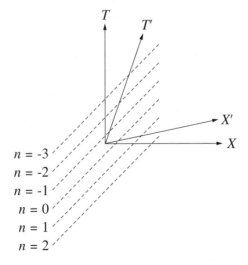

FIGURE 12.4
Lines of constant phase along a light wave: $x - t = \frac{n}{f}$.

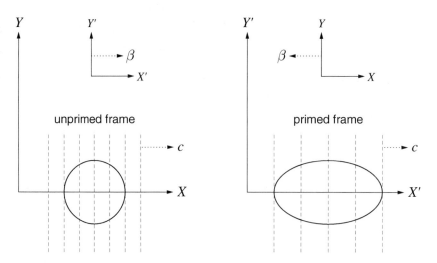

FIGURE 12.5
A light wave as seen by two observers in relative motion.

wave. The wave moves at c to the right in both reference frames. The observer on the right side of Figure 12.5 is moving in the direction of the wave, *relative* to the observer on the left. For the observer on the right, the wave appears to be "stretched out," and the distance between wave crests seems longer than it does to the observer on the left. (Compare the X and X' axes in Figure 12.4.) In the left frame, we draw an imaginary sphere of radius R enclosing part of the light wave. The sphere encloses five waves (along the X or X' axis). The center of the sphere coincides with one of the wave crests and also moves to the right at velocity c. A certain volume of physical space is enclosed by the sphere, which can be viewed by any observer. Since the waves are spaced further apart according to the primed observer, the sphere in the unprimed (left) frame becomes a prolate ellipsoid in the primed (right) frame. In the next section we will calculate the energy of the light contained within the physical volume of this imaginary sphere or ellipsoid. Here we will calculate exactly how the frequency of the light depends on the observer.

We can see from Figure 12.5, that an observer moving along with the light at a relative velocity $\beta < 1$, will see $f' < f$, because the wavelength $\lambda' > \lambda$. This makes intuitive sense: If one runs in the direction of a light wave, the waves are spaced further apart. Each observer will say that the nth crest of the wave obeys the equation (for $y = z = 0$)

$$(x - t)_n = \frac{n}{f}, \quad (x' - t')_n = \frac{n}{f'}. \tag{12.19}$$

Using the Lorentz transformation (12.10) to express x', t' in terms of x, t, we find

$$x' - t' = \sqrt{\frac{1+\beta}{1-\beta}} (x - t). \tag{12.20}$$

12.6 OBSERVING LIGHT WAVES

The Doppler shift formula for frequency transformation is therefore

$$f' = \sqrt{\frac{1-\beta}{1+\beta}}\, f. \tag{12.21}$$

This agrees with our intuitive guess that $f' < f$ if the primed frame moves with respect to the unprimed frame along the direction of the light wave.

→ *Note:* It is wrong to regard the left (unprimed) frame as "the rest frame" and the primed frame to be "the moving frame." Either frame is at rest, if you are an observer in that frame. It is the other frame that is moving. The situation is completely symmetrical. Both observers look at the same physical situation: a light wave moving to the right. The ratio of observed frequencies depends on the relative motion of the two observers.

QUESTION 8: *Intrinsic Frequency* Is it correct to say that an electromagnetic wave has no intrinsic frequency? Is the only difference between a radio wave and an x-ray beam in the viewpoint of the beholder? Would the same statement apply to sound waves in air from a train whistle? Explain.

QUESTION 9: *Frequency Transformation* Two observers are in relative motion with respect to each other, and both observe a light wave moving to the right. How would you determine the ratio of the observed frequencies? Consider how you might use formula (12.21) for all relative motion situations.

Transformation Law for Energy of a Light Wave

We are interested in proving that the energy in a light pulse transforms just like its frequency. We will use the classical theory of electric and magnetic fields created by charges and currents. Assume that the light passes through a vacuum. The energy in a light beam is proportional to $(\mathcal{E}_{\text{light}}^2 + \mathcal{B}_{\text{light}}^2)V$, where V is the volume, $\mathcal{E}_{\text{light}}$ is the electric field amplitude of the light wave, and $\mathcal{B}_{\text{light}}$ is the magnetic field amplitude.* In the vacuum, $\mathcal{B}_{\text{light}} = \mathcal{E}_{\text{light}}$, so the energy is proportional to $\mathcal{E}_{\text{light}}^2 V$. How do electric field amplitude $\mathcal{E}_{\text{light}}$ and volume V in a light pulse transform between inertial frames? This is no longer a question merely of kinematics, but involves dynamics, since electric fields cause forces on

---

* In previous chapters, we have used SI units for the electromagnetic fields. In this chapter, we find it more convenient to avoid the extra constants by using cgs Gaussian units. We hope this does not lead to unnecessary confusion for the reader. In the cgs Gaussian system, electric and magnetic fields have the same units. The Lorentz force exerted by a magnetic field is proportional to $\frac{\vec{v}}{c} \times \vec{B}$ rather than $v \times \vec{B}$, as in the SI or MKS system. The density of energy contained in the fields is $\frac{1}{8\pi}(\vec{\mathcal{E}} \cdot \vec{\mathcal{D}} + \vec{\mathcal{H}} \cdot \vec{B})$. The momentum density (Poynting vector) is $\frac{c}{4\pi} \vec{\mathcal{E}} \times \vec{B}$. Also, in a vacuum, $\vec{\mathcal{D}} = \vec{\mathcal{E}}$ and $\vec{B} = \vec{\mathcal{H}}$.

charged particles. Force lies at the root of dynamics, so we must discover how transverse and longitudinal forces will transform from one frame to another.*

How Do Electric and Magnetic Forces Transform?

The easiest way to discover how electromagnetic forces transform is to use a simple example – the transformation of forces produced by a parallel plate capacitor. We will need to make three assumptions in order to proceed. First, as a consequence of the first postulate, all forces transform in the same way, so we only need to find the transformation law for our capacitor example and then we can apply it to the light pulse. If all forces do not transform in the way that our example does, then we could set up a hypothetical situation where the electric force was exactly balanced in one reference frame by a mechanical force such as a spring. Let us transform both forces to a second frame, moving with respect to the first frame. If electric and mechanical forces transform differently, they would no longer balance when viewed by an observer in the second frame, so acceleration would result. This would provide a mechanism for determining one's absolute velocity – which is forbidden by the first postulate. Hence all forces must obey the same transformation law. The second assumption that we will make is that the electromagnetic force in any frame can be calculated from the Lorentz force

$$\vec{F} = e\left(\vec{\mathcal{E}} + \frac{\vec{v}}{c} \times \vec{B}\right). \tag{12.22}$$

We will prove this later in the chapter. The third assumption is that charge is a Lorentz invariant (i.e., does not change under Lorentz transformations). If it were otherwise, neutral matter would become charged while in motion.

Figure 12.6 shows the two cases we will consider. In both cases, the parallel plate capacitor has a charge $+Q$ on the top plate and $-Q$ on the bottom plate and is moving with velocity β in the $+x$ direction with respect to the unprimed (space) frame. The primed frame is attached to the capacitor so it is also moving with velocity β with respect to the unprimed frame. The electric field points in the $-y$ direction in the first case (perpendicular to the motion) and in the $+x$ direction in the second case (parallel to the motion).

First we will consider the case where the electric field is perpendicular to the direction of motion as shown in Figure 12.6A. Viewed in the primed frame, the electric field between the plates is a static field

$$\vec{\mathcal{E}}'_{\perp,\text{capacitor}} = -\frac{Q}{A'}\hat{j}' \tag{12.23}$$

with A' being the area of the plates in the primed frame, and $\hat{j}'$ being a unit vector along the Y' axis. Viewed from the unprimed frame, the Lorentz contraction has reduced the

---

* Einstein deduced the transformation of the electric and magnetic fields from the requirement that Maxwell's equations retain their form under Lorentz transformations (we will not explicitly do this here, however).

12.6 OBSERVING LIGHT WAVES

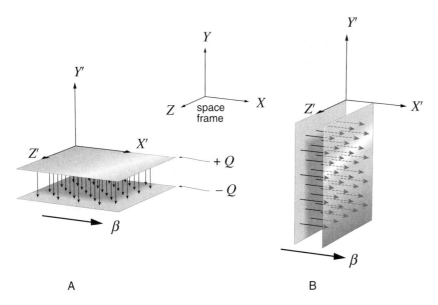

FIGURE 12.6
A) Perpendicular electric field. B) Parallel electric field.

length of the side in the x direction and hence the area of the plates by $A = \frac{A'}{\gamma}$. This means that the electric field, proportional to $\frac{Q}{A}$ in the unprimed frame, is increased by a factor of gamma when compared with the primed frame:

$$\vec{\mathcal{E}}_{\perp,\text{capacitor}} = \gamma \vec{\mathcal{E}}'_{\perp,\text{capacitor}}. \tag{12.24}$$

In the unprimed frame there is also an electric current caused by the fact that the charges on the plates of the capacitors are moving. The current is to the right ($+x$ direction) on the top plate and to the left ($-x$ direction) on the bottom plate. This causes a magnetic field in the $-z$ direction in between the plates, which can be calculated using the Biot–Savart law to obtain $\vec{\mathcal{B}} = \frac{\vec{v}}{c} \times \vec{\mathcal{E}}$:

$$\vec{\mathcal{B}}_{\perp,\text{capacitor}} = -\beta |\vec{\mathcal{E}}_{\perp,\text{capacitor}}| \hat{k}, \tag{12.25}$$

where $\hat{k}$ is a unit vector along the Z axis. Equation (12.25) relates the magnitude of the magnetic field to the magnitude of the electric field, both of which are in a frame (unprimed) in which the capacitor is moving.

We will use a test charge of magnitude e, a moving electron at rest with respect to the capacitors, to determine the electric force from the Lorentz force law (12.22). In the primed frame

$$\vec{F}'_{\perp} = e\vec{\mathcal{E}}'_{\perp,\text{capacitor}}. \tag{12.26}$$

And in the unprimed frame

$$\vec{F}_\perp = e\left(\vec{\mathcal{E}}_{\perp,\text{capacitor}} + \frac{\vec{v}}{c} \times \vec{B}_{\perp,\text{capacitor}}\right) = e(1-\beta^2)\vec{\mathcal{E}}_{\perp,\text{capacitor}}. \tag{12.27}$$

Using (12.24), we find

$$\vec{F}'_\perp = \vec{F}'_\perp \sqrt{1-\beta^2}. \tag{12.28}$$

According to Einstein's first postulate, transverse forces will always transform in this way.

Next we will consider the case where the electric field is parallel to the direction of motion as shown in Figure 12.6B. The electric field will be the same in the primed frame as in the transverse case (12.23) but in the x' direction. Since the capacitor plates are moving perpendicular to the direction of motion, the area of the plates must remain the same. Therefore, the electric field in the unprimed frame is the same as in the primed frame:

$$\vec{\mathcal{E}}_{\parallel,\text{capacitor}} = \vec{\mathcal{E}}'_{\parallel,\text{capacitor}}. \tag{12.29}$$

Similarly, the electric current in the unprimed frame caused by the moving plates will be the same as in the transverse case (in the $+x$ direction on the top plate and in the $-x$ direction on the bottom plate), but the net effect will be a cancellation of the magnetic field. Thus

$$\vec{B}_{\perp,\text{capacitor}} = \vec{B}'_{\perp,\text{capacitor}} = 0. \tag{12.30}$$

Finally, using the Lorentz force law (12.22), we see that

$$\vec{F}_\parallel = \vec{F}'_\parallel. \tag{12.31}$$

QUESTION 10: *Fields in Motion* Qualitatively explain why it is that for an electric field moving perpendicular to the motion, the moving charges add up to create a current that causes a magnetic field perpendicular to both the electric field and the direction of motion. Why does this not happen for an electric field parallel to the direction of motion?

How the Electric Field Transforms for a Light Pulse

In a light beam, observed from *any* inertial frame, the electric and magnetic fields are of equal magnitude; they are perpendicular to each other and to the direction in which the light beam is moving. Let us consider the force on a test charge *at rest* in the primed frame, which is moving with velocity $\vec{\beta}$ with respect to the unprimed frame, in the same direction as the light wave, as pictured in Figure 12.7. Since the charge is at rest in the primed frame, the only force comes from the electric field: $F'_y = e\mathcal{E}'_{\text{light}}$. Keeping in mind that $\vec{\mathcal{E}}_{\text{light}}, \vec{B}_{\text{light}}, \vec{\beta}$ are mutually perpendicular and that $\mathcal{E}_{\text{light}} = B_{\text{light}}$, the Lorentz force in the unprimed frame is $F_y = \vec{F}_\perp = e(1-\beta)\mathcal{E}_{\text{light}}\,\hat{j}$. From the transformation rule for transverse force (12.28), we can deduce the relation between the amplitude $\mathcal{E}_{\text{light}}$ of the

12.6 OBSERVING LIGHT WAVES

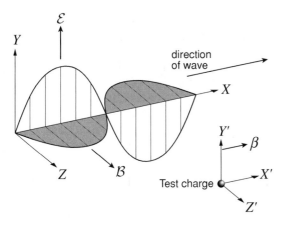

FIGURE 12.7
Electromagnetic wave showing electric and magnetic fields in space.

light wave in the unprimed frame and the amplitude $\mathcal{E}'_{\text{light}}$ in the primed frame:

$$\mathcal{E}'_{\text{light}} = \sqrt{\frac{1-\beta}{1+\beta}}\, \mathcal{E}_{\text{light}}. \tag{12.32}$$

This relation applies to light wave amplitudes only. It differs from the transformation of the static field of a capacitor owing to the presence of a magnetic field in both frames.

How the Volume Changes

As noted above, the energy of a light wave is proportional to the square of the amplitude times the volume V of the light pulse:

$$E \sim \mathcal{E}_{\text{light}}^2 V, \quad E' \sim \mathcal{E}_{\text{light}}'^2 V'. \tag{12.33}$$

To prove the transformation law for light pulse energy, all that remains is to find how a spherical volume enclosing part of the light wave in the unprimed frame is regarded in the primed frame. The equation of the surface of a sphere of radius R moving at velocity $c = 1$ down the X axis in the unprimed frame is

$$\frac{(x-t)^2}{R^2} + \frac{y^2}{R^2} + \frac{z^2}{R^2} = 1. \tag{12.34}$$

The volume of this sphere is $\frac{4\pi}{3} R^3$. In the primed frame, using the Lorentz transformation rules (12.20), the same sphere is an ellipsoid, with the equation

$$\left(\frac{1-\beta}{1+\beta}\right) \frac{(x'-t')^2}{R^2} + \frac{y'^2}{R^2} + \frac{z'^2}{R^2} = 1. \tag{12.35}$$

Contrary (perhaps) to one's intuition, the moving sphere becomes *elongated* along the direction of motion (see Figure 12.5). Calculating the volume of this ellipsiod we find

$$V' = \sqrt{\frac{1+\beta}{1-\beta}}\, V. \tag{12.36}$$

Putting (12.36, 12.32) together into (12.33), we finally arrive at the conclusion that the energy transforms like (12.21):

$$E' = \sqrt{\frac{1-\beta}{1+\beta}}\, E. \tag{12.37}$$

Note that Equation (12.37) is not applicable to energy in general. It *only* applies to light waves.

From the perspective of quantum theory, light consists of a stream of photons. All observers must count the same number of photons in a given beam of light. The formula $E = nhf$ is the Planck energy–frequency relation for n photons. The result we have obtained shows that Planck's constant must be a Lorentz invariant (i.e., the same for all observers).

12.7 WHAT IS MASS?

Thus far we have been concerned with lengths and times as viewed from different inertial frames. This is the kinematics of special relativity. It was Einstein who realized first that a new view on the transformation of space and time implies a new dynamics as well. At the heart of dynamics lies the concept of mass. We would like to review this fundamental concept here.

Prior to relativity, mass and force were introduced into physics by Newton's Second Law. Momentum is defined as the product of an intrinsic property of material bodies, called mass, and the velocity of that body in an inertial reference frame. Force is defined as the time derivative of momentum. Force is independent of any specific inertial frame, according to Newton. In a vague way, mass is related to "quantity of matter," but a more precise definition of mass is possible.

Say that by means of a weight and a pulley, two different masses m_1 and m_2 are subjected to the same force. It is observed experimentally that if the accelerations of two bodies, a_1 and a_2 respectively, acted on by the same force are compared, the ratio $\frac{a_1}{a_2}$ *does not depend on this applied force but is intrinsic to the bodies themselves.* It also is observed to be independent of any specific inertial frame. We can determine the mass ratio by taking the ratio of accelerations. By taking an arbitrary standard object for the unit mass (the standard kilogram in Paris, for example), we can determine the mass of any object. The force can be defined quantitatively as $F \equiv m_1 a_1 = m_2 a_2$. In this way of looking at

Newton's Second Law,* the experimental content is the independence of acceleration ratios on the reference frame and on the force applied – no matter whether it is a gentle constant force or an explosion lasting only a brief time. This is an experimental result, leading to the definition of force, a definition that agrees with our intuitive feeling that we need to apply as much force to accelerate two kilograms 1 m/s^2 as one kilogram 2 m/s^2. Classical pre-Einstein mechanics makes the additional assumption that if we combine two masses, the resultant object will have a mass that is the sum of the masses of its constituents. Mass is assumed to be an additive property of matter.

For masses moving at speeds close to the speed of light, this experimental result simply is not correct. Different bodies subject to identical forces will have accelerations that depend on their velocities. As a consequence, observers viewing this experiment from the vantage points of different inertial frames *will not agree on the ratio of the accelerations nor on the applied force*. Newton's Second Law is an approximation valid only for slowly moving objects. Also, in the Theory of Special Relativity, the mass of a combined object can be either more or less than the sum of the constituent masses. For example, in a fusion reaction the mass of the final product is less than the sum of the initial product. The energy emitted is the source of fusion energy. The nonadditivity of mass is below the limit of experimental observation for chemical reactions, but it can easily be observed in nuclear reactions, for example.

We modify the definition of mass by restricting our observation to systems in the limit of very small velocities. The mass ratio is defined as

$$\frac{m_1}{m_2} \equiv \lim_{v_1,v_2 \to 0} \frac{a_1}{a_2}. \qquad (12.38)$$

This will give us the mass of a system nearly at rest, which is referred to as the *rest mass*. This is independent of the motion of the body because it is defined in the rest frame of that body. In some textbooks and in the popular literature on relativity, one sees the statement that "relativistic mass" increases with velocity. *This corresponds to a different definition of "mass," which we will not use here.* By our definition, rest mass is a relativistic invariant (i.e., is the same for all velocities). Observers in all inertial frames will agree on the rest mass of a body. With this definition of mass, we cannot simply assume that momentum is $m\vec{v}$ but must discover the relativistic definition from more general principles. One bonus will be that momentum and energy conservation will be unified into a single conservation law, at the price of altering the definition of momentum. Another bonus will be a more general definition of force.

12.8 REST MASS IS A FORM OF ENERGY

By the end of the nineteenth century, many forms of energy were recognized, and the principle of energy conservation was believed to be one of the fundamental laws of

---

* This viewpoint is due to Mach, *The Science of Mechanics*. Mach had a great influence on Einstein.

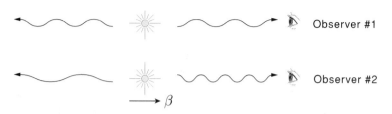

FIGURE 12.8
Atom emitting two light beams along $\pm X$ axis viewed by two observers in relative motion. Observer #1 is in the unprimed (rest) frame and observer #2 is in the primed frame.

physics. Even from 1775, the French Academy refused to consider further any proposed solutions to create perpetual motion machines. The steam engine was recognized by Mayer in 1842 as the creation of mechanical work due to the flow of heat energy from high to low temperature. Conservation of energy thus became elevated to a basic law: Energy can neither be created nor destroyed but only transformed. Mass also seemed (prior to 1905) to be indestructable, since matter itself could neither be created or destroyed, although it could be converted into different forms by chemical reactions or phase transitions such as freezing or boiling. Physicists at that time saw no connection between the two concepts of mass and energy other than that kinetic energy is proportional to mass, but kinetic energy was recognized to be only one of many possible forms of energy.

What Einstein showed is that the rest mass of a body that receives energy *in any form* will increase. The idea that the mass of a body could be changed by altering its energy content – heating it, for example – was to lead to a revolution in physics. This result is perhaps the most dramatic and the most important consequence of special relativity.

We will begin our derivation of relativistic dynamics with the hypothetical physical example originally proposed by Einstein.* Consider an atom emitting light in two directions along the $\pm X$ axes as shown in Figure 12.8. (You can think of two photons, but it isn't necessary to do so. The main idea is to avoid any momentum change in the atom (recoil) due to the light emission.) We will observe this atom both before and after it emits the light. In the first reference frame it remains at rest throughout the whole process. We will also observe the same emission process from a second reference frame, in which the atom moves at velocity β, both before and after emitting the light beams. (Note that this implies that the second observer moves with a velocity $-\beta$ with respect to the first observer.)

In the rest frame (observer #1), the total energy lost by the atom in the form of light is ΔE. Half of this energy goes to the right, half to the left. If the initial energy of the atom is E_0, by energy conservation, the final energy of the atom must be $E_1 = E_0 - \Delta E$.

Now view this same process from the second frame (observer #2), in which the atom has velocity β. The initial energy will now include some kinetic energy, which depends

---

* "Does the Inertia of a Body Depend upon Its Energy Content?," A. Einstein, *Annalen der Physik* 18 (1905), English translation in *The Collected Papers of Albert Einstein,* Volume 2, translated by A. Beck, pp. 172–174.

12.8 REST MASS IS A FORM OF ENERGY

on the velocity β. Call the initial energy $E'_0(\beta)$. The final energy is $E'_0(\beta) - \Delta E'$. We can express $\Delta E'$ in terms of β and ΔE using (12.37). The light emitted to the right will be Doppler-shifted towards the blue, because the atom is moving towards an observer on the right, along the X axis. Like the whistle of an oncoming train, the frequency or pitch grows higher. This half of the light energy is $\Delta E'_{\text{right}} = \sqrt{\frac{1+\beta}{1-\beta}} \frac{\Delta E}{2}$. Interchanging β and $-\beta$, we also calculate $\Delta E'_{\text{left}} = \sqrt{\frac{1-\beta}{1+\beta}} \frac{\Delta E}{2}$. Adding these two together, we obtain the total light energy in the primed frame:

$$\Delta E' = \Delta E'_{\text{right}} + \Delta E'_{\text{left}} = \frac{1}{\sqrt{1-\beta^2}} \Delta E. \tag{12.39}$$

QUESTION 11: Atoms Emitting Light Beams Check that the result in (12.39) is correct. Why is it that a single beam of light has a different energy transformation law than two oppositely directed light beams? Speculate on how this could be true, and give a physical reason. Hint: What is the total momentum carried by the oppositely directed light beams? Does a single beam of light carry momentum?

We can write the equations of energy conservation in the two frames and compare them:

$$E_1 = E_0 - \Delta E, \quad E'_1 = E'_0 - \gamma \Delta E. \tag{12.40}$$

Since energy conservation must apply to observers in any inertial frame, both equations in (12.40) must be true. Therefore

$$(E'_0 - E_0) - (E'_1 - E_1) = (\gamma - 1) \Delta E. \tag{12.41}$$

Except for a possible additive constant C, $(E'_0 - E_0)$ is the kinetic energy K'_0 of the atom moving at velocity β before the light is emitted, while $(E'_1 - E_1) = K'_1 + C$ is the kinetic energy of the same atom, still at rest in the unprimed system, after the light has been emitted. Hence the change in kinetic energy, viewed from the primed frame, is

$$K'_0 - K'_1 = (\gamma - 1) \Delta E. \tag{12.42}$$

Could (12.42) be true only for energy loss by light emission and not for other changes in energy? It is a critical point of the argument that, unless this transformation rule for energy changes were universally true (of any system, not just atoms), one could detect one's absolute velocity. An experiment could be conducted in which another kind of energy was added to the atom, then radiated as light. Two different observers would disagree as to whether the atom really returned to its original state unless *all energy transforms according to (12.42)* (only for zero net momentum, however).

Equation (12.42) was derived by using only the assumptions of energy conservation and the electromagnetic properties of light waves, which were well-known in Einstein's time. The physics comes in the interpretation of (12.42). It was Einstein's genius to recognize that the transformation of *all* kinds of energy changes is given by the universal rule (12.42) and that the simplest interpretation is then that energy *itself* must obey

$$E'_0 = \gamma E_0. \tag{12.43}$$

Similarly, $E'_1 = \gamma E_1$. Equation (12.43) then implies Equation (12.42). Recall that E'_0 contains the kinetic energy of the moving atom.

Let us look at the case when β is very small. We know then that the kinetic energy must take the Newtonian form $\frac{1}{2}mv^2 = \frac{1}{2}mc^2\beta^2$. For small β, expand γ in a power series:

$$\gamma \equiv \frac{1}{\sqrt{1-\beta^2}} = 1 + \frac{\beta^2}{2} + \frac{3\beta^4}{8} + O[\beta^6]. \tag{12.44}$$

For velocities encountered in everyday life, we can approximate γ by taking only the first two terms in (12.44). For small velocities, the kinetic energy before emission must be (from (12.43))

$$K'_0 = E'_0 - E_0 = \frac{1}{2}E_0\beta^2 + O[\beta^4]. \tag{12.45}$$

If we equate the term proportional to β^2 in (12.45) to $\frac{1}{2}m_0v^2$, it must be true that

$$E_0 = m_0 c^2. \tag{12.46}$$

It is hard to overestimate the revolutionary importance of (12.46): An atom at rest contains an energy proportional to its rest mass m_0. *Rest mass and rest energy are in fact identical, differing only by the numerical factor c^2.* Doing a similar calculation for K'_1, after emission one obtains $E_1 = m_1 c^2$. Thus we see that since the energy decreased in the rest frame of the atom, so does its rest mass. When we extend the theory to include molecules, the rest mass enables us to calculate binding energy. Thus Einstein's simple assumption helped to greatly increase the understanding of atomic physics. In 1905, it was beyond experimental verification to detect any increase in the mass of a body to which energy was added. In his original paper* Einstein says:

> Perhaps it will prove possible to test this theory using bodies whose energy content is variable to a high degree (e.g., salts of radium).
>
> If the theory agrees with the facts, then radiation transmits inertia between emitting and absorbing bodies.

---

* "Does the Inertia of a Body Depend upon Its Energy Content?," A. Einstein, *Annalen der Physik* 18 (1905), English translation in *The Collected Papers of Albert Einstein*, Volume 2, translated by A. Beck, pp. 172–174.

A proton (p) and an antiproton ($\bar{p}$) each have the same mass, about 1.67×10^{-24} g. These particles may sometimes annihilate into two gamma rays (high frequency light photons), which are pure energy.* In both cases, nothing of the original mass remains. According to Einstein's famous formula (12.46), for $\bar{p}p$ annihilation the energy of each photon is $1.67 \times 10^{-24}(3 \times 10^{10})^2 = 1.5 \times 10^{-3}$ ergs (cgs units) or 1.5×10^{-10} joules (mks units). Many experiments have measured this annihilation radiation. The existence of antimatter was unknown to Einstein in 1905.

From (12.43) and (12.46), we have found the connection between the *total* energy (rest plus kinetic) in an inertial frame and in the rest frame:

$$E'_0 = \gamma E_0 = \gamma m_0 c^2. \tag{12.47}$$

This result must be a general one for all massive particles. It must be valid for all matter and all forms of energy, except for those with zero rest mass (which we will discuss later).

12.9 HOW DOES MOMENTUM TRANSFORM?

Consider again the case of the atom that emits two oppositely directed beams of light, each with the same energy $\frac{\Delta E}{2}$ in the atom's rest frame as shown in Figure 12.8. We discovered that the mass changes from $m_0 = \frac{E_0}{c^2}$ to $m_1 = \frac{E_1}{c^2}$. What does this tell us about how momentum changes? In the frame in which the atom is initially at rest, since the two light beams are oppositely directed, the atom must remain at rest after the light is emitted. (This follows by a symmetry argument. Which way would it go?) Viewed from the primed frame (atom initially moving), assuming that total momentum is conserved, the two beams of light must carry off the momentum lost by the atom. Light transmits momentum as well as energy and thus can exert a force known as radiation pressure. Radiation pressure is a very small effect in the laboratory, although it can be detected. It is the reason the tails of comets point away from the Sun. (The same radiation pressure has been proposed as a means of "sailing" through space, by erecting a giant mirror of aluminized mylar.)

We must make one assumption:[†] that the momentum of a light wave[‡] is just its energy divided by c:

$$\Delta p_{\text{light}} = \frac{\Delta E_{\text{light}}}{c}. \tag{12.48}$$

---

* When the annihilation occurs, the emission of two photons is a relatively rare occurrence, since there are many other possibilities. For an electron–positron annihilation at rest, two photons occur 25% of the time. The total energy released is much less than for $\bar{p}p$ annihilation by the mass ratio of electron mass to proton mass – about $\frac{1}{1876}$.

[†] Note that we need not make this assumption if we use the Poynting vector from electromagnetic theory.

[‡] When we refer to the momentum carried by a light wave, we always mean a pulse of light moving in a certain direction, such as a beam of laser photons. A spherical wave can be thought of as made up of an infinite number of beams pointing in all directions; hence it carries energy but not momentum.

(Note: On dimensional grounds energy divided by c has the units of momentum. Up to a numerical factor, this must be the form of the light wave momentum. The fact that the numerical factor is one can be derived using Maxwell's equations and has to be taken on faith here.) Using this assertion, we can calculate the total momentum of the light as seen in the primed frame by taking the vector sum of the light momentum on the right and the left:

$$\Delta p'_{\text{total light}} = \Delta p'_{\text{light,right}} - \Delta p'_{\text{light,left}}$$
$$= \frac{\Delta E}{2c}\left(\frac{\sqrt{1+\beta}}{\sqrt{1-\beta}} - \frac{\sqrt{1-\beta}}{\sqrt{1+\beta}}\right) = \frac{\Delta E}{c}\frac{\beta}{\sqrt{1-\beta^2}}. \qquad (12.49)$$

This result must apply to the case of two-photon annihilation for a proton–antiproton system as well. In their common rest frame before annihilation, the total energy (keep them some distance apart, so that there is no interaction energy) is $2m_p c^2$, where m_p is the mass of a proton or an antiproton. The mass of the system m_0 is then $m_0 = 2m_p$. After the annihilation into two photons, all rest mass is destroyed. The two photons must have a total energy, by energy conservation, of $m_0 c^2$. From a moving observer's perspective, the total energy, either before or after the annihilation event is $\gamma m_0 c^2$. From (12.49), the total momentum of the two photons after the annihilation is $\gamma\beta m_0 c$.

Table 12.1 gives "before annihilation" and "after annihilation" momentum and energy. Assume that the unknown momentum of the material system before annihilation is p'_0. The second row is determined only by the known momentum–energy relations and transformation properties of light waves. The quantity $p'_0(\beta) - \gamma\beta m_0 c$ will equal zero if momentum is conserved as well as energy. Therefore the momentum along the X axis of the material system must be (generalized to any direction)

$$\boxed{\vec{p}'_0 = \gamma\vec{\beta}m_0 c.} \qquad (12.50)$$

This is the relativistic velocity dependence of momentum. However, note that we only considered how momentum changes in the direction of motion. How about the momentum transverse to the motion? In a homework problem it will be proved that the situation is analogous to the transformation of length – transverse momentum is not changed by a Lorentz transformation.

Combining Equation (12.47) and (12.50) we obtain $(E'_0)^2 - (p'_0 c)^2 = (m_0 c^2)^2$. This equation looks better if we revert to units in which $c = 1$ and drop the subscripts and

TABLE 12.1 ANNIHILATION AS SEEN BY A MOVING OBSERVER

System	Energy	Momentum
proton + antiproton	$\gamma m_0 c^2$	p'_0
two photons ($\pm x$ directions)	$\gamma m_0 c^2$	$\gamma\beta m_0 c$

primes. It becomes

$$E^2 - p^2 = m^2. \qquad (12.51)$$

The combination $E^2 - p^2$ is an invariant under Lorentz transformation and equals the square of the rest mass. Once again, we have proved a general result for all material objects, assuming only the properties of light waves. By this approach our theory is guaranteed to treat the two forms of energy and momentum in a consistent manner.

> **QUESTION 12:** *Momentum Conservation*   Verify the law of momentum conservation in the original Einstein example of atomic emission of two light beams, by using (12.50).

One might argue that the assumption of energy and momentum conservation in physical processes amounts to a third fundamental postulate, since it cannot be derived from the two postulates of Einstein. We do not wish to discuss this philosophical point further but only remark on it in passing. We could have made use of collisions between particles to derive the transformation laws for energy and momentum, without involving any properties of light waves. We have avoided doing this for two reasons: The algebra involved is somewhat messy, obscuring the physics, and secondly, the forces at the time of impact produce violent accelerations. Collisions of fast charged particles would not behave like ideal billiard ball collisions in real life. We will give some homework problems to work things out in this way.

Perhaps the most subtle part of the reasoning in the section above involves assuming that we can conclude the generality of the result from a particular choice of physical example. The student should consider what would happen if this were not true.

> **QUESTION 13:** *Determining Your Absolute Velocity*   Invent an experiment to determine your absolute velocity, assuming that the transformation law for energy applies only to the case considered – an atom emitting light waves – and not to other forms of energy as well. Hint: Consider an inelastic collision where two equal masses collide and then emit the original kinetic energy as light. Assume that mechanical kinetic energy does not transform like light wave energy. (Or you might invent another experiment.)

12.10 MORE THEORETICAL "EVIDENCE" FOR THE EQUIVALENCE OF MASS AND ENERGY

The following "gedanken" (thought) experiment was invented by Einstein after his initial publication.* Imagine a closed system as pictured in Figure 12.9. The system is

---

* See the article "Inertia and Energy," by M. von Laue, in *Albert Einstein: Philosopher-Scientist*, edited by P. A. Schlipp. This article contains a good discussion of the history of energy and momentum conservation in physical theory. The book also contains Einstein's autobiography.

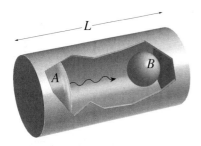

FIGURE 12.9
Cylinder of mass M.

a cylindrical cavity floating in empty space. The length is L, and the total mass is M. At each end are two small bodies, A and B, of negligible mass compared to the cylinder itself. These are free to move inside the cylinder without friction. Also, there is a mechanism inside the cylinder that can move B to A, such as a pulley system. No external forces act on this system, so it should be impossible by any internal processes to change the position of the center of mass.

Body A radiates energy ΔE, which is absorbed by body B. The time between emission and absorption is $\frac{L}{c}$. This time is assumed to be much longer than the time of emission or absorption. At the time of emission, the body A receives a momentum $-\frac{\Delta E}{c}$, recoiling from the emission of the light. Through A's connection to the cylinder, the momentum is transmitted to the cylinder, which recoils with a (nonrelativistic) velocity $-\frac{\Delta E}{cM}$ for a time $t = \frac{L}{c}$. The net displacement of the cylinder before the light is absorbed at B is towards the left, in the amount $\Delta x = -\frac{L}{M}\frac{\Delta E}{c^2}$. Body B absorbs the light. Body B is then displaced by the pulley system, using forces internal to the cylinder, until B touches A. The additional energy acquired by B is then transferred by any means (heat flow?) back into A. If the mass of B is m_1 just after it absorbs the light, the motion of B to A displaces the center of mass by $-\frac{Lm_1}{M}$. After transferring the extra energy back into A, the mass of B changes to m_0. When B is returned to its original position, the center of mass shifts by $+\frac{Lm_0}{M}$. The system is now back to its original condition. The net shift in the center of mass is found by adding up all of these changes:

$$\Delta x_{\text{cm}} = \frac{L}{M}\left(-\frac{\Delta E}{c^2} - m_1 + m_0\right). \tag{12.52}$$

It is not reasonable that a body subject to purely internal processes can shift its center of mass. The net displacement will be zero only if

$$m_1 - m_0 = \frac{\Delta E}{c^2}. \tag{12.53}$$

The rest mass of any body must increase by this amount if energy ΔE is added to the body.

This argument also shows that *any* transfer of energy from point B to A must involve a transfer of momentum as well. A flow of energy (mechanical from a rotating shaft, or heat conducted from *B* to *A*) must also give an impulse in the same direction. In a very profound sense, energy and momentum are two aspects of the same thing, just as x and t are not independent of each other when transforming from one frame to another. We will explore this in more mathematical terms in a later section.

This argument has the advantage of giving the equivalence of mass and energy without any use of transformation properties at all. It depends only on the momentum transmitted by a light pulse of a certain energy. This relation, (12.48), can be proved from Maxwell's equations without using special relativity explicitly. It is a very intuitive argument, since it is reasonable to assume that a body in empty space cannot shift its position by internal forces alone. However, it is not a rigorous argument, since it involves a number of approximations. We give it here as a beautiful example of the kind of "gedanken" experiment for which Einstein was famous. This type of very physical reasoning was the hallmark of his approach to theoretical physics.

12.11 MATHEMATICS OF RELATIVITY: INVARIANTS AND FOUR-VECTORS

We view a space–time event described in a specific reference frame as having four coordinates:

$$(x^0, x^1, x^2, x^3) \equiv (ct, x, y, z). \tag{12.54}$$

The superscripts do not denote exponents. Lower indices have another role to play, which we will explain. Denote the set of all four coordinates by an upper Greek letter index: The set of coordinates is written simply as x^μ, $\mu = 0, 1, 2, 3$. It is analogous to attaching an "arrow" above the symbol to indicate a vector quantity. The 0-th coordinate is the time coordinate, and the other three are the space coordinates. The Greek letters will always have the special significance of indices ranging from 0 to 3 unless otherwise noted. Notice that we define the four coordinates by listing ct first and then the spatial coordinates x, y, z. This is a convention usually adopted when writing the four coordinates together.

The *interval* between two events, such as the two supernova explosions discussed earlier, is the difference in the space–time coordinates of the two events: $\Delta x^\mu \equiv x_B^\mu - x_A^\mu$. Just as the distance between two points is invariant under space rotations, the interval between two space–time events has an invariant quantity which is the same for all observers connected by any combinations of Lorentz transformations and space rotations:

$$\Delta s^2 \equiv (\Delta x^0)^2 - (\Delta x^1)^2 - (\Delta x^2)^2 - (\Delta x^3)^2. \tag{12.55}$$

invariant quantity

The term Δs^2 can be positive, zero, or negative. If it is positive, we say the interval is *timelike*, like the interval between ticks of a clock located at a fixed position. If Δs^2 is a timelike interval, then there always exists a reference frame in which it corresponds to the time interval Δt^2 between two clock readings at the same space location.

> **QUESTION 14: Invariance of Δs^2** Check that an observer moving with respect to the clock will get the same value for Δs^2 for the interval between ticks, using the Lorentz transformation for a boost (Lorentz transformation) along the X axis.

If $\Delta s^2 = 0$, the two points A and B can be connected by a light wave.

> **QUESTION 15: Light Cone** Prove the above assertion.

Finally, if $\Delta s^2 < 0$, the interval is *spacelike*. If Δs^2 is a spacelike interval, then there always exists a reference frame in which two events occur at the same time, but in different places, like the supernova explosions. There can be no causal connection between events separated by a spacelike interval, that is, future and past depend on the observer.

> **QUESTION 16: Spacelike** Assuming that physical causes cannot be propagated faster than light, this means that events with spacelike separations are not causally related. Explain this statement.

If event A is placed at the origin of a space–time diagram in which we plot the X, Y axes in the horizontal plane and the time axis in the vertical direction (the Z axis is supressed for clarity), then we can plot a surface in this space for all points B that obey $\Delta s^2 = 0$ with respect to event A. This surface (really a three-dimensional solid) is called the *light cone*. An interval of space–time can lie inside, outside, or on the light cone, depending on whether Δs^2 is positive, negative, or zero. The situation is pictured in Figure 12.10.

Any set of four physical quantities that transforms under Lorentz transformations exactly like space and time (12.55) is called a *four-vector*. Energy and momentum ($\frac{E}{c}, \vec{p}$) form another important example of a four-vector* (there are many others). The invariant relation $m^2 = E^2 - p^2$ indicates that energy is the 0-th component and momentum the other three components of a four-vector. In the last formula, we simplify by assuming units such that $c = 1$. For a light wave, we can use the result that the momentum of a directed beam of light is $p = \frac{E}{c}$, which would imply that the mass of the light wave is zero, since $E^2 - p^2 = 0$.

There are physical quantities that behave like three-vectors (i.e., ordinary vectors) under space rotations, but which are *not* four-vectors. Electric and magnetic fields $\vec{E}$ and $\vec{B}$ are such quantities. They transform like vectors under (proper) space rotations,

---

* The coordinates of a four-vector must have the same units; therefore we need to divide the energy by c.

12.11 MATHEMATICS OF RELATIVITY: INVARIANTS AND FOUR-VECTORS

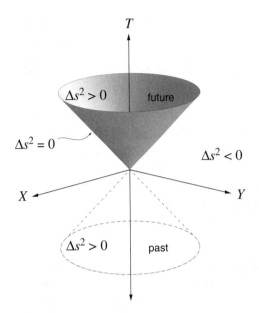

FIGURE 12.10
The light cone. For spacelike intervals, there is no causal connection – future and past depend on the observer. Therefore, future and past can only be distinguished inside the light cone.

but are in fact tensors, requiring two Greek indices, when Lorentz transformations are included.

The inverse of (12.55) is also an important theorem. Suppose that for four arbitrary physical quantities a^0, a^1, a^2, a^3, the combination

$$\mathbf{a} \cdot \mathbf{\Delta x} \equiv a^0 \Delta x^0 - a^1 \Delta x^1 - a^2 \Delta x^2 - a^3 \Delta x^3 \tag{12.56}$$

is known to be a Lorentz invariant (i.e., is the same in all inertial reference frames). (Here Δx^μ is an arbitrary interval of space–time. The notation of "dot product" of two four-vectors is extended from its three-dimensional meaning, being now defined by Equation (12.56).) We can use the invariance of the dot product (more properly known as the *scalar product*) to deduce that a^μ is a four-vector. Thus, if the dot product of a^μ with a four-vector is an invariant, a^μ is another four-vector.

To prove this fundamental theorem, first consider a situation with only time and x-space components for $\mathbf{a}, \mathbf{\Delta x}$. Let $\mathbf{a}', \mathbf{\Delta x}'$ be this same set viewed in the primed frame. By assumption

$$\mathbf{a} \cdot \mathbf{\Delta x} = a^0 \Delta x^0 - a^1 \Delta x^1 = (a')^0 (\Delta x')^0 - (a')^1 (\Delta x')^1 = \mathbf{a}' \cdot \mathbf{\Delta x}'. \tag{12.57}$$

The Lorentz transformation along $+x$ for the coordinate interval is

$$\Delta x^0 = \gamma((\Delta x')^0 + \beta(\Delta x')^1), \quad \Delta x^1 = \gamma((\Delta x')^1 + \beta(\Delta x')^0). \tag{12.58}$$

Substitute Equations (12.58) into Equation (12.57). Since the coordinate interval chosen is an arbitrary one, solve for **a'** in terms of **a**:

$$(a')^0 = \gamma(a^0 - \beta a^1), \quad (a')^1 = \gamma(a^1 - \beta a^0). \tag{12.59}$$

We have proved in this special case that $(a')^0$, $(a')^1$ are mixed together in a Lorentz transformation exactly like the components of Δx, except for a minus sign. This minus sign is to be expected since the observer in the primed frame sees the observer in the unprimed frame moving with velocity $-\beta$. Hence **a'** is a four-vector too, with components in the primed frame. We will drop the **a**, **a'** notation and just refer to a^μ for the components of this four-vector in an unspecified inertial frame. A more elaborate proof with all four components present is possible, but it will not be given here.

QUESTION 17: 3-D Invariance Theorem For ordinary vectors and three-dimensional space rotations, a similar theorem holds. Assume that $\vec{r}$ is a vector and therefore has known rotation properties. Prove that, if $ax + by + cz$ is invariant under all space rotations, then the quantities (a, b, c) are the components of a vector.

In summary,

INVARIANCE THEOREMS

1. The scalar product of a four-vector times another four-vector as defined in (12.55) is an invariant.
2. If the scalar product of four arbitrary quantities with a four-vector is an invariant, the four arbitrary quantities are the components of another four-vector (i.e., the inverse of (12.55)).
3. If a scalar times four arbitrary quantities is a four-vector, then the four arbitrary quantities are the components of another four-vector.

The third theorem is just an extension of such a property of three-vectors.

These results allow us to identify various quantities as four-vectors and hence as possessing the same transformation properties as coordinates under a rotation or a boost or any combination of these. We will demonstrate the use of the theorems above in the following examples:

⊃ Example 1:

Consider, as a trivial example, the mathematical identity

$$\frac{\partial(ct)}{\partial(ct)} + \frac{\partial x}{\partial x} + \frac{\partial y}{\partial y} + \frac{\partial z}{\partial z} = 4. \tag{12.60}$$

Equation (12.60) is true in any inertial frame. If we consider the four-vector to be (ct, x, y, z), the arbitrary quantities to be $\partial^\mu \equiv \frac{1}{c}\frac{\partial}{\partial t}, -\vec{\nabla}$, and the invariant to be the

12.11 MATHEMATICS OF RELATIVITY: INVARIANTS AND FOUR-VECTORS

number four, we can conclude from theorem 2 that the operator ∂^μ transforms like the components of a four-vector. (We introduce here the standard notation ∂^μ for the collection of four partial derivatives with respect to space and time defined above.)

⊃ **Example 2:**

The combination of second derivatives given by

$$\frac{1}{c^2}\frac{\partial^2}{\partial t^2} - \frac{\partial^2}{\partial x^2} - \frac{\partial^2}{\partial y^2} - \frac{\partial^2}{\partial z^2} = \frac{1}{c^2}\frac{\partial^2}{\partial t^2} - \vec{\nabla}\cdot\vec{\nabla} \qquad (12.61)$$

must, according to theorem 1, be a scalar (invariant) under the Lorentz group, since both quantities in the dot product are four-vectors as proved in example 1. Thus, if this operator acts on a function of (ct, x, y, z), it cannot change the transformation properties of the function being differentiated.

⊃ **Example 3:**

The conservation of charge requires the *continuity equation*, which connects partial time derivatives of the charge density ρ with the gradient of the current density $(\vec{j})$:

$$\frac{\partial\rho}{\partial t} + \vec{\nabla}\cdot\vec{j} = 0. \qquad (12.62)$$

This equation must remain true for all observers; otherwise the rate of change of the amount of charge contained in a little test volume ($\iiint \rho\, dx\, dy\, dz$) will not agree with the surface integral of the current density over a closed surface containing this volume ($\iint \vec{j}\cdot d\vec{S}$). We conclude immediately from theorem 2 that $j^\mu = (\rho, \frac{\vec{j}}{c})$ form the time and space components of a four-vector, because (12.62) can be rewritten as $\partial_\mu j^\mu$. Charge density in a static charged volume, with no current present, is increased by a factor of γ when viewed from a frame moving to the right at velocity $\vec{\beta}$. This is intuitively understood from the Lorentz contraction of the volume and the invariance of charge itself. The electric current density in the primed frame is, in this example, $\frac{\vec{j}'}{c} = -\gamma\vec{\beta}j^0$.

⊃ **Example 4:**

One knows from the theory of electromagnetic fields that it is possible to define a vector potential function $\vec{A}(x, y, z, t)$ and a scalar potential function $\Phi(x, y, z, t)$ which can then be used to calculate the electric and magnetic fields. Charges and currents act as sources for these potentials, obeying the equations (Gaussian units)

$$\frac{1}{c^2}\frac{\partial^2}{\partial t^2}\vec{A} - \vec{\nabla}\cdot\vec{\nabla}\vec{A} = \frac{4\pi}{c}\vec{j}, \quad \frac{1}{c^2}\frac{\partial^2}{\partial t^2}\Phi - \vec{\nabla}\cdot\vec{\nabla}\Phi = 4\pi\rho. \qquad (12.63)$$

(Equations (12.63) can be derived from Maxwell's equations. We are only interested in the transformation properties here, so we ask you to take them on faith if they are not familiar.) Since $(\rho, \frac{\vec{j}}{c}) = j^\mu$ was proved to be a four-vector, we see immediately from the scalar nature of (12.61) and theorem 3 that $(\Phi, \vec{A}) = A^\mu$. The electrostatic and vector potentials are the time and space components of a four-vector A^μ.

◯ **Example 5:**

Consider a light wave moving in the general direction denoted by the unit vector $\hat{n}$. The electric field associated with this light wave is proportional to an amplitude times a time and space–dependent phase factor, say $\mathcal{E}_0 \sin(k\hat{n} \cdot \vec{r} - \omega t)$. The phase of the wave is $\Psi \equiv \vec{k} \cdot \vec{r} - \omega t$. (Define $\vec{k} \equiv k\hat{n}$. $\vec{k}$ is called the *wave vector*; ω the *angular frequency*.*) For constant phase, along a particular wavefront, the point $\Delta\vec{r} = \frac{\omega}{k}\Delta t\, \hat{n}$ lies on the same wavefront as the point $\vec{r} = 0, t = 0$. The *phase velocity* of the wave is therefore $\frac{\omega}{k} = c$, and so the crests and nodes of the wave must be the same for all observers. This means Ψ is a Lorentz invariant quantity. We can thus see from theorem 2 that $(\omega, \vec{k})$ must constitute the components of a four-vector.

QUESTION 18: *Consequences of $(\omega, \vec{k})$ Four-Vector* Use this last fact directly to prove that the transformation equation (Doppler shift) of a directed light wave is given by Equation (12.21). If the energy is proportional to the frequency, show that the momentum is proportional to the wave vector and hence that the energy and momentum of a light beam constitute a four-vector.

QUESTION 19: *Phase Velocity and Invariant "Mass" of a Light Wave* What is the phase velocity and invariant "mass" of a light wave of frequency ω which passes through a medium of refractive index $n > 1$? Does a rest frame exist for this light wave? Why or why not?

An important consequence of the equivalence of all inertial frames is that if the left side of an equation transforms like a scalar or a four-vector, then the right side must also do this. Otherwise there would be a preferred frame in which the equation holds, and it would not be true in other inertial frames. If the transformation properties of right and left sides are known to be identical, we need only verify the equation in one inertial frame. It is then automatically valid for all such reference frames.

For example, suppose we know that light energy and momentum transform like four-vectors. By the conservation of momentum and energy we can write the energy and momentum change of an *atom* emitting light on the right side and the energy and momentum emitted as *light* on the left side of an equation (or rather four equations). This immediately tells us that the energy and momentum of a material object with mass also transform like a four-vector.

12.12 A SECOND LOOK AT THE ENERGY–MOMENTUM FOUR-VECTOR

An object with finite nonvanishing rest mass $m > 0$ has a particular inertial frame in which it is at rest. If the massive object is accelerated, there still may be an "instantaneous"

---

* If the wave travels in the $+x$ direction, and the phase velocity is $c = 1$, then $\Psi_n = 2\pi n = kx_n - \omega t_n$ or $(x - t)_n = \frac{2\pi n}{\omega} = \frac{n}{f}$, Equation (12.19).

12.12 A SECOND LOOK AT THE ENERGY–MOMENTUM FOUR-VECTOR

rest frame, a frame in which it is momentarily at rest with zero instantaneous velocity. A clock in that particular rest frame registers what is, by definition, *proper time* τ. Proper time is independent of the motion of the body, just like rest mass which we defined earlier in a similar way. If we consider the derivative of the four-vector x^μ with respect to proper time, we denote it by

$$u^\mu \equiv \frac{dx^\mu}{d\tau}. \tag{12.64}$$

The parameter u^μ is called the *four-velocity*. Notice that $u \cdot u = c^2$. Since proper time intervals $d\tau$ are Lorentz invariant by definition, u^μ is also a four-vector, *just by inspection*. In a different inertial frame, the time interval dt of a clock is lengthened relative to an interval of a clock in the rest frame by the time dilation factor: $dt = \gamma d\tau$. In this second frame, moving with respect to the rest frame at a velocity $-\beta$, the massive object has the velocity β:

$$u^\mu = \left(\frac{d(ct)}{d\tau}, \frac{dx}{d\tau}, \frac{dy}{d\tau}, \frac{dz}{d\tau}\right) = c\gamma(1, \beta, 0, 0). \tag{12.65}$$

Comparing (12.65) with the form $p^\mu \equiv (\frac{E}{c}, p)$, where we use (12.47, 12.50) and drop the subscript 0, we see that, for any massive object

$$\boxed{p^\mu = mu^\mu, \quad m > 0.} \tag{12.66}$$

The fact that energy and momentum comprise a four-vector means that we should talk about "energy–momentum" rather than energy and momentum separately. The p^μ in Equation (12.66) is often called the *four-momentum*. The conservation laws of energy and momentum are not separate laws but are components of the same law. The square of the rest mass is the invariant $m^2 = E^2 - p^2$.

Since rest mass is the proportionality factor between the kinematic quantity u^μ and the dynamical quantity p^μ, it is natural to ask why we didn't just write down u^μ, multiply it by the invariant rest mass, and use this immediately as the definition of four-momentum. The transformation law of energy and momentum is then explicit and obvious. The problem is that we do not a priori know the transformation properties of energy and momentum. A lucky guess that energy–momentum is a four-vector *would* be a possible way to discover this crucial transformation law of relativistic dynamics;* however, we think it is better and more convincing to show first by gedanken experiments that it is the only way energy and momentum conservation can be maintained for massive objects that interact with light

---

* One student remarked about this approach to relativistic mechanics: "It works, but you don't know *why* it works."

waves. This rather "Machian"* viewpoint is mainly used for pedagogical reasons. It runs rather contrary to the style of modern theoretical physics, which has become quite formalistic out of necessity.

In fact (12.66) cannot be taken as the fundamental definition of energy and momentum transformation, because it applies only to massive objects. Light waves obey the relation $(c = 1)$ $E^2 - p^2 = 0$, so photons have zero rest mass. There is no relationship to u^μ in this case (unless u^μ is taken to be infinite), and yet energy and momentum are well-defined quantities for light and other objects that (might) have zero mass, such as neutrinos.

QUESTION 20: *Transverse Momentum* Prove that transverse momentum is not changed by a boost along the X axis. Is this consistent with the laws for velocity addition?

⊃ Example: Particle Decay

A particle of mass M_0 and four-momentum p_0^μ at rest decays spontaneously into two other particles of masses M_1 and M_2 and four-momenta p_1^μ, p_2^μ respectively. By conservation of four-momentum we know that $p_0^\mu = p_1^\mu + p_2^\mu$. It is often easiest to solve for quantities by using conservation of four-momentum along with the invariance of the scalar products of four-vectors. For example, rewrite the conservation of momemtum equation as $p_2^\mu = p_0^\mu - p_1^\mu$, and then square this to enable us to solve for the individual scalar products:

$$\left(p_2^\mu\right)^2 = \left(p_0^\mu\right)^2 - 2 p_0^\mu p_1^\mu + \left(p_1^\mu\right)^2. \tag{12.67}$$

We can write out the components of the four-momenta for each particle in the rest frame of the decaying particle as

$$p_0^\mu = \left(\frac{E_0}{c}, 0\right), \quad p_1^\mu = \left(\frac{E_1}{c}, \vec{p}_1\right), \quad p_2^\mu = \left(\frac{E_2}{c}, \vec{p}_2\right). \tag{12.68}$$

Then, by performing the scalar products[†] in (12.67) using (12.68), and noting that the scalar product of the four-momentum is the mass, we can reduce (12.67) to be

$$M_2^2 = M_0^2 + M_1^2 - 2\frac{E_1 E_0}{c^2}. \tag{12.69}$$

Finally, noting that $E_0 = M_0 c^2$ since the decaying particle is at rest, we can solve for the energy of the first particle in the rest frame of the decaying particle:

$$E_1 = \frac{M_0^2 + M_1^2 - M_2^2}{2 M_0}. \tag{12.70}$$

---

* Ernst Mach resisted very strongly the introduction of concepts not directly related to experience and measurements into physics. Unfortunately, Mach's beliefs led him to oppose the atomic hypothesis. One wonders what he would have felt about quantum mechanics, quantum field theory, quarks, etc.
† Remember to use the form of (12.55) to find the scalar products of the four quantities.

By repeating the calculation using $p_1^\mu = p_0^\mu - p_2^\mu$, we find that the energy of the second particle in the rest frame of the decaying particle is

$$E_2 = \frac{M_0^2 + M_2^2 - M_1^2}{2M_0}. \tag{12.70}$$

12.13 WHY ARE THERE BOTH UPPER AND LOWER GREEK INDICES?

The minus sign in the formation of a Lorentz invariant scalar (12.55) means that there is an asymmetry in the way time and space are treated in calculations. To avoid the appearance of this, we can, for purely mathematical and esthetic reasons, introduce a *metric tensor*:

$$g_{\mu\nu} \equiv \begin{pmatrix} 1 & 0 & 0 & 0 \\ 0 & -1 & 0 & 0 \\ 0 & 0 & -1 & 0 \\ 0 & 0 & 0 & -1 \end{pmatrix}. \tag{12.71}$$

The order of indices indicates that $\mu = 0, 1, 2, 3$ denotes the rows and $\nu = 0, 1, 2, 3$ denotes columns. The invariant scalar product of two four-vectors can then be written using the Einstein summation convention, in which μ, ν are summed over all four possible values:

$$a \cdot b = g_{\mu\nu} a^\mu b^\nu. \tag{12.72}$$

This gives us a more symmetrical treatment of time and space, since we can avoid writing the components separately. The metric tensor can also be used to define lower indices (called *covariant components*) in terms of the upper indices (called *contravariant components*):

$$a_\mu \equiv g_{\mu\nu} a^\nu. \tag{12.73}$$

The Einstein summation convention is used again in Equation (12.73). If we think of a^ν as a four-vector transforming like $(t, \vec{r})$, then a_μ transforms like the four-vector $(t, -\vec{r})$. The metric tensor can be used to convert one type (contravariant) of four-vector into another (covariant). It can be used to raise or lower the indices. By using a metric tensor with upper indices $g^{\mu\nu} \equiv g_{\mu\nu}$, we can "raise" the indices:

$$a^\nu \equiv g^{\mu\nu} a_\mu \tag{12.74}$$

Using either (12.73) or (12.74), we can write the invariant scalar product (12.72) as the multiplication of lower with upper indices:

$$a \cdot b = a_\mu b^\mu = a^\nu b_\nu. \tag{12.75}$$

What we have just described are mathematical conventions as far as special relativity is concerned. *We avoid the original Minkowski notation in which he uses "ict" ($i = \sqrt{-1}$) as the fourth component of a four-dimensional Euclidean vector and considers Lorentz transformations as rotations in this partly imaginary space. Instead, we prefer to introduce the metric tensor above, and keep all coordinates as real quantities.* The importance of this notation is not really apparent until one studies general relativity. Then it becomes more than esthetics: The metric tensor is not necessarily diagonal nor are its elements constants. In fact the functions appearing in the metric tensor determine the motion of all objects that can be affected by gravitational fields. Now $g^{\mu\nu} = (g_{\mu\nu})^{-1}$ (i.e., they are inverses). Thus there is a distinction between contravariant and covariant components of a four-vector that is more than just the minus sign on the space components in special relativity.

The term $g_{\mu\nu} = g^{\mu\nu}$ is not the only 4×4 tensor. It can be shown that the electric and magnetic fields taken together transform like the six independent components of an antisymmetric four-tensor:*

$$F^{\mu\nu} \equiv \begin{pmatrix} 0 & -\mathcal{E}_x & -\mathcal{E}_y & -\mathcal{E}_z \\ \mathcal{E}_x & 0 & -\mathcal{B}_z & \mathcal{B}_y \\ \mathcal{E}_y & \mathcal{B}_z & 0 & -\mathcal{B}_x \\ \mathcal{E}_z & -\mathcal{B}_y & \mathcal{B}_x & 0 \end{pmatrix}. \tag{12.76}$$

The most general second-rank tensor will have $16 = 4 \times 4$ components. It transforms like the 16 possible products $x^\mu x^\nu$ transform to under Lorentz transformations ("boosts") and ordinary space rotations.

12.14 RELATIVISTIC LAGRANGIAN MECHANICS

We begin by assuming that there exists a relativistic scalar quantity called the action. If we minimize the action, the invariance of this scalar guarantees that all observers will agree that it is a minimum for a given path. When we minimized the action in Chapter 2, we focused on minimizing a certain integral over time along different paths in space. The choice of the difference of kinetic and potential energy in the integral may have seemed rather arbitrary and contrived. But now the choice of what to minimize is forced upon us by the requirement that the result be a Lorentz invariant. The actual description of the physical path will depend on the coordinate system, but the minimized action itself will depend neither on the observer nor on the choice of coordinates.

In Figure 12.11, an electron with nonzero rest mass is located at the origin $x = y = z = t = 0$ for all observers. This point is denoted by the space–time event A in the diagram. At a later time, the electron moves to another point, B. Each observer records the position and time of the electron – say if it emits a flash of light when at point B. The time and space coordinates of B will depend on the observer. What is the path in

---
* See Jackson, *Classical Electrodynamics*, 2d ed., p. 550.

12.14 RELATIVISTIC LAGRANGIAN MECHANICS

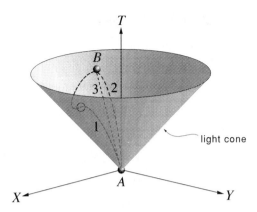

FIGURE 12.11
Possible paths in space–time connecting event A with event B.

space–time that connects the physics events A and B? As we said above, we seek a way to formulate this problem which is manifestly Lorentz-invariant. All observers must calculate the same value for the action.

If no forces act upon this particle, in its rest frame, the particle would seem to "move" straight up the time axis. In other moving frames, the particle appears to have an initial velocity, and so it is displaced in space as time increases. (The light cone is the same for all observers, according to Einstein's second hypothesis.) In Figure 12.11, three space–time paths (called *world lines*) are shown. Which path is the correct physical path? We can immediately exclude path 1 as being physically impossible. At the point on this curve marked by the small circle, the slope becomes less than 1 (units are $c = 1$) (i.e., the velocity at that point is greater than that of light). One way to see that this is true is to draw a miniature light cone emanating from the curve at the small circle. This is the locus of all light waves that could be emitted by the electron at that point. The curve moves into a region that is a spacelike interval from the vertex of this small cone. This is impossible. Curves 2 and 3 are physically possible in Figure 12.11.

The only two Lorentz invariants that exist for a particle not acted on by forces are its rest mass and proper time τ. Let us assume that the action $\mathcal{S}$ is given by

$$\mathcal{S} \equiv \lambda \int_A^B d\tau. \qquad (12.77)$$

In this way, the action is proportional to the time recorded on a clock in the rest frame. Here λ is an unknown constant, which we will choose so that the action agrees with the nonrelativistic limit, and $d\tau$ is the differential of proper time. All observers will agree on the action to be calculated for any given path, such as paths 2 or 3 in Figure 12.11, since τ is a scalar under Lorentz transformations. We are asking that a clock located on the particle record the minimum possible time when the particle moves along a world line from event A to event B.

In a particular inertial frame, the connection between proper time and local time for events at the same location in the rest frame is given by the time dilation factor: $dt = \gamma \, d\tau$.

For an observer in this particular frame, the variational principle becomes

$$\delta S = 0 = \lambda \delta \left(\int_{t_A}^{t_B} \frac{dt}{\gamma} \right) = \lambda \delta \left(\int_{t_A}^{t_B} \sqrt{1 - \beta^2}\, dt \right). \tag{12.78}$$

By analogy with the definition of a Lagrangian in nonrelativistic mechanics (2.16), we *define* the relativistic Lagrangian L by the equation

$$S \equiv \int_{t_A}^{t_B} L\, dt. \tag{12.79}$$

Note that L and dt depend on the observer although S does not.

If the particle is nearly at rest in the particular inertial frame we are considering, we can expand the square root in (12.78) in powers of the velocity β: $\sqrt{1-\beta^2} \approx 1 - \frac{1}{2}\beta^2 + O[\beta^4]$. If we ignore an insignificant constant term independent of β, and also drop the higher-order terms, assuming that β is sufficiently small, we obtain

$$\delta S \equiv \delta \left(\int L\, dt \right) = \delta \int \left(-\frac{1}{2} \lambda \beta^2 \right) dt. \tag{12.80}$$

For (12.80) to agree with the nonrelativistic limit of $\frac{1}{2}mv^2$, $\lambda = -mc^2$. This choice for λ is only a convention, since the Lagrangian can always be changed by an additive or multiplicative constant without affecting the equations of motion.

The relativistic Lagrangian for a free particle is thus (by comparing (12.78) with (12.79), and using $\lambda = -mc^2$)

$$\boxed{L = -mc^2 \sqrt{1 - \frac{v^2}{c^2}}.} \tag{12.81}$$

relativistic Lagrangian for a free particle

Proceeding in the usual way, define the canonical momenta $\vec{p} = (p_x, p_y, p_z) \equiv (\frac{\partial L}{\partial v_x}, \frac{\partial L}{\partial v_y}, \frac{\partial L}{\partial v_z})$. Computing the partial derivatives with respect to velocity components of (12.81), we obtain the canonically conjugate momentum to $q = x, y, z$:

$$\vec{p} = \frac{m\vec{v}}{\sqrt{1 - \frac{v^2}{c^2}}}. \tag{12.82}$$

This is identical to our previous expression for the relativistic momentum (12.50). The Euler–Lagrange equations for a free particle are

$$\frac{d\vec{p}}{dt} = 0. \tag{12.83}$$

The momentum $\vec{p}$, and hence $\vec{v}$, are constants for a free particle in accord with (12.83) and with what we expect.

Using the definition of the Hamiltonian,

$$H \equiv E \equiv \vec{p} \cdot \vec{v} - L, \tag{12.84}$$

rewrite (12.84) multiplied by dt as $-d\mathcal{S} = -L\,dt = H\,dt - \vec{p} \cdot d\vec{x}$. By the previously proven invariance theorem 2, $H = E$, since $(\frac{H}{c}, \vec{p})$ must be a four-vector and thus must equal $(\frac{E}{c}, \vec{p})$. This *is* the total energy in the relativistic theory (we don't have "$T + V$" to use here). Because there is no explicit time dependence, it should be conserved.

The reader should verify that, by this definition (12.84)

$$E = \frac{mc^2}{\sqrt{1-\beta^2}} = \gamma mc^2. \tag{12.85}$$

We have the previously obtained expression for the total energy (12.47), *including the energy of the rest mass.*

The derivation of the canonical momenta and the Hamiltonian remains valid even when the particle is subject to forces.

QUESTION 21: *Total Energy* Prove Equation (12.85) follows from (12.84).

Path 3 is the correct physical choice, since only on this path is the slope and hence the velocity of the electron constant. Elapsed proper time, and hence the action, are minimized on path 3.

Starting from the inescapable fact that, *if* a variational principle governs the dynamical behavior of a free particle, the only possible choice is $\delta \int d\tau = 0$, we have proved that we can get the correct expressions for momentum and *total* energy in just a few lines. It follows immediately and inevitably once we are committed to the existence of a Lorentz-invariant variational action principle for dynamics. We haven't done anything except trace the consequences of this one theoretical assumption. However, Einstein's gedanken experiments not only established Equations (12.82, 12.85), but also demonstrated the consistency of his hypotheses with emitting and absorbing light, as well as the transformation properties that proved $p^\mu = (\frac{E}{c}, \vec{p})$ is a four-vector.

12.15 WHAT IS THE LAGRANGIAN IN AN ELECTROMAGNETIC FIELD?

What could we write for the Lagrangian in case of a particle of charge e in an electromagnetic field? The field will accelerate the particle – an accelerated charged particle will produce electromagnetic radiation – and there will be a "radiation reaction force" back on the particle in order to conserve energy and momentum. *As a first approximation, we neglect any effect on the particle caused by that radiation.* We require that the new Lagrangian must reduce to the proper form in the nonrelativistic limit. The electromagnetic addition to the free particle Lagrangian must therefore be linear in $A^\mu = (\Phi, \vec{A})$, in order to give equations of motion that are linear in A^μ. The action $dS = L\, dt$ must remain a scalar, so we must form a scalar product for A^μ with something else. Since p^μ and four-velocity u^μ differ only by the constant factor m, we can assume, up to a constant, that u^μ is the only four-vector we have available besides x^μ (provided we do not consider derivatives of the vector potential – see below).

We can write the most general form consistent with these assumptions and with Lorentz invariance of the action differential dS:

$$dS = \underbrace{-mc^2\, d\tau}_{\text{free}} + \underbrace{(ax_\mu A^\mu + bu_\mu A^\mu)}_{\text{linear in electric field}} d\tau, \tag{12.86}$$

where a and b are unknown constants at this point. The summation convention is assumed for the scalars $x_\mu A^\mu$ and $u_\mu A^\mu$. We can immediately conclude that the constant $a = 0$, because it would give a term that would depend on the choice of coordinate system origin. To find the constant b, evaluate the electromagnetic part of (12.86) in a convenient frame. We choose a frame where the particle is at rest. To simplify even further, we assume there is only an electric field present in this frame, so $A^\mu = (\Phi, 0)$. (To find A^μ in any other inertial frame, make a Lorentz transformation that mixes space and time coordinates just as for $t, \vec{r}$. To find the magnetic field in this new frame, take the curl of $\vec{A}$ in that frame.) With this choice of frame,

$$d\tau = dt, \quad u_\mu A^\mu = c\Phi. \tag{12.87}$$

For a particle of charge e, the potential energy in an electrostatic field is $V = e\Phi$, so we expect a term $-e\Phi$ in the Lagrangian; hence $b = -\frac{e}{c}$. With this choice, both terms in Equation (12.86) above have the dimensions of energy multiplied by time (i.e., dimensions of action). The electric charge e is Lorentz-invariant, since all observers will agree on the total charge of any object isolated from its surroundings. Within the assumption of Lorentz invariance and linearity in the fields, we obtain the action differential

$$\frac{dS}{d\tau} = -mc^2 - \frac{e}{c} u_\mu A^\mu. \tag{12.88}$$

Using (12.79), the Lagrangian, written in *any* inertial frame, becomes

$$\boxed{L = -mc^2 \sqrt{1 - \frac{v^2}{c^2}} - e\Phi + e\frac{\vec{v}}{c} \cdot \vec{A}.} \tag{12.89}$$

Lagrangian in an electromagnetic field

12.16 DOES A CONSTANT FORCE CAUSE CONSTANT ACCELERATION?

In the electromagnetic part of (12.89) there is a factor of γ from u_μ that cancels $\sqrt{1-\beta^2}$ from the time dilation.

The ease with which we are led to this unique form demonstrates yet again the power achieved when the action principle and Lorentz invariance are combined. We have above the complete and most general form of the Lagrangian in an electromagnetic field, assuming that radiation reaction is neglected. (For extremely large fields, and hence large accelerations, these reaction terms must be included, but we will not discuss this possibility further.) Most problems of motion in various electromagnetic fields can be solved by using Equation (12.89) as the Lagrangian.

12.16 DOES A CONSTANT FORCE CAUSE CONSTANT ACCELERATION?

In nonrelativistic mechanics a constant force means a constant acceleration, which means that the particle will eventually exceed the velocity of light – a physical impossibility.

QUESTION 22: Accelerating to Light Speed If a test object is subjected to a constant gravitational field, with acceleration $g = 9.8$ m/s², how many seconds will it take for the object to reach the velocity of light, if it starts from rest? Over what distance will it have fallen? Use nonrelativistic mechanics to answer this question.

Suppose that a test particle of mass m is charged and is therefore accelerated by a constant electric field $\mathcal{E}$ which exists everywhere in space. The particle is assumed to start from rest. The observer in the rest frame (defined as frame 1) sees only an electric field – let's say it is directed along the X axis. A moving observer will see both electric and magnetic fields, but we will focus on the acceleration in only two frames: a) the original frame 1 in which the particle starts at rest and b) an *accelerated* (i.e., noninertial) frame 2 in which the particle remains always at rest. *We will call frame 2 the instantaneous rest frame.* Recall that one can use any coordinate system – even noninertial frames – to describe motion with Lagrangians, but the Lagrangian must be evaluated in an inertial frame. The use of a variational principle grants total freedom in the choice of coordinate systems to describe the motion.

In frame 1, there is no magnetic field. The electrostatic potential is $\Phi = -\mathcal{E}x$, so that minus the gradient of the potential will give a uniform electric field directed along the X axis. The four-vector $A^\mu = (\Phi, \vec{A}) = (-\mathcal{E}x, 0)$.

Use the Lagrangian (12.89) together with the specific form of the vector potential in frame 1:

$$L = -mc^2\sqrt{1 - \frac{v^2}{c^2}} + e\mathcal{E}x. \tag{12.90}$$

The problem is solved in the standard way. The Euler–Lagrange equations are only nonzero in the x direction:

$$\frac{d}{dt}\left(\frac{\partial L}{\partial v_x}\right) = \dot{p}_x = \frac{\partial L}{\partial x} = e\mathcal{E}. \tag{12.91}$$

The x component of the momentum increases without limit proportional to the elapsed time: $p_x = e\mathcal{E}t$. This is exactly the same as the nonrelativistic formula. The velocity is given by the general formula (using (12.47, 12.50, 12.51))

$$\beta = \frac{pc}{E} = \frac{ec\mathcal{E}t}{\sqrt{(mc^2)^2 + (ec\mathcal{E}t)^2}}. \tag{12.92}$$

We see that the velocity does not increase indefinitely, but instead $\beta \to 1$ as $t \to \infty$. The acceleration is, by definition, $a \equiv \frac{dv}{dt}$:

$$a(t) = \left(\frac{e\mathcal{E}}{m}\right)\left(1 + \left(\frac{e\mathcal{E}t}{mc}\right)^2\right)^{-\frac{3}{2}}. \tag{12.93}$$

At sufficiently short times, the acceleration takes on its nonrelativistic value $\frac{e\mathcal{E}}{m}$, but at times such that $\frac{e\mathcal{E}t}{mc} \geq 1$, the acceleration is reduced from the nonrelativistic value. This becomes significant when the momentum $e\mathcal{E}t$ has increased beyond mc. In terms of the velocity, we can show that $\beta \geq \frac{1}{\sqrt{2}}$, which might be called the boundary between relativistic and nonrelativistic motion. An electron must be accelerated to a kinetic energy of $\sqrt{2}-1 = .41$ times its rest mass energy for the velocity to exceed this amount.

From this example, we can see why our definition of mass, Equation (12.38), had to be taken in the limit where the velocity is small. We have also learned that $\vec{F} \equiv m\vec{a}$ is no longer a convenient definition of force for relativistic motion. $\vec{F} \equiv \frac{d\vec{p}}{dt}$ is a more accurate definition. One can save the form $\vec{F} = m\vec{a}$ by defining something called "relativistic mass," which varies with velocity. We regard "relativistic (velocity-dependent) mass" as a misleading concept, since longitudinal forces depend on velocity differently than transverse forces, so we will not use this idea. Mass for us is always rest mass. The concept of force is not really of such critical importance to us if we formulate everything in terms of Lagrangian mechanics. If we refer to "force," however, we must be very careful about what we call "momentum," as will be illustrated in the Section 12.17.

Calculating Proper Time for a Particle in a Constant Force

We will make a Lorentz transformation to frame 2 – the "instantaneous rest frame" – to see how things look to the electron. In a linear accelerator, the electron rides a wave that has a phase velocity reduced somewhat from c and therefore is synchronous with the electron. To a good first approximation, we can consider that the electron sees a constant electric field as it moves along with the wave, much like a surfboard moving along the crest of an ocean wave. The electric field is neither constant in space nor in time, but it remains nearly constant in the vicinity of the electron.

In the Stanford accelerator SLAC for example, electrons are accelerated by an electric field of about 18 MeV/m,[*] many times the rest energy of $mc^2 = 0.511$ MeV. The electron,

---

[*] 1 electron volt (eV) = 1.6×10^{-12} erg = 1.6×10^{-19} joules.

even only after one meter, has $\beta = 0.999597$, if it starts from rest. Over an accelerator length of just under two miles, the electron would be accelerated to an energy of about 60 GeV.* After acceleration, $\gamma = \frac{60 \times 10^9}{.511 \times 10^6} \approx 10^5$, and β differs from unity by only 5.2×10^{-11}. With the approximation that the electron travels at c over the length of the accelerator, it will take just over 10 microseconds to traverse the entire two mile machine, when viewed from a frame where the accelerator is at rest (i.e., the laboratory frame 1).

What would a clock fixed with respect to this accelerated electron (clock is in frame 2) read when the electron finishes the trip down the full length of the accelerator? We will assume that we can still use the Lorentz transformation to calculate the passage of proper time:

$$\tau = \int d\tau = \int \frac{dt}{\gamma(t)}. \tag{12.94}$$

To simplify, assume† that $\gamma(t) = 1 + at$. Then for $\gamma(10 \ \mu s) = 10^5$, $a = 10^{10} \ s^{-1}$. The elapsed proper time for the electron to fly down the accelerator is $\tau = \frac{1}{a} \ln \gamma_{\max} = 1.1 \times 10^{-9}$ s. For the electron, the time down the accelerator is just over one nanosecond, almost 10,000 times less than is measured by clocks remaining at rest in the laboratory. One can also calculate from the Lorentz contraction that the accelerator appears to be very short, about one foot long from the perspective of the electron.

This can be applied to muons, heavy electrons that are produced in the upper atmosphere by decay products of nuclear reactions between the air and incoming cosmic rays. They have a lifetime $\approx 2 \ \mu s$. Were it not for the time dilation factor, very few of them would actually reach the ground in a typical time that is 10–20 times longer than their lifetime at rest. The very existence of cosmic rays at ground level is proof of time dilation.

Notice that we have assumed that we can use a Lorentz transformation in a noninertial reference frame (Equation (12.94)). We have not given any proof that this procedure was correct, but we believe it is.

12.17 DERIVATION OF THE LORENTZ FORCE FROM THE LAGRANGIAN

Starting from the Lagrangian (12.89), we allow the vector potential to be a function of x, y, z, t, so it is completely general for any electromagnetic field. As always, the motion is given by the Euler–Lagrange equations. We must first find‡

$$\frac{\partial L}{\partial \vec{r}} = -e \vec{\nabla} \Phi + \frac{e}{c} \vec{\nabla} (\vec{v} \cdot \vec{A}), \tag{12.95}$$

$$\vec{p} = \frac{\partial L}{\partial \vec{v}} = m \gamma \vec{v} + \frac{e}{c} \vec{A}. \tag{12.96}$$

---

* In practice, this rate of acceleration is not maintained over all of this distance, and the final energy is closer to 46 GeV.
† $\frac{dE}{dx}$ is really what is constant here. Since $\frac{dE}{dt} = \frac{dE}{dx} v$, if we assume $v \simeq c$ (which is true for most of the time the electron is accelerated), we can assume a constant rate of energy gain too.
‡ $\frac{\partial L}{\partial \vec{r}}$ stands for the vector $(\frac{\partial L}{\partial x}, \frac{\partial L}{\partial y}, \frac{\partial L}{\partial z})$.

The Euler–Lagrange equations (there are three of them) are

$$\frac{d}{dt}\left(\frac{\partial L}{\partial \vec{v}}\right) = \frac{\partial L}{\partial \vec{r}}. \tag{12.97}$$

The total time derivative on the left side of (12.97) gives

$$\frac{d}{dt}(m\gamma\vec{v}) + \frac{e}{c}\frac{d\vec{A}}{dt}. \tag{12.98}$$

The potential $\vec{A}$ changes with time for two reasons: a) there can be an explicit time dependence and/or b) the particle moves through space and thus changes the vector potential. By the chain rule then, we have $\frac{d\vec{A}}{dt} = \frac{\partial \vec{A}}{\partial t} + (\vec{v} \cdot \vec{\nabla})\vec{A}$. Using (12.98) in the fundamental Equation (12.97) together with (12.95) on the right side, and rearranging terms, we obtain

$$\frac{d}{dt}(m\gamma\vec{v}) = \underbrace{-e\left[\vec{\nabla}\Phi - \frac{1}{c}\frac{\partial \vec{A}}{\partial t}\right]}_{e\vec{\mathcal{E}}} + \underbrace{\frac{e}{c}[\vec{\nabla}(\vec{v}\cdot\vec{A}) - (\vec{v}\cdot\vec{\nabla})\vec{A}]}_{\frac{e}{c}\vec{v}\times\vec{B}}. \tag{12.99}$$

The well-known vector identity $\vec{a} \times (\vec{b} \times \vec{c}) = (\vec{a}\cdot\vec{c})\vec{b} - (\vec{a}\cdot\vec{b})\vec{c}$ can be used to show that the part in the right-hand brackets in Equation (12.99) is $\vec{v} \times \vec{B}$, since $\vec{\nabla}$ acts only on $\vec{A}$ and not on $\vec{v}$, and $\vec{B} \equiv \nabla \times \vec{A}$.

Many people will recognize the expression on the right-hand side as the Lorentz force:

$$\boxed{\vec{F} = e\left(\vec{\mathcal{E}} + \frac{\vec{v}}{c} \times \vec{B}\right).} \tag{12.100}$$

Lorentz force

We only agree with (12.100) if we adopt a definition of *kinetic momentum* $\vec{\pi} \equiv \gamma m\vec{v}$ and then take the definition of force to be $\vec{F} = \dot{\vec{\pi}}$. This is a definition; it does not affect in any way the equations of motion we have already derived. Kinetic momentum and canonical momentum differ only in the presence of a vector potential (see (12.96)).

Since all observers will follow the same procedure and arrive at Equation (12.100), even though the velocities and fields each of them observes differ from what the others observe, we say this is a *Lorentz-invariant* or a *covariant* procedure. The action is minimized (actually an extremum) on the path taken, and all observers will agree on a unique physical path, although not on the coordinate description of it. The Lorentz invariance of the description can be made more explicit by using the covariant form of (12.100):

$$\boxed{\frac{d\pi^\mu}{d\tau} = \frac{e}{c}F^{\mu\nu}u_\nu.} \tag{12.101}$$

Here $F^{\mu\nu}$ is the antisymmetric tensor defined in (12.76), and $\pi^\mu \equiv (m\gamma c, m\gamma \vec{v})$, the kinetic momentum. This is an equation with four-vectors on both right and left sides, so it is certainly a Lorentz-invariant or covariant equation. Let us see if this works out to give us the Lorentz force. Recall $u^\nu \equiv c(\gamma, \gamma\vec{\beta})$ so the lower index $u_\nu = c(\gamma, -\gamma\vec{\beta})$. The sum over indices in (12.101) can be written as matrix multiplication of a 4×4 matrix $(F^{\mu\nu})$ with a column vector u_ν to give another column vector

$$F^{\mu\nu} u_\nu = c \begin{pmatrix} 0 & -\mathcal{E}_x & -\mathcal{E}_y & -\mathcal{E}_z \\ \mathcal{E}_x & 0 & -\mathcal{B}_z & \mathcal{B}_y \\ \mathcal{E}_y & \mathcal{B}_z & 0 & -\mathcal{B}_x \\ \mathcal{E}_z & -\mathcal{B}_y & \mathcal{B}_x & 0 \end{pmatrix} \begin{pmatrix} \gamma \\ -\gamma\beta_x \\ -\gamma\beta_y \\ -\gamma\beta_z \end{pmatrix}. \quad (12.102)$$

Carrying out the matrix multiplication, we obtain four equations for the four components of this four-vector:

$$\frac{d\pi^\mu}{d\tau} = e \begin{pmatrix} \gamma\vec{\beta}\cdot\vec{\mathcal{E}} \\ \gamma(\vec{\mathcal{E}} + \vec{\beta}\times\vec{\mathcal{B}}) \end{pmatrix}. \quad (12.103)$$

Since the differential of proper time $d\tau = \frac{dt}{\gamma}$, if we divide both sides of (12.103) by γ we obtain (12.100) for the $\mu = 1, 2, 3$ components. Evaluating the $\mu = 0$ component of (12.103) we get

$$\gamma \frac{d\frac{E}{c}}{dt} = e\gamma\vec{\beta}\cdot\vec{\mathcal{E}} \quad (12.104)$$

and thus

$$\boxed{\frac{dE}{dt} = e\vec{v}\cdot\vec{\mathcal{E}}.} \quad (12.105)$$

Equation (12.105) is the equation of energy conservation: The increase in energy is due to the work done by the electric field. The magnetic field can never do work; hence it does not appear in (12.105).

12.18 RELATIVISTIC CIRCULAR MOTION

Uniform circular motion can be produced by a constant magnetic field pointing in a direction perpendicular to the plane of the motion. If we have a magnetic field of magnitude B_0 in the z direction, and neglect radiation as always, Equation (12.100) takes the form

$$\frac{d\vec{\pi}}{dt} = \frac{e}{c}(\vec{v}\times\vec{B}). \quad (12.106)$$

By taking the vector dot product of (12.106) with $\vec{\pi}$, we see that the magnitude of kinetic momentum $|\vec{\pi}|$ is constant if there is no electric field present. The left side is proportional to the time derivative of $\vec{\pi} \cdot \vec{\pi}$, and the right side is zero, since $\vec{\pi}$ is proportional to $\vec{v}$. Since the magnitude of the kinetic momentum remains constant, assume the motion is circular, with $\vec{\pi} = |\vec{\pi}|(\cos \omega t, \sin \omega t, 0)$, where ω is an unknown constant. Substituting this form for π into (12.106), we find

$$\omega = \frac{eB_0}{\gamma m c} \tag{12.107}$$

ω is known as the relativistic cyclotron frequency.

Because the motion lies on a circle in a plane perpendicular to the magnetic field, we can obtain the radius of this circle by the relation $\omega = \frac{v}{r}$. Substituting this into (12.107), we obtain the important relativistic relation

$$eB_0 r = |\vec{\pi}|c = mc^2 \gamma \beta. \tag{12.108}$$

The product of the charge times the magnetic field strength times the radius of curvature equals the kinetic momentum of the moving charged particle. Once again, we must be careful to distinguish between kinematic momentum, which is always proportional to velocity, and canonical momentum, which is not proportional to velocity if a vector potential is present. (The Hamiltonian formalism deals only with canonical momentum. Working out Hamilton's equations for relativistic particles is completely straightforward. We leave this to the homework problems.)

QUESTION 23: Units Formula (12.108) is of great practical use. Prove, by making the correct unit conversions, that:

$$\begin{aligned} \mathcal{B}[\text{gauss}] \, r[\text{cm}] &= \frac{1}{3} \times 10^4 \, (|\vec{\pi}|c) \, [\text{MeV}] \quad \text{(cgs system)}, \\ \mathcal{B}[\text{tesla}] \, r[\text{m}] &= \frac{10}{3} \, (|\vec{\pi}|c) \, [\text{GeV}] \quad \text{(mks system)}. \end{aligned} \tag{12.109}$$

We have assumed that the charge is the same as the electronic charge. The formulas above will then be true for electrons, positrons, protons, antiprotons, etc. One often sees the momentum of a particle quoted in units of "MeV/c" or "GeV/c."

SUMMARY OF CHAPTER 12

- Einstein's postulates are: a) all inertial frames are equivalent and b) the speed of light is the same for all observers.

PROBLEMS

- Space–time diagrams and the two Einstein hypotheses were used to find the form of the Lorentz transformation from one inertial reference frame to another:

$$x' = \gamma(x - \beta t), \quad t' = \gamma(t - \beta x),$$
$$y' = y, \quad z' = z. \tag{12.110}$$

($\gamma \equiv \frac{1}{\sqrt{1-\beta^2}}$, $\beta \equiv \frac{v}{c}$, v is velocity.)

- Two events that appear simultaneous for one observer will not be simultaneous for a second observer in relative motion to the first observer.
- Moving clocks slow down (time dilation), and moving lengths appear to shorten (Lorentz contraction).
- Rest mass is a form of energy. $E = mc^2$.
- Momentum and energy transform together like space and time. Relativistic momentum is $m\gamma\vec{v}$ and relativistic energy is γmc^2.
- All quantities that transform like space and time are called four-vectors. They are denoted by a^μ, with the index $\mu = 0, 1, 2, 3$, the first index representing time and the last three, the space coordinates.
- A metric tensor $g^{\mu\nu}$ can be used to raise and lower indices. The invariant scalar product of two four-vectors is written as $a^\mu b_\mu$. The Einstein summation convention is implied.
- The relativistic Lagrangian for a free particle can be found from the nonrelativistic limit of the action integral, the integral of the proper time along a world line. A variational principle gives the equations of motion just as for nonrelativistic mechanics:

$$L = -mc^2\sqrt{1 - \frac{v^2}{c^2}}. \tag{12.111}$$

- The Lagrangian in an electromagnetic field represented by the vector potential A^μ is

$$L = -mc^2\sqrt{1 - \frac{v^2}{c^2}} - e\Phi + e\frac{\vec{v}}{c} \cdot \vec{A}. \tag{12.112}$$

- The Lorentz force on a charged particle in an electromagnetic field can be derived from the Lagrangian (12.112). It has the same form for all observers in inertial frames.

PROBLEMS

Kinematics

Problem 1: *(Finding τ)* Find a relationship for the quantity τ from Section 12.1 in terms of β and L'. τ is the space–time distance in the unprimed frame of the time in the primed frame $t' = \frac{L'}{2}$.

Problem 2: *(Relative velocity)* Two particles are moving along orthogonal axes (say X and Y). Each particle moves at a speed β with respect to the same inertial frame. What is the relative velocity of the particles?

Problem 3: *(Length of a stick)* A stick measured to be length L' in its rest frame (call it the primed one) is seen to move by at a velocity of $.5c$ in another frame (call it the unprimed one). A ball is seen to move by at a velocity of $.5c$ in the unprimed frame but in the opposite direction.

a) How long does it take (according to the observer in the unprimed frame) for the ball to pass the stick?

b) The relative velocity of the ball and stick as seen in the unprimed frame is c. Does this violate the principle of relativity? Solve the problem in part a) using this relative velocity (if you have not already done so).

Problem 4: *(Does the pole fit in the barn?)* A person carrying a pole of rest length 10 meters is running on level ground and holding the pole in a horizontal position. The person runs at a speed of $\frac{3}{5}c$. The person runs through the open door of a barn of rest length 8 meters. The doors close at the instant the end of the pole reaches the far end of the barn. The people who shut the doors argue that the pole is Lorentz contracted by a factor $\frac{1}{\gamma} = \frac{4}{5}$ and therefore does fit into the barn. The person carrying the pole argues that the barn is contracted by the factor $\frac{4}{5}$ and hence that the pole does *not* fit into the barn.

a) Does the pole fit or not? Draw a space–time diagram to show the sequence of events, which should reveal what happens unambiguously. Where is the fallacy in the incorrect argument? Assume that the pole is strong enough to crash through the barn doors.

b) Derive a formula for the length (at rest) of the longest pole that can be run into a barn of length L as a function of γ.

Problem 5: *(Alternate derivation of relativistic velocity transformation)* The velocity transformation rules can be derived from the invariance of the speed of light. An observer in his own rest frame sees a train of length L moving with velocity v. He also sees a particle and a photon start a race down the length of the train. He sees the particle move at velocity $w < c$ and the photon move at velocity c. After a time T, he sees the photon reach the end of the train where it is reflected back. After a second time interval T', he sees the photon meet the particle. When this occurs, he observes that the particle has traveled a distance fL, with f being a dimensionless number less than one. The total elapsed time since the race started is $T + T'$.

a) Prove that

$$f = \frac{(c+v)(c-w)}{(c-v)(c+w)} \qquad (12.113)$$

from the point of view of the observer.

b) The fraction of the train's length the particle has traveled when it meets the reflected photon is certainly a number on which all observers will agree – it is an invariant quantity. Now consider the same sequence of events from the point of view of a passenger on the train. Assume that the passenger measures the particle's velocity to be u and the light's velocity to still be c. We can use the same formula, (12.113), and the invariance of c and f to claim that

$$f = \frac{(c-u)}{(c+u)}. \tag{12.114}$$

((12.114) is (12.113) with $w \to u$ and $v \to 0$.) Set the two expressions for f equal and thus prove the relativistic law for addition of velocities:

$$w = \frac{u+v}{1+\frac{uv}{c^2}}. \tag{12.115}$$

(From N. D. Mermin, *American Journal of Physics*, **51**, (12), pp. 1130–31 (1983).)

Problem 6: *(Four-acceleration)* The four-velocity was defined as $u^\mu \equiv \frac{dx^\mu}{d\tau}$, with τ equal to the proper time. Show that, if we define four-acceleration as $a^\mu \equiv \frac{du^\mu}{d\tau}$, the general expression for a^μ in an arbitrary inertial frame for a body moving at velocity $\vec{v} = c\vec{\beta}$ with respect to that frame is

$$a^\mu = (a^0, \vec{a}) = c\left(\gamma^4 \vec{\beta} \cdot \frac{d\vec{\beta}}{dt},\ \gamma^2 \frac{d\vec{\beta}}{dt} + \gamma^4 \left(\vec{\beta} \cdot \frac{d\vec{\beta}}{dt}\right)\vec{\beta}\right). \tag{12.116}$$

The formula (12.116) may be used to relate acceleration in one inertial frame to acceleration in another inertial frame. For motion in which the acceleration is perpendicular to the velocity, prove that $a^\mu = (0, \gamma^2 \vec{a})$, and if the acceleration is parallel to the velocity, prove that $a^\mu = \gamma^4(\beta|\vec{a}|, \vec{a})$. Finally, use these relations to prove that the relation between the laboratory acceleration of a particle undergoing circular motion at constant speed and the acceleration in the instantaneous rest frame of the particle is $a_{\text{rest}} = \gamma a_{\text{lab}}$. (The instantaneous rest frame for this case is an inertial frame moving tangent to the circle with the instantaneous velocity of the particle at the point of tangency.) For linear acceleration prove that the relation between laboratory and instantaneous rest frame acceleration is $a_{\text{rest}} = \gamma^3 a_{\text{lab}}$.

Problem 7: *(Time dilation for accelerated muons)* In an experiment done at the CERN laboratory in Geneva, Switzerland, muons were stored in a constant magnetic field causing the particles to move in a circle of radius 7 m (known as a storage ring). The momentum of the muons was 1.3 GeV/c. The muon mass is 105.7 MeV, and the charge is the same as for an electron.

a) Assuming a uniform magnetic field, what was the field in tesla?
b) What was the mean lifetime of these muons in the laboratory? (The mean lifetime of muons at rest is 2.2 μs.) The decay of these muons was observed over a time up to about five mean lifetimes. Perfect agreement with Einstein's time dilation was found, within very small experimental errors, despite the fact that the moving "clock" was not an inertial frame.
c) Prove that the "proper" (instantaneous rest frame) acceleration of the muons is $\approx 2 \times 10^{17} g$ ($g = 9.8$ m/s^2). This shows that the rate of moving clocks does not depend on acceleration but only on velocity (i.e., on γ).

Problem 8: *(Tachyons)* Hypothetical particles whose velocity is greater than the speed of light are known as tachyons. Imagine a situation where a person on Earth wishes to send a message via tachyons to another person who is far from the Earth. Assume the tachyons travel at a velocity $v_t > c$ with respect to the source of the tachyons.

a) If the receiving observer is at rest a distance L' away and immediately sends a tachyonic reply, how much time will have elapsed before the reply is received by the Earth-bound observer?
b) Now consider the same situation but with the receiving observer moving with a velocity $v_0 < c$ directly away from the Earth-bound observer. If the receiving observer is at a distance L' away exactly when the tachyonic message is obtained and the reply sent, how much time will have elapsed before the Earth-bound observer receives the reply? (Assume you can use the relativistic velocity transformation formulas.)
c) Show that if

$$v_0 > \frac{2v_t}{1 + \frac{v_t^2}{c^2}} \tag{12.117}$$

the Earth-bound observer could receive the reply before the original tachyonic message was sent.
d) Make a space–time diagram of the events of part b), for the case that the velocity v_0 satisfies (12.117). Explain how this diagram shows that the reply is received before the message is sent.

(From Lightman et al., *Problem Book in Relativity and Gravitation*.)

Problem 9: *(Deriving the Lorentz transformation; rapidity)* The hypothesis that the speed of light appears the same to all observers means

$$t'^2 - x'^2 = 0 \quad \text{implies} \quad t^2 - x^2 = 0 \tag{12.118}$$

(we assume $c = 1$ here for simplicity). In other words, points on the light cone for one observer in an inertial frame are also on the light cone for all other observers in inertial

PROBLEMS

frames. If we also assume that the correct transformation law between unprimed and primed observers is a linear transformation, a consequence of the invariance of $t^2 - x^2 = 0$ is that $t^2 - x^2$ is an invariant everywhere.* The linearity of the transformation laws means that there is no natural unit of length/time – doubling the units of length and time simultaneously for one observer means doubling them both for all observers.

a) Let $x_\pm \equiv t \pm x$. Einstein's hypothesis about the velocity of light (12.118) can be stated succinctly: $x_+ x_- = 0$ implies and is implied by $x'_+ x'_- = 0$ for all points on the light cone. If the primed x', t' coordinates are related by a linear transformation to the unprimed x, t coordinates, show that the most general form for $x'_+ x'_-$ is $A(\beta) x_+ x_+ + B(\beta) x_- x_- + C(\beta) x_+ x_-$, where β is the relative velocity between primed and unprimed frames, and $A(\beta), B(\beta), C(\beta)$ are undetermined functions of β. Explain why $A = B = 0$ if (12.118) is correct; hence

$$x'_+ x'_- = C(\beta) x_+ x_-. \qquad (12.119)$$

b) We can go from $x', t' \to x, t$ by the substitution $\beta \to -\beta$. Show that this implies $C(\beta) C(-\beta) = 1$. $\beta \to -\beta$ if the X axis is reflected, and (12.119) must hold under the transformation $x \to -x$. Why does this mean that $C(-\beta) = C(\beta)$. Hence $C^2 = 1$. $C = -1$ would give nonsense in the limit $\beta \to 0$, so it must be true that

$$x'_+ x'_- = x_+ x_- \qquad (12.120)$$

everywhere, not just on the light cone.

c) Show that (12.120) means that

$$\frac{t' + x'}{t + x} = \frac{t - x}{t' - x'} = f(\beta). \qquad (12.121)$$

($f(\beta)$ is a still unknown function of β. Redefine $f \equiv e^{-y}$. $y(\beta)$ is called the *rapidity*.)

d) Prove that, in terms of the rapidity, the transformation law must be

$$t' = \cosh[y] t - \sinh[y] x, \quad x' = \cosh[y] x - \sinh[y] t. \qquad (12.122)$$

Require that the point $x = \beta t$ corresponds to the origin $x' = 0$. Then we have the connection between rapidity and velocity:

$$\beta = \tanh[y]. \qquad (12.123)$$

Also prove that (12.123) is correct and that then the transformation (12.122) takes the usual form of the Lorentz transformation as a function of β.

---

* This proof is due to Prof. Kurt Gottfried of Cornell.

e) Show by using (12.121) that, for two successive "boosts" (Lorentz transformations) in the same direction, with rapidities y_1 and y_2, the rapidity of a single equivalent boost is $y = y_1 + y_2$. In other words, rapidity is additive.

Energy and Momentum

Problem 10: *(Doppler shifting)* One observer sees blue light (400 nm). How fast and in which direction do you need to be moving with respect to that observer so that you see red light (700 nm)?

Problem 11: *(Photon disintegration)*
a) Can a photon spontaneously disintegrate into an electron and a positron (both of mass m)? Why or why not?
b) How about if a stationary nucleus of mass M is nearby? Why or why not? What would be the minimum energy a photon could have to undergo disintegration?

(From Rindler, *Introduction to Special Relativity*, 2d ed.)

Problem 12: *(Exotic particle decays)* A B^0 meson with a mass (in energy units) of 5279.0 MeV is produced at rest in an electron–positron colliding beam storage ring. This meson sometimes decays according to the decay mode

$$B^0 \rightarrow J/\Psi + K^0, \qquad (12.124)$$

where J/Ψ is the name of a particle. The mass of J/Ψ is 3096.9 MeV. The K^0 meson subsequently decays into two π mesons: $K^0 \rightarrow \pi^+ + \pi^-$. The mass of a K^0 meson is 497.67 MeV and of π mesons is 139.57 MeV.

a) What is the momentum (in MeV/c) of the K^0 in the original B^0 decay? What is γ for this meson?
b) What momentum does each of the π mesons have in the rest frame of the K^0 after it decays?
c) The decay of the K^0 is observed as a "vee" some distance from the original production point of the B^0. What is the maximum possible opening angle of this $\pi^+\pi^-$ pair in the laboratory (rest frame of the B^0)?

Problem 13: *(The best way to make a Z_0)* High energy physicists originally were satisfied with accelerating a beam of electrons or protons to the highest possible energy and then bombarding a target in order to produce new particles. Eventually it was realized that most of the energy was being wasted on the motion of the center of mass of the beam and target particles. Assume that you want to produce the famous Z_0 vector boson, which is responsible for a part of the *weak force* in nature. The mass of this particle, in energy units, is $M_{Z_0} = 91.19$ GeV. You have three possible choices for experiments:

a) Bombard electrons in a target with high energy positrons. The reaction is $e^+ + e^- \to Z_0$. What is the minimum energy of the positrons you will need to do this?

b) Bombard a proton target with high energy electrons. Assume the final state contains a Z_0, a proton ($M_{\text{proton}} = 0.938$ GeV) and an electron ($M_{\text{electron}} = 0.0005$ GeV). What is the minimum energy required for the electron beam?

c) Make a colliding beam accelerator to collide a beam of electrons of energy E with a beam of positrons of the same energy E. In some cases a perfectly inelastic collision takes place, producing a Z_0 boson at rest. What is the necessary energy of your electron and positron beams? Compare this result with parts a) and b). Which way would you choose to make a Z_0?

Problem 14: *(Is energy always conserved in relativistic collisions?)* In ordinary non-relativistic mechanics when two masses collide inelastically and stick together, we say that mechanical energy is not conserved. Usually, kinetic energy is converted into heat. The description in relativistic mechanics is rather different. In all cases, the total energy is conserved, and any heat or other forms of energy generated contributes to the rest mass of the composite object.

Imagine two equal masses, each of mass m, with velocities β and $-\beta$ along the X axis with respect to the unprimed reference frame. These masses collide and stick together, forming a single massive object at rest, as can be seen from the symmetry (and also by momentum conservation). Now view this collision from an inertial (primed) frame moving to the left with velocity β. In the primed frame, the right-hand mass appears now at rest to this observer ($\beta'_1 = 0$). The left-hand mass has its velocity increased, according to the formula for velocity addition, to $\beta'_2 = \frac{2\beta}{1+\beta^2}$. After the collision, the composite mass must be moving to the right with velocity β, since it was at rest in the first reference frame.

a) Prove that the rest mass m^* of the composite particle is $\frac{2E}{c^2}$, where E is the *total* energy of either one of the two incident masses before the collision in the unprimed frame.

b) Using the known relativistic transformation properties of total energy and momentum, prove that energy and momentum are conserved in either reference frame. Could you *instead* assume energy and momentum conservation in both reference frames and deduce the velocity dependence of energy and momentum from this?

Problem 15: *(Using collisions to find relativistic form of momentum)* Assume that relativistic momentum is of the form

$$\vec{p} = \mathcal{P}(v)\vec{v}, \tag{12.125}$$

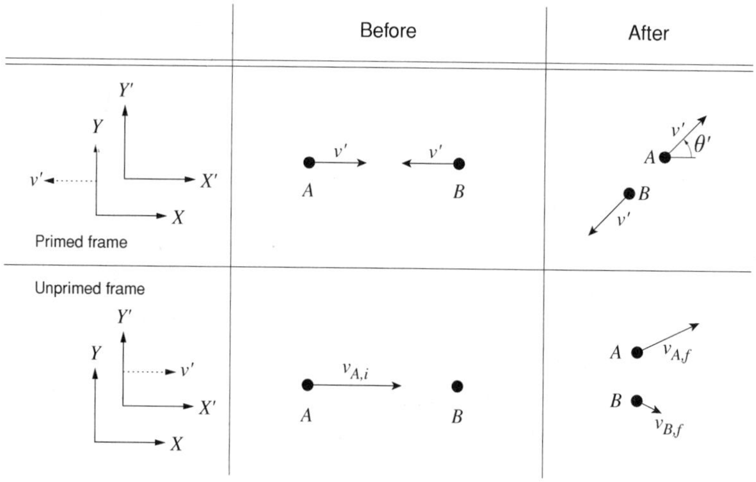

FIGURE 12.12

where $\mathcal{P}(v)$ is a function of the magnitude of the velocity. $\mathcal{P}(v \to 0) = m$ so that (12.125) reduces to its nonrelativistic form. Let's investigate a collision and see what $\mathcal{P}(v)$ must be for momentum conservation to hold in the relativistic case.

Assume we have a collision as drawn in Figure 12.12. In a particularly symmetric primed frame, particle A has velocity $\vec{v}'_{A,i} = \vec{v}'$ in the x direction before the collision and particle B has velocity $\vec{v}'_{B,i} = -\vec{v}'$ also in the x direction. After the collision, particle A has $\vec{v}'_{A,f} = \vec{v}'$ making an angle θ' with respect to the X axis, and particle B has velocity $\vec{v}'_{B,f} = -\vec{v}'$ making an angle $\theta' + \pi$ with respect to the X axis. Momentum is seen to be conserved in this case without knowing the form of $\mathcal{P}(v')$.

Now let's transform to the frame in which B is at rest before the collision, as shown in the unprimed frame in Figure 12.12.

a) Using the velocity transformation rules, find the x and y components of $v_{A,i}$, $v_{B,i}$, $v_{A,f}$, $v_{B,f}$ in terms of v', θ', and c only. Then rewrite them in terms of $\beta = \frac{v'}{c}$.

b) Using conservation of momentum in the y direction, find the ratio $\frac{\mathcal{P}(v_{A,f})}{\mathcal{P}(v_{B,f})}$. Then find what the ratio would be in the limit that $\theta' = 0$. You should be able to find the ratio $\frac{\mathcal{P}(v_{A,i})}{\mathcal{P}(0)}$ from this. Hint: When $\theta' = 0$, $\vec{v}_{A,f} = \vec{v}_{A,i}$ and $\vec{v}_{B,f} = 0$.

c) Prove that

$$\frac{1+\beta^2}{1-\beta^2} = \frac{1}{\sqrt{1-\frac{v_{A,i}^2}{c^2}}}. \tag{12.126}$$

Use this to find $\mathcal{P}(v_{A,i})$. Check that you obtain the expected form for momentum of $\vec{p} = \gamma m \vec{v}$.

(From Jackson, *Classical Electrodynamics*, 2d ed.)

Problem 16: *(Transverse momentum)* Prove, using the same collision as in the previous problem (see Figure 12.12), that when you make a Lorentz transformation into a

PROBLEMS

moving frame, the relativistic momentum transverse to the motion between the frames does not change. For example, in the previous problem we transformed between frames moving in the x direction. Therefore, the individual (not total here) y momentum of both particle A and particle B should be the same in either frame. Make sure to use the relativistic form for momentum here; otherwise the y momentum will change.

Problem 17: *(Compton scattering)* Individual photons behave like particles of rest mass equal to zero. When a photon scatters from a free electron, the electron can recoil, taking up both energy and momentum. There is a unique relationship between the energy of the scattered photon, E'_γ and the scattering angle from the incident photon, θ'. This relation also depends on the electron rest mass m_e and the incident photon energy E_γ. See Figure 12.13. If the four-momentum of the incident and scattered photons are denoted by k^μ, k'^μ respectively, the statement of energy and momentum conservation is $k^\mu + p_e^\mu = k'^\mu + p'^\mu_e$, where p_e^μ and p'^μ_e are the initial and final energy–momentum four-vectors of the electron.

a) Prove that the conservation laws of energy and momentum require that

$$E'_\gamma(\theta) = \frac{E_\gamma}{1 + \frac{E_\gamma(1-\cos\theta')}{m_e c^2}}. \tag{12.127}$$

Hint: Set $c = 1$, solve the four-momentum conservation equation for $p'_{e,\mu}$, and then "square" $p'^\mu_e p'_{e,\mu} = m_e^2$. In these units the momentum of the photon equals its energy. You may also want to use $k_\mu k^\mu = k'_\mu k'^\mu = 0$ and $p_{e,\mu} p_e^\mu = p'_{e,\mu} p'^\mu_e = m_e^2 c^4$.

b) From (12.127), prove that the maximum possible photon energy for backscattered photons, regardless of the incident photon energy, is $\frac{m_e c^2}{2} = 255$ KeV.

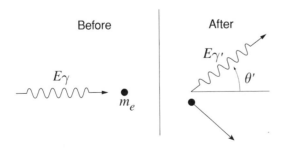

FIGURE 12.13
Compton scattering. (Notice that here the prime indicates the after collision quantities rather than quantities in another frame.)

Invariance

Problem 18: *(Transformation properties of $\mathcal{E}$ and $\mathcal{B}$)* We will use the invariance of $\partial^\mu \equiv (\frac{\partial}{\partial t}, -\vec{\nabla})$ to deduce the transformation properties of $\mathcal{E}$ and $\mathcal{B}$ fields from the postulated invariance of Maxwell's equations in vacuum (here written in cgs Gaussian units):

$$\vec{\nabla} \cdot \vec{\mathcal{E}} = 4\pi\rho, \quad \vec{\nabla} \cdot \vec{\mathcal{B}} = 0,$$
$$\vec{\nabla} \times \vec{\mathcal{E}} = -\frac{1}{c}\frac{\partial \vec{\mathcal{B}}}{\partial t}, \quad \vec{\nabla} \times \vec{\mathcal{B}} = \frac{4\pi}{c}\vec{J} + \frac{1}{c}\frac{\partial \vec{\mathcal{E}}}{\partial t}. \tag{12.128}$$

a) First derive a second-order equation for electric and magnetic fields separately from the first-order equations above. For example, take the curl of the $\vec{\nabla} \times \vec{\mathcal{E}}$ equation and use the vector calculus identity $\vec{\nabla} \times (\vec{\nabla} \times \vec{\mathcal{E}}) = \vec{\nabla}(\vec{\nabla} \cdot \vec{\mathcal{E}}) - \nabla^2 \vec{\mathcal{E}}$. By combining this result with another Maxwell equation, obtain the equation

$$\left(\frac{1}{c^2}\frac{\partial^2}{\partial t^2} - \nabla^2\right)\vec{\mathcal{E}} = -4\pi\left(\frac{1}{c}\frac{\partial}{\partial t}\left(\frac{\vec{J}}{c}\right) + \vec{\nabla}\rho\right). \tag{12.129}$$

b) The left side of (12.129) involves only the components of the electric field operated on by a Lorentz invariant, second-order, differential operator, which can also be written $\partial_\mu \partial^\mu$. The transformation properties of the right and left sides have to be the same if Maxwell's equations are to be Lorentz invariant. On the right side, if $j^\mu \equiv (\rho, \frac{\vec{J}}{c})$, we have an expression that is proportional to the $\mu = 0, \nu = i$, $i = 1, 2, 3$ components of the antisymmetric tensor $\partial^\mu j^\nu - \partial^\nu j^\mu$. Verify this last statement. Then prove that this result implies that $\vec{\mathcal{E}}$ must be three components of a tensor $F^{\mu\nu}$.

c) Check that the other components (for $\vec{\mathcal{B}}$) of the antisymmetric tensor arise from the same type of equation and supply the right side of the covariant equation below:

$$\left(\frac{1}{c^2}\frac{\partial^2}{\partial t^2} - \nabla^2\right)F^{\mu\nu} = ? \tag{12.130}$$

Relativistic Lagrangian

Problem 19: *(Positronium acceleration)* When electrons (e^-) and positrons (e^+) annihilate into pure electromagnetic energy, at first a compound called "positronium" is formed. Depending on the way the spins of the two particles are oriented, two reactions are then possible, each with its own lifetime, before annihilation takes place: a) 25% of the time, $e^+ + e^- \to 2$ photons with a mean lifetime of 0.125 nanoseconds or b) 75% of the time $e^+ + e^- \to 3$ photons, with a mean lifetime of 138.6 nanoseconds.

Suppose that you wish to accelerate the long-lived state of positronium to an energy of 1 TeV (10^{12} electron volts). Assume, as in Section 12.16, that $\gamma(t) = 1 + at$, where a is a constant, and t is the time measured in the laboratory frame. To keep from

PROBLEMS

losing the positronium during the acceleration, you want the elapsed proper time to be equal to or less than one mean lifetime. Approximating the mass of positronium by that of two electrons, what is the minimum value of the constant a that will accomplish this goal? Convert your answer to an acceleration in units of MeV/m, assuming the velocity of the positronium is close to the velocity of light. What is the minimum length of the accelerator in kilometers? (Note: There is no known way to accelerate positronium, so this question is a purely hypothetical one.)

Problem 20: *(Relativistic Lagrangian)* Prove that terms proportional to $u_\mu u_\nu F^{\mu\nu}$ and $u_\mu x_\nu F^{\mu\nu}$ do not enter into the action differential (12.86). Hint: Use symmetry arguments for the former, and translation invariance for the latter. Challenge: What about a term like $\partial^\mu A_\mu$? Hint: Look at the nonrelativistic limit of this term – does it have any physical meaning?

Problem 21: *(Relativistic harmonic oscillator)* As a somewhat contrived problem, solve the 1-D relativistic harmonic oscillator. The Lagrangian is defined to be

$$L \equiv -mc^2\sqrt{1-\beta^2} - V(x), \quad V(x) \equiv \frac{1}{2}kx^2. \quad (12.131)$$

Find the total energy, the equation of motion, and the period expressed as an integral from $x=0$ to the maximum amplitude $x=x_0$. In your integral, evaluate the total energy at the maximum amplitude. Express your integral in terms of the dimensionless variable $u \equiv \frac{x}{x_0}$ and possible other dimensionless constants. If the amplitude is extremely small, the oscillator will be nonrelativistic, with a period of $2\pi\sqrt{\frac{m}{k}}$. Prove that for larger amplitudes, the period depends on the amplitude. Show that the correction factor for the period is $1 + \frac{3}{8}\frac{V(x_0)}{mc^2}$. (Neglect terms of order higher than x_0^2.) Thus the period of a relativistic harmonic oscillator is lengthened as the amplitude grows larger. Give a physical reason for this effect.

Problem 22: *(Relativistic Hamiltonian for the electromagnetic field)* Prove that the Hamiltonian for a relativistic charged particle in an electromagnetic field with arbitrary space and time dependence is given by

$$H(\vec{r}, \vec{p}, t) = e\Phi + c\sqrt{(\vec{p}-e\vec{A})^2 + m^2c^2}. \quad (12.132)$$

In the formula above, $\vec{r}$ is the position and $\vec{p}$ the *canonical* momentum of the charged particle, which has charge e; $\Phi = \Phi(\vec{r}, t)$ is the electrostatic potential and $\vec{A} = \vec{A}(\vec{r}, t)$ is the vector potential.

Experiments Leading to Relativity

Problem 23: *(Michelson–Morley experiment)* Eighteen years before Einstein published the Theory of Special Relativity, Michelson and Morley did a famous experiment that was designed to detect the Earth's motion through the "ether" – the medium

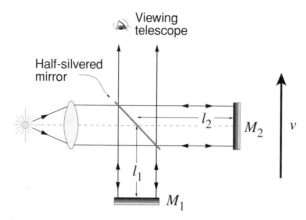

FIGURE 12.14
Michelson–Morley experiment.

in which it was thought that light traveled at velocity c. The time difference of round trips made by two light beams was measured by an extremely sensitive interferometric technique, as sketched in Figure 12.14. The interferometer shown is assumed to be moving through the ether at velocity v in the vertical direction.

a) Using Newtonian kinematics, prove that the time t_1 for the trip from the half-silvered mirror to M_1 and back and the time t_2 for the trip from the half-silvered mirror to M_2 and back are given by the formulas

$$t_1 = \frac{2l_1}{c} \frac{1}{1-\beta^2}, \quad t_2 = \frac{2l_2}{c} \frac{1}{\sqrt{1-\beta^2}}. \quad (12.133)$$

Let $\Delta t \equiv t_1 - t_2$ and $\Delta t' \equiv t_1' - t_2'$, where the prime means that the apparatus has been rotated 90° with respect to the velocity v as shown in Figure 12.14. Prove that

$$c(\Delta t' - \Delta t) = \frac{2}{\sqrt{1-\beta^2}}(l_1 + l_2)$$

$$\times \left(1 - \frac{1}{\sqrt{1-\beta^2}}\right) \approx -(l_1 + l_2)\beta^2. \quad (12.134)$$

b) If the formulas (12.133, 12.134) were correct, the observed interference fringes would shift when the apparatus rotated by a number given by the ratio of $\frac{c(\Delta t' - \Delta t)}{\lambda}$, where λ is the wavelength of the light used. Michelson and Morley looked at possible fringe shifts as a periodic function of time throughout the year, because the Earth's velocity reverses its direction during that time. An upper limit of about $\frac{1}{3}$ of the Earth's orbital velocity, or 10 km/s, was placed on the velocity through the ether. Subsequent experiments with more modern and quite different techniques have placed the limit at 5 cm/s.

c) The null result of this experiment was "explained" by Fitzgerald and later Lorentz who advanced the hypothesis that moving lengths shrink in the direction of their motion through the ether. The velocity dependence would be $l = l^0\sqrt{1 - \beta^2}$, where $\beta = \frac{v}{c}$ and v is the velocity parallel to the length. Explain why this would cancel the effect in the original version of the Michelson–Morley experiment. The Lorentz–Fitzgerald contraction hypothesis means that no fringe shift will be observed upon rotation of the apparatus through 90°: $c\,\Delta t' = c\,\Delta t$.

d) According to the Lorentz–Fitzgerald hypothesis, the position of the fringes *does* shift as the velocity changes in magnitude if the lengths of the two arms are not equal. Prove that, if moving lengths shrink parallel to their motion,

$$c\,\Delta t = \frac{2(l_1^0 - l_2^0)}{\sqrt{1 - \beta^2}}. \tag{12.135}$$

Subsequent experiments done with unequal length arms in the interferometer showed no such fringe shift.

Problem 24: *(Velocity of light in a moving medium)* Starting in 1851, Fizeau performed a series of experiments in which he measured the velocity of light in a column of water moving parallel to the light at velocity v. The velocity of light in water at rest is $c_{\text{water}}(0) \equiv \frac{c}{n}$, where c is the velocity of light in a vacuum and $n > 1$ is the index of refraction. He observed that, if the water moves at velocity v,

$$c_{\text{water}}(v) = c_{\text{water}}(0) + kv, \quad \text{where } k = 1 - \frac{1}{n^2}. \tag{12.136}$$

His explanation for this was that the ether is "dragged" such that (12.136) is true. "Ether-dragging" in this strange way struck Einstein as a most unlikely explanation of this experimental result and influenced his search for a different explanation. Show that if you use the relativistic velocity addition formula for the addition of velocities in the direction of motion, then (12.136) is a consequence of this formula. Assume that $v \ll c$.

Problem 25: *(Stellar aberration)* Careful observations of the stars reveal that they appear to move in very small ellipses, which depend in size on the elevation of the star above the plane in which the Earth and the Sun rotate (ecliptic). The maximum effect is an ellipse with a semimajor axis of angular width 20.6 seconds of arc and a period of 1 year (for stars directly overhead, at the zenith of the ecliptic). This "stellar aberration" is due to the motion of the Earth around the Sun.

a) In a desperate attempt to explain the null result of the Michelson–Morley experiment (see Problem 24), it was hypothesized that the Earth dragged the ether with it, so that the Earth was always at rest with respect to the local ether. If this were true, explain why we would not observe stellar aberration.

b) In principle we could test the theory of relativity by measuring this stellar aberration very precisely. In a pre-relativity theory with a stationary ether, the relation

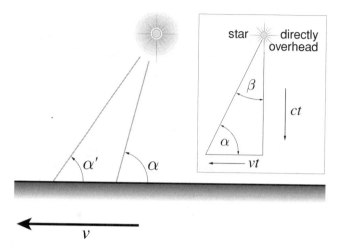

FIGURE 12.15
Stellar aberration – lines show path of light from star. α corresponds to the angle if the Earth didn't move and α' to the angle actually observed. The inset shows the relationship if the star was directly overhead.

between the angles α and α' is (see Figure 12.15)

$$\tan \alpha' = \frac{\sin \alpha}{\beta + \cos \alpha}. \qquad (12.137)$$

Show by using the relativistic form of the velocity addition formulas for vertical and horizontal components of the velocity of a photon (or a classical light wave) that the correct *relativistic* formula for stellar aberration is

$$\tan\left(\frac{\alpha'}{2}\right) = \sqrt{\left(\frac{1-\beta}{1+\beta}\right)} \tan\left(\frac{\alpha}{2}\right). \qquad (12.138)$$

c) Check that the magnitude of the effect is correct for stars that are at the zenith, $\alpha = 90°$. The angle from the vertical is $\approx \beta$ to first order in β in either case. The only difference occurs in second order. In practice, the difference between (12.137) and (12.138) is too small to measure, but the very existence of the observed effect rules out an ether dragged along with the Earth as a possible explanation of the null result of the Michelson–Morley experiment.

APPENDIX

THE TWIN PARADOX

Identical twins part company on their twentieth birthday. Twin A climbs into a rocket ship and leaves the Earth on an extended voyage into space. Twin B, who remains on the

APPENDIX THE TWIN PARADOX

Earth, sees A move away. After a few years of travel at very high speeds, B sees A reverse his direction of motion and head back towards Earth. A's description is exactly the same, except B moves with a equal but opposite relative velocity. When they meet, one of the twins is just over 27 years old. The other twin has a biological age of almost 32. *Viewed from the perspective of twin A, the speed of twin B would seem to change in the same way as when twin B describes the motion of A.* How do we decide which twin is older when they meet again? This is known as the "Twin Paradox."

The purpose of the trip is to explore the nearest star, α Centuri, for undiscovered planets.* This star lies at a distance of 4.3 light years from the Earth. To simplify the problem, during the first quarter of the trip, A accelerates at a constant rate. At a point halfway to the star, A's rocket reverses its acceleration and the rocket slows down. At the moment it reaches α Centuri, the rocket is nearly at rest with respect to Earth. In the return trip, the acceleration is constant again until halfway to Earth, then reversed so that the rocket arrives at rest as it lands on Earth. The alert reader will notice that there *is* an asymmetry here: Twin A is accelerated, while twin B is not. This is the key to resolving the twin paradox.

Denote the time and distance measured by B on Earth t and x. Twin A also has a clock that registers his local time τ, also called A's proper time. It is useful to calculate the time $t(\tau)$ and distance $x(\tau)$ of twin A in B's reference frame in terms of τ. Once we have this, we can make a space–time diagram that shows clearly how the intervals of A's clock are related to intervals of B's clock.

First we need to obtain a four-vector proportional to acceleration, by differentiating the four-velocity (12.65) is with respect to the scalar τ:

$$a^\mu \equiv \frac{du^\mu}{d\tau} = c\left(\gamma^3 \vec{\beta} \cdot \frac{d\vec{\beta}}{d\tau},\ \gamma^3\left(\vec{\beta} \cdot \frac{d\vec{\beta}}{d\tau}\right)\vec{\beta} + \gamma \frac{d\vec{\beta}}{d\tau}\right). \qquad (12.139)$$

Equation (12.139) was obtained by noting that the β^2 in the definition of γ can be written as $\vec{\beta} \cdot \vec{\beta}$ before differentiating. We can greatly simplify (12.139) by restricting ourselves to the case where $\frac{d\vec{\beta}}{d\tau} \parallel \vec{\beta}$. In that case, we can assume that all motion takes place along the X axis, writing the four-vector a^μ as (a^0, a^1). Then we have

$$a^\mu = c\frac{d\beta}{d\tau}(\gamma^3 \beta, \gamma^3). \qquad (12.140)$$

(Obtaining (12.140) from (12.139) makes use of the identity $\gamma^2 \beta^2 + 1 = \gamma^2$.)

In the frame of the accelerated twin A, $\gamma = 1$, $\beta = 0$, $c\frac{d\beta}{d\tau} \equiv g(\tau)$, where g is the acceleration twin A reads on his accelerometer. It is a function of his time τ since he wants to stop to examine the planets, before returning to Earth. The form of a^μ in A's frame is, explicitly,

$$\text{frame of } A: (a^0, a^1) = (0, g). \qquad (12.141)$$

Since the four-acceleration is a four-vector, we can obtain a second expression for it by Lorentz transforming (12.141) into B's frame. The Lorentz transformation is the same for

---

* For a report on the discovery of planets outside the solar system, see Alan P. Boss, "Extrasolar Planets," *Physics Today*, September 1996.

all four-vectors:

$$\text{frame of } B: (a^0, a^1) = (\gamma\beta g, \gamma g). \tag{12.142}$$

Equating (12.142) to (12.140), we obtain

$$\gamma^2 \frac{d\beta}{d\tau} = \frac{g}{c}. \tag{12.143}$$

Equation (12.143) is a differential equation for $\beta(\tau)$. Assuming that twin A starts with zero velocity, we can integrate both sides of this equation to obtain

$$\int_0^\beta \gamma^2 \, d\beta = \frac{1}{c} \int_0^\tau g(\tau) \, d\tau \equiv \theta(\tau). \tag{12.144}$$

Equation (12.144) defines $\theta(\tau)$ (a dimensionless parameter) for any arbitrary acceleration g as a function of time in A's rest frame. The integral on the left side of (12.144) is $\frac{1}{2} \ln \frac{1+\beta}{1-\beta}$. With this result, we have the solution

$$\beta(\tau) = \tanh\theta, \quad \gamma(\tau) = \cosh\theta. \tag{12.145}$$

To obtain x and t as functions of τ, integrate the four-velocity with respect to τ, using (12.145):

$$x(\tau) = c \int_0^\tau \sinh\theta(\tau)\, d\tau, \quad t(\tau) = \int_0^\tau \cosh\theta(\tau)\, d\tau. \tag{12.146}$$

From the expression for t in (12.146), we can see immediately that $t \geq \tau$, since $\cosh\theta \geq 1$. Twin B will be older than twin A unless $g = 0$. For constant acceleration in A's rest frame an analytic solution to (12.146) is possible.

If g is a constant, then $\theta = \frac{g\tau}{c}$. This gives

$$x = \frac{c^2}{g}(\cosh\theta - 1), \quad t = \frac{c}{g} \sinh\theta. \tag{12.147}$$

If we assume that g equals its value on Earth (for the comfort of the passengers), we find from (12.147) that $\theta_{\frac{1}{4}} = 1.836$ after the first quarter of the round trip to the distant star. Converting $\theta_{\frac{1}{4}}$ to the time τ observed by A, he reverses his acceleration after 1.781 years. When he lands back on Earth, twin A is 7.1 years older than when he started. Assuming twin A started at the age of 20, he arrives back here on Earth at the age of 27. How old then is twin B? Using (12.147) to calculate the time elasped for twin B we obtain 11.9 years, and he is now almost 32 years old. Had twin A undertaken an even longer journey, the age difference would have been even more. It is a way to travel into the future and search

APPENDIX THE TWIN PARADOX

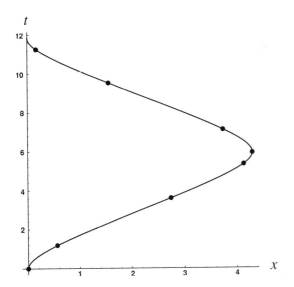

FIGURE 12.16
B's view of A's trip. The Y axis is twin B's time in years; the X axis is distance between twin A and twin B, as measured by B, expressed in light years.

for other planetary systems at the same time! At the halfway distance while outbound to the star, A's $\gamma = 3.216$. His maximum velocity relative to B, divided by c, differs from 1 by only .05. This high speed is reached even with a relatively gentle acceleration equal to that of gravity on Earth, an acceleration that could be tolerated for an indefinite period. Unfortunately, one would need to carry many times the rest mass of the human cargo and convert it all to kinetic energy, making such a trip highly impractical.

The whole trip is depicted in the space–time diagram in Figure 12.16. The diagram is seen from twin B's perspective. The lack of symmetry between the world lines for the twins* is rather clear from the diagram. The difference in the clock rates for the twins is largest at the midpoints of the journey out and back, as can be seen from the separation of the dots on the curve, which are years as recorded by twin A.

---

* Twin B's world line is straight up the time axis.

BIBLIOGRAPHY

CHAPTERS 1–3

Elementary and Intermediate-Level Textbooks on Classical Mechanics

Most of the books listed here have a review of particle kinematics and standard vectorial Newtonian mechanics.

Baierlein, R., *Newtonian Dynamics*, McGraw-Hill, 1983.
Corben, H. C., and Stehle, P., *Classical Mechanics*, second edition, John Wiley & Sons, 1960.
Fowles, G. R., and Cassiday, G. L., *Analytical Mechanics*, fifth edition, Saunders Coolege Publishing, Harcourt Brace College Publishers, 1993.
Kittel, C., Knight, W. D., and Ruderman, M. A., *Mechanics*, Berkeley Physics Course, vol. 1, McGraw-Hill, 1965.
Kleppner, D., and Kolenkow, R. J., *An Introduction to Mechanics*, McGraw-Hill, 1973.
Knudsen, J. M., and Horth, P. G., *Elements of Newtonian Mechanics*, Springer, 1995.
Landau, L. D., and Lifshitz, E. M., *Mechanics*, third edition, Pergamon Press, reprinted 1986.
Marion, J. B., and Thornton, S. T., *Classical Dynamics of Particles and Systems*, fourth edition, Harcourt Brace, 1995.
Slater, J. C., and Frank, N. H., *Mechanics*, McGraw-Hill, 1947.
Sommerfeld, A., *Mechanics*, Academic Press, 1964.
Symon, K. R., *Mechanics*, third edition, Addison-Wesley, 1971.

Graduate-Level Textbooks

Arnold, V. I., *Mathematical Methods of Classical Mechanics*, Springer-Verlag, 1978.
Fetter, A. L., and Walecka, J. D., *Theoretical Mechanics of Particles and Continua*, McGraw-Hill, 1980.
Goldstein, H., *Classical Mechanics*, second edition, Addison-Wesley, 1980.
Whittaker, E. T., *A Treatise on the Analytical Dynamics of Particles and Rigid Bodies*, fourth edition, Cambridge Univ. Press, 1959.

Mathematics and Programming Textbooks

Kaplan, W., *Advanced Calculus*, third edition, Addison-Wesley, 1984.
Leon, S. J., *Linear Algebra*, second edition, Macmillan Publ. Co., 1986.
Percival, I., and Richards, D., *Introduction to Dynamics*, Cambridge Univ. Press, 1982.
Press, W. H., et. al., *Numerical Recipes in C*, second edition, Cambridge Univ. Press, 1992.

Other References

Galilei, G., translated by H. Crew, A. de Salvio, *Dialogues Concerning Two New Sciences*, Prometheus Books, 1991.
Lanczos, C., *The Variational Principles of Mechanics*, Univ. of Toronto Press, 1970.
Mach, E., *The Science of Mechanics*, ninth German edition, sixth American edition, Open Court Publishing Co., 1960, reprinted 1989.
Pippard, A. B., *The Physics of Vibration*, combined volume, Cambridge Univ. Press, 1989.
Smith, D. E., *A Source Book in Mathematics*, Dover Press, 1959.

CHAPTERS 4–6

History

Alioto, A. M., *A History of Western Science*, second edition, Prentice Hall, 1993.
Chandrasekhar, S., *Newton's Principia for the Common Reader*, Clarendon Press, 1995.
Cohen, I. B., *The Birth of a New Physics*, Penguin Books, 1987, reprinted 1992.
Drake, S., *Galileo at Work: His Scientific Biography*, Univ. of Chicago Press, 1978.
Koyré, A., translated by R. E. W. Maddison, *The Astronomical Revolution: Copernicus–Kepler–Borelli*, Cornell Univ. Press, 1973.
Lindberg, D. C., *The Beginnings of Western Science*, Univ. of Chicago Press, 1992.

Astronomy

Anderson, H. L., ed., *AIP 50th Anniversary Physics Vade Mecum*, American Institute of Physics, 1981.
Hetherington, N. S., ed., *Cosmology: Historical, Literary, Philosophical, Religious, and Scientific Perspectives*, Garland Publishing, Inc., 1993.
Maran, S. P., ed., *The Astronomy and Astrophysics Encyclopedia*, van Nostrand Reinhold and Cambridge Univ. Press, 1992.
Sagan, C., *Cosmos*, Random House, 1980.
Sciama, D. W., *Modern Cosmology and the Dark Matter Problem*, Cambridge Univ. Press, 1993.

CHAPTERS 7 AND 8

Chandrasekar, S., *Ellipsoidal Figures of Equilibrium*, Dover Press, 1987.
Imbrie, J., and Imbrie, K. P., *Ice Ages: Solving the Mystery*, Harvard Univ. Press, 1979.
Munk, W. H., and Macdonald, G. J. F., *The Rotation of the Earth*, Cambridge Univ. Press, 1960.
Sciama, D. W., *The Physical Foundations of General Relativity*, Doubleday & Co., 1969.

CHAPTERS 9 AND 10

Brillouin, L., *Wave Propagation in Periodic Structures*, Dover, 1953.
Hayashi, C., *Nonlinear Oscillations in Physical Systems*, Princeton Univ. Press, 1985.
Mickens, R. E., *An Introduction to Nonlinear Oscillations*, Cambridge Univ. Press, 1981.
Nayfeh, A. H., *Introduction to Perturbation Techniques*, John Wiley & Sons, 1981.

Pauli, W., *Theory of Relativity*, Pergamon Press, 1959.
Sanders, J. A., and Verhulst, F., *Averaging Methods in Nonlinear Dynamical Systems*, Springer-Verlag, 1985.

CHAPTER 11

Popular and Semipopular

Gleick, J., *Chaos, Making a New Science*, Penguin Books, 1988.
Peterson, I., *Newton's Clock. Chaos in the Solar System*, W. H. Freeman & Co., 1993.
Schroeder, M., *Fractals, Chaos, Power Laws. Minutes from an Infinite Paradise*, W. H. Freeman & Co., 1991.
Stewart, I., *Does God Play Dice? The Mathematics of Chaos*, Blackwell Publishers, 1989.

Introductory Textbooks and Articles

Baker, G. L., and Gollub, J. P., *Chaotic Dynamics. An Introduction*, second edition, Cambridge Univ. Press, 1996.
Bergé, P., Pomeau, Y., and Vidal, C., *Order Within Chaos. Towards a Deterministic Approach to Turbulence*, John Wiley & Sons, 1984.
Berry, M. V., "Regular and Irregular Motion," reprinted in *Hamiltonian Dynamical Systems – a Reprint Collection*, see below.
Helleman, R. H. G., "Self-Generated Chaotic Behavior in Nonlinear Mechanics," review article published in *Fundamental Problems in Statistical Mechanics*, vol. V, pp. 165–233, 1980.
Tabor, M., *Chaos and Integrability in Nonlinear Dynamics: An Introduction*, John Wiley & Sons, 1989.
Thompson, J. M. T., and Stewart, H. B., *Nonlinear Dynamics and Chaos*, John Wiley & Sons, 1986.

Graduate-Level Textbooks

Gutzwiller, M. C., *Chaos in Classical and Quantum Mechanics*, Springer-Verlag, 1990.
Jackson, E. A., *Perspectives of Nonlinear Dynamics*, two volumes, Cambridge University Press, 1990.
Lichtenberg, A. J., and Lieberman, M. A., *Regular and Chaotic Motion*, Springer-Verlag, 1992.
Marek, M., and Schreiber, I., *Chaotic Behaviour of Deterministic Dissipative Systems*, Cambridge University Press, 1991.
McCauley, J. L., *Chaos, Dynamics and Fractals. An Algorithmic Approach to Deterministic Chaos*, Cambridge University Press, 1993.
Moon, F. C., *Chaotic Vibrations. An Introduction for Applied Scientists and Engineers*, John Wiley & Sons, Inc., 1987.
Ott, E., *Chaos in Dynamical Systems*, Cambridge University Press, 1993.
Reichl, L. E., *The Transition to Chaos in Conservative Systems: Quantum Manifestations*, Springer-Verlag, 1992.
Schuster, H. G., *Deterministic Chaos. An Introduction*, second revised edition, VCH Publishers, 1988.

Reprint Collections and Other References

Civitanović, P., *Universality in Chaos*, second edition (reprints), The Institute of Physics Publishing, reprinted 1993.

Ferrax-Mello, S., ed., *Chaos, Resonance and Collective Dynamical Phenomena in the Solar System*, Kluwer Academic Publishers, 1992.

MacKay, R. S., and Meiss, J. D., *Hamiltonian Dynamical Systems – A Reprint Collection*, Adam Hilger, 1987.

Richter, P. H., and Scholz, H. J., "Chaos in Classical Mechanics: The Double Pendulum," *Stochastic Phenomena and Chaotic Behaviour in Complex Systems*, ed. by P. Schuster, pp. 86–97, Springer-Verlag, 1984.

CHAPTER 12

Beck, A., translator, *The Collected Papers of Albert Einstein*, vol. 2, Princeton Univ. Press, 1989.

French, A. P., *Special Relativity*, Norton & Co., 1968.

Geroch, R., *General Relativity from A to B*, Univ. of Chicago Press, 1978.

Jackson, J. D., *Classical Electrodynamics*, second edition, John Wiley & Sons, 1975.

Kenyon, I. R., *General Relativity*, Oxford Univ. Press, 1990.

Pais, A., '*Subtle is the Lord...*,' Oxford Univ. Press, 1982.

Rindler, W., *Introduction to Special Relativity*, second edition, Oxford Univ. Press, 1991.

Schlipp, P. A., ed., *Albert Einstein, Philosopher-Scientist*, Tudor Publishing Co., 1951.

Schutz, B. F., *A First Course in General Relativity*, Cambridge Univ. Press, 1985.

Schwinger, J., *Einstein's Legacy*, W. H. Freeman & Co., 1986.

Taylor, E. F., and Wheeler, J. A., *Spacetime Physics*, W. H. Freeman & Co., 1966.

REFERENCES

Boss, A. P., "Extrasolar Planets," *Physics Today*, **49**, pp. 32–38, 1996.
Brans, C., and Dicke, R. H., "Mach's Principle and a Relativistic Theory of Gravitation," *Phys. Rev.*, **124**(3), pp. 925–935, 1961.
Case, W. B., "The Pumping of a Swing From the Standing Position," *American J. of Phys.*, **64**(3), pp. 215–220; Case, W. B., and Swanson, M. A., "The Pumping of a Swing from the Seated Position," *American J. of Phys.*, **58**(5), pp. 463–467; and Burns, J. A., "More Pumping on a Swing," *American J. of Phys.*, **38**(7), pp. 920–922.
Einstein, A., "Does the Inertia of a Body Depend upon its Energy Content?," *Annalen der Physik* 18 (1905), English translation in *The Collected Papers of Albert Einstein*, volume 2, translated by A. Beck, Princeton Univ. Press, 1989, pp. 172–174.
Feigenbaum, M. J., *J. Stat. Phys.*, **19**, p. 25ff, 1987.
Feigenbaum, M. J., *J. Stat. Phys.*, **21**, p. 699ff, 1979.
Gehrels, T., ed., *Asteroids*, Univ. of Arizona Press, 1979.
Gibbons, A., "Putting Einstein to the Test – In Space," *Science*, Nov. 15, 1991, pp. 939–941.
Grmela, M., and Marsden, J., eds., "On the Prevalence of Aperiodicity in Simple Systems," *Global Analysis*, Springer-Verlag, 1979.
Jung, K., "Figur der Erde," Geophysics I, vol. XLVII, *Encyclopedia of Physics*, Springer-Verlag, 1956.
Laplace, P. S., *A Philosophical Essay on Probabilities*, translated by F. W. Truscott and F. L. Emory, Dover, 1951.
Laskar, J., *Celestial Mechanics and Dynamical Astronomy*, **64**, pp. 115–62, 1996.
Lighthill, J., *Proc. Roy. Soc. (London)*, **A407**, p. 35, 1986.
Lightman, A. P., et. al., *Problem Book in Relativity and Gravitation*, Princeton University Press, 1975.
Mermin, N. D., "Relativistic Addition of Velocities Directly from the Constancy of the Velocity of Light," *American Journal of Physics*, **51**(12), pp. 1130–1131, 1983.
Pais, A., *'Subtle Is The Lord...' The Science and Life of Albert Einstein*, Oxford Univ. Press, 1982, p. 152.
Poinsot, L., *Théorie Nouvelle de la Rotation des Corps*, Bachelier, 1851.
Rindler, W., *Introduction to Special Relativity*, second edition, Oxford University Press, 1991.
Rucker, R., *Infinity and the Mind*, Birkhauser, 1982.
Schiff, L. I., "Observational Basis of Mach's Principle," *Reviews of Modern Physics*, **36**(2), pp. 510–511, 1964.
Sussman, G. J., and Wisdom, J., *Science*, **257**, p. 56, 1992.

Tremaine, S., "Is the Solar System Stable?," in *Frontiers of Astrophysics, Proceedings of the Rosseland Centenary Symposium*, P. B. Lilje, et. al., eds. 1995.

Von Laue, M., "Inertia and Energy," in *Albert Einstein: Philosopher-Scientist*, ed. by P. A. Schlipp, Tudor Publishing Co., 1951.

Wisdom, J., "Chaotic Behavior in the Solar System," in *International Conference on the Physics of Chaos*, M. Duong-Van, ed., 1987.

Wolf, A., Swift, J. B., Swinney, H. L., and Vastano, J. A., "Determining Lyapunov Exponents from A Time Series," *Physica* **16D**, pp. 285–317, 1985.

INDEX

accelerated systems, Hamiltonian dynamics. *See* Hamiltonian dynamics, in accelerated systems
acceleration
   and relation to mass, 512–513
   block on inclined plane, 2–3
   caused by a constant force, 535–536
   elevator, 20
   transformation of coordinate systems, 266
action functional S, 51, 178, 220
action principle. *See* variation principle
   extended. *See* extended action
   relativistic. *See* relativistic action
   variable, 230–231
action-angle variables, 230–235, 237, 238, 246–247 (prob)
   pendulum. *See* pendulum, action-angle variables
adiabatic invariants, 118, 233–235
airplane wings, 397
Almagest, 130
almost periodic, 238
amplifiers, parametric. *See* parametric amplifiers
analytical mechanics, 23
angle variable, 230, 232
angular frequency, 526
angular momentum, 137–139, 170–172, 283, 245 (prob)
   intrinsic. *See* angular momentum, rotational
   rotational, 292
   total, 291–292
   translational, 292
angular momentum barrier, 140–141
angular velocity, 256–257, 264–266, 283, 293, 304, 306, 276–277 (prob)
anharmonic oscillator, 398–401
   driven, 401–411
   period of, 401
annihilation radiation, 517
antisymmetric four-tensor, electric and magnetic fields, 530, 539
antisymmetric matrix, 264, 304
aphelion, ellipse. *See* ellipse, aphelion
apparent force, 267
approximate solutions to non-analytic problems, 383–411
apsides, ellipse. *See* ellipse, apsides
arc length, 138

area-preserving map. *See* map, area-preserving
Aristarchus of Samos, 317
Aristotle, 317
Arnol'd. *See* KAM theorem
Arnold diffusion, 451, 470, 472
asteroid belt, 469
asteroid, perturbed by Jupiter (prob), 417
astrophysical data, table of, 167–169
asymmetric top, 288, 331 (prob)
   kinetic energy, 306
   torque free, 305–312
attractive force, hyperbolic orbits. *See* hyperbolic orbits, attractive force
attractor, 452 (*see also* phase space trajectories)
   periodic, 425
   strange. *See* strange attractor
Atwood machine, double (prob), 31
axial vector. *See* pseudovector
axions, 155

band theory, of solids, 390
basins of attraction, 453
bead on wire (prob), 30, 34, 114
beam stability, 390
beat frequency. *See* frequency, beat
Bernoulli, Daniel, 45, 86, 344
Bernoulli, Jacques, 4
Bernoulli, Jean, 44–46, 344
Bertrand's theorem, 154
Bertrand's theorem (prob), 167
bicycle (prob), 27
bifurcation diagrams, 459–460, 484–485, 491
   pitchfork, 459, 483–484, 486
Big Bang, 325, 456
binding energy, 516
Biot-Savart law, 509
Birkhoff, 439 (*see also* Poincaré-Birkhoff theorem)
black body radiation, 325
black hole M87, 155
block on inclined plane, example, 1–7
body cone, 297, 298
body coordinate system, 259–264, 284, 300
boldface notation, 262
boost. *See* Lorentz transformations

bouncing ball (prob), 475
boundary conditions, matching. *See* matching boundary conditions
boundary curve for Poincaré section of double pendulum, 431
bowling ball, 38–39, 35 (prob)
box-counting dimension. *See* dimension, box-counting
brachistochrone, 44–46, 70 (prob)
Brahe, Tycho, 131
Britain, dimension of coastline. *See* dimension, of coastline of Britain
broken symmetry. *See* symmetry breaking
brown dwarfs, 155
bug crawling on a phonograph turntable, Hamiltonian and Lagrangian, 190–192
butterfly effect, 425

calculus of variations. *See* variational calculus
cannonball example, 52, 77
canonical transformations, 207–213, 215–218, 237, 250, 241–244 (prob)
canonically conjugate momentum, 22, 75, 170, 172, 175, 181, 538, 34–35 (prob)
Cantor set, 464–465
capacitor (prob), 116
capacity dimension. *See* dimension, box-counting
Cassini, 339–340
catenary curve, 62
causal Green's function. *See* Green's function, causal
cavity, cylindrical. *See* gedanken experiment, cylindrical cavity
center of mass, 134, 136–137
  motion, 283, 285, 292, 298, 160–161 (prob)
center of momentum. *See* center of mass
central force, 24, 133–141, [161–162, 198, 413] (prob)
  (*see also* Kepler problem)
  motion
    effective potential, 140–141
    energy diagram, 142
    Kepler's Laws, 132,139
    orbit equation, 142–143
    stability of circular orbit, 384–388
centrifugal force, 142, 267
centripetal acceleration, 40
Ceres, 148
cgs units, 507, 540
chain example, 61
chain rule, 6
Chandler wobble, 299–300
Chandler, S., 300
chaos, 237, 475 (prob)
  in conservative systems, 379, 425, 428–452, 472
  in dissipative systems, 425, 452–468
chaotic dynamics, 423–474
characteristic exponent, 393
  multiplier, 444
  polynomial, 337
charge conservation, 188
circular hoop with 3 point masses connected by massless springs, 365–367
  motion, relativistic. *See* relativistic circular motion
  orbits, 144, 385
classically forbidden, 126
clockwork universe, 423
coastline. *See* dimension, of coastline of Britain
cofactors, 355, 362, 380–382

comets (prob), 163
complex numbers, 111
  solutions, 89
Compton, A. H., 252, 280
Compton generator (prob), 280
Compton scattering (prob), 549
configuration space, 36–37, 229
conic section, 144
conjugate momentum. *See* canonically conjugate momentum
conservation of angular momentum, 170–172
  of charge. *See* charge conservation
  of energy, relativistic (prob), 547
  of momentum. *See* momentum conservation
conservative forces, 18, 41–43
conserved function, 21
  quantity, 174
constant of the motion, 21, 22, 138, 170, 172, 196 (prob)
constraint force, 2, 57
  generalized, 60–61
constraints, 12–13
  holonomic, 13, 57–58, 83, 85, 33 (prob)
  nonholonomic, 36–40, 62–64, 35 (prob)
  nonintegrable, 36
  rheonomic, 13, 22
  scleronomic, 13, 22, 85
contact transformation, 208, 211
continued fraction expansion, 436–437
continuity equation, 188, 525
continuous transformations. *See* transformations, continuous
contravariant components, 529
control parameter, 481
coordinate system
  body. *See* body coordinate system
  rotating. *See* rotating reference frames
  space. *See* space coordinate system
  polar
    cylindrical, 10, 19, 24–25
    elliptical, 309–310
    spherical, 10, 51, 138, 226
coordinates, generalized. *See* generalized coordinates
Copernicus, Nicolaus, 130–131
Coriolis force, 267
cosinelike, 390
cosmic rays, 537
Coulomb potential. *See* hyperbolic orbits, repulsive force
coupled pendulum. *See* pendulum, two coupled
covariant components, 529
  procedure, 538–539
Cramer's rule, 380
critical damping (prob), 117, 121
critically damped. *See* oscillator, damped simple harmonic
cross section (prob), 165
cubic polynomials, 315
curvature, of potential energy, 85–86
cyclic coordinates. *See* ignorable coordinates
cyclotron frequency, relativistic, 540

d'Alembert, 4, 322
d'Alembert's Principle, 5
damped driven pendulum, 425, 453–463, 467–468
  oscillator. *See* oscillator, damped
dark matter, 154–156

INDEX

day, length of in age of dinosaurs (prob), 282
degeneracy, 339, 343, 353
   modes, 365–367
degrees of freedom, 2, 8, 10–12, 36, 86, 27 (prob)
   rotational, 283
delta function, 99–100
derivative
   directional. *See* directional derivative
   Lie. *See* Lie derivative
determinant, 203, 352
deterministic chaos, 424
diagonalization, 392
   of symmetric matrices, 336–339
   of **t** and **v** matrices, 364–365
diatomic molecule (prob), 377
diffeomorphism, 55
differential equation
   linear, inhomogeneous, 95
      ordinary, second-order, and homogeneous, 88–89
      superposition, 88–89
   second order, 24
differential equations of motion. *See* equations of motion
dimension, 463–468
   box counting, 463–466
   fractional, 463–464 (*see also* fractal)
   Hausdorff, 463
   Lyapunov, 467
   magnification, 466
   of coastline of Britain, 465–466
   of strange attractor of damped driven pendulum, 468
Dirac delta function. *See* delta function
Dirac, P. A. M., 44
directional derivative, 236
discontinuity, in Green's function, 101–102
discrete transformations. *See* transformations, discrete
displaced axis theorem, 289
displacement ratio. *See* mode displacement ratio
dissipative flow, 189
divergence, 192
   exponential of trajectories. *See* characteristic exponent
Doppler shift, 505, 507, 515, 546 (prob)
dot cancellation, 14
dot product. *See* scalar product
double pendulum
   exact, 425, 426–428, 430–433, 435, 440–442, 451–452
   Hamiltonian, 426–427
   Lagrangian, 349
   small angle approximation, 348–355, 358–359, 377 (prob)
double-well potential (prob), 157
driven oscillator. *See* oscillator, driven
driven support of pendulum. *See* pendulum, with driven support
DSHO. *See* oscillator, damped simple harmonic
Duffing oscillator, 384, 398–401, [158, 419] (prob)
   driven, 401–411
   period of, 401
dumbbell, inertia tensor, 290–291
dynamical equilibrium. *See* equilibrium, dynamical
   instability. *See* instability, dynamical
   stability. *See* stability, dynamical
   variable, 7

Earth. *See* solar system, stability
   as noninertial reference frame, 267–275
   equatorial bulge. *See* equatorial bulge

eccentric anomaly, 148–150
eccentricity, 323, 470–473
   ellipse, 144
ecliptic, 317
effective potential, 140–142, 385
eigenvalues, 302–303, 336–339, 392, 444, 449
   generalized, 353, 359
eigenvectors, 302–303, 337–339, 392, 444, 449
   generalized, 353
Einstein, A., 154, 493–495, 508, 514, 516, 519, 533
Einstein summation convention, 253, 529
electric and magnetic field, antisymmetric four-tensor. *See* antisymmetric four-tensor, electric and magnetic field
electric field, 493, 508–510, 522
   charged particle. *See* particle in electric and magnetic fields
   transformation law for light pulse. *See* transformation law, electric field for light pulse
electric force, transformation law. *See* transformation law, electric and magnetic forces
electromagnetic field (prob), 199–200, 550, 551
electromagnetic field, relativistic Lagrangian. *See* relativistic Lagrangian, electromagnetic field
electron capture, 185
electron-positron annihilation, 517
electrostatics (prob), 67
elevator, 20, 200 (prob)
ellipse, 144–145
   aphelion, 145
   apsides, 145
   eccentricity, 144
   perihelion, 145
   semimajor axis, 145–146
   semiminor axis, 145–146
ellipsoid
   equation of, 306
   inertia, 306
   inertia tensor (prob), 327
   momental, 306
   oblate, 289
   prolate, 289
elliptic fixed points. *See* fixed points, elliptic
   orbits, 145, 385, 162 (prob)
elliptical motion
   period of, 145–147
   time dependence, 147–150
energy, 21 (*see also* Hamiltonian)
   of light wave, transformation law. *See* transformation law, energy of light wave
   binding. *See* binding energy
   relativistic, 515–517, 533
   stored in oscillator. *See* oscillator, stored energy
energy-momentum four-vector. *See* four-vector, energy-momentum
entrainment (prob), 480
EOM. *See* equations of motion
epicyles, 130–131
equant, 130–131
equations of motion, 17–19, 23–24, 51, 54
equator, 339–340
equatorial bulge
   Earth's, 339–342, 332 (prob)

equilibrium, 82–87, 114 (prob)
   dynamical, 384
   neutral, 85
   pendulum, 82–83
   stable, 82, 125
   unstable, 82, 125
equinoxes
   nutation of (prob), 333
   precession of, 317–323
ether, 493
Euler angles, 284, 300–302
Euler equation, 46–49, 65
Euler force, 267
Euler's equations, 283, 292–293, 297–299,
      329–330 (prob)
Euler, Leonard, 23, 46, 299
Euler-Lagrange equations, 19, 23–24, 51, 54
   matrix form, 352
   normal coordinates, 359
exotic particle decay (prob). *See* particle decay (prob)
exponent, characteristic. *See* characteristic
      exponent
extended action, 76
extremum, 46, 49

Falkland Islands battle (prob), 279
Feigenbaum constant, 458–459, 487, 490
Feigenbaum, M. J., 457, 485, 486
Fermat, 55
Fermat's principle, 55, 75, 229, 68 (prob)
fictitious force, 25, 253, 267, [279–282, 290] (prob)
fig tree, 485, 486
filamentation, 186
fixed point, 481–482
   elliptic, 431, 434–435, 440–441, 445
   hyperbolic, 435, 440–442, 446–448
   neutral, 428–483, 485
   stable, 482, 485
   superstable, 482–486
   unstable, 482–382, 485
Fizeau, M., 493
flip hyperbolic fixed point, 446
Floquet matrix, 390–392
Floquet Theory, 390–395, 397, 445
fluid, incompressible. *See* incompressible fluid
forbidden, classically. *See* classically forbidden
force, causing constant acceleration, 535–536
forced oscillator. *See* oscillator, driven
Foucault pendulum, 253, 272–275, 324
Foucault pendulum (prob), 281
four-acceleration, 555, 543 (prob)
four-momentum, 527
four-vector, 522
   energy-momentum, 502, 526–528
four-velocity, 527
Fourier analysis, 86, 403–404
   decomposition, 390
   series, 402
fractal, 425, 453, 463–468
frame-dragging, 324–325
frames. *See* reference frames
free particle, relativistic Lagrangian. *See* relativistic
      Lagrangian, free particle
free solution. *See* oscillator, transient solution
frequency
   beat, 348

cyclotron. *See* cyclotron frequency
mode. *See* mode frequency
of oscillator, 88
resonant, 108
frequency of light wave, transformation law. *See*
      transformation law, frequency of light wave
friction, 38, 90
full width at half max, 109
functional, 44, 46
functional calculus. *See* variational calculus
fundamental theorem of symmetric real matrices,
      336

Galilean Invariance Principle, 54
Galilean transformation laws, 504
Galileo, G., 87, 105, 131–132, 494
Gauss, 148
Gauss-Jordan elimination, 337
Gaussian units. *See* cgs units
gedanken experiment, cylindrical cavity, 519–521
general relativity, 44, 154, 494, 530,
      420–422 (prob)
generalized coordinates, 10, 350
   equations of motion, 17–18
   forces, 14–15, 18–19, 82
      central force, 24–25
      kinetic energy, 16–17
   momentum. *See* canonical momentum
   velocities, 14
generating function, 207, 210–217, 240–244 (prob)
geodesic curve, 51, 66–67 (prob)
Gibbs, J. W., 253
global chaos, 442, 451
golden mean, 437
governor (prob), 32
gradient vector, 307
grandfather's clock. *See* pendulum, nonlinear
gravitational constant, 339–342
gravitational field, particle in, Hamiltonian-Jacobi
      equation, 220–222, 224–228
   potential energy of, 318
gravity, 134, 141–142, 159 (prob)
Gravity Probe B, 317
Green's function, 96, 102, 399, 120 (prob)
   causal, 102
   drive with arbitrary force, 103–105
   simple harmonic oscillator, 99–102, 103, 105
Gutzwiller, M. C., 237, 442
gyroscope, 323, 330 (prob)

hairy ball theorem. *See* Poincaré-Hopf theorem
Hale-Bopp, 163
Halley, Edmond, 132–133, 147–148
Hamilton, 48
Hamilton's Characteristic Function, 222, 230
   equations of motion, 180–181, 207, 218, 248–249
   Principal Function, 219
   Principle, 46, 51, 53–55, 75, 178–186, 220, 72 (prob)
      cannonball example. *See* cannonball example
Hamilton-Jacobi equation, 207, 218–230,
      246–247 (prob)
   particle in a gravitational field, 220–222, 224–228
Hamiltonian, 21–22, 175, 178, 180, 211, 234, [29–30,
      551] (prob)
   bug crawling on a phonograph turntable. *See* bug
      crawling on a phonograph turntable

INDEX

chaos. *See* chaos, in conservative systems
charged particle in electric and magnetic fields, 192
double pendulum, exact. *See* double pendulum, exact Hamiltonian
dynamics, 175–184, 197–198 (prob)
  in accelerated systems, 190–195
invariance. *See* invariance, Hamiltonian
particle on a parabolic wire. *See* particle on a parabolic wire, Hamiltonian
relativistic. *See* relativistic Hamiltonian
simple harmonic oscillator, 88, 182
spherical pendulum. *See* spherical pendulum, Hamiltonian
handedness, 456
harmonic analysis, 402–404
harmonic oscillations, 383
Hausdorff dimension. *See* dimension, Hausdorff
heavy symmetric top. *See* symmetric top, heavy
heliocentric, 130–132, 317
Henon-Heiles Hamiltonian (prob), 378, 478
Hermitian matrices, 337
herpolhode, 308, 311–312
heteroclinic point, 447
  tangle, 448
Hill equation, 383, 389–395, 409, 411
Hill, G. W., 389
Hipparchus of Rhodes, 317
holonomic constraints. *See* constraints, holonomic
homoclinic point, 447
  tangle, 446–448
homogenous differential equation. *See* differential equation
Hooke, Robert, 132
Hooke's Law, 154
Hubble telescope, 155
hurricanes, 253, 271–272, 281 (prob)
Huygens, C., 87, 125, 480
Hyakutake, 163
hyperbola, 144, 150
hyperbolic fixed points. *See* fixed points, hyperbolic
  orbits, 145, 150–154, 164 (prob)
    attractive force, 150–151, 153
    repulsive force, 150–154
hyperplane, three-dimensional, 429
hysteresis, 402, 408–409

ice ages, 323
ice cream cone rolling on a table, 257–259
identity matrix, 262
identity transformation. *See* transformations, identity
ignorable coordinates, 22–23, 138, 226
impact parameter (prob), 165
impulse force, 99
  arbitrary force as superposition of, 104–105
  response of oscillator to, 100–101
inclination, 472–473
inclined plane, example, 1–7
incompressible fluid, 188
index of refraction (prob). *See* refractive index (prob)
inertia ellipsoid. *See* ellipsoid, inertia
inertia tensor, 283, 286–291, 339, 327–328 (prob)
  dumbbell, 290–291
inertial reference frame, 20, 252, 323–324, 494, 502, 535
infinitesimal rotation. *See* rotation, infinitesimal

infinity, orders of, 465
inhomogenous differential equation. *See* differential equation
initial conditions, 7–8, 89
instability, 392
  dynamical, 383, 386
instantaneous angular velocity. *See* angular velocity
  axis of rotation. *See* rotation axis
  rest frame, 526–527, 535–536
integrable, 425, 447–448, 451–452
integrable systems, 207, 235–237
intermittency, 457
invariable line, 308–309
invariable plane, 308
invariance, 28–29 (prob)
  Hamiltonian, 211
  Lagrangian, 171, 173, 209
  rotational, 170–172
  translational, 135–136
  theorems. *See* Lorentz invariant, invariance theorems
  transformation. *See* transformation, invariance
invariant tori, 237–239, 424, 431–433, 435
invariants, adiabatic. *See* adiabatic invariants
  Lorentz. *See* Lorentz invariant
inverse square force, 134
invertible mapping, 55
involution, 235–236
irrational winding number. *See* winding number, irrational
irregular motion, 425, 447, 451
iteration methods, 384, 405

Jacobian, 203, 249–251, 432, 443–444, 452, 454, 241 (prob)
  of Lorentz transformation, 501
Jupiter, perturbation effects. *See* asteroid, perturbed by Jupiter (prob)
  See solar system, stability

KAM barriers, 451
  theorem, 435–439
  tori, 433–439, 448, 451, 469
Kepler ellipses, precession of (prob), 420–422
Kepler, J., 131–132
Kepler orbit (prob), 413
  problem, 130–133, 141–150, 196 (prob), 226, 228 (*see also* central force problem)
  perturbation of, 384–388
Kepler's equation, eccentric anamoly, 149
  Laws, 131–132, 139–140, 147
kinematics, 252
kinetic energy, [27, 328] (prob)
  asymmetric top. *See* asymmetric top, kinetic energy
  block on inclined plane, 4
  function of both coordinates and velocities, 16–17
  matrix, Taylor series near equilibrium, 350
  moving constraints, bead on rotating wire example, 7–9
  relativistic change, 515–516
  rigid body, 284–286
  rotational, 285–287
    symmetric top. *See* symmetric top, rotational kinetic energy
  sliding ladder example, 16
  total, 285
  translational, 285
kinetic momentum, 538–539

Kirkwood, 469
Kirkwood gap, 469–470
Kohlrausch, 493
Kolmogorov. *See* KAM theorem
Kovalevskaya, S. V., 451
Kronecker delta, 286, 307

Lagrange multipliers, 56–65, 73 (prob)
  chain example, 61
  for problems with explicit holonomic constraints, 57–58
  linear pendulum example, 58–59
  nonintegrable nonholonomic constraints, penny example, 62–64
Lagrange, Joseph Louis, 1, 23, 86
Lagrangian, 1, 19, 51, 54–55, 83–86, 209, [28, 30–34, 199–200] (prob)
  bug crawling on a phonograph turntable. *See* bug crawling on a phonograph turntable
  cannonball. *See* cannonball
  central force. *See* central force
  central force, spherical polar coordinates, 138
  charged particle in electric and magnetic fields, 192
  equation of motion, 19
  Foucault pendulum. *See* Foucault pendulum
  heavy symmetric top. *See* symmetric top, heavy, Lagrangian
  in center of mass frame, 137
  invariance. *See* invariance, Lagrangian
  linear pendulum. *See* pendulum, linear
  matrix form, 351
  mechanics, relativistic. *See* relativistic Lagrangian mechanics
  momentum space. *See* momentum space Lagrangian
  nonlinear pendulum. *See* pendulum, nonlinear
  normal coordinates, 359
  penny. *See* penny
  simple harmonic oscillator, 88
Lao-Tze, 437
Laplace, 423, 468
Laplace-Runge-Lenz vector (prob), 196
Larmor frequency, 193
Larmor's Theorem, 192–193
Laskar, J., 471–474
least action. *See* Hamilton's Principle
Legendre transformation, 175–180, 211, 213, 197 (prob)
Leibnitz, 45, 75
length contraction. *See* Lorentz contraction
Lenz vector, 228
LHC, 452
libration, of pendulum. *See* pendulum, nonlinear, libration
Lie derivative, 236
light cone, 522
light pulse, electric field transformation law. *See* transformation law, electric field for light pulse
light wave energy, transformation law. *See* transformation law, energy of light wave
light
  bending (prob), 68–69
  emitted by atom, 514–517
  minimal optical path, 55
  speed of. *See* speed of light
Lighthill, Sir James, 474
Lindstedt-Poincaré perturbation theory, 384, 398–401, 419–420 (prob)

linear differential equation. *See* differential equation
  oscillator. *See* oscillator, simple harmonic
  pendulum. *See* pendulum, linear
  system, 343
linearizing, 383, 385–386
linearly accelerated reference frames. *See* reference frames, linearly accelerated
Liouville's theorem, 184–189, 202–204, 250, 391, 444, 449, 452
local chaos, 442, 451
locomotive (prob), 267
logistic map, 425, 481–492, 474 (prob)
Lorentz, 494
  contraction, 500
  force, 199, 508, 537–539
  invariant, 521–526, 538
    action, 530–532
    charge, 508, 534
    invariance theorems, 524
    mass, 519, 527
    Planck's constant, 512
    proper time, 527
  transformation, 498–501, 503–504, 555–556, 544 (prob)
Lorentzian line shape, 109
Lorenz equations (prob), 477
Lyapunov dimension. *See* dimension, Lyapunov exponents
  conservative systems, 449–450
  dissipative systems, 458–459, 467, 471–472
  logistic map, 483, 485–486
Lyapunov time, 471

Mach, E. 324, 513, 528
MACHOS, 155
Mach's principle, 323–325
Maclaurin, 342
magnetic field, 493, 509–510, 522, [30, 74, 277, 550–551] (prob)
  charged particle. *See* particle in electric and magnetic fields
  relativistic particle. *See* particle in a magnetic field, relativistic
magnetic force, transformation law. *See* transformation law, electric and magnetic forces
magnification. *See* dimension, magnification
manifold
  stable, 440, 446
  unstable, 440, 446
map
  area-preserving, 431–432
  logistic. *See* logistic map
  Poincaré section, 434, 439–440
  standard (prob). *See* standard map (prob)
  tangent. *See* stability matrix
marginal stability. *See* stability, marginal
Mars. *See* solar system, stability
mass
  definition of, 512–513
  inertial, 324
  reduced, 136–137
  rest, 513–517
matching boundary conditions, 96–98
Mathematical Principles of Natural Philosophy, 133
Mathieu equation, 383, 395–398
  functions, 397

matrix
    antisymmetric. *See* antisymmetric matrix
    Hermitian. *See* Hermitian matrix
    identity. *See* identity matrix
    mondronomy. *See* stability matrix
    notation, 262
    orthogonal. 263
    stability. *See* stability matrix
    transformation, 265
    unitary. *See* unitary matrix
Maupertuis, 55, 339
    cannonball trajectory, 77
Maupertuis' Principle, 75–80, [74, 162] (prob)
Maxwell, J., 493
Mayer, 514
m-cycle, logistic map, 481
Mechanique Analytique, 23, 86
Mercury. *See* solar system, stability
    precession of, 154
method of matching boundary conditions, 96–98
    of successive approximations. *See* successive
        approximations, method of, 356
metric tensor, 51, 529
Michelson-Morley experiment (prob), 551
Milankovitch, M., 323
minimal action principle of Maupertuis. *See* Maupertuis'
    Principle
minimal optical path for light, 55
Minkowski, 495, 501
mixer, 402
mks units. *See* SI units
mode. *See* normal mode
    amplitude ratio. *See* mode displacement ratio
    displacement ratio, 343, 346, 352, 353–355, 357, 361,
        363–367
    frequency, 343, 353, 357, 362–366, 374
    vector, 352, 355, 362–367
mode-locking. *See* phase locking
molecule
    diatomic (prob). *See* diatomic molecule (prob)
    triatomic linear. *See* triatomic molecule, linear
moment of inertia tensor. *See* inertia tensor
momental ellipsoid. *See* ellipsoid, momental
moments, principal. *See* principal moments
momentum
    canonically conjugate. *See* canonically conjugate
        momentum
    relativistic, 517–519, 547 (prob)
    transverse, relativistic, 517, 532, 548 (prob)
momentum conservation, 519
    space, 189–190, 201–202 (prob)
        Lagrangian, 190
mondronomy matrix. *See* stability matrix
Moon. *See* solar system, stability
Moser. *See* KAM theorem
multiplicative ergodic theorem. *See* Oseledic's
    multiplicative ergodic theorem
mylar, aluminized, 517

neutral equilibrium. *See* equilibrium, neutral
neutral fixed point. *See* fixed point, neutral
neutrinos, 155, 528
Newton, Isaac, 45, 132–133, 284, 299, 324, 339–342,
    423, 468, 493
Newtonian mechanics, review with sliding block example,
    1–3

Newton's Second Law, 2, 5, 17, 512–513
    Third Law, 3
Noether, Emmy, 172
Noether's theorem, 172–175, 245
nonconservative force, 90
    example of a rubber band, 42–43
nonconstraint force, 5
nondissipative chaos. *See* chaos, in conservative
    systems
nonholonomic constraints. *See* constraints, nonholonomic
noninertial reference frames. *See* reference frames,
    noninertial
nonintegrable, 36, 63, 207, 237, 305, 425, 442, 447–448,
    451–452
nonlinear oscillator. *See* oscillator, nonlinear
    pendulum. *See* pendulum, nonlinear
    resonance, 425
        double pendulum, 440–442
normal coordinates, 343, 359
    mode, 343, 352–355
    mode frequency. *See* mode frequency
normalization, 338, 356–357
North Pole, 339–340
N-torus. *See* torus
nuclear magnetic resonance, 277
nutation, 316, 333 (prob)

oblate ellipsoid. *See* ellipsoid, oblate
obliquity, 317
    chaotic of Earth, 474
ODE. *See* differential equation, ordinary
one-dimensional systems, 123–125, 158 (prob)
orbital period, 147
orbits, of Kepler problem, 143–151
orders of infinity. *See* infinity, orders of
ordinary differential equation. *See* differential equation,
    ordinary
orthogonal matrix, 263, 277–278 (prob)
    modes, 355–356
    transformation, 288, 302, 336
oscillations
    harmonic. *See* harmonic oscillations
    small. *See* small oscillations
    subharmonic. *See* subharmonic oscillations
oscillator
    anharmonic. *See* anharmonic oscillator
    damped, 81
    damped simple harmonic, 90–94, 103, 106–110,
        116–120 (prob)
        complex driving force, steady state, 108
        critically damped, 92
        overdamped, 92
        response to sinusoidal driving force, relative phase,
            110–113
        underdamped, 91
    driven, 81, 94–113
        damped simple harmonic (prob), 120–121
        simple harmonic (prob), 119
    Duffing. *See* Duffing oscillator
    linear. *See* oscillator, simple harmonic
    nonlinear, 125–127, 121–122 (prob)
    simple harmonic, 81, 88–90, 124, 212–213, 218,
        237–239, [115, 202, 240, 246, 551] (prob)
        general solution with Green's function, 99–102, 105
        Hamiltonian. *See* Hamiltonian, oscillator, simple
            harmonic

simple harmonic response to sinusoidal driving force, 106
steady-state solution, 95, 97–98, 107–108
step function drive force, 96–99
stored energy, 94, 109
transient solution, 95, 97–98, 107–108
Oseledic's multiplicative ergodic theorem, 449
overdamped. *See* oscillator, damped simple harmonic

parabola, 144
parabolic orbits, 145, 164 (prob)
parallel axis theorem, 289
parallel plate capacitor, 508–510
parametric amplification, 384
   amplifiers, 390
   resonance, 384, 388–398, 414–419 (prob)
      unstable, 397
particle decay, 528, 546 (prob)
   in a gravitational field, Hamiltonian-Jacobi equation, 220–222, 224–228
   in a magnetic field, [30, 74] (prob)
      relativistic, 539–540
   in electric and magnetic fields
      Lagrangian and Hamiltonian, 192
   on a parabolic wire
      Hamiltonian, 182–183
particular solution. *See* oscillator, steady-state solution
pendulum. *See also* oscillator, [33–34, 116–117, 373, 415, 419] (prob)
   action-angle variables, 231–232
   damped driven. *See* damped driven pendulum
   double. *See* double pendulum
   equilibrium, 82–83
   linear, 58–59
   nonlinear, 125–130
      libration, 126
      period, 128–130
      phase portrait, 127–128
      rotation, 126
   spherical. *See* spherical pendulum
   spring. *See* spring pendulum
   upside-down. *See* upside-down pendulum
   with driven support, 388–398, 414 (prob)
   two coupled, 344–348, 376–377 (prob)
pendulum (prob), physical. *See* physical pendulum (prob)
pendulum (prob), triple. *See* triple pendulum (prob)
penny example, 62–64, 73 (prob)
perihelion, ellipse. *See* ellipse, perihelion
period
   anharmonic oscillator. *See* anharmonic oscillator, period of
   Duffing oscillator. *See* Duffing oscillator, period of
   nonlinear pendulum. *See* pendulum, nonlinear
   of elliptical motion. *See* elliptical motion, period of
   orbital. *See* orbital period
period doubling, 456–458, 483–485
   cascade, 457
perpetual motion machine, 514
perturbation, 383, 384–388, 433, 439, 442
perturbation theory, 398, 404–408
   Lindstedt-Poincaré. *See* Lindstedt-Poincaré perturbation theory
phase advance, 390, 393
phase portrait (prob), 156–158
phase portrait, pendulum. *See* pendulum, nonlinear, phase portrait

phase space, 127, 203–204, 248
   density, 186–189
phase trajectories, 127–128, 424
   of damped drive pendulum, 454–457, 461–462
phase velocity, 526
phase-locking, 425, 461
phenomenological, 453
photon disintegration (prob), 546
physical pendulum (prob), 32, 34, 157, 332
pitchfork bifurcation. *See* bifurcation, pitchfork
Planck energy-frequency relation, 512
planet, perturbed by Jupiter (prob). *See* asteroid, perturbed by Jupiter (prob)
planetary orbits (prob), 162–164
Pluto. *See* solar system, stability
Poincaré, 423–424, 426, 428, 433, 439, 447–448, 468, 494
   Recurrence Theorem, 204–206
   section, 393, 428–436, 442–443, 445, 451, 459
      damped driven pendulum, 454–457, 462–463
      double pendulum, 430–433, 435, 441, 451
   section map. *See* map, Poincaré section
Poincaré-Birkhoff theorem, 439–442
Poincaré-Hopf theorem, 235
Poinsot construction, 283, 307–312
Poinsot, L., 306
point mass sliding on a bowling ball, 38–39
masses
   N connected by a string, 367–371, 379 (prob)
   transformation, 208, 216
Poisson brackets, 207, 217–218, 236–237, 251, 243–245 (prob)
Polaris, 317
polhode, 308, 311–312
positive definite, 288, 337, 351
positronium acceleration (prob), 550
postulates, of special relativity. *See* special relativity, two postulates
potential energy, 18–19, 42, 85–86. (*See also* conservative forces)
   for the earth in a gravitational field, 318
   matrix, Taylor series near equilibrium, 350
potential, effective. *See* effective potential
Poynting vector, 507, 517
precessing orbits, 154
precession, 273, 295–298, 315–316, 321
   of equinoxes. *See* equinoxes, precession of
   of Foucault pendulum. *See* Foucault pendulum
   of Kepler ellipses (prob). *See* Kepler ellipses, precession of (prob)
   of Mercury, 154
   of torque free top. *See* symmetric top, torque free, asymmetric top, torque free
   of symmetric heavy top, 313–316
   retrograde, 317
pressure, radiation. *See* radiation pressure
primed frame, 260
principal axis, 283, 292, 306, 339
   transformation, 288
principal moments, 288
Principia, 133, 340, 474
Principle of Least Action. *See* Hamilton's Principle
   of Relativity, 493–494

INDEX

prolate ellipsoid, *See* ellipsoid, prolate
proper time, 527, 531
   for a particle in a constant force, 536–537
proton-proton annihilation, 517–518
pseudovector, 257, 277 (prob)
Ptolemy, 130, 317
pumping a swing (prob), 416
Pythagorean theorem, 50

Q, *See* quality factor
quadratic form, 22, 351, 29 (prob)
quadrature, 22, 125
quality factor, 90, 93–94, 453
quantum mechanics, 44, 229–230, 67 (prob)
quasiperiodic, 238
   route to chaos, 428

radiation pressure, 517
radiation reaction force, 533
radiofrequency resonant cavity, 185
rank, of tensor, 333
rapidity (prob), 544
rational winding number. *See* winding number, rational
reduced echelon form, 337
reduced mass, 137
reference frames, 495
   body, 260 (*see also* body coordinate system)
   inertial. *See* inertial reference frames
   linearly accelerated, 193–195
   noninertial, 20
   noninertial (prob), 200
   rotated. *See* rotated reference frames
   rotating. *See* rotating reference frames
   space, 260. *See also* space coordinate system
reflection, 257 (*see also* transformations, discrete)
   hyperbolic fixed point, 446
refractive index (prob), 69
regular motion, 425, 447, 451
relative phase, 110–113
relativistic action, 530–532
   circular motion, 539–540
   energy. *See* energy, relativistic
   Hamiltonian, 533
   Lagrangian, 550–551 (prob)
      electromagnetic field, 534–536
      free particle, 532
   Lagrangian mechanics, 530–540
   mechanics, 494 (*see also* special relativity)
   momentum. *See* momentum, relativistic
   velocity. *See* velocity transformation rules, relativistic
relativity
   general. *See* general relativity
   special. *See* special relativity
   of simultaneity. *See* simultaneity
renormalization, 486–490
repulsive force, hyperbolic orbits. *See* hyperbolic orbits, repulsive force
resonance, 105–110
   nonlinear. *See* nonlinear resonance
   parametric. *See* parametric resonance
resonant cavity, radiofrequency. *See* radiofrequency resonant cavity
   frequency. *See* frequency, resonant
   terms, 400 (*see also* secular terms)
rest frame, instantaneous. *See* instantaneous rest frame
rest mass, 513–517

restricted three-body problem (prob). *See* three-body problem, restricted (prob)
retrograde precession. *See* precession, retrograde
rheonomic constraints. *See* constraints, rheonomic
rigid body, 254–255
   as collection of mass points 11
   kinetic energy, 284–286
   motion, 283–325
   torque free motion. *See* Euler's equations
rotated reference frames, 259–263
rotating reference frames, 190–192, 263–264, 266
rotating turntable (prob), 243
rotation matrix, 259–263
   axis, 255, 257–259
   infinitesimal, 171, 254–259
   number. *See* winding number
   of hurricanes. *See* hurricanes
   of pendulum. *See* pendulum, nonlinear
rotational angular momentum. *See* angular momentum, rotational
   invariance. *See* invariance, rotational
   kinetic energy. *See* kinetic energy, rotational
rotations, 261–263, 302
   simultaneous. *See* simultaneous rotations
   three-dimensional, 239, 261
rotator, 289
Routhian, 23, [35, 197, 331, 377] (prob)
row elimination, 347, 355, 380
   reduction. *See* Gauss-Jordan elimination
rubber band, 42–43
Rutherford, 151
   scattering, 151–152, 165–167, 165 (prob)

saddle hyperbolic fixed point, 446
saddle point, 53, 87
satellite (prob), 200
scalar, 333–334
   potential, 525, 534
   product, 254, 334, 523
scaling function, universal. *See* universal scaling function
scaling transformation, 487
scattering angle, 152
Schrodinger equation, 229, 384, 67 (prob)
scleronomic constraints. *See* constraints, scleronomic
second-order differential equation. *See* differential equation, second-order
secular terms, 400, 404–405
self-similarity, 487–488
semi-axes, 306
semimajor axes, 472
   ellipse, 145–146
semiminor axis, ellipse, 145–146
separability, 208, 225–226
separation constants, 221
   of variables, 221, 228
separatrix, 128, 185, 232, 435, 442
sgrad, 236
SHO. *See* oscillator, simple harmonic
SI units, 507, 540
similarity transformation, 487
simultaneity, 494, 500, 501–503
simultaneous rotations, 258
sinelike, 390
skew gradient, 236
SLAC, 536

sleeping top, 316
small oscillations, theory of, 343–371, 373–374 (prob)
  vibrations. *See* small oscillations
small-angle approximation, 349
Snell's Law (prob), 71
solar system, stability, 424, 468–474
solenoid lens for electrons (prob), 199
Southern cross, 317
space cone, 297, 298
  coordinate system, 259–264, 284, 300
spacelike separation, 522
space-time areas
  diagrams, 495–498, 501–503
  event, 521
  transformation of, 501
special relativity, 493–540, [542, 546–548, 550, 553] (prob)
  two postulates, 494
speed of light, 494
spherical pendulum, 27 (prob)
  Hamiltonian, 183–184
spin, 293, 296–297, 313
  of electron in magnetic field (prob), 277
spring pendulum (prob), 27
square pulse, 100–101, 104
stability, 392, 413–414 (prob)
  matrix, 443–446, 449, 458, 485
  dynamical, 383, 386
  marginal, 474
  of many-dimensional linear systems near equilibrium, 359–360
  solar system. *See* solar system, stability
stable equilibrium. *See* equilibrium, stable
  fixed point. *See* fixed point, stable
  manifold, *See* manifold, stable
Staeckel conditions, 226
standard map (prob), 476
standing waves on a string, 371
state vector, 350, 355, 357
stationary action, path for, 53
steady-state solution. *See* oscillator, steady-state solution
stellar aberration (prob), 553
step function, 96–97
Stoke's theorem, 231
strange attractor, 425, 453, 461–463
subharmonic oscillations, 383, 409–411
successive approximations, method of, 270–271
summation convention for indices, 253
supercycle, 484, 486–487
superposition, 89, 344, 120 (prob)
superstable fixed point. *See* fixed point, superstable
surface of section. *See* Poincaré section
Sussman, G. J., 471–472
swing. *See* pumping a swing (prob)
symmetric matrices
  diagonalization, 336–339
  top, 289
    heavy, 284, 313–317 (*see also* gyroscope), 331 (prob)
    Lagrangian, 313
    rotational kinetic energy, 305
    torque free, 293–300, 374 (prob)
symmetry breaking, 455–456
symplectic, 248–251, 431–432

tachyons (prob), 544
tangent map. *See* stability matrix
Taylor series, 482
  expansion, 24, 47–48, 83–84, 128
tensor, 286, 333–336
  first-rank, 333–334
  inertia. *See* inertia tensor
  second-rank, 286, 319, 333–336
  zero-rank, 333–334
tether ball (prob), 162
three-body problem, 226–235
  restricted (prob), 478
three-dimensional rotations. *See* rotations, three-dimensional
tides, 322, 163 (prob)
time derivative
  transformation of coordinate systems, 266
  vector. *See* vector, time derivative
time dilation, 499–500, 543 (prob)
timelike separation, 521–522
tippe top (prob), 331
top. *See* symmetric top, asymmetric top
topology, 235
tori, KAM. *See* KAM tori
torus, 235, 429. *See also* invariant tori
tower of Pisa problem, 253, 267–271
trace, 204, 303, 392, 445
transformation law
  electric and magnetic forces 508–510
  electric field for light pulse, 510–511
  energy of light wave, 507–512
  frequency of light wave, 505–507
  Galilean. *See* Galilean transformation laws
  volume, 511–512
transformation
  canonical. *See* canonical transformations
  contact. *See* contact transformation
  continuous, 171
  discrete, 171
  identity, 171, 215
  invariance, 174
  Lorentz. *See* Lorentz transformation
  matrix. *See* matrix transformation
  of coordinates, 173
  orthogonal. *See* orthogonal transformation
  point. *See* point transformation
  principal axis. *See* principal axis, transformation
  scaling. *See* scaling transformation
  similarity. *See* similarity transformation
  velocity. *See* velocity transformation rules
transient solution. *See* oscillator, transient solution
translation invariance. *See* invariance, translation
translational angular momentum. *See* angular momentum, translational
  kinetic energy. *See* kinetic energy, translational
transpose, 249–250, 262
transverse momentum, relativistic. *See* momentum, transverse, relativistic
Tremaine, S., 471
triatomic molecule, linear, 360–363, 377 (prob)
triple pendulum (prob), 373
true anomoly, 148
tune, 445
turning point, 125

INDEX

turntable (prob), 279
twin paradox, 554–557
two-body problem, 134

uncountable, 465
underdamped. *See* oscillator, damped simple harmonic
unitary matrix, 263
universal scaling function, 488–490
unprimed frame, 260
unstable equilibrium. *See* equilibrium, stable
   fixed point. *See* fixed point, unstable
   manifold. *See* manifold, unstable
   motion. *See* instability
upside-down pendulum, 397–398, 415 (prob)

Van der Pol equation (prob), 420
variation $\delta y(x)$, 47
variation in the integral I, 48
   principle, 54–55
variational calculus, 44, 51 (*see also* Euler equation), 66–71 (prob)
   shortest distance between two points, 49–50
   derivative, 53
   principle, 532–533
varied curve, 47
vector, 253–254, 333–334
   field, 235
   potential, 525, 534, 538
   time derivative, 253
vectorial mechanics, 23

velocity
   generalized. *See* generalized velocity
   relativistic, 504–505, 542 (prob)
   transformation of coordinate systems, 264–265
   transformation rules
Venus. *See* solar system, stability
vibrational state, complex. *See* state vector
vibrations, small. *See* small oscillations
virtual displacement, 4–5, 14–15
   work, 4–5, 9, 14–15, 27 (prob)
Voltaire, 75, 340
volume, transformation law. *See* transformation law, volume
voyager, 163

water molecule (prob), 375
wave equation, 371
   vector, 526
Weber, 493
Whittaker, E. T., 283, 308–309
width, full at half max, 109
Wigner, E. P., 44
winding number, 237–239, 425, 427–432, 434, 461
   irrational, 238–239, 436–437
   rational, 238–239
Wisdom, J., 469–472
WKB limit, 230
work, definition of, 4–5
world line, 496, 531
Wren, Sir Christopher, 132
WWI, Falkland Islands battle (prob). *See* Falkland Islands battle (prob)